Alexander of Aphrodisias
and the Text of
Aristotle's *Metaphysics*

CALIFORNIA CLASSICAL STUDIES

NUMBER 4

Editorial Board Chair: Donald Mastronarde

Editorial Board: Alessandro Barchiesi, Todd Hickey, Emily Mackil, Richard Martin, Robert Morstein-Marx, J. Theodore Peña, Kim Shelton

California Classical Studies publishes peer-reviewed long-form scholarship with online open access and print-on-demand availability. The primary aim of the series is to disseminate basic research (editing and analysis of primary materials both textual and physical), data-heavy research, and highly specialized research of the kind that is either hard to place with the leading publishers in Classics or extremely expensive for libraries and individuals when produced by a leading academic publisher. In addition to promoting archaeological publications, papyrological and epigraphic studies, technical textual studies, and the like, the series will also produce selected titles of a more general profile.

The startup phase of this project (2013–2015) is supported by a grant from the Andrew W. Mellon Foundation.

The current volume has been selected by the Editorial Board as the winner of the 2014 CCS competition to identify distinguished work by junior scholars.

Also in the series:

Number 1: Leslie Kurke, *The Traffic in Praise: Pindar and the Poetics of Social Economy*, 2013

Number 2: Edward Courtney, *A Commentary on the Satires of Juvenal*, 2013

Number 3: Mark Griffith, *Greek Satyr Play: Five Studies*, 2015

ALEXANDER OF APHRODISIAS AND THE TEXT OF ARISTOTLE'S *METAPHYSICS*

Mirjam E. Kotwick

Berkeley, California

© 2016 by Mirjam E. Kotwick.

California Classical Studies
c/o Department of Classics
University of California
Berkeley, California 94720–2520
USA
http://calclassicalstudies.org
email: ccseditorial@berkeley.edu

ISBN 9781939926067

Library of Congress Control Number: 2015953938

CONTENTS

Acknowledgments ix
Sigla and Abbreviations xi

1. Introduction 1
2. The Transmission of Alexander's *Metaphysics* Commentary 20
   2.1 The authentic part of the commentary (books A–Δ) 20
   2.2 The Greek manuscripts and the modern editions of the commentary 23
   2.3 The Latin translation by Sepúlveda 26
   2.4 The so-called *recensio altera* 28
   2.5 The Arabic fragments of the commentary on book Λ 29
3. Alexander's Commentary as a Witness to the *Metaphysics* Text 33
   3.1 Preliminary considerations: How many *Metaphysics* exemplars did Alexander use? 34
   3.2 The evidence in the lemmata 38
   3.3 The evidence in the quotations 50
   3.4 The evidence in the paraphrase and the critical discussion 55
   3.5 Alexander's sources for the *Metaphysics* Text 60
      3.5.1 Aspasius? Others? 60

3.5.2 Did Alexander know readings from $\omega^{\alpha\beta}$?		70
3.5.2.1 Alex. *In Metaph.* 354.28–355.5 on Arist. *Metaph.* Δ 3, 1014a26–31		70
3.5.2.2 Alex. Fr. 12 Freudenthal (Averroes, Lām 1481) on Arist. *Metaph.* Λ 3, 1070a18–19		75
3.5.2.3 Alex. *In Metaph.* 137.2–5; 138.24–28 on Arist. *Metaph.* α 1, 993a29–b2		78
3.5.2.4 Alex. *In Metaph.* 169.4–11 on Arist. *Metaph.* α 3, 995a12–19		83
3.6 Alexander's distinction between variants and conjectures		89
4. Alexander's Text (ω^{AL}) and the Direct Transmission ($\omega^{\alpha\beta}$)		99
4.1 Separative errors in $\omega^{\alpha\beta}$ against ω^{AL}		99
4.1.1 Alex. *In Metaph.* 174.5–6; 25–27 on Arist. *Metaph.* α 3, 995a12–20		101
4.1.2 Alex. *In Metaph.* 264.28–35; 265.6–9 on Arist. *Metaph.* Γ 3, 1005a19–23		105
4.1.3 Alex. *In Metaph.* 220.1–4 on Arist. *Metaph.* B 4, 1000a26–32		112
4.1.4 Alex. *In Metaph.* 204.23–31 on Arist. *Metaph.* B 3, 998b14–19		121
4.2 Separative errors in ω^{AL} against $\omega^{\alpha\beta}$		124
4.2.1 Alex. *In Metaph.* 11.3–6 on Arist. *Metaph.* A 2, 982a19–25		124
4.2.2 Alex. *In Metaph.* 167.7–14 on Arist. *Metaph.* α 3, 994b32–995a3		126
4.2.3 Alex. *In Metaph.* 273.20–26; 34–274.2 on Arist. *Metaph.* Γ 4, 1006a18–24		130
4.2.4 Alex. *In Metaph.* 228.29–229.1 on Arist. *Metaph.* B 5, 1001b26–28		134
4.3 ω^{AL} as criterion for priority in cases of divergence between α and β		138
4.3.1 Separative errors in α against β + ω^{AL}		140
4.3.1.1 Alex. *In Metaph.* 299.5–9 on Arist. *Metaph.* Γ 4, 1008b12–19		140
4.3.1.2 Alex. *In Metaph.* 419.25–420.3 on Arist. *Metaph.* Δ 22, 1022b32–36		146
4.3.1.3 Alex. *In Metaph.* 257.7–16 on Arist. *Metaph.* Γ 2, 1004a31–b3		153
4.3.2 Separative errors in β against α + ω^{AL}		157
4.3.2.1 Alex. *In Metaph.* 292.13–16 on Arist. *Metaph.* Γ 4, 1007b 29–1008a2		157
4.3.2.2 Alex. *In Metaph.* 182.32–38 on Arist. *Metaph.* B 2, 996a 29–996b1		161

4.3.2.3 Alex. *In Metaph.* 303.23-29 on Arist. *Metaph.* Γ 5, 1009a22-28 — 164

4.3.3 Reconstruction of an $\omega^{\alpha\beta}$-reading from ω^{AL} and two differently corrupted readings in α and β: Alex. *In Metaph.* 329.33-330.8 on Arist. *Metaph.* Γ 7, 1011b35-1012a1 — 167

5. Contamination of the Direct Transmission by Alexander's Commentary — 178

   5.1 Contamination of $\omega^{\alpha\beta}$ by Alexander's comments — 178

   5.1.1 Alex. *In Metaph.* 206.9-12 on Arist. *Metaph.* B 3, 998b22-28 — 178

   5.1.2 Alex. *In Metaph.* 438.14-17 on Arist. *Metaph.* Δ 30, 1025a21-25 — 187

   5.1.3 Alex. *In Metaph.* 372.10-17 on Arist. *Metaph.* Δ 7, 1017a35-b6 — 191

   5.1.4 Alex. *In Metaph.* 164.15-165.5 on Arist. *Metaph.* α 2, 994b21-27 — 198

   5.1.5 Alex. Fr. 12 Freudenthal (Averroes, Lām 1481-82) on Arist. *Metaph.* Λ 3, 1070a13-19 — 200

   5.2 Contamination of β by Alexander's comments — 207

   5.2.1 Alex. *In Metaph.* 421.7-15 on Arist. *Metaph.* Δ 23, 1023a 17-21 — 208

   5.2.2 Alex. *In Metaph.* 285.32-36; 286.2-6 on Arist. *Metaph.* Γ 4, 1007a20-23 — 212

   5.2.3 Alex. *In Metaph.* 262.37-263.5 on Arist. *Metaph.* Γ 2, 1005a2-8 — 219

   5.2.4 Alex. *In Metaph.* 144.15-145.8 on Arist. *Metaph.* α 1, 993b19-23 — 224

   5.2.5 Alex. *In Metaph.* 31.27-32.9 on Arist. *Metaph.* A 3, 984b8-13 — 230

   5.2.6 Alex. *In Metaph.* 295.29-32 on Arist. *Metaph.* Γ 4, 1008a18-27 — 235

   5.3 Contamination of α by Alexander's comments — 241

   5.3.1 Alex. *In Metaph.* 26.14-18 on Arist. *Metaph.* A 3, 983b33-984a3 — 242

   5.3.2 Alex. *In Metaph.* 38.5-7 on Arist. *Metaph.* A 5, 985b23-29 — 245

   5.3.3 Alex. *In Metaph.* 33.17-19; 23-26 on Arist. *Metaph.* A 4, 985a4-10 — 249

   5.3.4 Alex. *In Metaph.* 67.20-68.4 on Arist. *Metaph.* A 8, 989a22-26 — 254

   5.3.5 Alex. *In Metaph.* 380.25-30; 381.1-4 on Arist. *Metaph.* Δ 10, 1018a20-25 — 259

   5.4 Contamination of β by ω^{AL} or of α by Alexander's report of a *varia lectio*? — 266

   5.4.1 Alex. *In Metaph.* 347.19-25; 348.5-8 on Arist. *Metaph.* Δ 1, 1013a17-23 — 266

   5.4.2 Alex. *In Metaph.* 145.8-12; 19-146.4 on Arist. *Metaph.* α 1, 993b19-23 — 271

6. Results — 279

Appendices

A. A Diagram of the Ancient Greek Tradition of the *Metaphysics* — 282
B. Lemmata in Alexander's commentary — 283
C. Quotations from the *Metaphysics* in Alexander's commentary — 293
D. Alexander's paraphrase in cases of α-/β-divergences — 313

Bibliography — 323
Index Locorum — 335
General Index — 337

ACKNOWLEDGMENTS

This study is a translated (from the German) and revised version of my dissertation, which was accepted in February 2014 by the Ludwig-Maximilians-Universität Munich, Germany.

First of all I want to thank my doctoral adviser, Professor Oliver Primavesi, who introduced me to the study of the transmission of Aristotle's *Metaphysics* and generously shared with me his broad experience, his deep learning, and his masterful skill. The training I received through participation in his project of a new critical edition of *Metaphysics* A was tremendous, yet the learning I acquired through my time as his assistant reached far beyond that.

During my work on this study I could draw from a variety of sources that Professor Primavesi generously made available to me. The Greek text of Aristotle's *Metaphysics* used in the present study is based on the new collations that were produced by Dr. Pantelis Golitsis and Ingo Steinel (Aristoteles-Archiv, Berlin) on behalf of Professor Primavesi and funded by the Deutsche Forschungsgemeinschaft. The Greek text of Alexander's commentary I used in my dissertation was based on the editions by Bonitz (1847) and Hayduck (1891), the manuscript Laurentianus plut. 85,1 (= **O**), and the Latin translation by Sepúlveda (1527). A digital copy of ms. **O** was made available to me by Professor Primavesi and through the funding of the Deutsche Forschungsgemeinschaft. Dr. Pantelis Golitsis, who is currently preparing a new critical edition of the authentic parts of Alexander's commentary, checked against all extant Greek manuscripts the Greek text of all those passages that function as evidence in this study (in other words, those passages that are quoted in Greek and followed by an English translation in the body of the page). In accordance with the evidence presented in his forthcoming paper (Golitsis 2016), I included in the apparatus the readings of the three independent manuscripts **A**, **O**, and **P**[b].

The text of Michael Scotus's Latin translation of the first five books of the Arabic version of the *Metaphysics*, which I compared with the Greek text of the relevant *Metaphysics* passages, was kindly made available to me by Professor D. N. Hasse and Dr. Stefan Georges (Würzburg), who are currently preparing a critical edition of the text. I am also indebted to Andreas Lammer (Munich), who went with me through the text of those passages of Alexander's commentary that are preserved only in Arabic.

All English translations of Aristotle's *Metaphysics* are based on Ross 1984, but have been modified. The English translations of Alexander's commentary on the *Metaphysics* are based on Dooley 1989 (book A), Dooley 1992 (book α), Madigan 1992 (book B), Madigan 1993 (book Γ) and Dooley 1993 (book Δ), but have been modified.

I would like to express my gratitude to Professor Christof Rapp (Munich), who was the second reader of my dissertation and greatly supported my work in a number of ways. I am also indebted to Professor Claudia Wiener (Munich), who acted as the third reader of my dissertation. She always had her door open for me and was ready to offer helpful advice on matters both academic and non-academic.

I would also like to express thanks to my former colleagues at the Munich School of Ancient Philosophy (Musaph), in particular Professor Peter Adamson, Dr. Andreas Anagnostopoulos, Dr. Cordula Bachmann, Dr. Laura Castelli, Antonio Ferro, Dr. Matteo Di Giovanni, Dr. Rotraud Hansberger, Peter Isépy, Mareike Jas, Annika von Lüpke, Dr. Christopher Noble, and Christian Pfeiffer. They all in different ways and at different occasions contributed to the completion of this book.

Various parts of this work were presented at workshops and conferences in Munich, Berlin and Lille, France. I am very thankful to the organizers, participants and audiences at these occasions. In particular, I would like to thank Dr. Andreas Anagnostopoulos, Professor Stephen Menn and Dr. Gweltaz Guyomarc'h.

For their tremendous help with my English, I would also like to thank Joshua Crone, Hugo Havranek, and Edmond Kotwick.

I am much indebted to the two anonymous referees who made a large number of extremely thoughtful and stimulating comments and to the editorial board chair of California Classical Studies, Professor Donald Mastronarde, whose diligence, consideration, and kindness made the preparation of the manuscript a most pleasurable experience. I would also like to thank the editorial assistant Anna Pisarello.

Finally, my deepest gratitude goes to my husband Edmond Kotwick, who not only read both my German dissertation and my English manuscript, but also listened to my arguments over and over, never tired of criticizing them, and even shared with me my excitement about αὐτοῦ. I thank him for his enduring patience and abundant support. To him this book is dedicated.

Mirjam E. Kotwick
Ann Arbor, November 2015

SIGLA AND ABBREVIATIONS

Ω	Archetype of Aristotle's *Metaphysics* (first century BC edition)
α	Hyparchetype of the *Metaphysics* (E J V^d E^s J^b L^c H N E^b I^b P^b H^a W)[1]
β	Hyparchetype of the *Metaphysics* (Y A^b M V^k C)
ω^{αβ}	Ancestor of α and β
ω^{AL}	The *Metaphysics* text Alexander used
Al.^{l}	Lemma in Alexander's commentary
Al.^{c}	Alexander's quotation from the *Metaphysics*
Al.^{p}	Alexander's paraphrase (and/or critical discussion) of the *Metaphysics*
ω^{ASP1}	The *Metaphysics* text Aspasius used
ω^{ASP2-n}	Possible further texts Aspasius (or other commentators) used
Ascl.^{l}	Lemma in Asclepius's commentary
Ascl.^{c}	Asclepius's quotation from the *Metaphysics*
Ascl.^{p}	Asclepius's paraphrase (and/or critical discussion) of the *Metaphysics*
Ar^u	The Arabic translation of the *Metaphysics* by Ustāth
Ar^i	The Arabic translation of the *Metaphysics* by Ishāq ibn Hunayn
Ar^m	The Arabic translation of the *Metaphysics* by Abū Bišr Mattā

Manuscripts of Aristotle's *Metaphysics* mentioned in this study:[2]

α

E	Parisinus gr. 1853
J	Vindobonensis phil. gr. 100
V^d	Vaticanus gr. 255
I^b	Parisinus Coisl. 161

[1] For a complete *stemma codicum* see Harlfinger 1979.
[2] On very few occasions hyparchetypes γ (= J and hyparchetype δ) and ζ (V^k and C) are mentioned in the apparatus. See the *stemma* in Harlfinger 1979.

β
Y   Parisinus Suppl. 687
A^b Laurentianus plut. 87,12
M   Ambrosianus F 113 sup.
V^k Vaticanus gr. 115
C   Taurinensis B VII 23
J^c Vindobonensis phil. gr. 189

Manuscripts of Alexander's *Metaphysics* commentary mentioned in this study:
A   Parisinus gr. 1876
O   Laurentianus plut. 85,1
P^b Parisinus gr. 1878
M   Monacensis gr. 81
C   Parisinus Coisl. 161 (= *Metaphysics* ms. I^b)
V   Vaticanus Reg. gr. 115
B   Ambrosianus 115³
L   Laurentianus plut. 87,12 (= *Metaphysics* ms. A^b)
F   Ambrosianus F 113 sup. [363] (= *Metaphysics* ms. M)

S   The Latin translation of Alexander's commentary by Sepúlveda (1527)

For the key to editions cited by editor's name only in the apparatus accompanying Greek excerpts, see Bibliography.

---

³ = **D** in Golitsis 2016.

To my husband

Alexander of Aphrodisias
and the Text of
Aristotle's *Metaphysics*

CHAPTER 1

Introduction

Aristotle's *Metaphysics* was written in the fourth century BC. But our testimonies about the transmission of Aristotle's writings suggest that the earliest date of an edition containing the 14 books known to us, in the order known to us, is the first century BC. Worse still, our manuscript tradition containing Aristotle's *Metaphysics* begins with the transliteration process in the ninth century AD: *Metaphysics* manuscripts of an earlier date did not survive. This means that our direct access to the *Metaphysics* begins about 1200 years after it was written.

One might readily ask: is there not another way to access the *Metaphysics* text before the ninth century AD? It would be of great value to know what happened to the *Metaphysics* text before then—was the text evolving and shaped under the conditions of the transmission process?

Luckily, our knowledge about the textual history of Aristotle's *Metaphysics* is not restricted to the direct manuscript tradition. There exists an *indirect* transmission of the text, constituted by references to and quotations of the *Metaphysics* text in other works, most importantly, of course, works that were written before the ninth century AD. In this study I am going to analyze the earliest and most important indirect witness to the *Metaphysics*, the commentary by Alexander of Aphrodisias. I will investigate how Alexander's commentary can function as witness to the *Metaphysics* text, what it tells us about the ancient history and transmission of this text, and how it hence can help us to improve the current state of the text of Aristotle's *Metaphysics*.

There are four different sources of evidence that either directly (i) or indirectly (ii–iv) give us access to Aristotle's *Metaphysics*:

(i) The Greek manuscripts containing either parts, or the whole, of the *Metaphysics*. The manuscripts that are available to us all derive from either of two versions called α and β,[1] whose ancestor I call $\omega^{\alpha\beta}$.

(ii) The *Metaphysics* versions available through the ancient and late ancient

---

[1] For the 53 manuscripts and a complete *stemma codicum* see Harlfinger 1979.

commentaries of Alexander of Aphrodisias, Syrianus, Asclepius of Tralles, and Michael of Ephesus.[2]

(iii) The Arabic translation of the *Metaphysics* preserved in Averroes' *Long Commentary* on the *Metaphysics*.[3]

(iv) The partially or completely preserved Medieval Greco-Latin translations of the *Metaphysics*.[4]

Although the commentaries preserve the *Metaphysics* text only indirectly, that is, through the medium of their quotations, paraphrases, and comments on the text, they can give us access to a much older and more authentic version of the *Metaphysics* than the version our direct manuscript transmission can. The earliest and most important of these commentaries is the one by Alexander of Aphrodisias (ca. AD 200), itself preserved in the Greek original for the first five books (A–Δ) and in Arabic fragments for parts of book Λ.

The present study analyzes Alexander's commentary as a textual witness to Aristotle's *Metaphysics*. It thereby pursues two main objectives, which correspond to two different ways in which Alexander's commentary provides information on the *Metaphysics* text. The first objective is to analyze how the *Metaphysics* text Alexander used when composing his commentary relates to the versions of the direct transmission, α and β, and to their common ancestor $ω^{αβ}$. A clear picture of how these versions interrelate will enable us to use the readings we can extract from Alexander's commentary more effectively. The second objective is to investigate the effects that Alexander's commentary had on the transmission of the *Metaphysics* text. Alexander's impact on the *Metaphysics* text can be identified through words or phrases present in the *Metaphysics* text that were not actually written by Aristotle but were adopted into the text from Alexander's commentary. Such traces of contamination reveal to us the dynamics that shaped the text we read today, and hence can improve our understanding of the textual history of the *Metaphysics*.

---

[2] A further commentary is the so-called '*recensio altera* of Alexander,' an anonymous revision of the first two books of Alexander's commentary (see 2.4).

[3] The Arabic version of the text goes back (probably via a Syriac intermediate version) to a Greek exemplar that was written before the ninth century AD. The Arabic version is transmitted through Averroes' *Commentum magnum* (*Tafsīr Mā ba'd at-Tabī'at*). Averroes' *Commentum magnum* has been edited by Bouyges 1938–52. In his 1957 edition of the *Metaphysics*, Jaeger made sporadic use of the textual information contained in the Arabic version (see Jaeger 1957: xx and Primavesi 2012b: 402). Walzer 1958 then analyzed more carefully the textual evidence contained in the Arabic version of books A, α, and Λ of the *Metaphysics*. Primavesi 2012c provided a detailed consideration of the Arabic version in his edition of *Metaphysics* A (cf. Primavesi 2012b: 399–403). For the present study, I evaluated the Arabic tradition through the Latin translation of the Arabic text provided by Michael Scotus (13th century) for those *Metaphysics* passages that are relevant to my concern.

[4] Gudrun Vuillemin-Diem provided the first critical edition of the four extant Medieval Greco-Latin translations of the *Metaphysics* (published between 1970 and 1995, in the series *Aristoteles Latinus*). For a brief overview of these translations see Primavesi 2012b: 403–406.

Both objectives have been aspired to before in some way or other by previous scholarly investigations. Regarding the first, *Metaphysics* editors since Brandis (1823) have become increasingly aware of the fact that sometimes the direct transmission is corrupt and Alexander's commentary alone witnesses to the correct (or at any rate preferable) reading of the text. Investigations into the second objective have only been undertaken more recently. In his study on the first book of the *Metaphysics*, Primavesi (2012) argues that the β-version of book A underwent an editorial revision,[5] for which Alexander's commentary was used as a source for emendations and interpolations. Other scholars like Freudenthal (1885) and Rashed (2007) have mentioned in passing the possibility that Alexander's comments could have had an impact on our version of the *Metaphysics*.[6]

The present study offers the first systematic investigation into all preserved parts of Alexander's commentary as a witness to the *Metaphysics* text. It will furthermore show how many of the results attained by previous scholars fit into a larger picture of the ancient transmission of the *Metaphysics*, a picture I establish on the basis of all the evidence on Alexander's commentary that is available.

In the course of this introduction I will do four things. First, I will give a brief overview of the direct transmission of the *Metaphysics*. Second, I will offer a short historical survey of previous scholarly explorations of Alexander's commentary as a textual witness to the *Metaphysics*. Third, I will give an overview of the present study's methods, agenda, and scope. Fourth, I will justify the time-frame (first century BC to ninth century AD) relevant for this study.

THE TRANSMISSION OF THE *METAPHYSICS*

All extant Greek manuscripts, which contain either completely or partially the text of Aristotle's *Metaphysics*, can be traced back to two versions of the text, the α- and the β-versions.[7] The direct transmission of the *Metaphysics* begins in the

[5] Primavesi 2012b: 409–39. Frede/Patzig (1988: 13–14), in their study of book Z, had also suggested that the β-version was the work of a reviser, who improved the readability of the text (see below).

[6] Rashed 2007: 315 n. 861 and Freudenthal 1885: 87 n. 2. For the former see 5.1.4; for the latter see 5.1.5.

[7] On the manuscripts and a complete *stemma* see Harlfinger 1979. For a concise review of the current scholarship concerning the two families see Primavesi 2012b: 387–99. Cf. (regarding book Γ) Hecquet-Devienne 2008: 3–53 and the short sketch in Golitsis 2013.

Bonitz had early on drawn attention to the significant differences between codex A^b (*Laurentianus Plut.* 87,12) and the other *Metaphysics* manuscripts (Bonitz 1848: XV–XVI). Wilhelm v. Christ based his 1886 edition solely upon the two codices E (α) und A^b (β) (cf. Christ 1853: 2–3). Alfred Gercke recognized that we are in fact dealing here with two independent *families* of the *Metaphysics* text. Gercke identified in 1892 the Viennese ms. *Vindobonesis phil. gr.* 100 (= J) as a second, independent α-manuscript (Gercke 1892: 147). In 1979, Harlfinger showed that also C (*Taurinensis* VII. B. 23) and M (*Ambrosianus* F 113 sup.) are, with A^b, independent witnesses of the β-branch, and that A^b ceases to be a descendant of the β-family from Λ 7, 1073a1 onwards. (Concerning the question as to where precisely

ninth century AD.⁸ It was during this Photian Renaissance (or Macedonian Renaissance) that ancient texts were copied out of the hitherto typical majuscule script and into a new space- and time-saving minuscule script,⁹ a process known as *transliteration* (μεταχαρακτηρισμός). In the case of Aristotle's *Metaphysics*, two exemplars were transliterated.¹⁰ These exemplars can be reconstructed using the extant manuscripts that are derived from these two copies.¹¹

Two pieces of evidence suggest that both the α- and the β-version of the *Metaphysics* already existed as two distinct versions of the *Metaphysics* before their transliteration in the ninth century.¹² First, as Wilhelm v. Christ 1886a pointed out, there are certain scribal errors, typical of the majuscule script, which distinguish **A**ᵇ (β) from **E** (α).¹³ These separative errors show that the two extant versions of the *Metaphysics* had split *before* the text was converted to the minuscule script. Second, there are indications that α and β had separated even prior to AD 400. In the manuscripts of the β-version, the first words of books Δ, Θ, and Κ appear twice. Books Γ, H, and I were each the last book of a papyrus roll, and at the ends of these rolls were included the first words of the succeeding books Δ, Θ, and Κ, which were written on new papyrus rolls.¹⁴ These catchwords, or

---

Aᵇ ceases to be a descendant of the β-family, see Fazzo 2010, Fazzo 2012b: 113–18, and the critique of Fazzo's thesis in Golitsis 2013.)

⁸This period of renaissance in Byzantine scholarship after two dark centuries (see Wilson 1983: 61–78) is closely associated with Photios, the patriarch of Constantinople (810–91). See Irigoin 1962 and Wilson 1983: 79–119; see also Reynolds/Wilson 2013: 58–66; Gastgeber 2003: 14–18; 28–29. For the transmission of the Aristotelian corpus see Harlfinger 1971: 36–52.

⁹Space conservation was an important factor, as papyrus was a rare commodity and parchment was expensive. The new script, together with the import of paper from the Orient (at the end of the eighth century AD), was the answer to the dearth of papyrus material (see Wilson 1983: 63–67).

¹⁰For many works of Greek literature only one transliteration-exemplar can be reconstructed; this is due to the fact that, at the time of the transliteration, only one majuscule-exemplar was available. Wilson 1983: 67: "It often happens that all extant copies of a text seem to derive from a single archetype, and the fact may be due not so much to the unwillingness of scribes to use different capital letter exemplars as to the survival of only one such exemplar in a conveniently accessible library."

¹¹That the *stemma codicum* for Aristotle's *Metaphysics* is twofold is not uncommon for Aristotle—a twofold *stemma* is given for several of Aristotle's works. Two *hyparchetypi* can be reconstructed, for instance, for *MA* (see the new edition that is presently under preparation by Oliver Primavesi) or *Cael.* (see also Moraux 1965: CLXVIII). Cf. also Rashed 2001a: 315–38 for the question of whether there are two or three families reconstructable for *GC*. On the question of a preponderance of bipartite *stemmata* in general ('Bédier's paradox') see Reeve 2011.

¹²See Primavesi 2012b: 390–93.

¹³See v. Christ 1886a: VII: *Scripturae continuae codicis archetypi haud pauca vestigia relicta sunt, [...]; eiusdem libri archetypi quadratam litteraturam testantur nonnulli errores hoc modo facile explicandi, velut δεῖ (ΔΕΙ) pro ἀεί (ΑΕΙ) p. 998b, 17. 1016a, 15. 1026a, 21, σύνοδος (ΣΥΝΟΔΟΣ) pro σύνολος (ΣΥΝΟΛΟΣ) [...].* Due to the new collations of Golitsis and Steinel, it is possible to identify these *errores separativi* in **A**ᵇ as errors in β. They are also found in **M** and (where extant) in **C**.

¹⁴Primavesi 2012b: 391.

reclamantes, facilitated the correct ordering of rolls.[15] Wilhelm v. Christ (1886a: VII) first brought attention to these duplicated words in A^b, and in 1912 Werner Jaeger drew the conclusion relevant for our purposes: "die Kustoden führen uns in die Zeit der Buchrollen zurück, in die Zeit vor der Umschrift der Texte in codices."[16] Following Alexandru 2000,[17] Primavesi 2012b points to the fact that the "Kustoden" preserved in A^b are a feature of the entire β-version.[18] Accordingly, the β-version can most likely be traced back to an edition of the *Metaphysics* in a papyrus roll. Now, the use of codices began to be implemented in the second century AD, and until the end of the third century AD codices and papyrus rolls existed side-by-side.[19] By the end of the fourth century AD, however, the papyrus roll was obsolete.[20] The *reclamantes* in the β-version, therefore, take us back to a papyrus edition from before AD 400. The α-version, on the other hand, does not contain such *reclamantes*.

This *terminus ante quem* of the β-version (as separate from the α-version) becomes even more likely given the following considerations.[21] Let it be supposed that both the β- and the α-version of the text contained these *reclamantes*, and that while the β-version retained its *reclamantes*, the α-version of the text lost its own some time after AD 400, yet still before the transliteration. On this supposition, the presence of *reclamantes* could not be used to date the split of the two texts. There is, however, evidence that indicates that the β-text underwent a revision process.[22] This strengthens the claim that the two texts separated before AD 400. For, if the revision had occurred after AD 400, at which time papyrus rolls were no longer in use, the superfluous *reclamantes* would have been eliminated. Yet the *reclamantes are* in the β-text. Therefore the revision must have been made for an edition on a papyrus roll, for which the *reclamantes* were still useful. Therefore the revision took place before AD 400.

Werner Jaeger maintained in his *praefatio* (1957) that both versions, which he designated as Π (α) and A^b (β), already existed with most of their characteristic features in the first century AD, and that Alexander knew of and used both versions of the text.[23] However, as Primavesi 2012b argues, Alexander's commentary

---

[15] See Schironi 2010: 31–35 and 74–75.

[16] Jaeger 1912: 181. See also Jaeger 1957: ix–x.

[17] Alexandru 2000: 13–14.

[18] Primavesi 2012b: 393: "So it seems that Jaeger was right in claiming that our β-text goes back to an ancient edition which precedes not only the transliteration of ancient Greek texts from uncial script to minuscule (which took place during the ninth century), but also the replacement of the papyrus scroll by the codex (which emerged gradually during the two centuries before and after AD 300)."

[19] Dorandi 1997: 7.

[20] See Bülow-Jacobsen 2009: 25.

[21] I am indebted to conversations with Oliver Primavesi on these issues.

[22] On this β-revision see below.

[23] Jaeger 1957: ix–x. Jaeger even speculates that the two versions represent two different sets of Aristotle's lecture notes. On Jaeger's view, and its grounding in his concept of *Entwicklungsgeschichte*,

can be used as *terminus post quem* for the β-revision, and hence also for the β-text and most of its characteristic features. According to Primavesi, the β-version did not exist in the form in which it has been transmitted to us until *after* a revision occurred, during which phrases from Alexander's commentary (as well as other authorities) were placed into the *Metaphysics* text.[24] In the course of the present study, further evidence will be provided that clearly speaks against Jaeger's assumption that Alexander could have used both versions of the *Metaphysics* (see 5.1).

Frede/Patzig 1988, in their study on Book Z, first suggested that the β-version had been revised.[25] The hypothesis of such a revision of the text can explain why in many cases both versions of the text offer divergent, yet nevertheless viable, readings.[26] All the divergences between α and β cannot be due merely to the scribal errors that each version suffered during the transmission process. The fact that the β-text seems to offer the 'smoother' text[27] invites the supposition that it underwent a revision that did not occur in the α-version.[28] Cassin/Narcy 1989[29] share this view, as their edition of book Γ displayed. Primavesi 2012b introduced Alexander as a reference point concerning the puzzle about the divergence of α or β in book A, and on that basis showed that the β-version is the work of a reviser who used, besides other authoritative models, Alexander's commentary.[30]

It is still an open question whether the β-version underwent a revision process in *all* books of the *Metaphysics*. The answer can only be based on a full assessment of all divergences between α and β.[31] But even if future research on the *Metaphys-*

---

see Kotwick 2016.

[24] Primavesi 2012b: 424–39.

[25] Frede/Patzig 1988: 13–14. Cf. also Bonitz 1848: XVI: *non desunt loci (scil. in A^b), ubi interpretis potius quam simplicis librarii manum agnoscere tibi videaris.*

[26] Ross 1924: clxi: "In very many passages A^b on one side, EJ on the other have divergent readings between which there is little or nothing to choose from the point of view of sense, style, or grammar."

[27] Frede/Patzig 1988: 14. "Diese hypothetischen Eingriffe lassen sich in drei Gruppen einteilen, freilich mit der üblichen Unbestimmtheit hinsichtlich von Grenzfällen: (i) Normalisierung der Texte durch Tilgung grammatischer Besonderheiten, (ii) Glättung des Textes infolge tatsächlicher oder vermuteter sachlicher Unstimmigkeiten, (iii) Regulierung des Textes durch Tilgung unverstandener oder mißverstandener Ausdrücke."

[28] Frede/Patzig 1988: 16. Yet, from Jaeger's (1917: 481–82 and 1957: vi–vii) point of view it is Π (equivalent to our α-version) that offers a text that has been revised by Byzantine scholars, whereas A^b (β-version) often preserves the rougher, but more authentic text. Concerning Aristotle's writings other than the *Metaphysics*, see Dreizehnter 1962, who finds that one of the two transmitted versions of the *Politics* goes back to a revision of the text ("Vereinfachung des Textes," 42) in Byzantine time.

[29] Cassin/Narcy 1989: 111. See also Bydén 2005: 106. Cf. Hecquet-Devienne 2008: 5.

[30] Primavesi 2012b: 457–58.

[31] As we will see in 5.2, contamination of β by Alexander's commentary can be found in all books for which the commentary is extant. This seems to speak in favour of an affirmative answer to the question. The evidence of $ω^{AL}$ that can be found in Alexander's commentary provides additional information about the status of β: Alexander's text agrees slightly more often with α than with β (*lemmata*:

ics will affirm the revision thesis, this ought not to overshadow the fact that both versions, α and β, were exposed to other modifications during the time of the transmission, intentional or otherwise.[32]

The present investigation into Alexander's commentary as witness to the *Metaphysics* and the conclusions drawn from it on the one hand confirm that intentional changes based on Alexander's comments were made to the β-version, and on the other argue that Alexander's comments left traces in *all* versions of the *Metaphysics* that we can reconstruct. More often than not, these traces do not need to be attributed to a careful revision of the text, but rather seem to result from an accidental incorporation of glosses containing the upshot of Alexander's comments on a passage.

Primavesi 2012b (439–56) also identifies phrases or passages in the α-version as later additions to the text. These 'α-supplements' are distinguished by the fact that they are absent from the β-version *and* unknown to Alexander.[33] These instances corroborate the point already made that both versions appear to have undergone changes, albeit of a different kind. The α-supplements should also make us aware that influences on the text of the *Metaphysics* may have a source other than Alexander's commentary (cf. 5.4).[34]

ALEXANDER'S COMMENTARY AS A WITNESS TO THE TEXT OF THE *METAPHYSICS*

To say that most of the ancient commentators on Aristotle's works had access to copies that are today lost would be to state the obvious. In the case of the *Metaphysics*, and starting with Brandis (1823), editors recorded with an increasing degree of thoroughness the readings that Alexander attested to. Speaking again of Aristotle's works in general, recent attention has been drawn to the long-known

---

agreements with α 61 vs. agreements with β 51; *quotations* agreements with α 126 vs. agreements with β 82; *paraphrase*: agreements with α 198 vs. agreements with β 143; see appendices B–D). This indicates that the α-version preserves the readings in $ω^{αβ}$ slightly more faithfully than the β-version.

[32] Primavesi 2012b: 409 acknowledges this fact, yet in his edition of book A, he tends to follow the α-reading whenever possible. The fact that scribal errors occurred in both versions means also that not every α-reading that is rougher than the corresponding β-reading should be preferred as the *lectio difficilior*; cf. also the analysis of α-supplements in Primavesi 2012b: 439–56. Cassin/Narcy 1989, however, disregard this rule and follow the α-reading in all cases where α and β differ (see Hecquet-Devienne 2008: 5). Here are some passages in book Γ for which they should not have followed the α-reading, as the α-reading can be shown to be the result of a later intervention: Γ 4, 1008b15 (see 4.3.1.1) and Γ 2, 1004a32 (see 4.3.1.3). Cf. also 4.3.3 on Γ 7, 1011b35–1012a1.

[33] In 5.3.3 I show that one of the phrases that Primavesi identifies as α-supplements stems in fact from Alexander's commentary.

[34] On the question whether the commentary by Asclepius of Tralles influenced the α-version of the *Metaphysics* see Kotwick 2015.

implications of the commentaries for a scholar interested in reconstructing the ancient text of an Aristotelian treatise. Jonathan Barnes writes:[35]

> ... and the history of Aristotle's text is far more twisted—and rather more exciting ... It is precisely here that the evidence of the ancient commentators is invaluable; for the commentaries are themselves far earlier than our earliest manuscripts of Aristotle's text, and they thus testify—in principle and under certain conditions—to the state in which that text found itself in several centuries before the scribes whose ink we now read rolled up their cuffs.

Barnes is here referring to his work on Aspasius's commentary on the *Nicomachean Ethics*. Speaking of the *Metaphysics*, Frede and Patzig point to the importance of and the room for further investigation into the ancient commentators for the constitution of the text:

> [Wir sind] der Meinung, daß in der Textkonstitution der aristotelischen ‚Metaphysik' durchaus noch wichtige Verbesserungen möglich sind, wozu u.a. eine stärkere Beachtung der antiken Kommentare und der arabischen Überlieferung beitragen könnte.[36]

Primavesi confirmed this opinion by introducing a new way of treating the evidence presented in Alexander's commentary. Previous editors tended to consult Alexander's commentary either with the intention of confirming or disconfirming a particular reading, or in order to find a reading alternative to the directly transmitted text. They did not—or at least did not sufficiently—use Alexander's commentary to judge the age and the value of the two transmitted versions of the text as a whole. In 2012 Primavesi introduced Alexander's commentary as a criterion to decide whether α or β contains a revised version of the original text and to date the emergence of α and β as two separate versions.

My study of all books for which we have Alexander's commentary will follow and further extend the route taken by Primavesi by evaluating Alexander's commentary as a source for establishing the textual history of the *Metaphysics* in antiquity. I will treat the following two aspects of the evidence that Alexander's commentary offers: first, the access it provides to a text or texts much older than the text we find in our manuscripts, and second, the active role it played in the transmission process of the *Metaphysics* text.

Before I embark on this project, I would like to look briefly back to the beginning and development of the evaluation of ancient commentaries as textual witnesses and specifically to the evaluation of Alexander's commentary and its relation to the *Metaphysics*. The starting point of any exploration of the commentaries on Aristotle is the *Commentaria in Aristotelem Graeca* (CAG) editions which were edited at the behest of the Königlich-Preußische Akademie der Wissenschaften,

---

[35] Barnes 1999: 34.
[36] Frede/Patzig 1988: 13.

first under the direction of Adolf Torstrik and then of Hermann Diels.[37]

The first edition made available was Diels's 1882 edition of the first part of Simplicius's commentary on Aristotle's *Physics*, and in the following years editions of most of the extant commentaries were—often for the first time—made available. Diels was then the first to make editorial use of Simplicius's *Physics* commentary as a witness to the text of Aristotle's *Physics*. In his article "Zur Textgeschichte der Aristotelischen Physik"[38] he showed that the *Physics* text used by Simplicius was independent of the directly transmitted text.[39] In several passages, he restored the *Physics* text on the basis of the evidence in Simplicius's commentary. Diels furthermore concluded, first, that the *archetypus* of the direct transmission contained marginal glosses, and second, more generally, that we should be aware of the fact that during the transmission process, copies of the *Physics* were regularly made on the basis of multiple manuscripts and with the ancient commentaries at hand.[40] In the following, clear parallels will become evident between the results that Diels attained concerning the transmission of the *Physics* in light of Simplicius's commentary, and those attained in the present study on the transmission of the *Metaphysics* in light of Alexander's.

Alexander's testimony has always been of interest to editors of the *Metaphysics* text. The editorial history of the *Metaphysics* therefore also comprises the history of the use of Alexander's commentary as a textual witness. In his edition of 1831, Bekker recorded variant readings found in Alexander's commentary in his apparatus using the siglum F^b. The readings he thus labelled are drawn mainly from the lemmata and citations in Alexander's commentary. F^b stands for the commentary manuscript *Parisinus* 1876 = **A**.[41] Bekker thereby gave equal weighting to a manuscript of Alexander's commentary and the manuscripts of the *Metaphysics*. Schwegler (1847) adopted Bekker's text,[42] but advocated a more thorough exploration of Alexander's commentary,[43] and indeed recorded in greater detail the evidence available there. He took over Bekker's stock of evidence in F^b, but in numerous places added the evidence stored in Alexander's comments (using the

---

[37] For the history of the editorial enterprise of the CAG see Usener 1892: 197–201; see also 2.2.

[38] Diels 1882. Diels confines his study to the first four books of the *Physics* and Simplicius's commentary, the latter of which he had recently edited (CAG, Bd. IX, 1882). For an evaluation of Diels's results see Ross 1936: 103–108.

[39] Diels 1882: 5–7.

[40] Diels 1882: 19–20; also Freudenthal 1885: 46, who has Diels in mind when he says: "die Schreiber haben den aristotelischen Text nach ihrem Gutdünken geändert und bisweilen Conjecturen der Commentatoren aufgenommen, die so zur Vulgata geworden sind." Cf. also Bonitz 1848: XVI; Jaeger 1917: 486, 491; Ross 1924: clxii and Ross 1936: 103–106.

[41] See 2.2.

[42] Schwegler 1847a: IV.

[43] Schwegler 1847a: IX: "Natürlich musste der Commentar Alexanders genau verglichen werden, um den Aristotelischen Text, den dieser Ausleger vor sich gehabt hat, constatiren und wiederherstellen zu können. Freilich ist diess keine so ganz leichte und einfache Aufgabe."

abbreviation "Alex."),⁴⁴ which several times disagrees with **F**ᵇ.⁴⁵

In the same year in which Schwegler published his edition of the *Metaphysics*, Bonitz brought forward a new edition of Alexander's commentary,⁴⁶ which he regarded as preparatory work for his own edition of the *Metaphysics* (text: 1848, commentary: 1849). Alexander's commentary had clearly come to hold central importance as a textual witness,⁴⁷ and Bonitz read the commentary carefully when he set out to edit the *Metaphysics*.⁴⁸ As Schwegler before him, Bonitz criticized Bekker's consideration of Alexander's commentary as insufficient,⁴⁹ but went even further than Schwegler and distinguished three types of evidence in his evaluation of the commentary: the lemmata in **A**,⁵⁰ designated as **F**ᵇ; the text that was supposedly read by Alexander, designated as "Alex."; and the variant readings recorded by Alexander, designated as "γρ Alex."

In his 1886 edition, Wilhelm v. Christ combined the different types of testimony found in Alexander's commentary, bringing them all under the abbreviation "Alex." This parallels Christ's reduction of the direct textual witnesses to a single manuscript of each branch of the transmission, that is, to the manuscripts **E** and **A**ᵇ. Christ had in fact already made use of Alexander's commentary in 1853 in order to correct the directly transmitted *Metaphysics* text.⁵¹

Werner Jaeger made extensive use of Alexander's commentary in his 1917 and 1923 studies of the *Metaphysics* text, as well as in his 1957 edition of the text. At the beginning of his study in 1917, he writes:

> Die nähere Untersuchung des Verhältnisses von **A**ᵇ zu Alexander und beider zu der byzantinischen Recension Π [= α], die einer andern Stelle überlassen bleiben soll, beweist den hohen Wert Alexanders als Quelle für die antike, an Varianten reiche Überlieferung, und die Notwendigkeit, das Vorurteil von der schönen Einheitlichkeit unsres Textes aufzugeben. (...) Die nächste Aufgabe der Kritik wird sein, Alexander sorgfältig durchzuinterpretiren und den Bestand seiner Lesarten aufzustellen, keine ganz einfache Sache ...⁵²

---

⁴⁴Schwegler (1847a: XI–XII) based his information about Alexander's commentary on Brandis's edition of the *scholia* on the *Metaphysics* (1836). For this edition see 2.2.

⁴⁵Schwegler furthermore suggests conjectures for the Aristotelian texts that he justified by the reading found in Alexander. For instance, in 1004a12 Schwegler conjectured ἢ <γὰρ> ἁπλῶς λέγομεν based on Al. 253.1-2; Bonitz, Christ, Ross, and Jaeger follow (β, followed by Bekker, reads ἢ ἁπλῶς λεγομένη, α reads ἢ ἡ ἁπλῶς λεγομένη; **V**ᵈ ἢ ἁπλῶς λεγομένη).

⁴⁶See 2.2.

⁴⁷See already Bonitz 1842: 84–131. See also the *praefatio* of the commentary edition 1847: IV–V.

⁴⁸Bonitz 1848: IX: *accuratissime investigavi*. We will encounter Bonitz's intimate familiarity with Alexander's commentary on several occasions during the present study.

⁴⁹Bonitz 1848: VII and IX.

⁵⁰Bonitz notes some cases where the lemma **A** (*Paris.* 1876) and **M** (*Monac.* 81) disagree.

⁵¹See Christ 1853: 2–3.

⁵²Jaeger 1917: 482.

From Jaeger's point of view, his predecessors—especially Bonitz—had not yet fulfilled this task. Jaeger demands (i) that a complete inventory of readings ("Bestand seiner Lesarten") preserved in Alexander be taken, and—although Jaeger does not say it specifically—(ii) that these readings be evaluated with respect to the question of how Alexander relates to the direct transmission of the *Metaphysics*. Here, in 1917, Jaeger presents this task as a future project ("einer andern Stelle überlassen"). But he would never publish the study he envisions here.

Be that as it may, Jaeger did undertake the second task in his *praefatio* to the 1957 edition. According to Jaeger 1957: x, the two versions of the *Metaphysics* that came down to us via the direct transmission were not only already extant at the time at which Alexander wrote his commentary, but they were even *used* by him: *Al certe suo usus iudicio utramque versionem adhibuit*.[53] Jaeger's view is based on his own conception of an *Entwicklungsgeschichte* of the *Metaphysics* and its text.[54] He contends that the two versions of the text go back to two versions of the Peripatetic school.[55] This contention has the effect of diminishing the importance of the evidence in Alexander: suppose that Alexander, when composing his commentary, could have chosen his readings of the *Metaphysics* from an α- or a β-text; his testimony to a particular reading would not necessarily lead us to the older reading, but only to Alexander's personal preference.[56] Nevertheless, as Jaeger himself also points out, most explicitly in his last, unfinished article,[57] Alexander may still be the only witness to an authentic reading that in the other witnesses has since been corrupted.[58]

---

[53] See also xii. Cf. Jaeger's earlier comments in Jaeger 1917: 503. In the course of the present study we will see that it is most unlikely that Alexander used more than one textual exemplar of the *Metaphysics* when writing his commentary (see 3.1). This fact weakens Jaeger's claim that Alexander used α and β, yet it does not constitute proof that the split of the two versions happened after Alexander. Proof of the latter can be attained by identifying traces of contamination in ω^{αβ} by Alexander (on which see 5.1).

[54] Jaeger 1912.

[55] Cf. Jaeger 1956: x–xii.

[56] For further discussion of Jaeger's view see Kotwick 2016.

[57] Jaeger 1965: 408–409.

[58] Just as his editorial forerunner Ross, Jaeger presents the information found in Alexander's commentary according to the standard Diels had set up in his study on the *Physics* and Simplicius's commentary (Diels 1882: 4 n. 1: "Ich bezeichne diese Lemmatavarianten des Simplicius mit *Sl*, seine wörtlichen Citate mit *Sc*, die paraphrasierenden Textanführungen (innerhalb des Commentars) mit *Sp*."). Jaeger used the abbreviations Al (= *Alexandri Aphrodisiensis Commentarius in Aristotelis Metaphysica*), Al^c (= *Alexandri citatio*), Al^l (= *Alexandri lemma*), Al^p (= *Alexandri interpretatio vel paraphrasis*). See Jaeger 1957: xxii. Ross, however, does not distinguish between Al^p and Al^l; Al^l subsumes Al^p in his edition. Both Jaeger and Ross in their apparatus do not differentiate between the authentic part of Alexander's commentary (A–Δ) and the inauthentic part written by Michael of Ephesus (E–N) (see, however, Jaeger's note in his apparatus at Λ 1, 1069a32). Cf. Bydén 2005: 105–106 and my 2.1. This is surprising, because Ross (1924: clxi–clxiii) in his *praefatio* clearly distinguishes Alexander from Michael ("pseudo-Alexander") when addressing their status as textual witnesses for the *Metaphysics*. In his 1923 article, Jaeger refers to Ps.-Alexander interchangeably by the names "Alexander" and "Ps.-Al-

Ross shares Jaeger's view that the separation of the two versions of the *Metaphysics* text happened at a time before Alexander.[59] Ross, however, did not go so far as to claim that Alexander had both versions at his disposal. He more modestly observes:[60]

> Where EJA^{b}Al. do not agree, the usual alternatives are either A^{b}Al. right, EJ wrong, or EJAl. right, A^b wrong. Each of these alternatives was elected approximately equally often.

From this he formulates the following rule for the editor of the *Metaphysics* text:[61]

> We shall do well, generally speaking, to treat the consensus of any two of them as taking us as near as we can hope to get to the text of Aristotle.

The case studies I will examine in the following offer an occasion to discuss this rule and its implications (see 4.3). In general, Ross's rule holds true only when "Al" signals a reading in the *Metaphysics* exemplar Alexander used, but even here we have to be mindful of possible exceptions.

Most recently, Primavesi 2012 has provided evidence against Ross's and Jaeger's hypothesized dating of the two versions α and β, and for the fact that the defining characteristics of the β-version came about after Alexander wrote his commentary.[62] In the introduction to his edition of book A of the *Metaphysics*, Primavesi argues that the β-version is the product of a reviser, who used authoritative models for his revision of the text. Alexander's commentary was one such model. In 5.2, I discuss evidence found in the books subsequent to A that confirms Primavesi's discovery of β's contamination by Alexander's commentary. In 5.1, I adduce another argument that corroborates the dating of α and β to a period after Alexander.

OVERVIEW OF THE PRESENT STUDY: METHODOLOGY, AGENDA, AND SCOPE

The heart of the present study is my analysis of how the text Alexander used, his commentary itself, and the text of our manuscripts interrelate. The results of this analysis are presented in the diagram given on p. 282 (appendix A). In order to establish the relationships and influences among these different versions I follow

---

exander" (Jaeger 1923: 260 and 271–72).

[59] Ross 1924: clxiii: "The facts point to the existence in Alexander's time of three texts of approximately equal correctness, represented now by EJ, A^b, and Alexander's commentary." Cf. also Ross 1924: clxi: "Alexander (fl. 200 AD) represents a tradition intermediate between the two."

[60] Ross 1924: clxi. Cf. also Jaeger 1957: ix.

[61] Ross 1924: clxiii.

[62] Primavesi 2012b: 388 and 457–58.

the basic rules of textual criticism, most concisely set out by Paul Maas in his treatise *Textual Criticism*.[63]

The interrelation between our witnesses (including manuscripts, quotations, commentaries)[64] and the original text is determined on the basis of indicative errors (*errores significativi*), which these witnesses either share or do not share. In order to prove that a witness B is independent of a witness A, i.e. that B is not a copy of A, witness A has to contain at least one 'peculiar error' that B does not share.[65] In order to find such errors and determine the relationship of the direct and indirect Greek witnesses to the *Metaphysics*, I explore Alexander's commentary in close comparison to the readings preserved in both manuscript families, α and β. On the basis of separative errors in the text used by Alexander and the ancestor of α and β, $ω^{αβ}$, I will be able to determine that the text Alexander used and $ω^{αβ}$ are two independent witnesses to the *Metaphysics*.

There is another factor in the transmission of manuscripts that is relevant to the present study. In the simplest cases, a manuscript is copied from one single manuscript, called its *exemplar* or *Vorlage*. However, it is often the case that the copy is produced not from just one *Vorlage*, but several, and that it is additionally influenced by other sources such as commentaries or alternative readings written in the margins of a *Vorlage*. A manuscript derives from this variety of sources. This means that most manuscripts do not just bear witness to one manuscript; they also include readings from the other manuscripts that the scribe might have consulted and from the margins of the *Vorlage* where *variae lectiones* or corrections were noted. The phenomenon that a manuscript might derive from several sources is called 'contamination'[66] or 'horizontal transmission.'[67] In this study, I apply the term 'contamination' to describe the influence that Alexander's commentary exerted on different versions of the *Metaphysics*. In the context of this study, then, contamination means that information originally given by Alexander in his commentary was later incorporated into the *Metaphysics* text, where it features as Aristotle's own words.

---

[63] Maas's *Textkritik* was originally published in 1927 as part of the *Einleitung in die Altertumswissenschaft* (edited by A. Gercke and E. Norden). See also Pasquali's review and extensive discussion in Pasquali 1929. The fourth and final edition of Maas's work is Maas 1960. When I employ terms or concepts from Maas I follow the English translation by Barbara Flower (Maas 1958). For further discussions and developments of Maas's theory see Timpanaro 2005; Erbse 1979; Pöhlmann 2003b; Reeve 2007; Reeve 2011.

[64] Maas 1958: 3–4.

[65] Maas 1958: 42: the error in A has to be "so constituted that our knowledge of the state of conjectural criticism in the period between A and B enables us to feel confident that it cannot have been removed by conjecture during that period." On the relevance of this condition for the present study see pp. 137–38.

[66] Maas 1958: 3; 7–8.

[67] This term was introduced by Pasquali 1962: 140–141 in order to free the phenomenon from negative connotation.

My analysis of Alexander's commentary begins with a brief discussion both of the state of preservation of Alexander's commentary as well as of the various editions of it (chapter 2). I will then discuss how many texts Alexander used, how his "philological" work should be assessed, and what sources he made use of (chapter 3). I will argue that he used one *Metaphysics* text, ω^{AL}, which can be partially reconstructed from his commentary, and that he had occasional knowledge of some *variae lectiones* and conjectures, which he acquired from marginal notes in his *Metaphysics* exemplar or from commentaries. I will then explore the authenticity, and thereby the reliability, of the following four types of evidence for ω^{AL} found in Alexander's commentary (3.2–3.4): (i) *lemmata*; (ii) *quotations*; (iii) Alexander's *paraphrase* of Aristotle's argument; and finally (iv) Alexander's *critical discussion* of Aristotle's thought.

After my critical assessment of Alexander's commentary as a textual source for the *Metaphysics*, I will undertake a systematic exploration of how ω^{AL} relates to the two directly transmitted versions, α and β, as well as to their common ancestor ω^{αβ}. I will analyze separative errors in ω^{αβ} (4.1) and ω^{AL} (4.2) and the consequences these have for the evaluation of Alexander's commentary as textual witness to the *Metaphysics* (4.3). Thereafter, I explore how Alexander's commentary itself relates to the versions ω^{αβ}, α, and β (5). Primavesi has pointed to traces of Alexander's commentary in the β-text of *Metaphysics* A,[68] which prompts the following questions: are traces of Alexander's commentary also identifiable in the β-version of books α–Δ (5.2)? Furthermore, are traces of the influence of Alexander's commentary identifiable (in the text of A–Δ and parts of Λ) in the other versions of the *Metaphysics* text, that is, in the α- or even the ω^{αβ}-version (5.1 and 5.3)?

I should remark briefly on the scope of the present study. The case studies from which I draw my conclusions about the interrelations of ω^{AL}, ω^{αβ}, α, and β have been selected from a wide array of possible cases. I do not discuss all possible cases in which Alexander's commentary might be relevant for the named texts and the relationships they hold to one another. Some of the criteria by which I have selected the cases that I discuss vary according to the purpose of the particular analysis in question, but the following three apply in all cases. First, the textual differences between the versions of the *Metaphysics* under discussion must be substantive. In other words, I am not primarily interested in word order or other minor divergences. This leads to the second criterion: the differences between the *Metaphysics* versions must be such that it makes sense to ask which of the two (or more) available readings is more likely to have been written by Aristotle. In other words, it must be possible to assess the divergent readings on the basis of Aristotle's philosophy or diction. Third, the reading that I identify as the reading in ω^{AL} has to be reconstructed on the basis of at least two types of evidence in Alexander's commentary (see p. 60 for this rule). Any more specific selection criteria will be

---

[68] Primavesi 2012b: 424–29 and 457–58.

indicated and explained in the relevant sections of this study.

Since the cases examined below are highly specific in their features, this study does not provide the reader with a discussion of all cases and passages for which Alexander's commentary offers evidence helpful to the reconstruction of the *Metaphysics* text. On the contrary, the aim of this study is not to discuss exhaustively all available evidence, but first of all to determine, guided by the rules of textual criticism, the relationship of Alexander's *Metaphysics* text as well as his commentary to the direct transmission of the *Metaphysics*. I base my analysis on those cases that, as I found after having worked through the entire commentary,[69] offer the *most reliable* evidence. The conclusions I draw from the extensive analysis of these select cases provide the set of all possible textual relationships that might hold between Alexander's commentary and the direct transmission of the *Metaphysics*. This will offer any future editor or reader of a critical edition of the *Metaphysics* a schema whereby each and every piece of evidence found in Alexander's commentary can be efficiently ascertained and evaluated. The goal of the present study is not to complete this task, but to lay the groundwork for its completion.

As a necessary means for drawing the above conclusions and for preparing the way for the completion of the above task, this study develops methods for analyzing Alexander's commentary with a view to improving the textual evidence of the text of Aristotle's *Metaphysics*. These methods could in principle be applied to the remaining indirect witnesses of the *Metaphysics* text, namely, the commentaries by Syrianus, Asclepius of Tralles,[70] and Michael of Ephesus (and possibly also the Arabic and Latin translations of the text).

THE TIME-FRAME

The time-frame of my analysis is defined, at its one extremity, by the first century BC edition of the *Metaphysics*, often referred to as the Roman edition, which most likely consisted of 14 books. At the other extremity, it reaches as far as the transliteration process of the ninth century AD.[71] There are divergent ancient reports (particularly by Strabo[72] and Plutarch[73]) about the disappearance of (parts of) Aristotle's esoteric writings after the death of his student Theophrastus (ca. 287 BC) and their later reappearance in the first century BC. The reliability of these reports and especially the role of Andronicus of Rhodes (first cent. BC), whom Plutarch mentions as the editor of the hitherto inaccessible Aristotelian works, has been

---

[69] A concise and inevitably compressed overview of the evidence available in Alexander's commentary is offered in the appendices. Furthermore, I offer more specific lists throughout the study itself.
[70] Cf. Kotwick 2015.
[71] On the transliteration process see above pp. 3–4.
[72] Strabo, *Geographica* XIII,1,54 = Radt 2004: 602–605.
[73] Plutarch, *Sulla* 26; 468B.

judged quite differently among modern scholars.[74]

Beyond dispute, however, is the fact that in the only surviving catalogue of Aristotelian works from Hellenistic times,[75] preserved in two different versions by Diogenes Laertius[76] and Hesychius,[77] several important works, among them the *Metaphysics*, are missing.[78] The Arabic author Ptolemy al-Gharīb preserves a later catalogue, however, which lists almost all of the titles of our *Corpus Aristotelicum*.[79] The temporal divergence between the Hellenistic and the Arabic catalogues is confirmed by a fact to which Primavesi drew attention in 2007. In the older catalogue, the books of Aristotle's works are numbered according to the Hellenistic system of labeling books, whereas in the catalogue given by Ptolemy, the numbering is pre-Hellenistic, which is the way in which Aristotle's works have come down to us.[80]

Ptolemy names Andronicus of Rhodes as the author of a comprehensive catalogue (*Pinakes*) of Aristotle's works.[81] Furthermore, both Plutarch and Porphyry describe Andronicus as editor of Aristotle's writings.[82] Therefore it seems reasonable to conclude that, in the first century BC, Andronicus compiled and edited works of Aristotle that had been hitherto inaccessible.[83] Whether Andronicus's work included "text-critical initiatives" is a matter of debate.[84] It seems that Andronicus, while editing and organizing, that is, while labeling the books of Aristotle's works, preserved the pre-Hellenistic numbering system present in the re-

---

[74] See Moraux 1973: 3–44; Gottschalk 1987: 1083–97; Barnes 1997: 2–31; Primavesi 2007; cf. also Hatzimichali 2013: 11–27.

[75] Concerning the question whether the catalogue goes back to the library of Alexandria see Primavesi 2007: 58–59.

[76] Diogenes Laertius (5.22–27; ll. 257–409 Dorandi); see also Moraux 1951: 15–193.

[77] Hesychius of Miletus (text: Düring 1957: 83–89). See also Moraux 1951: 195–209.

[78] See Moraux 1951: 73 and Primavesi 2011c: 60–61.

[79] See Moraux 1951: 287–309 and Hein 1985: 424–29, who presents the Arabic text with translation. See also Primavesi 2011c: 62 and Gottschalk 1987: 1090.

[80] Primavesi 2007: 63–70 and Primavesi 2011c: 60–63. The pre-Hellenistic system consisted of 24 letters (Α–Ω), whereas the Hellenistic system consisted of 27 letters (including ϛ = 6, ϟ = 90, and ϡ =900). Note, however, that the differentiation between the two systems was not always sharp. For example, the (pre-Hellenistic) numbering of Homer's works in 24 books remained stable throughout. See also the examples given in Lapini 1997.

[81] Hein 1985: 417–19. Hatzimichali 2013: 15–20 works out the precise implications of the editorial work credited to Andronicus by Plutarch and Porphyry.

[82] Plutarch, *Sulla* 26; 468B; Porphyry, *De vita Plotini*, 24,2–11.

[83] It is likely that Andronicus drew on editorial work done by Tyrannio. See Hatzimichali 2013: 16. For Tyrannio see also Moraux 1973: 33–44 and Barnes 1997: 17–20; concerning editorial work done after Andronicus's death see Gottschalk 1990: 67.

[84] See most recently Hatzimichali 2013: 27: "A scrutiny of our sources has shown that it was the [i.e. Andronicus's] processes of cataloguing, canon-formation and corpus-organisation that had the greatest impact on the texts we now read, and not the appearance of new 'editions' and text-critical initiatives." Barnes is "cutting Andronicus down to size" (1997: 59); Fazzo 2012a: 54–55 endorses Barnes's general doubt on the impact of Andronicus on the Aristotelian writings.

discovered writings and, what is more, even carried the pre-Hellenistic method of book-labeling over to those writings of Aristotle that had already been accessible during Hellenistic times and had consequently been adjusted to the Hellenistic numbering system. It is this pre-Hellenistic system of book numbering that is preserved in the present-day *Corpus Aristotelicum*. Andronicus's edition, which was a contributing factor to the renaissance of Aristotelianism of the first century BC, therefore serves as the starting point of the textual history of the *Metaphysics* that I am going to explore.[85]

The precise dating of Andronicus's engagement with Aristotle's work is no easy matter. Barnes reasons that Cicero's death in 43 BC can be taken as *terminus post quem* for Andronicus's activity, since Cicero knew neither of Andronicus nor of his work on the Aristotelian corpus.[86] Gottschalk was able to date Andronicus's work to as early as the 60s of the first century BC because he claimed that Andronicus was active in *Athens* at that time.[87]

How much do we know about the *Metaphysics* version of Andronicus's edition?[88] Ptolemy's catalogue, presenting the post-Andronican state of Aristotle's works, lists a 'Metaphysics' in 13 and *not* (as we would expect) in 14 books.[89] The Hellenistic catalogue preserved through Diogenes Laertius and Hesychius includes only book Δ of the *Metaphysics*.[90] (The entry of a Metaphysics in 10 books in Hesychius's catalogue is a later inauthentic addition.[91]) Can we infer from the fact that Ptolemy was acquainted with the *Metaphysics* in 13 books that it was Andronicus who enlarged an earlier (10-book?) version and hence gave shape to the *Metaphysics* as we know it today? Even Barnes seems partial to this explanation.[92] But why then does this Metaphysics not (yet) contain the *14* books we find in our *Metaphysics*? Is the difference in the number of books merely due to the fact that the ancients did not count the so-called second book of our *Metaphysics*, α ἔλαττον, as an independent book, but saw it rather, as the letter α indicates, as an appendix to book A?[93] If so, our 14-book *Metaphysics* and the ancient 13-book *Metaphysics* would in fact be identical. In itself this line of reasoning is plausi-

---

[85] See Gottschalk 1987: 1095 and Hatzimichali 2013: 17.

[86] Barnes 1997: 21–24. Moraux 1973: 45–58, however, holds the view that Andronicus became head of the Peripatos in Athens already around 80–78 BC.

[87] Gottschalk 1987: 1093–96 and 1990: 79.

[88] See Gottschalk 1987: 1086–97; Barnes 1997: 28–66.

[89] Hein 1985: 429.

[90] Book Δ appears here under the title Περὶ τῶν ποσαχῶς λεγομένων ἢ κατὰ πρόσθεσιν (Diog. Laert. 5, 23; l. 293 Dorandi; Hesychius, Nr. 37; Düring 1957: 84).

[91] See Hesychius, Nr. 111; Düring 1957: 86 with note in 90: "must be a later addition" and Nr. 154 (in the *appendix Hesychiana*); Düring 1957: 87 and 91. See Jaeger 1912: 177–80, who asserts that the 10-book *Metaphysics* lacked books Δ, K, Λ and α. See also Primavesi 2007: 70

[92] Barnes 1997: 62: "Perhaps, then, it was Andronicus who first produced the *Metaphysics*." See also Primavesi 2007: 70. Hatzimichali 2013: 25–26.

[93] Cf. Jaeger 1912: 178 and Primavesi 2007: 70. See also 3.5.2.3–4.

ble. And as a matter of fact there is an ancient witness who knows and regards α ἔλαττον as Aristotelian and as part of the *Metaphysics*,[94] but does not count it as an independent book: Alexander of Aphrodisias. Alexander explicitly calls book B the *second* book of the treatise.[95]

Nicolaus of Damascus (born 64 BC) in his *De philosophia Aristotelis*[96] provides evidence for the fact that α ἔλαττον was already part of the *Metaphysics* in the first century BC.[97] This shows that the 14 book *Metaphysics* known to us existed in the first century BC. If that is the case, then it is, for my purposes, an almost insignificant detail whether it was Andronicus himself (who did not count α ἔλαττον as independent book) or someone else who added α ἔλαττον to Andronicus's *Metaphysics* edition.[98] The next testimony that the *Metaphysics* contained our 14 books comes from Alexander of Aphrodisias. We know that Alexander held a chair of Aristotelian philosophy in Athens[99] in the years AD 189 to 209,[100] and so we can date his commentary to approximately AD 200. The commentary on the preserved books A–Δ (and the fragments of book Λ in Arabic) bears witness to the fact that Alexander's *Metaphysics* exemplar, which has to be dated to the second century AD, and the text transmitted to us, had the same number and order of books.[101]

The present study analyzes the textual history of Aristotle's *Metaphysics* from the first century BC until the ninth century AD. In the first century BC, there existed an edition of the *Metaphysics* containing our 14 books (Ω). Among the copies that

---

[94] The very title α ἔλαττον suggests that the compiler (Andronicus?) regarded it as part of the *Metaphysics*. He seems to have taken it as an introduction, yet seeing that the *Metaphysics* already had an introduction (A major) he called it 'little α' (see also 3.5.2.3).

[95] Alex. *In Metaph.* 257.10–16; 264.31. Cf. also 137.2–9.

[96] Recently Silvia Fazzo (Fazzo 2008; Fazzo/Zonta 2008) raised doubts concerning the attribution of the work in question to Nicolaus of Damascus and therefore also about the dating of the work to the first century BC. She considers the real author of the work to be Nicolaus of Syria, who lived in the fourth century AD.

[97] The fragments of the work, preserved only in Arabic (F 21; Drossaart Lulofs 1965: 76, 137–39), contain a periphrastic excerpt of the passage in α 1, 993b9–11.

[98] Cf. Jaeger 1912: 178, who takes the title α ἔλαττον to indicate that the book was a later addition to the canonical *Metaphysics* in 13 books. Drossaart Lulofs 1965: 30 suggests that Nicolaus himself could have added the book. Cf. also Gottschalk 1990: 67.

[99] That the chair was at Athens is confirmed by a 2001 discovery of an inscription in Aphrodisias. See Chaniotis 2004 and Sharples 2005.

[100] Alexander dedicates his treatise *De fato* to the emperors Septimius Severus and Antoninus Caracalla (*De fato* I, 164.3–6 Bruns), who appointed him professor of Aristotle's philosophy. See Sharples 1987: 1177. For the precise dating of Alexander, see Sharples 2005.

[101] See Alexander's remarks on book α in 137.5–138.23. See Di Giovanni/Primavesi 2016 on the status and origin of the complete list of the books of the *Metaphysics* in 1395–1405 Bouyges (Genequand 1986: 60–65), in which the order of A and α is reversed and book K is declared missing. Alexander also testifies to the title (μετὰ τὰ φυσικά): 137.2; 169.22; 171.6. For the age and origin of the title μετὰ τὰ φυσικά see Jaeger 1912: 179–80 (arguing for a pre-Andronican origin) and Fazzo 2012a: 56 (arguing for an Andronican origin).

were produced of this version and which then began to circulate from the first century BC onwards, we know of, first, the version Alexander used when writing his commentary in AD 200 and which I call ω^{AL}; and second, the version, $\omega^{\alpha\beta}$, which became the ancestor of our direct transmission, represented by the two ninth-century AD versions α and β. How these versions interrelate, to what era $\omega^{\alpha\beta}$ can be dated, and whether or not we are able to recognize even further versions of the *Metaphysics*—these topics will all be addressed in the course of this study.

CHAPTER 2

The Transmission of Alexander's *Metaphysics* Commentary

2.1 THE AUTHENTIC PART OF THE COMMENTARY (BOOKS A–Δ)

The scope of this study is confined to books A–Δ, which are the parts of Alexander's commentary that are regarded as authentic.[1] The view that the commentary on books E–N that had been transmitted under Alexander's name was not in fact written by Alexander of Aphrodisias was already widely accepted by the time of Juan Ginés de Sepúlveda (1490–1573), Spanish Humanist and translator of Alexander's commentary.[2] In the preface to his translation of the commentary on books E–N, Sepúlveda describes this opinion as *famam illam vulgo absque auctore dissipatam*.[3] Sepúlveda wrote the preface to the later books of the commentary as an epistle to his sponsor Pope Clement VII, to whom he dedicated his translation of the commentary.[4] In this epistle, Sepúlveda attempts to justify the inclusion of books E–N in his translation by refuting the assumption or "rumor" that the later books of the commentary are inauthentic. Sepúlveda undertakes to show that this assumption is based on shaky evidence[5] and that the later books are in fact authentic. To that end, he focuses on the following four criteria: *inscriptionum antiquitas, dicendi character, opinionum constantia, ratioque testimoniorum.*[6]

Regarding the first criterion, Sepúlveda states that the inscriptions or captions in all four of his commentary manuscripts name Alexander as the author.[7] Re-

---

[1]On the Arabic fragments of Alexander's commentary to book Λ see 2.5.

[2]On Sepúlveda as translator of works by Aristotle and Alexander see Coroleu 1995 and Coroleu 1996. On his translation of Alexander's commentary see section 2.4 below.

[3]Sepúlveda 1527: f. A.i.r. (*Ad Clementem. vii. Pont. Max. Io. Genesii Sepuluedae praefatio in alex. aphr. enarrationem posteriorum librorum Arist. de prima philosophia*).

[4]See also Sepúlveda's dedicatory epistle introducing the whole of the commentary.

[5]Sepúlveda 1527: f. A.i.r.

[6]Sepúlveda 1527: f. A.i.v.

[7]Sepúlveda 1527: f. A.i.v: *illud tamen testari possum, quattuor antiquissima exemplaria, quorum fidem sum in conversione secutus, alexandri nomine sine ulla distinctione inscripta esse atque notata.*

garding the second, he argues that the diction is largely homogeneous through both parts of the commentary,[8] and as concerns the third, he points out that the content of the second part is congruent with Alexander's opinions attested elsewhere.[9] Sepúlveda gives particular attention to the fourth criterion, because he sees in it the origin of doubts about the second part's authenticity.[10] The explicit mentions of "Alexander of Aphrodisias" that occur repeatedly in the second part of the commentary[11] seem to rule out Alexander as the author. However, as Sepúlveda argues, these mentions can be explained by the peripatetic custom of using the names of famous persons such as Socrates or Plato in examples. If Alexander mentions Alexander, Sepúlveda reasons, it is because Alexander preferred to use his own name in his examples.[12]

Sepúlveda's arguments were apparently unable to lay to rest once and for all doubts about the authenticity of the commentary on books E–N. These doubts would resurface three hundred years later in the first modern edition of Alexander's commentary in Greek. Edited by Christian August Brandis (1836), the second part of the commentary appeared only in the form of extracts. Brandis names the Byzantine commentator Michael of Ephesus, who is given as the author of book E in manuscript **A** (*Parisinus* 1876),[13] as a possible author, but he does not commit to a solution.[14] A year later, Félix Ravaisson (1837) argues for the authorship of Michael on the basis of a remark that Michael made about his own commentary on *Metaphysics* Z–N.[15]

This information furthermore provides us with a valuable criterion for identifying which commentary manuscripts Sepúlveda used. Since manuscript **A** names Michael of Ephesus as the author of book E we can rule out that this manuscript was among those used by Sepúlveda, unless one argues that the ascription in **A** is an addition that came into the text after the 16th century.

[8]Sepúlveda 1527: f. A.i.v: *Nam dicendi character, seu mavis Latino vocabulo dictionem nuncupari, tam est in utraque parte similis, ut, quod aiunt, lac non sit lacti similius.*

[9]Sepúlveda 1527: f. A.i.v: *Cum non solum, quae in hoc opere ab ipso disputantur, utrobique sint consentientissima, sed etiam quaedam alexandri dogmata, quae ab averroi ceterisque peripateticis celebrantur, et in aliis ipsius operibus apparent, in his potissimum libris, de quibus quaeritur, habeantur.*

[10]Sepúlveda 1527: f. A.ii.r: *cum ipse in libro sexto de alexandro aphrodisieo philosopho, id est, de se ipso mentionem faciat, quod mihi opinari saepe in mentem venit, caput fuisse atque fontem, unde totus error emanavit.*

[11]See Ps.-Alex. (Michael Ephesius) *In Metaph.* 466.17; 524.6–11; 532.8–19; 663.2–4.

[12]Sepúlveda 1527: f. A.ii.r: *Ut autem haec caeteris sunt familiaria in exemplis vocabula, sic alexandro, ego et alexander.*

[13]On this manuscript see 2.2.

[14]Regarding the scholia on book E–N Brandis says: *Ad libros seqq. Metaphysicorum non integros dedi Alexandri, qui feruntur, commentarios, sed scholia tantum ex iis excerpta, cum mihi dubium non sit falso eos Aphrodisiensis nomen prae se ferre, sive Michaelis Ephesii sunt, quemadmodum cod. Reg. Par. 1876 autumat* (Μιχαὴλ τοῦ Ἐφεσίου σχόλια εἰς τὸ ε' τῶν Μετὰ τὰ Φυσικὰ τοῦ Ἀριστοτέλους), *sive alius cuiusdam similis notae scholiastae.* On this ascription to Michael see Hadot 1987: 242–45 ("Note supplémentaire à la note 12").

[15]Ravaisson 1837: 65 n. 1. Michael's remark is found in his commentary on *Parva Naturalia, in PN*

Bonitz (1847) also questions the authenticity of the second part of Alexander's commentary,[16] yet at the same time he stresses the importance of an edition of the commentary that includes the second part even if it proves to be spurious. According to Bonitz, the second part still contains valuable information on the text of the *Metaphysics*. Acknowledging the positions held by Brandis and Sepúlveda,[17] Bonitz investigates further into the matter of the second part's authenticity.[18] He observes that the commentaries by Syrianus and Asclepius do not offer any reliable evidence for determining the identity of the author of the second part of Alexander's commentary. Pseudo-Philoponus's commentary, however, names Michael as the author of the commentary on book E.[19] Bonitz further probes the second part of Alexander's commentary with a view to references to other works, internal congruity, interpretation, language, and diction, the last of which he finds especially suspicious.[20] Even after a thorough investigation and extensive examination of the evidence, Bonitz concludes that he cannot give a definite answer. So he concludes tentatively: Alexander is the original author of the second part, but it was later reworked by someone else, possibly Michael of Ephesus.[21]

When Jakob Freudenthal published his dissertation in 1885 he brought new evidence to the discussion about the authorship of the second part of the commentary. Freudenthal claims to prove, first, that Alexander is not the author of the preserved second part and, second, that its real author did not even use Alexander's original commentary on the later books. Freudenthal points to the fact that the fragments of Alexander's commentary on book Λ, which are preserved in Averroes' *Metaphysics* commentary (see 2.5), are incompatible with the directly preserved commentary on book Λ.[22] Freudenthal does not hold that Michael of Ephesus authored the inauthentic second part of the commentary. He dates the inauthentic part to the fifth or sixth century AD.[23]

In his 1906 review of Hayduck's edition of Michael's commentaries *in PA*,[24] Karl Praechter was the first to adduce other works of Michael as evidence to show

---

149.14 15 Wendland: γέγραπται δέ μοι καὶ εἰς τὰ Μετὰ τὰ φυσικὰ ἐξ αὐτοῦ τοῦ ζῆτα ἕως τοῦ νῦ. See also Rose 1854: 147–50. On the fact that Michael mentions only books Z–N (instead of E–N) and the possible implications of this, see Golitsis 2014b: 220–23.

[16] Bonitz 1847: IV–V.

[17] Bonitz 1847: XIV–XVIII. Bonitz speaks remarkably positively (cf. Freudenthal 1885: 10) about the acumen demonstrated by Sepúlveda, whose arguments he nonetheless refutes in detail.

[18] Bonitz 1847: XVIII–XXVII.

[19] On the commentary of Ps.-Philoponus see Alexandru 1999. Alexandru ascribes the commentary of Ps.-Philoponus to George Pachymeres (1242–1310).

[20] Bonitz 1847: XXII–XXVII, on diction esp. XXV–XXVI.

[21] Bonitz 1847: XXVII.

[22] Freudenthal 1885: 3–64.

[23] Freudenthal 1885: 53–55. See also Praechter 1906: 882–83.

[24] This edition by Hayduck (1904) includes Michael's commentaries on *de Partibus Animalium*, *de Animalium Motione* and *de Animalium Incessu*.

that Michael is the author of the commentary on books E–N of the *Metaphysics*.[25] His evidence comes down to language and diction: the writing style of the inauthentic part agrees with the writing style in Michael's commentaries on *PA, MA*, and *IA*. Concetta Luna 2001 agrees with Praechter's conclusion.[26] She adduces further evidence by showing that Michael drew from the commentary of Syrianus and not, as is sometimes suspected, the other way around.[27]

2.2 THE GREEK MANUSCRIPTS AND THE MODERN EDITIONS OF THE COMMENTARY

The Greek text of Alexander's *Metaphysics* commentary was not printed until the 19th century.[28] In 1836 Christian August Brandis edited large extracts from Alexander's commentary in the fourth volume of the Berlin-academy edition of Aristotle's work (1831–1870).[29] Brandis's *scholia in Aristotelem* comprise extracts from various commentaries on works of Aristotle.[30] From Alexander's *Metaphysics* commentary[31] Brandis includes the first five books completely and the commentary on books E–Λ only partially. For the first five books Brandis relies primarily on manuscript **A** (*Parisinus gr.* 1876).[32] For the text of A, α, Γ 4–8, and Δ he draws additionally on manuscript **M** (*Monacensis gr.* 81),[33] and for the text of B and Γ

---

[25] Praechter 1906: 863 n. 3 and 882–907. Cf. Rose 1854: 147–52. On Michael of Ephesus as a commentator see Praechter 1909: 533–37 and Mercken 1990: 429–36.

[26] Golitsis 2014b recently questioned the attribution of the entire second part (E–N) to Michael. He argues that the commentary on book E might actually belong to the anonymous so-called *recensio altera* (6th century AD). On the *recensio altera* see 2.4.

[27] Luna 2001: 53–71. On the opposite view, according to which Ps.-Alexander (not Michael) influenced Syrianus, see Luna 2001: 37–53.

[28] There was, however, a Latin translation of Alexander's commentary by the Spanish Humanist Juan Ginés de Sepúlveda, published as early as 1527 (see 2.3).

[29] On the editorial history of the Aristotelian corpus see Primavesi 2011b: 57–59. On Brandis's edition see Usener 1892: 1004–1005.

[30] Olaf Gigon's outline of Brandis's edition, which Gigon placed at the beginning of the second edition of the fourth volume of the Berlin edition of Aristotle (1961: X–LI), provides a useful overview of the commentaries included by Brandis. On the extracts from commentaries on the *Metaphysics* see XLII–LI and 518a14–942b27.

[31] Brandis includes the following material on the *Metaphysics*: extracts from the commentaries by Syrianus (edited by Usener) and Asclepius as well as scholia ("cod. Reg.") from manuscript **E** (*Parisinus gr.* 1853). On the scholia see Jaeger 1957: vii, who dates most of the scholia to the 15th century, some to the 10th century, and Golitsis 2014a, who dates them all to the 12th century.

[32] Bonitz (1847: VII) describes **A** as *optimum et certissimum*. **A** is to be dated to the 13th century. Hayduck 1891: VII; Harlfinger 1975: 18; Golitsis 2014b: 219: "[A, O] qui datent à mon avis des années 1270–1290." This codex contains the commentary *in plena pagina*. Mondrain 2000: 17–18 provides a description of the manuscript. See also Hadot 1987: 242–43. For the identification of the scribe ('Anonymus Aristotelicus') see Rashed 2001a: 230–32.

[33] According to Bonitz (1847: VII–VIII), this 16th-century manuscript stems from the same exem-

1–3 additionally on manuscript **C** (*Parisinus Coisl.* 161).[34] Moreover, Brandis occasionally uses manuscript **V** (*Vaticanus Bibl. Reginae* 115), **L** (*Laurentianus* 87,12)[35] and Asclepius's commentary.[36]

The first edition of the whole commentary transmitted under Alexander's name was provided by Hermann Bonitz in 1847. Bonitz edited Alexander's commentary as he was preparing a new edition of Aristotle's *Metaphysics* (edition 1848, commentary 1849). Bonitz justifies his decision to include all of the commentary ascribed to Alexander with the following argument: the commentary, even if parts of it were not written by Alexander himself, provides valuable information on the text of the *Metaphysics*, and so it is in its entirety of vital importance for establishing the text of the *Metaphysics*. Even if Alexander is not the author of the commentary on books E–N, this part of the commentary is still old enough to be of relevance to the *Metaphysics* editor.[37] Compared to Brandis's edition, Bonitz's offers an occasionally corrected text of the first five books and an entirely new edition of the later books E–N.[38] On the basis of new collations of manuscripts **A** and **M**,[39] Bonitz thoroughly evaluates Sepúlveda's Latin translation (= **S**).[40] As far as manuscripts **C**, **L**, and **V** are concerned, Bonitz adopts the testimony of their

---

plar as **A**, but was copied less carefully. In **M** the commentary is written *in plena pagina*.

[34] According to Harlfinger, this 14th-century manuscript is a copy of **A** (Harlfinger 1975: 19; Mondrain 2000: 20: "une copie du *Parisinus gr. 1876*"). **C** offers the commentary *in margine*, surrounding a text from the α-version of the *Metaphysics* (= **I**ᵇ), which is contaminated by the β-version (Harlfinger 1979: 27; see also Harlfinger 1971: 55–56 and Rashed 2001a: 229–30, who both argue that *Paris. Coisl.* 161 represents one of the most influential editions of Aristotle of the 14th century. Cf. also Hadot 1987: 242–45).

[35] See 2.4.

[36] Brandis 1836: 518 n. Cf. also Brandis 1836: 734a. Brandis, just as the editors after him, occasionally relies on the indirect evidence for Alexander's commentary that is provided in Asclepius of Tralles' commentary. Asclepius bases his commentary on the lectures of his teacher Ammonius Hermiae (see Luna 2001: 99–106). Asclepius's commentary is preserved for books Α–Ζ; it contains excerpts from Alexander's commentary on books Α–Γ (Luna 2001: 107–41).

[37] Bonitz 1847: IV–V. From today's perspective and on the supposition that Ps.-Alexander is to be identified with Michael of Ephesus one might question the importance of the textual information available in the commentary on E–N. First, Michael did not have access to the original commentary by Alexander. Second, he wrote three centuries later than our earliest *Metaphysics* manuscripts, the oldest of which is from the ninth century (**J** = *Vindobonensis phil. gr.* 100), were produced. Nonetheless Marwan Rashed and Thomas Auffret in a recent paper (delivered at the workshop "The Text-History of Aristotle's *Metaphysics*" in Berlin, June 2014) accredit new importance to the role of Michael for the constitution of the *Metaphysics* text; they argue that Michael is especially relevant for settling the question about the stemmatic shift of **A**ᵇ from the β-branch to the α-branch from book Λ 7 (1073a1) onwards (on this shift see 1; pp. 3–4 n. 7).

[38] Bonitz 1847: V–VI; VII.

[39] Bonitz's complete evaluation of **M** is confined to the second part of the commentary (Bonitz 1847: VIII).

[40] Bonitz 1847: VIII–IX. Brandis consulted the Latin translation occasionally. On Sepúlveda's translation see 2.3.

readings from Brandis's apparatus.⁴¹

In 1874 Eduard Zeller initiated at the Prussian Academy in Berlin the major editorial project of editing—in many cases for the first time—the ancient and late-ancient *Commentaria in Aristotelem Graeca* (CAG). Between the years 1882 and 1909, 23 volumes were published under the editorship of Hermann Diels. The first volume, published in 1891, contains Alexander's commentary on the *Metaphysics*, which was edited by Michael Hayduck.⁴² In his *praefatio* Hayduck declares that the work of Jakob Freudenthal (see 2.1) refuted Bonitz's hypothesis that the author of books E–N of the commentary had extended access to Alexander's authentic commentary on these books. Still, since Hayduck leaves open the case about who wrote the second part of the commentary, he can hold on to Bonitz's view that also the inauthentic part provides valuable evidence to the *Metaphysics* editor. Hayduck builds significantly on Bonitz's work and his evaluation of the manuscripts, but he consults two other manuscripts that Bonitz had not taken into account: **L** and **F** (see 2.4). Hayduck further checks, though occasionally, the reading in **B** (*Ambrosianus* D 115),⁴³ a manuscript that had not been considered previously.⁴⁴

In 1975 Dieter Harlfinger edited those sections of Alexander's commentary on book A in which Alexander cites from the otherwise lost Aristotelian treatise *De ideis*.⁴⁵ Harlfinger based his edition on the manuscripts **A**, **C**, **L**, and **F** and on the previously disregarded manuscript **O** (*Laurentianus plut.* 85,1), also called 'Oceanus.'⁴⁶ Harlfinger regards **O** as a most important witness to the text of Alexander's commentary, especially in view of the possibility that the important manuscript **A** is actually dependent on **O**.⁴⁷ This immense codex contains a collection of commentaries on Aristotle's works. Although it was evaluated for several editions of the CAG series,⁴⁸ Hayduck did not consult it for his edition of Alexander's commentary.

---

⁴¹Occasionally Bonitz relies on the indirect evidence in Asclepius's or Syrianus's commentaries (on Syrianus's usage of Alexander's commentary see Luna 2001: 72–98).

⁴²Cf. 1. Alexander's commentary is the first volume of the series, yet it was not the first volume that was published. The first published volumes are Diels's edition of the first part of Simplicius's commentary on the *Physics* (vol. IX, 1882) and Hayduck's edition of Simplicius's commentary on *De anima* (vol. XI, 1882).

⁴³On this manuscript and Hayduck's confused denomination see below.

⁴⁴Hayduck, just as his predecessors, occasionally relies on the indirect evidence in Asclepius's or Syrianus's commentaries.

⁴⁵See Leszl 1975: 22–54. On the identification of the fragments from *De ideis* see Wilpert 1940.

⁴⁶Just as **A**, the manuscript **O** is from the 13th century. Cacouros 2000 speaks of the second half of the 13th century. Golitsis 2014b: 219: "1270–1290." Cf. Moraux 1976: 275: "13.–14. Jh." and Harlfinger 1975: 18. This codex contains the commentary *in plena pagina* (Moraux 1976: 275–76).

⁴⁷Harlfinger 1975: 19. Harlfinger collated for his edition the six manuscripts that are dated to a time before 1400. The manuscript *Marcianus* Z.255 (coll. 872) reveals itself to be an *apographon* of **O** and can therefore be disregarded (Harlfinger 1975: 18–19).

⁴⁸See e.g. the editions of Simplicius, *In Physicorum libros quattuor priores commentaria*, ed. H.

The present study is based on my dissertation, in which I used the following sources for the text of Alexander's commentary: Hayduck's edition, manuscript **O** (*Laurentianus plut.* 85,1), and the Latin translation by Sepúlveda (**S**), which Hayduck did take into account, yet sometimes insufficiently. For the present study, Pantelis Golitsis kindly checked all of the Alexander passages I analyze extensively in all extant manuscripts. Golitsis is currently preparing a new edition of the authentic part of Alexander's commentary, and in his forthcoming article he argues that *three* independent manuscripts of Alexander's commentary are extant: **A**, **O**, and **P**^{b}.[49] According to Golitsis, **A** and **O** together represent one family and **P**^{b} a second one.[50] The codex **P**^{b} (*Parisinus gr.* 1878)[51] has never been used for the constitution of Alexander's commentary before. **P**^{b}, Golitsis argues, represents a new family, which we previously knew only indirectly through the Latin translation **S** and in extremely rare places where Hayduck followed the readings of **B**.[52] Given his conclusion that the three independent manuscripts are **A**, **O**, and **P**^{b}, it is the readings of these manuscripts that I include in my apparatus. Furthermore, I include the evidence I found in **S** and the *recensio altera* (**L** and **F**)[53] whenever they provide information relevant to my investigation.

2.3 THE LATIN TRANSLATION BY SEPÚLVEDA

In 1847 Bonitz drew attention to the fact that the four manuscripts used by Sepúlveda (1527) for his Latin translation of Alexander's commentary[54] occasionally provide readings independent from the manuscripts available to Bonitz himself.[55] Whether or not Sepúlveda's manuscripts are independent also from all the other Greek manuscripts that are available to us today[56] and that exceed by far

---

Diels 1882 (= CAG IX) (see *suppl. praefationis* XII, where **O** is named **B**), Philoponus, *In De generatione et corruptione*, ed. H. Vitelli 1897 (= CAG XIV.2) (see *praefatio* VIII, where **O** is named **S**), and Olympiodorus, *In Meteora*, ed. G. Stüve 1900 (= CAG XII.2) (see *suppl. praefationis* XIII, where **O** is named **II**).

[49] Golitsis 2016.

[50] There are, however, conjunctive errors (*Bindefehler*) between **A** and **P**^{b} in the passages from Alexander's commentary that are under consideration here (see *In Metaph.* 165.3–4; 299.6–8; 330.7–8; 354.28). In private correspondence, Golitsis ascribed these conjunctive errors to coincidence.

[51] Golitsis dates **P**^{b} to about 1440.

[52] Hayduck gives the *siglum* **B** to the 'Ambrosianus B 115,' which, as Golitsis pointed out to me, is actually the Ambrosianus D 115, hence in Golitsis 2016 "**D**."

[53] For the status of **L** and **F** see 2.4.

[54] Sepúlveda says in his preface: *innumera librariorum errata, quae passim scatebant, quatuor exemplaribus conferendis, per laboriosum examen mihi fuerunt castiganda.*

[55] Bonitz 1847: IX. Sepúlveda's translation does not have the large lacuna that **A**, **O**, and **M** have in 318.21–319.27. Hayduck could supplement the Greek text of the passage on the basis of **L** and **F** (see 2.4).

[56] In 113.13, where **A** and **M** have a lacuna (and **L** and **F** read the text of the *recensio altera*), **S** offers a text that can now be confirmed on the basis of **O**. In **O** (*f.* 708r 24–25), I read between the words ἀρχῶν

the number of Bonitz's manuscripts, is a matter currently investigated by Pantelis Golitsis (see my remarks in 2.2). Given the precision of Sepúlveda's Latin translation[57] it is often possible to determine which Greek text he is translating, especially when it matches with the evidence in (one of) our Greek manuscripts.[58] The possibility, however, remains that Sepúlveda's translation is misleading simply because he misunderstands what Alexander says.[59]

Independent of Bonitz's and Hayduck's[60] records of the evidence in S, I checked the commentary passages relevant for my study in the 1527 edition of Sepúlveda's translation. In those cases where the Latin text cannot be identified as a translation of the Greek text available to us, I note it in my apparatus or add a note in my text. The faithfulness with which Sepúlveda testifies to the four Greek commentary manuscripts is limited when it comes to lemmata or quotations from the *Metaphysics*. The limited reliability of Sepúlveda's reports of lemmata and quotations is readily explainable from the specific layout of Sepúlveda's edition. Sepúlveda's edition includes a *complete Metaphysics* text in Latin that is printed in sections, each of which is followed by Alexander's comments on it. This Latin version of the *Metaphysics* is based on a *Metaphysics* text that was available to Sepúlveda in the 16th century and which therefore can diverge widely from the *Metaphysics* text presupposed in Alexander's commentary. Given this editorial situation it does not come as a surprise that Sepúlveda in many places either adjusted the text of Alexander's lemmata (and quotations) to the *Metaphysics* text that precedes the commentary in his edition or simply abbreviated lemmata and quotations to avoid repetition.[61] This procedure might result in blatant inconsistencies between Alexander's paraphrase and the *Metaphysics* text that appears in the lemma and

---

... πῶς (113.13) the following: τὰ γὰρ μεταξὺ τινῶν κατὰ κοινωνίαν τινὸς καὶ οἰκειότητα μεταξύ ἐστι. ταῦτα δὲ ἑτέρου γένους καὶ ἑτέρας ὄντα φύσεως.

[57] See Coroleu 1995: 182 on Sepúlveda's method of translation: "According to him [Sepúlveda], this [clarity] can be obtained in two ways: by avoiding literalness and blind loyalty to word-by-word method, and on the other hand, by not falling into an extreme liberty which turns the version into a mere explanation of the text."

[58] Bonitz 1847: VIII–IX.

[59] For instance, Sepúlveda adds a sentence that betrays his misunderstanding of Aristotle's words: in 167.20 Alexander writes οὕτως ἀξιοσπούδαστός ἐστιν ("so much is it worthy of utmost devotion"), referring to people who believe in mythical stories about their region and who therefore are ready to fight for their land (167.15–20). Sepúlveda, however, translates this expression as *atque ad hunc quidem modum Aristoteles dicto suo fidem facere conatur* (f. i.iv.r).

[60] Hayduck occasionally records Bonitz's information on S imprecisely, e.g. in 59.2.

[61] See e.g. in Sepúlveda f. m.vi.r (220.2–3 Hayduck, see also 4.1.3). In Sepúlveda f. h.iii.r. the lemma reads the β-version whereas Alexander's commentary shows that he must have read the α-version (144.15–16 Hayduck), see 5.2.4. Sepúlveda does not repeat a quotation but just says *idem: f.* h.iii.r (145.7–8 Hayduck and 5.2.4). Yet it does also happen that Sepúlveda writes out the lemma in its authentic form without minding the repetition of Aristotle's text: e.g. 11.3–5 Hayduck vs. Sepúlveda, *f.* a.iii.r (see also 4.2.1).

quotations.[62] Therefore, the credibility of the evidence that Sepúlveda provides for the lemmata and quotations in Alexander's commentary is questionable and so the text in these lemmata and quotations might not testify to the actual reading in the four Greek manuscripts available to Sepúlveda.[63]

2.4 THE SO-CALLED *RECENSIO ALTERA*

The manuscripts Hayduck used for his edition offer two different versions of the commentary. The authentic commentary (on Α–Δ) of Alexander of Aphrodisias, represented by **A, O, M,** and **C**, differs from the version that is preserved in **L**[64] and **F**.[65] Hayduck calls the version of **L** and **F** *recensio altera* and presents its text in a separate apparatus. The differences between the authentic version and the *recensio altera* become apparent especially in the first two books of the commentary.

According to Golitsis 2014, who offers the first extensive treatment of the text preserved in **L** and **F**, the *recensio altera* is a commentary that heavily depends on Alexander's commentary. It also shows an influence of Asclepius's commentary. The latter fact gives us a *terminus post quem* in the sixth century AD.[66] This dating squares well with my research into the *Metaphysics* version(s) used by the author of the *recensio altera* and which suggests that the author was familiar with both versions α and β.[67] Golitsis furthermore suspects the anonymous author to have

---

[62] For example in Al. 273.20-25 (=*f.* q.ii.v), see also 4.2.3 below or in Al. 228.29-229.1 (= *f.* n.iii.v), see also 4.2.4.

[63] Moreover, Sepúlveda sometimes wishes to expand Alexander's short remarks on variant readings. The short notification ἄμεινον γεγράφθαι (68.3), which hints at a conjecture, Sepúlveda translates with the following sentence: *quanquam nescio an rectius sit, quod in quibusdam exemplaribus legitur ad hunc modum*. On this see 5.3.4 below. Relatedly, Sepúlveda translates Alexander's description of Aristotle's wording (περὶ τοῦ κακοῦ ἡμῖν προσθεῖναι κατέλιπε 33.26) not verbatim, but as he interprets its meaning (*omisit mentionem de malo*). See 5.3.3.

[64] Manuscript **L** (*Laurentianus* 87,12) contains the *Metaphysics* text (= **A**[b]) and Alexander's commentary in the margins. The *Metaphysics* text until Λ 7, 1073a1 is from the 12th century (see Golitsis 2014b: 219-20 on Cavallo's proposed dating to the 11th century), the rest from the 14th century (Moraux 1976: 302-304). The marginal text consisting of the commentary was written in the 12th century (Harlfinger 1979: 32).

[65] Manuscript **F** (Ambros. F 113 sup. [363]), a codex from the 14th century (Harlfinger 1979: 32-33; Golitsis 2014b: 216-17), also contains the commentary written *in margine*, surrounding the *Metaphysics* text of (the *Metaphysics* manuscript) **M**. Both **L** and **F** come with a *Metaphysics* text of the β-version. According to Golitsis (2014: 217) the two manuscripts **L** and **F** are two independent witnesses to the *recensio altera*.

[66] Golitsis 2014b: 214-16. Already Hayduck (1891: IX) supposes that the author of the *recensio altera* used Asclepius's commentary.

[67] See *rec alt.* app. 10 (apart from the quoted τίθεσθαι [β] the author also knows the alternative πείθεσθαι [α]); *rec. alt.* app. 67 (vs. Al. 68.3-4; see also 5.3.4). Cf. also *rec. alt.* app. 138 (vs. Al. 138.26-28) and *rec. alt.* app. 132 (vs. Al. 132.16-133.4): here the distinction is not between α and β, but rather between Alexander's text and ω^{αβ}. Cf. also Golitsis 2014b: 208.

been a professor with a Christian and a Platonic background.⁶⁸

The divergence between the authentic version and the *recensio altera* is apparent only in certain passages in the commentary's first two books (A, α) and the beginning of the third book (B). Only in those passages does it make sense to speak of a separate version that diverges from the authentic one.⁶⁹ By contrast, in books B–Δ we encounter only minor differences between the authentic and the alternative version.⁷⁰ So it seems as though the anonymous commentator restricted his re-composition of Alexander's commentary mainly to the first two books.

Nevertheless, the textual evidence of the *recensio altera* preserved in L and F can be of vital importance for restoring the text of Alexander's authentic commentary. Since the *recensio altera* often faithfully copies the text of Alexander's commentary, it can become a crucial witness to the original text of Alexander in those passages where all manuscripts of the authentic version are corrupt.⁷¹ For instance, the beginning of the commentary (1.3–2.3) as presented in the editions by Bonitz and Hayduck is based solely on the evidence of the *recensio altera* in manuscript F. The present study focuses on the authentic commentary of Alexander of Aphrodisias, and also occasionally references relevant information available in the *recensio altera* as preserved by the manuscripts L and F.

2.5 THE ARABIC FRAGMENTS OF THE COMMENTARY ON BOOK Λ

Ibn Rushd (1126–1198) or, in Latin, Averroes, is the most important Arabic commentator on Aristotle.⁷² He wrote three commentaries on the *Metaphysics*, which the Latin tradition classed into a short, middle, and long commentary.⁷³ My consideration of Averroes is limited to the Long Commentary on the *Metaphysics* (*Tafsīr Mā baʾad at-Tabīʾat*),⁷⁴ and specifically to the commentary on book Λ (*Lām*).⁷⁵ Averroes was destined to share with Alexander the title 'the commentator'; he shared with him additionally the distinction of interpreting Aristotle not

⁶⁸Golitsis 2014b: 214–16.

⁶⁹Golitsis 2014b: 216–17 n. 20, however, insists on the autonomous character of the *recensio altera* and calls it a 'selective' rather than a 'partial' re-composition of Alexander's commentary.

⁷⁰With the exception of one passage in book Δ where the *recensio altera* diverges considerably from the vulgate version: instead of Alexander's comments in 431.10–437.2 the *recensio altera* reads an extract from Asclepius's commentary (*In Metaph.* 352.26–354.5 Hayduck).

⁷¹Cf. Hayduck 1891: VIII.

⁷²For an overview of the Arabic commentary tradition see Adamson 2012, and for the relationship between Averroes and Alexander in particular see 648 and 653.

⁷³On this standard classification see Gutas 1993: 41–42 and 55.

⁷⁴On the name of *Tafsīr* see Gutas 1993: 33.

⁷⁵The edition of the text is by Bouyges 1948. I cite Averroes by the page numbers of Bouyges's edition.

through a Neoplatonic lens, but as faithfully as possible.⁷⁶ It is no great surprise then that Averroes in his Long Commentary refers to and quotes Alexander extensively. As Averroes makes clear at the beginning of his commentary (1393), he had access only to Alexander's commentary on book Λ, and even here only to two thirds of it.⁷⁷ Averroes says the following about his use of Alexander's commentary (1393): "It seemed to me best to summarize what Alexander says on each section of it as clearly and briefly as possible."⁷⁸ So Averroes' excerpts of Alexander do provide us access, however limited and abbreviated, to Alexander's authentic commentary on book Λ of the *Metaphysics*.⁷⁹

Jakob Freudenthal was the first to complete a study (with German translation) on the Λ-fragments of Alexander's commentary.⁸⁰ Apart from Freudenthal's translation of the fragments into German, there is an English translation of book Λ of Averroes' commentary by Genequand (1986) and a French translation by Martin (1984). For my study of the Λ-fragments I will rely not only on these modern translations but also take into account the Latin translation by Michael Scotus († about 1235).⁸¹ The Latin translation⁸² is especially important since the Arabic text is transmitted by one manuscript only.⁸³ Additionally, there is a Hebrew ver-

---

⁷⁶Adamson 2012: 653: "Significantly, for Averroes the most important previous commentator on Aristotle was the Peripatetic Alexander, whereas the Bagdad school was influenced primarily by the late Alexandrian Neoplatonist school (though they too read and made use of Alexander). Averroes' reading of Aristotle is, in short, a far cry from the Platonizing and harmonizing approach of al-Kindī."

⁷⁷Genequand 1986: 59 translates: "No commentary by Alexander or by the commentators who came after him has been found on the books of this science, nor any compendium, except on this book; we have found a commentary by Alexander on about two thirds of the book and by Themistius a complete compendium on it according to the sense." The limited access to Alexander's commentary on book Λ is connected to the fact that the Arabic translation of the *Metaphysics* by Mattā, which was written in the same manuscript as Alexander's commentary, was also unavailable from line 1072b16 onwards (up to 1073a13) (1613). The last Alexander-fragment (fr. 32 Freudenthal) is in 1623. That Alexander's commentary and Mattā's translation do not stop at exactly the same point could be due to the material condition of the manuscript (Genequand 1986: 7).

⁷⁸Genequand 1986: 59.

⁷⁹Recently Di Giovanni/Primavesi 2016 questioned the authenticity of the Alexander-fragments preserved by Averroes, arguing that the commentary Averroes used was a *revised* version of Alexander's commentary.

⁸⁰Freudenthal counts 36 fragments in total. On Freudenthal's discovery that the fragments in Averroes are incompatible with the inauthentic *Metaphysics* commentary on book Λ see 2.1.

⁸¹I thank Dag N. Hasse and Stefan Georges (Würzburg) for providing me access to their forthcoming edition of Scotus's translation. On Scotus's translation see also Primavesi 2012b: 401. On the Latin tradition of Averroes' commentary in general see Bouyges 1952: LXVI-LXXXIV.

⁸²Freudenthal 1885: 121-23 speaks rather dismissively of the Latin rendering of the fragments of Alexander.

⁸³Bouyges 1952: LXVI. On Cod. *Leid. or. 2074* (**B**) see Bouyges 1952: XXVII-XXXVIII; on the erroneously inserted folia in *Leid. or. 2075* (**C**) see Bouyges 1952: XXXVIII-XLII. A first description of the manuscript was offered by Fränkel 1885.

sion of the Arabic text.[84]

The form of Averroes' commentary differs from Alexander's in that the *Tafsīr* contains the complete text of the *Metaphysics*, which is placed as *textus* in front of each commentary section. Averroes himself read the *Metaphysics* in Arabic translation,[85] and in the case of book Λ Averroes used multiple translations.[86] He read the translation by Abū Bišr Mattā,[87] which Averroes used for most of book Λ (1069a18[beginning of Λ]–1072b16; 1073a14–1076a4[end of Λ]), in a manuscript that also contained Alexander's commentary. How were the *Metaphysics* text and Alexander's commentary arranged in this manuscript? Bertolacci (2005: 245 n. 9) suspects that the manuscript contained Alexander's commentary with lemmata that, when combined, formed a complete text of the *Metaphysics*, but it is also possible that Alexander's commentary was written *in margine* around the *Metaphysics* text. The evidence in the Fihrist (AD 988), the great index of Arabic literature, does not help to settle the matter. There we read: "Abū Bišr Mattā translated treatise 'L'—namely the eleventh letter—with Alexander's commentary into Arabic."[88] In addition to Mattā's translation Averroes references other translations in his commentary, as, in the case of book Λ, the translation by Ustāth.[89]

Given that Averroes had access to the words of Aristotle and Alexander through an Arabic translation (that had been translated from a Syriac intermediary) and I in the present study rely on modern and mediaeval translations of Averroes commentary,[90] one might raise doubt whether it is at all possible to draw from Averroes information about the *Greek text* of Alexander's commentary and Alexander's *Metaphysics* text.[91] Yet, as I hope to show in the case studies 3.5.2.2 and 5.1.5, it remains true that, in some places, Averroes' information about

---

[84] See Bouyges 1952: LXXXIV–XCIX. In fact, Freudenthal at first based his translation on the Hebrew text only. After the discovery of the Arabic text in the Leiden codex, S. Fränkel checked Freudenthal's translation against the Arabic original and Freudenthal adjusted his translation accordingly (Freudenthal 1885: 66).

[85] For extensive information on the Arabic translations of the *Metaphysics* see Bertolacci 2005.

[86] Bertolacci 2005: 253: "Λ is the book for which Averroes uses the highest number of translations."

[87] Mattā most likely based his translation on a Syriac version. Bouyges 1952: CLXXVII–CLXXVIII (also Walzer 1958: 221). Genequand 1986: 5 however has doubts about this. From Mattā's other translations one can infer that he usually worked with Syriac texts (cf. Bertolacci 2005: 245 n. 9).

[88] Bertolacci 2005: 244. Averroes' remark in his commentary (1537), "I found the section which I transcribed first in the manuscript of Alexander blended with the text of Alexander," refers most likely not to his usual practice, but highlights an exception. When Averroes speaks of the *Metaphysics* text that is "blended with" Alexander's comments he probably refers to the quotations of the Aristotelian text within Alexander's commentary.

[89] Bertolacci 2005: 244, 251, and 253. The translation that appears for the other parts of book Λ (1072b16–1073a13) in textus and citation is the one by Ustāth.

[90] Andreas Lammer (Munich) kindly checked my statements about the Arabic text against the Arabic original.

[91] Cf. Ross 1924 II: 347–48.

Alexander's commentary allows us to draw conclusions about the text Alexander used and his comments on it.

CHAPTER 3

Alexander's Commentary as a Witness to the *Metaphysics* Text

There are many obstacles to the evaluation of Alexander's commentary as a witness to the *Metaphysics* text. These result from the special conditions in which the evidence of the *Metaphysics* text appears within Alexander's commentary. One might, for instance, raise doubt about the accuracy with which Alexander quotes the *Metaphysics* text. Additionally, the commentary itself is a product of a transmission process (cf. 2). Furthermore, given that the reception of a work and the reception of its commentary are typically simultaneous, the question arises whether the transmission of the commentary and the transmission of the *Metaphysics* text were at some points intertwined[1] and hence subject to mutual influence. As it happens it seems that quotations from the *Metaphysics* in the commentary were likely adjusted to the *Metaphysics* text in cases where they differed.[2] If such contamination of Alexander's commentary is to be expected we are in need of a criterion that will help us to determine which passages in Alexander's commentary faithfully represent Alexander's *Metaphysics* text.

The search for an adequate criterion has to start from the following questions, which I am going to discuss in the subsequent sections: what kind of textual evidence of the *Metaphysics* did Alexander have when he wrote his commentary? Did he use more than one *Metaphysics* exemplar (see 3.1)? We must also ask: how does Alexander's commentary provide us access to his *Metaphysics* texts? We can distinguish four different types of evidence in Alexander's commentary: (i) lemmata, that is, citations from the *Metaphysics* text that introduce a commentary section (see 3.2); (ii) quotations within a commentary section (see 3.3);

---

[1]We encounter the phenomenon of a transmission community in the β-family. The commentary of the *recensio altera* is written in the margins of **L** (= *Metaph.*-ms. **A**[b]) and **F** (= *Metaph.*-ms. **M**). Furthermore, manuscript **C** (= *Metaph.*-ms. **I**[b]) contains the authentic version of the commentary *in margine* surrounding a *Metaphysics* text of the α-version, which is, however, contaminated by the β-version.

[2]For contamination in the opposite direction (i.e. from the commentary to the *Metaphysics* text) see 5.

(iii) Alexander's paraphrase of Aristotle's argument; and (iv) Alexander's critical discussion of it (see 3.4).[3]

Alexander's knowledge of the *Metaphysics* text is not restricted to the exemplar he used when writing his commentary. Alexander reports variant readings that differ from his own *Metaphysics* exemplar. From where does he have this information (see 3.5)? One source Alexander explicitly names is the early second-century commentator Aspasius (see 3.5.1).

3.1 PRELIMINARY CONSIDERATIONS: HOW MANY *METAPHYSICS* EXEMPLARS DID ALEXANDER USE?

Alexander wrote a detailed continuous commentary on the *Metaphysics*[4] and through his commentary we gain access to the *Metaphysics* exemplar he used. This access is, however, limited because the reconstruction of Alexander's *Metaphysics* text from the commentary depends on the way in which Alexander presents this text. Before we look at how exactly Alexander's comments allow us to reconstruct the text he used the following question has to be discussed: did Alexander use one or more *Metaphysics* exemplars? The answer to this question is not readily apparent, because Alexander does not comment on the type[5] and condition of his *Metaphysics* exemplar(s).[6] The fact that Alexander now and then refers to variant readings by no means implies that he *himself* found these in another manuscript.

Where in his commentary could Alexander have provided a clue about his text? His introduction is a likely candidate, but we do not have the introduction to the first book of the *Metaphysics* commentary. Moraux has hypothesized that there once existed an extensive introduction to the whole of the commentary.[7] Should we expect Alexander to have commented on his *Metaphysics* exemplar in this lost introduction? When we look at the preserved introductions to other commentaries by Alexander we see that this question can be answered in the negative.[8] Looking at the introductions to the subsequent books of the *Metaphysics*

---

[3]Cf. Barnes's (1999: 36) classification of the evidence in Aspasius's commentary on *EN*: "lemmata, … citations, … paraphrases, … the commentary itself, … passages in the commentary where Aspasius explicitly discusses textual points." Diels 1882: 4 n. 1 classifies Simplicius's *Physics* commentary in terms of lemma, citation and paraphrase.

[4]For the continuous philosophical commentary see Hadot 2002; for Alexander's use of it and its later adaptations see Fazzo 2004: 8–9 and D'Ancona 2002: 206–26.

[5]Alexander read the text from a papyrus scroll, which was the usual medium of such texts in the second century AD (see the chart in Bülow-Jacobsen 2009: 25).

[6]Cf. Busse 1900: 74 in respect to Ammonius's commentary on *Int*. (fifth to sixth century AD): "Aber von einer Mitteilung über seine Handschrift findet sich in dem Kommentare nicht eine Spur. Alle in uns auftauchenden Fragen nach dem Ursprung, dem Alter, der Beschaffenheit derselben bleiben unbeantwortet."

[7]Moraux 2001: 431–32 with n. 26.

[8]In the following commentary introductions no comments are made on the type and condition of

(α, Γ, and Δ), we see that these usually contain discussions on topics such as the authenticity, composition, and position of the respective books,[9] and so it is likely that in the lost introduction to book A and the whole commentary Alexander had raised the very same issues with respect to the whole of the *Metaphysics*.[10] Since in the introductions to the subsequent books of the *Metaphysics* Alexander never says a word about the number or, more generally, the material condition of his *Metaphysics* copy, we may conclude that such questions were outside the scope of Alexander's interests.

Should we then infer from Alexander's silence that he had but one *Metaphysics* text in front of him when composing his commentary?[11] None of the available evidence contravenes the assumption that he had only one, and three considerations speak in favor of it. First, Alexander sporadically mentions *variae lectiones* (see 3.6) that he judges favorably and even prefers to the reading he finds in his own text. He therefore distinguishes clearly between his *own*, standard text and the variant but preferred reading of *another* source. If he generally had two or even three different texts at his disposal it would not make sense for him to designate the preferable reading as a *variant*. If he had worked with multiple texts, why would he not have simply taken the (in his view) correct reading as the starting point of his comments and proceed from there?[12]

Second, it seems unreasonable to suppose that Alexander himself actively sought variant readings in other manuscripts. Moraux (2001: 429) points to Galen as a clear counterexample. Galen, a commentator on Hippocrates and contemporary of Alexander,[13] explicitly discusses the acquisition of old manuscripts that contain more accurate readings.[14] Had Alexander engaged in this kind of re-

---

the exemplars: commentary on *De Sensu* (CAG III.1; 1.3–2.24 Wendland), commentary on the *Topics* (CAG II.2; 1.3–7.2 Wallies), commentary on *Analytica Priora* (CAG II.1; 1.3–9.2 Wallies); see also the beginning of the commentary on the *Meteorologica*, which does not include an introduction in the proper sense (CAG III.2; 1.3–4.11 Hayduck).

[9] In his introduction to book α, Alexander extensively discusses the question of the status and position of the book (136.8–17; 137.1–138.4). See also 3.5.2.3 and 4.1.1. The authenticity of book B is discussed in 196.20–24. In the introduction to book Γ (237.3–239.3), Alexander situates the book's content within its context in the *Metaphysics*: 237.8–238.3. In his introduction to book Δ (344.2–345.20), Alexander addresses concerns about the book's authenticity: 344.2–7; questions about the composition are discussed in 344.20–345.1; for the completeness of the book see 345.4–11.

[10] Moraux 2001: 431–32.

[11] Cf. Busse 1900: 74–75 concerning Ammonius: "wir können [...] kaum den Argwohn unterdrücken, dass er in eine andere Handschrift überhaupt keinen Blick geworfen hat. [...] Dass also Ammonius irgend eine andere Handschrift zu Rate gezogen haben sollte, [...] ist höchst unwahrscheinlich." See also Ilberg 1890: 112 on the pre-Galen commentaries on the Hippocratic treatises.

[12] For this argument see Flannery 2003: 125 n. 25.

[13] For mutual references in the writings of Galen and Alexander see Sharples 2005: 50–51, with further literature.

[14] See, for example, the beginning of his treatise *In Hippocr. libr. de officina med. comm.* (XVIII, II, pp. 630–31 Kühn). What is remarkable is that Galen speaks in the first person: he himself searched for other manuscripts (ὥσπερ τὰ παρ' ἡμῖν ἐν Περγάμῳ ... ὅπως ἐκ τῶν πλείστων τε καὶ ἀξιοπιστοτάτων

search, we could expect him to have mentioned it in his work. Alexander does refer to variant readings in other manuscripts, but he never gives a clue about where and how he found these variants.[15] In addition, Alexander consistently introduces *variae lectiones* with the prosaic formula "some [manuscripts] read" (ἔν τισι(ν) φέρεται / γράφεται). The plural ἔν τισι(ν) is strikingly anonymous.[16] It is simply a standard formula for reporting variant readings (see also 3.6). By no means does it imply that Alexander independently collated different manuscripts.[17]

Another helpful counterexample is Simplicius, a sixth-century Neoplatonic commentator who in many ways stands in the tradition of Alexander.[18] Simplicius's presentation of variant readings and his attitude towards them is very different from Alexander's.[19] For example, he proclaims emphatically and repeatedly that he *himself* found another reading in another manuscript:[20] *In Phys.* 377.24–26: οὐχ οὕτως ἔχει ἡ γραφὴ <u>τῶν ἐμοὶ συνεγνωσμένων</u> ἀντιγράφων πάντων, ἀλλ' οὕτως … and *In Phys.* 1317.6–7 <u>ἐγὼ</u> μέντοι ἔν τισιν ἀντιγράφοις ἀντὶ τοῦ 'ἡ φορὰ' 'ἡ περιφορὰ' γεγραμμένον <u>ηὗρον</u>.[21] Whether or not it is true that Simplicius personally compared different manuscripts is another question. The important thing is that Simplicius thought it worth mentioning and that this activity was within the realms of possibility for him as a commentator. There is no passage in Alexander's commentary where he claims to have found another reading in another manuscript. There is no mention of his involvement in the discovery of a variant version of the text.[22] It seems that Alexander simply did not have a genuine inter-

εὕροιμεν τὰς γνησίας γραφάς). Galen even mentions the different formats of the manuscripts: codices, papyrus scrolls, or single sheets (τὰ μὲν ἔχοντες ἐν τοῖς βιβλίοις, τὰ δὲ ἐν τοῖς χάρτοις, τὰ δὲ ἐν διαφόροις φιλύραις)!

[15] See Barnes 1999: 41 in respect to Aspasius's (first to second century AD) commentary on the *Ethics*.

[16] Cf. Busse 1900: 74–75 about Ammonius's commentary on *Int.*: "und wir vermuten, dass das gespreizte ἐν τοῖς πλείστοις ἀντιγράφοις nichts weiter als eine hohle Redensart ist." See also Fazzo 2012a: 62 n. 35: "Please note that the expression ἔν τισι <ἀντιγράφοις>, is always in the plural, so that it is not possible to judge whether one or more manuscripts are mentioned."

[17] Moraux's premature inference (2001: 429: "Dennoch lässt sich feststellen, dass er bemüht ist, den Text nicht nach einem beliebigen Manuskript zu erklären, sondern verschiedene Lesarten zu ermitteln.") is surprising given the fact that elsewhere he underlines the contrast between Alexander and Galen. Hecquet-Devienne (2008: 11) also seems to be too quick in her assessment of the matter.

[18] On the question how Simplicius worked with Alexander's commentary see Baltussen 2008: 107–35.

[19] In *In Phys.* 395.20–21, Simplicius states as a general fact that many variants of the text of Aristotle's *Physics* have been transmitted. In *In Phys.* 450.32–36, he also takes into consideration that a corruption could have been caused by the insertion of a marginal gloss into the text. See Baltussen 2008: 33–42.

[20] See Golitsis 2008: 79. For the question how Simplicius worked with his sources see Golitsis 2008: 66–71 and Baltussen 2008: 31–53.

[21] See also Simp. *In Phys.* 1093.5–6 Diels; Simp. *In Cael.* 291.24–25 and 521.25–26 Heiberg.

[22] Alexander does show personal interest now and then in discussing how a passage might be im-

est in the matter—he was a philosopher, not a philologist.²³

Yet Alexander's attitude is by no means unusual among the commentators. As Barnes puts it concerning Aspasius's *Ethics* commentary from the early second century AD:²⁴

> Like all the extant commentaries on Aristotle. The later commentaries all indeed contain philological notes, just as Aspasius' commentary does; but such notes are generally sparse, and they do not constitute the intellectual centre of the works.

Simplicius seems to be an exception here. It has been noted that due to its scholarly character, his commentary draws on an enormous number of sources *in addition to* the text on which he is primarily commenting.²⁵ The situation is different with Alexander.

This leads to the third consideration. On the whole, Alexander's references to *variae lectiones* play a small role in his commentary work.²⁶ Many of the *variae lectiones* that Alexander reports bear no relation to his interpretation of the relevant passage. Also, Alexander often does not decide which of two possible readings he prefers or regards as original.²⁷ Like other ancient commentators he shows astonishingly little interest in the search for the one and only *original* reading. He does not raise questions as to what error might have occasioned a certain unsatisfactory reading.²⁸ After all, collating different manuscripts in the form of papyrus scrolls

---

proved, but these conjectures are not connected to a variant reading. Rather these are interpretative endeavors that are based on a discussion of the Aristotelian argument. In these contexts we sometimes find formulations that point to Alexander himself: e.g. δοκεῖ δέ μοι αὕτη ἡ λέξις … τὴν τάξιν ἔχειν (267.14–17; see also 3.6). Alexander also shows dedication when discussing questions like the authenticity and order of the books: e.g. 137.7: μοι δοκεῖ. These matters are of interest to Alexander, but the collation of other manuscripts is not.

²³ Jaeger 1923: 32 calls Alexander in passing 'philologically naïve' ("er [Alexander], der in philologisch-kritischer Hinsicht freilich naiver war, als man es zu seiner Zeit zu sein brauchte").

²⁴ Barnes 1999: 24 n. 67.

²⁵ See Baltussen 2008: 21–53; 211–15 and Golitsis 2008: 65–79. This has been connected with historical circumstances that made Simplicius a teacher without a classroom. On this see Golitsis 2008: 18–22 and Baltussen 2008: 48–51.

²⁶ This is not contradicted by the fact that later commentators referred to Alexander as a witness for other readings or textual matters (see Golitsis 2008: 66–70; Baltussen 2008: 127–29; Kupreeva 2012: 113). This is just a natural part of the commentary tradition. A commentary is a place where variant readings are mentioned and thereby preserved. Early commentaries in a sense also served the role of a modern apparatus.

²⁷ Alexander often contrasts several possible interpretations and sometimes even variant readings. See e.g. 13.9–17; 21.14–31; 27.15–25; 37.6–12; 41.21–32; 42.20–21; 50.24–51.25; 63.25–27; 100.22–30; 169.9–17; 186.25–187.6. See also Moraux 2001: 438 with n. 60.

²⁸ Diels 1882: 30 n. 3: "Am meisten fehlt den Interpreten die Einsicht, dass doch nur eine Lesart richtig sein kann." See also Wittwer 1999: 78 concerning Aspasius: "And, finally, note also that, surprisingly, Aspasius does not share with us a quite fundamental principle according to which only one reading can be the true one. […] The reason, I think, is not that he did not know which reading to prefer and therefore suspended judgment, it is rather that he was happy to have both readings at the

requires immense effort, not at all comparable to the crosschecking of different editions today. Were someone to make this effort, we would reasonably expect to see a clearer sign of it than Alexander's scant references to variant readings.

Still, since Alexander now and then mentions variant readings he must have some sort of access to information on the text of the *Metaphysics* beyond what his exemplar could supply. The most likely explanation is that he gained this information from earlier commentators (see 3.5) or from marginal notes in his own exemplar.[29] Alexander seems simply to be reporting those variant readings that he knows of. As there is no intrinsic connection between a variant reading he introduces and his interpretative interest in the text, we may assume that he used just one exemplar when composing his commentary. This does not, of course, exclude the possibility that Alexander had available in his school other *Metaphysics* exemplars. The evidence suggests that Alexander based his commentary mostly on *one* text, his working exemplar, which most likely contained variant readings in the margins. This text I call ω^{AL}. I assume that Alexander used this text for the lemmata (see 3.2), and that whenever he quotes directly from the Aristotelian text without designating the text as coming from *another* version he takes it from ω^{AL}.[30] Furthermore, I take it that his paraphrase, explanation, and critical discussion are based on ω^{AL}, unless otherwise noted.

3.2 THE EVIDENCE IN THE LEMMATA

A lemma is a citation from the text being commented on that stands at the head of a commentary section and indicates the passage under examination.[31] It is syntactically independent of the preceding or subsequent comments and is marked by textual layout and punctuation.[32] The length of the lemma in Alexander's commentary ranges from a few words[33] to multiple lines.[34] The investigation into the

---

same time."

[29] It was a common practice among (late) ancient *philologists* and also booksellers to check one edition against another and mark down corrections or variant readings (see Erbse 1979: 548–49, who refers to Strabo, *Geographica* XVII 1, 5 / 790 C.24–27 = IV, 420 Radt and XIII 1, 54 / 609 C.18–22 = III, 602–604 Radt), so it is likely that Alexander's exemplar of the *Metaphysics* contained variant readings.

[30] Clearly this does not imply that what we find in the lemmata and citations of our commentary manuscripts necessarily represents the readings in ω^{AL}.

[31] An introduction to the lemma in ancient commentaries is provided by Wittwer 1999: 51–58.

[32] The typographical marking in modern editions has ancient roots. Wittwer 1999: 52: "We also have evidence that lemmata of Aristotelian commentaries were indeed written in ἔκθεσις, that is, that their lines projected into the left margin by one or two letters." See McNamee 1977: 33–36: papyri (also from the third century AD and especially relevant for my present purpose) show that lemmata were marked either by ἔκθεσις, or by a paragraphos ("a short horizontal stroke of the pen, written at the left of a column usually under the line in which the new quotation begins," 34–35).

[33] The lemmata in 29.5 (984a18) and 31.6 (984b3) each contain only three words.

[34] Extensive lemmata (three to four lines of text) are given e.g. in: 19.21–23 (983a24–26), 64.13–15

origin and authenticity of the lemmata in Alexander's *Metaphysics* commentary could commence from either of two different starting points:[35] (i) the investigation could begin with what was original to Alexander's commentary and ask *whether* Alexander used lemmata in his commentary at all, and if so, *why*; (ii) or the investigation could start at the opposite end in the history of the commentary and raise the question of *how reliably* the lemmata, as printed in Hayduck's edition, bear witness to Alexander's own *Metaphysics* exemplar, a question that scholars tend to answer in the negative.[36] In the following, I will follow both approaches to investigating the lemmata, and I do this in order to get a firm grip on the status and value of the lemmata in Alexander's commentary.

The question of whether Alexander put lemmata into his commentary can be answered in the affirmative. There is evidence in his commentary as well as external testimony indicating that he equipped his commentary with lemmata.[37] I will first make some general remarks on the type of commentary to which Alexander's belongs, then look at some pieces of external evidence, and finally focus on the characteristics of Alexander's commentary and the internal evidence for the genuineness of its lemmata.

Alexander's commentary on the *Metaphysics* is a continuous or 'running' commentary: Alexander stays close to the Aristotelian text and comments on almost every sentence in Aristotle. Since the practice of setting the commentary in the margins of the source text emerged only after the codex had replaced the roll,[38] it is safe to assume that Alexander designed his commentary to be used alongside Aristotle's text, but nevertheless as a standalone piece of writing. The reception of Aristotle's *Metaphysics* took place within the context of philosophical teaching. Alexander held a chair of philosophy in Athens,[39] and so it is most likely in this academic setting that his commentary was composed[40] and subsequently stud-

(988b22–24), 111.1–3 (991b22–25), 123.15–18 (992b9–13), 158.1–3 (994b6–9).

[35] For a full list of all extant lemmata in Alexander's commentary see appendix B.

[36] See e.g., Barnes 1999: 37 and Primavesi 2012b: 407–408.

[37] Bloch 2003: 24 concerning Alexander's commentary on *De Sensu*: "It can be established beyond reasonable doubt that Alexander must have used some sort of lemmata." This assertion is taken for granted by D'Ancona 2002: 209–11. Wittwer 1999: 62–67 comes to the same conclusion concerning Aspasius and his commentary. Cf. also Baltussen 2008: 114–16.

[38] Pfeiffer 1968: 218. See also Hoffmann 2009: "A major phenomenon of the history of commentaries in antiquity was the transition from the practice of putting the text commented upon and the commentary in separate books (rolls) to the practice of reuniting the commentary with text receiving commentary in the same book and on the same page—parceling the commentary out in the margins or encircling the text commented upon." See also Schironi 2012: 410.

[39] As testified by the newly discovered inscription at Aphrodisias: see Chaniotis 2004 and Sharples 2005.

[40] Sluiter 1999: 173: "The existence of a commentary on any given text is evidence that that text was used in teaching." See also Hadot 2002: 183–85; Fazzo 2004: 5–7 and (regarding Neoplatonic commentaries) Hoffmann 2009: 615–16. This, however, does not automatically mean that Alexander's commentary (in the form in which it came down to us) was a lecture script. The detailed discussion

ied alongside Aristotle's work for centuries to come. It therefore is reasonable to assume that the typical recipient of Alexander's commentary also had direct access to the text of the *Metaphysics*. These considerations immediately reveal the function of the lemma within the commentary: The lemma guides the student to the passage in Aristotle to which the comments pertain. Additionally, the lemma makes it easy for the student to find comments pertaining to a *Metaphysics* passage of interest to him.[41] Finally, lemmata divide the commentary into different sections, thus providing a helpful structure. All told, the lemmata seem to be a genuine part of this type of commentary.

These general thoughts are supported by specific evidence given, on the one hand, by Alexander's *Metaphysics* commentary itself, and on the other, by the commentator Simplicius. Simplicius writes in his commentary on *De caelo* A 12, 282a25–26:[42]

> Simplicius *In Cael.* 336.29–31 Heiberg
>
> Ὁ μέντοι Ἀλέξανδρος, καίτοι ἐν τῇ τῆς λέξεως ἐκθέσει γράψας ὁμοίως δὲ εἰ καὶ ἄφθαρτον, ὂν δέ, ἐν τῇ ἐξηγήσει ὡς οὕτως ἔχουσαν τὴν γραφὴν ἐξηγεῖται ὁμοίως δὲ εἰ καὶ ἀίδιον, ὂν δέ.
>
> However, although Alexander writes in his lemma[43] "and similarly if it is indestructible and existent," in his exegesis he expounds the passage as though it read "and similarly if it is eternal and existent."[44]

Simplicius's remark indicates that his exemplar (from the fifth or sixth century AD) of Alexander's commentary on *De caelo* contained lemmata.[45] If this is the case with Simplicius's exemplar of Alexander's *De caelo* commentary then it might not be too far-fetched to assume that it also holds for Alexander's *Metaphysics* commentary. But this is not the only information we can extract from this passage. Simplicius's surprise over the discrepancy between the reading in the lemma (ἐν τῇ τῆς λέξεως ἐκθέσει) and the paraphrase (ἐν τῇ ἐξηγήσει) shows not only that the lemmata in Alexander's commentary had suffered occasional

---

of the Aristotelian text rather suggests that the commentary is either an elaborate version of teaching notes or was designed as a work on which a teacher could draw for teaching purposes. Cf. Sharples 1990: 95–97.

[41] Lemmata were already being used by Alexandrian scholars. See Lamberz 1987: 7 with n. 24; Pfeiffer 1968: 218. Cf. Wittwer 1999: 69 with n. 63.

[42] Wittwer 1999: 52 n. 6 also draws attention to this important passage. See also Resigno 2004: 112 with n. 213 and 228; Golitsis 2008: 58–59, esp. n. 73.

[43] Wittwer 1999: 52: "The truth is that there is no name for lemmata in Greek. ... Most of the few [references to lemmata in ancient texts] I have found refer to them ... as 'what is said in ἔκθεσις.'" See also Wildberg 1993: 191.

[44] The English translation is by Hankinson 2006, but has been modified.

[45] A collection of fragments from Alexander's commentary on *De caelo* can be found in Resigno 2004. On the question of how precisely Simplicius references his sources see Baltussen 2008: 31–53.

corruption prior to or in the sixth century AD, but also that Simplicius takes it for granted that the lemmata display Alexander's *very own Metaphysics* text.[46] Simplicius says γράψας ("he [*sc.* Alexander] has written [in the lemma]," 336.29), which plainly indicates that Simplicius thinks the lemmata came from Alexander.

Further evidence can be found in the anonymous commentary on Plato's *Theaetetus* preserved on papyrus from the second century AD.[47] The lemmata contain partial or complete sentences from Plato's text, and they are highlighted in the mise en page.[48] And just as in the transmitted text of Alexander's commentary, they do not contain the whole passage to which the comments refer. While the anonymous commentary on the *Theaetetus* brings us closer to the time of Alexander but not to an exemplar of Alexander's commentary, Simplicius's remark brings us closer to an exemplar of Alexander's commentary, though only to about 300 years after that commentary was written. Still, the testimony corroborates our initial suspicion.

I turn now to the internal evidence in Alexander's commentary that speaks for the authenticity of the lemmata. The lemmata in Alexander's commentary are not just syntactically independent headlines. Rather, they are clearly anchored in the structure of the commentary.[49] This can be seen by the fact that a lemma marks a caesura in the text, which means that the commentary subsequent to the lemma begins for the most part *asyndetically*.[50] Additionally, Alexander often starts off his comments with a short summary of what was said in the preceding part of the commentary. He situates the new passage within the logic of the argument by introducing his summary of what Aristotle had previously said with an aorist participle (εἰπών..., λαβών ..., δείξας ..., διελών).[51] The main verb (...(ἑξῆς) δείκνυσιν..., φησί...) then directs attention to the given topic: "After having said such and such ... Aristotle now shows that ..." (cf. 3.4).

Apart from this, there are many cases in which Alexander begins a commentary section by immediately referring back to the words cited in the lemma. This shows that the lemma is a genuine part of the commentary. Here are some examples: in 194.8–11, the lemma contains the text of B 2, 997a25–6. This is the fifth of the aporiae treated in book B: ἔτι δὲ πότερον περὶ τὰς οὐσίας ἡ θεωρία μόνον

---

[46] Cf. Wittwer 1999: 54 n. 9.

[47] See Diels/Schubart 1905: VIII; XX–XXIV and the new edition of the papyrus by Bastianini and Sedley with extensive introduction: Bastianini/Sedley 1995: 227–61.

[48] Bastianini/Sedley 1995: 240–41.

[49] Cf. Bloch 2003: 24.

[50] Cf. Wittwer 1999: 67 concerning Aspasius's commentary and Lamberz 1987: 9–10 concerning the Neoplatonic commentaries. Lamberz considers whether the exceptional cases in which the first sentence contains a particle are due to later 'corrections.' In the case of Alexander's commentary, there are also other occasions when we encounter an asyndetical beginning: e.g. when introducing a *varia lectio* (46.23; 58.31–59.1; 91.5; 194.3). In these cases the asyndeton marks a new thought: see Kühner/Gerth II: § 546e, p. 346.

[51] See Moraux 2001: 437–38 and Luna 2003: 251.

ἐστὶν ἢ καὶ περὶ[52] τὰ συμβεβηκότα ταύταις ("Further, does our investigation deal with substances alone or also with their attributes?"). After the lemma Alexander begins by saying: καὶ αὕτη ἡ ἀπορία … ("This aporia too …"). Here he directly refers back to the aporia quoted in the lemma.[53] This back reference would not make sense if the aporia were not quoted in the lemma immediately preceding it.

We find a parallel case in 188.7–9. The lemma quotes B 2, 996b35: εἴπερ οὖν ὁμοίως μὲν ὁποιασοῦν ἐστίν, ἁπασῶν δὲ μὴ ἐνδέχεται.[54] This is the protasis of a statement made to address the second aporia (*Does wisdom bear on substances only, or does it also consider the principles of demonstration?*). Alexander continues his commentary by asking: Διὰ τί μὴ ἐνδέχεται ἁπασῶν;[55] In this way Alexander directly and immediately questions the assumption that is implicit in Aristotle's words in the lemma. Without the lemma Alexander's question would be incoherent, even unintelligible.

In 332.1–2, Alexander begins with a lemma quoting Γ 7, 1012a9 (ἔτι ἐν ὅσοις γένεσιν ἡ ἀπόφασις τὸ ἐναντίον ἐπιφέρει, καὶ ἐν τούτοις / "Again, in all classes in which the negation of an attribute means the assertion of its contrary, even in these…"), before continuing in the commentary section (332.3) with the words ὃ λέγει, τοιοῦτόν ἐστιν ("What he means is as follows."). This subsequent explanation of Aristotle's words (332.3–5) makes it clear that the relative pronoun ὃ refers to the content of the lemma,[56] which is therefore indispensable to Alexander's comments.

Further examples can be found in Alexander's comments on chapters Γ 6 and 7 (on the principle of non-contradiction), which often exhibit the following characteristics: in the lemma, Alexander quotes the Aristotelian argument against the denial of the principle of non-contradiction, and then begins the commentary with the formula ἡ ἐπιχείρησις τοιαύτη ("The argument is as follows.").[57] This is comparable to the situation in book A 9, where Aristotle enumerates several arguments against the Platonic theory of Forms. In 128.10–11, for example, Alexander quotes one of the arguments in the lemma and prefaces his comments by simply saying ὁ λόγος τοιοῦτος (128.12).

There are not only back references to the preceding lemma, but also references

---

[52] The word περί (997a26) is missing in the lemma of Alexander's commentary.

[53] Cf. also 203.1–3; 203.12–14; 204.23–24.

[54] The complete sentence reads (996b35–997a2): εἴπερ οὖν ὁμοίως μὲν ὁποιασοῦν ἐστίν, ἁπασῶν δὲ μὴ ἐνδέχεται, ὥσπερ οὐδὲ τῶν ἄλλων οὕτως οὐδὲ τῆς γνωριζούσης τὰς οὐσίας ἴδιόν ἐστι τὸ γιγνώσκειν περὶ αὐτῶν. / "If then it belongs to every science alike, and cannot belong to all, it is not peculiar to the science which investigates substances, any more than to any other science, to know about these topics." On the transmission of the word εἴπερ in Alexander's commentary see Hayduck's *app. ad loc.*

[55] *In Metaph.* 188.9: "Why can it not belong to all sciences?"

[56] Cf. the parallel case in 119.14 (ὃ δὲ λέγει τοιοῦτόν ἐστιν) and 198.33 (ὃ ἀπορεῖ τοιοῦτόν ἐστιν).

[57] *In Metaph.* 329.5–7; cf. 330.19 (συντιθέντι τὸν πάντα νοῦν ἡ ἐπιχείρησις τοιαύτη.); 331.9–10 (ἡ ἐπιχείρησις βραχέως μὲν εἴρηται καὶ αὕτη, ἔστι δὲ ὁ νοῦς αὐτῆς τοιοῦτος); 332.16–18 (ἡ ἐπιχείρησις αὕτη ἐστίν). See also the cases in 104.19–105.2; 294.22–23.

at the end of a commentary section which point forward to the subsequent lemma. In 200.31 Alexander writes ὑπὲρ γὰρ τοῦ ταῦτα δεῖξαι παρέθετο ("In order to prove this Aristotle goes on to say [lemma quoting Aristotle's words]"). Here Alexander segues from the end of a commentary section into the next lemma[58] by marking the quotation in the lemma explicitly as Aristotle's words (παρέθετο / "*he* goes on to say"). Compare the case in 281.37: here Alexander concludes the paragraph by saying δείξας δὲ ταῦτα ἐπιφέρει τοῖς δεδειγμένοις ("Having shown these things, he adds to what he has shown: [lemma]") as a transition to the subsequent lemma. In both cases the commentary after the lemma begins with an asyndeton.[59]

If we take these examples as evidence that Alexander used lemmata in his commentary, the next question is whether the precise format of the preserved lemmata is authentic.[60] Many commentaries from different centuries show lemmata in abbreviated form. In these cases the lemma contains the first and last word(s) of a given passage connected by ἕως (τοῦ) ("up to," "until").[61] Is this also the original format of Alexander's lemmata?[62] Is the format in which the lemmata are transmitted in our manuscripts the product of a later revision that expanded the text of the lemmata? This is not likely. There is just one instance in Alexander's commentary where the lemma is abbreviated by ἕως: in 37.4-5.[63] The lemma belongs to the passage in A 5, 985b23–26 and reads: ἐν δὲ τούτοις καὶ πρὸ τούτων ἕως ἐπεὶ

---

[58] This lemma contains B 2, 998a7-9: εἰσὶ δέ τινες οἵ φασιν εἶναι μὲν τὰ μεταξὺ ταῦτα λεγόμενα τῶν τε εἰδῶν καὶ τῶν αἰσθητῶν, οὐ μὴν χωρίς γε τῶν αἰσθητῶν ("Now there are some who say that these so-called intermediates between the Forms and the perceptible things exist, not apart from the perceptible things, however...").

[59] One might object that what appears as a lemma in these cases was originally a quotation (cf. 3.3) *within* the commentary that had later been marked as a lemma.

[60] Bloch 2003: 25 in regard to Alexander's commentary on *De sensu*: "We might ask if the lemmata were originally full quotations or just a few words as a reference to Aristotle's text." Bloch then concludes (27): "This being the case, the discrepancies [between the text in the lemma and the citation in the commentary] are certainly better explained, if Alexander did not himself write the full lemmata."

[61] Cf. Wittwer 1999: 70-72. Such abbreviated lemmata are given, for instance, in Aspasius's *Ethics* commentary, and Simplicius's commentary on *Cael*. Against this background Bonitz's method of abbreviating the lemma in his edition of Alexander's *Metaphysics* commentary (1847) appears as a modern construction that is not sustained by manuscript evidence. In his edition, Bonitz prints the first and last *word* of the sentence(s) transmitted in the lemma.

[62] Cf. Wittwer 1999: 67-73 on the abbreviated lemmata in Aspasius's commentary. Wittwer compares the evidence in Aspasius with other commentaries. Regarding the commentaries contemporary to Aspasius which are preserved on papyri the answer is clear (70): "In all of them the text of the lemmata is fully written out." After taking into consideration the manuscript tradition, Wittwer says (71): "Later on, most of the commentaries abbreviate the text of the lemmata either by quoting only the beginning of the text or by using lemmata of the ἕως τοῦ form." For the transmission of the lemmata in Proclus's commentaries see Lamberz 1987: 6-13.

[63] The other extant commentaries by Alexander do not show ἕως τοῦ-abbreviations in the lemmata either.

δὲ τούτων οἱ ἀριθμοὶ φύσει πρῶτοι.⁶⁴ Is this only an exception to the rule, or is it rather the sole remnant of Alexander's original format?⁶⁵

Is there further evidence within the commentary that speaks in favor of the assumption that Alexander himself wrote the lemmata in *unabridged* form? In 220.1 Alexander quotes the following text in the lemma (B 4, 1000a27–28): δόξειε δ' ἂν οὐδὲν ἧττον καὶ τοῦτο γεννᾶν. ("But this [i.e. *Strife*]⁶⁶ too would appear to beget no less...").⁶⁷ In the subsequent sentence, Alexander refers back to the wording in the lemma by explicating the meaning of the verb γεννᾶν, which Aristotle uses in Empedocles' parlance to denote "begetting." The commentary text begins with τουτέστι γεννητικὸν εἶναι καὶ ποιητικόν / "That is, generative and productive" (220.2). This kind of direct back reference introduced by τουτέστι would be unintelligible if the lemma were written in the ἕως-format, that is, if only the first words of this sentence (without the verb γεννᾶν) were quoted, followed by ἕως and the quotation of the last sentence of the relevant passage. The same holds for 42.18–21. In the lemma we read (986b8–9) τῶν μὲν οὖν παλαιῶν καὶ πλείω λεγόντων τὰ στοιχεῖα τῆς φύσεως ("Of the ancients who said the elements of nature were more than one ... "). The commentary section starts immediately with an explanation of the words πλείω and τὰ στοιχεῖα, which likely would have been skipped in an abridged lemma format.⁶⁸ Again, the same is true of 114.20–22. The lemma reads (992a2–3): ἔτι δὲ πρὸς τοῖς εἰρημένοις, εἴπερ εἰσὶν αἱ μονάδες διάφοροι. ("Again, besides what has been said, if the units are diverse..."). Alexander immediately goes on to comment on the word διάφοροι: ἂν ᾖ ἀδιάφοροι γεγραμμένον, τοιοῦτον ἂν εἴη τὸ λεγόμενον ("If the text were written, 'not diverse' what Aristotle would be saying is the following"). There are many such direct back references to words quoted in the lemma. If the lemmata had originally been abbreviated with the relevant words skipped over these back references would have been unintelligible.⁶⁹

Finally, we should consider the possibility that Alexander's lemma quoted not

---

⁶⁴This is the reading in **A** (checked in digital copy), **O** (checked in digital copy), and **M** (checked in digital copy). For the reading in **LF** see *app.* in Hayduck. Even the Latin translation preserves this feature in the lemma: *Inter autem hos et ante hos usque ad illud: Sed quoniam horum numeri sunt primi natura.* Dooley 1989: 62 n. 120, however, grossly misinterprets the word ἕως (followed by Lai 2007: 259 n. 334): "The first [of the two combined] text has *heôs* after *pro toutôn*; *heôs* is not found in Aristotle's text, and makes little sense in this combination. It may be a variant for *pro*."

⁶⁵Alexander abbreviates a quotation using ἕως τοῦ once *within* a commentary section: 104.20–21. Here Alexander draws attention to the fact that the passage was dropped from another version of the text (cf. 3.5.2 and also 267.14–19). See also Wittwer 1999: 71 n. 72.

⁶⁶Strife is one of Empedocles' principles.

⁶⁷For my analysis of this passage see 4.1.3.

⁶⁸*In Metaph.* 42.20–21: τουτέστι τὰ τῶν φυσικῶς γιγνομένων στοιχεῖα. πλείω δὲ λέγει ἤτοι τὰ ὑλικά, ἢ πλείω τὰ στοιχεῖα ἀντὶ τοῦ τὰ αἴτια.

⁶⁹See also the examples in *In Metaph.* 8.19–20, 49.16–17, 90.3–5, 106.7–9, 117.20–23, 286.25–26, 296.22–23.

only the first sentence of the passage under examination, but also the remainder of the passage. This method was used in the Arabic commentary tradition as evidenced by Averroes' *Tafsīr*. The *textus* of this commentary on the *Metaphysics* contains the complete *Metaphysics* text (see 2.5).[70] This, however, was not the method Alexander employed, as concluded from the examples I gave above. Alexander's back references to a single word within the lemma would be imprecise if the lemma contained the complete text of the relevant *Metaphysics* section, and it would have rendered the commentary impractical. It could take the reader quite some time to search an entire section of the *Metaphysics*[71] for the word to which Alexander's comments refer.

The arguments I have provided so far speak in favor of the view that Alexander designed his commentary to include lemmata and that these lemmata consisted of unabridged, roughly sentence-long quotations from the beginning of the *Metaphysics* section to which his subsequent comments refer. This brings me to the second part of my investigation into the lemmata, in which I ask how reliably the lemmata in our manuscripts of the commentary preserve the readings of ω^{AL}.

What happened to Alexander's lemmata during the transmission process? Do the lemmata presented in Hayduck's edition and in the manuscripts he used contain the exact words that Alexander wrote down?[72] Is it possible that the lemmata were removed at some point in the transmission, only to be restored later from another *Metaphysics* text? Primavesi 2012b holds that the lemmata in Alexander's commentary do not bear witness to Alexander's actual text. As justification Primavesi points to the fact that in the course of the transmission the commentary was sometimes fitted around a complete *Metaphysics* text, thus rendering the lemmata superfluous.[73] Furthermore, the lemmata might have been contaminat-

---

[70] Sepúlveda inserted the complete text of the *Metaphysics* in his Latin translation (1527) of Alexander's commentary. This text regularly differs from the text Alexander used (ω^{AL}).

[71] The size of the text segments treated under one lemma varies considerably in Alexander's commentary. For example, in the commentary on A 4 and 5 the span between the lemmata regularly measures between 20 and 30 Bekker-lines (e.g. from 984b33 to 985a21; from 985a22 to 985b19; from 985b27 to 986a13; from 986a13 to 986b8).

[72] Barnes (1999: 37) is skeptical. His conclusions concerning the authenticity of the lemmata in commentaries on Aristotle are based on his study of Aspasius's commentary on *EN*: "But one thing is plain: we may not assume that the lemmata which stand in our manuscripts of the commentaries—or which stood in the archetypes of those manuscripts—represent the text of Aristotle which the commentator had open before him." Wittwer 1999: 53–55, however, offers a more nuanced view on the matter and speaks against an overall dismissal of the readings in the lemmata.

[73] Primavesi 2012b: 407–408: "But the relationship between the lemmata and the main body of Alexander's commentary is, unfortunately, a very loose one: in the course of its transmission, the commentary underwent phases when lemmata were altogether superfluous since the text was fitted around a *complete* text of the *Metaphysics*." See also Barnes 1999: 37: "It follows that we cannot use the lemmata as evidence for the state of Aristotle's text at the date of the commentary." Hadot 2002: 184–85 is also skeptical. Although it is granted that lemmata in individual cases can preserve readings that are older than our *Metaphysics* manuscripts, these scholars argue against the view that the wording in the lem-

ed by the *Metaphysics* text, often transmitted side by side with the commentary. Or, Alexander's own comments, especially his paraphrases and reformulations of Aristotle's text, could have been misinterpreted as evidence for ω^{AL}, resulting in adjustments of the lemmata.

The possibility that lemmata could indeed be omitted during the transmission process becomes a reality upon closer examination of Alexander's commentary: from 376.13 (commentary on Δ 9) to 439.13, where the authentic part of Alexander's commentary ends, the lemmata are, with one exception (407.16), absent.[74] In 62.1–2 (lemma containing 988a34–35) we find that a lemma had been put in the wrong place in the commentary:[75] Here, the lemma comes eight lines too early.[76]

In what form are the lemmata transmitted in the manuscripts?[77] Is Hayduck's way of presenting them justified? Bonitz's edition abbreviates the lemmata by giving only the first and the last word of each lemma.[78] The manuscripts that have the commentary *in plena pagina* (**A O M**) as well as those that have the commentary *in margine* (**C L**) preserve (apart from **F**)[79] the lemmata in unabridged form.[80] This evidence speaks against the view that whenever the commentary was written *in margine* around the full *Metaphysics* text the lemmata were regarded as superfluous and omitted subsequently.[81] At least for the time period to which the manuscripts give us access we can say with confidence that the lemmata were understood as a genuine part of Alexander's commentary. Furthermore, the fact that in ms. **L** (= *Metaphysics* ms. **A**^{b}) the wording in the lemmata by no means always agrees with the *Metaphysics* text surrounded by the commentary[82] speaks against the suspicion that the lemmata of the commentary have generally been adjusted to the neighboring *Metaphysics* text. It therefore is plausible to conclude that the manuscript testimony of the lemmata on the whole is not weaker than the manuscript testimony of the commentary itself. Hayduck's presentation of the lemmata, we conclude, is more authentic than that of Bonitz.

---

mata and Alexander's own exemplar are directly related.

[74] It is an open question whether there is any connection between the lack of lemmata at the end of book Δ and the loss of Alexander's commentary altogether from book E onwards.

[75] See Dooley 1989: 94 n. 197.

[76] Bonitz (1847) prints the lemma as it is presented in the manuscripts (including **O** and **S**).

[77] That is, the manuscripts used by Bonitz and Hayduck and also **O**.

[78] For example, Bonitz 1847: 4.15: φύσει … ζῷα.

[79] Manuscript **F** has blank spots instead of lemmata. In these spots is placed a siglum that refers to the corresponding passage in the *Metaphysics* text.

[80] Markers on the margins of the commentary text can be found in both types of commentaries (*in plena pagina*: **A** and **O**; *in margine*: **L** and **C**).

[81] Primavesi 2012b: 407–408.

[82] The following are a few examples: 982a6–7 τοῦ σοφοῦ α Al.¹ 9.18 (**L** =**A**^{b} f.6r) : τοὺς σοφούς β (**A**^{b} = **L** f.6r); 995a24 ἐπιζητουμένην α Al.¹ 171.3 (**L** =**A**^{b} f.63v) : ζητουμένην β (**A**^{b} = **L** f.63v); 995b19 ἡ θεωρία μόνον α β (**A**^{b} = **L** f.65v) : μόνη ἡ θεωρία Al.¹ 176.17 (**L** =**A**^{b} f.65v); 1003a21 τούτῳ α β (**A**^{b} = **L** f.100r) : αὐτῷ Al.¹ 239.4 (**L** =**A**^{b} f.100r).

One may, however, ask about the fate of the lemmata in the centuries before our manuscript tradition began. Had the lemmata's original form been preserved throughout that period? Their syntactical independence from the rest of the commentary proper would seem to facilitate textual corruption. Further, since the lemmata are quotations from the *Metaphysics* text, they might well have been influenced by the transmission of the *Metaphysics*.[83] As the remark by Simplicius quoted above indicates, the lemmata in Alexander's commentary could already have been corrupted by the fifth or sixth century AD.[84]

On the basis of these considerations it seems easy to question the authenticity of the lemmata in general. Yet, there are clear indications that a general condemnation of the lemmata as they have come down to us would be inappropriate. There are several cases in which a lemma has a reading that disagrees with the manuscript tradition of the *Metaphysics* but agrees with the subsequent commentary, and so is confirmed as the reading of ω^{AL}.[85] In 228.29–30, the lemma quotes lines 1001b26–28 from the twelfth aporia (τούτων δ' ἐχομένη ἀπορία πότερον οἱ ἀριθμοὶ καὶ τὰ σώματα <u>καὶ τὰ ἐπίπεδα</u> καὶ αἱ στιγμαὶ οὐσίαι τινές εἰσιν ἢ οὔ).[86] However, the words καὶ τὰ ἐπίπεδα, which are transmitted by all of our manuscripts, are missing from the lemma. These words very likely have been dropped due to *saut du même au meme* (καὶ ... καὶ). Furthermore, from Alexander's comments (228.31–229.1) it is plain to see that he did not read these words in his own text. For he says that Aristotle's words καὶ τὰ σώματα καὶ αἱ στιγμαὶ also include the terms καὶ ἐπιφάνειαι καὶ γραμμαί, which makes it clear that these words were not present in Alexander's text as the quotation in the lemma indicated. The lemma, then, attests to the actual reading in ω^{AL}. It was not subject to later correction

---

[83] It seems reasonable that a student or a teacher who reads Alexander's commentary alongside the *Metaphysics* would be inclined to 'correct' the reading of a lemma whenever it disagrees with his *Metaphysics* text. Cf. Lamberz 1987: 7–10.

[84] Cf. Wittwer 1999: 54 n. 9.

[85] Lemmata containing *peculiar* readings (i.e., readings differing from ω^{αβ}) that are explicitly *confirmed* by the evidence in the commentary are: Al.^{l} 8.6, Al.^{p} 8.7–8, text 981b27; – Al.^{l} 11.3–4 (cf. 4.2.1), Al.^{p} 11.5–8, text 982a21; – Al.^{l} 54.21, Al.^{p} 54.15; 18, text 987b31; – Al.^{l} 71.10–11, Al.^{p} 71.14, text 989b29–30; – Al.^{l} 95.4, Al.^{p} 96.2, text 990a9; – Al.^{l} 106.7, Al.^{p} / Al.^{c} 106.9, text 991b3–4; – Al.^{l} 132.9, Al.^{p} 132.12, text 993a2–3; – Al.^{l} 196.29, Al.^{p} 196.31, text 997b5; – Al.^{l} 204.8, Al.^{p} 204.11, text 998b11–12; – Al.^{l} 228.30, Al.^{p} 228.32–229.1, text 1001b26–28; – Al.^{l} 239.5, Al.^{p} 239.7, text 1003a21–22; – Al.^{l} 245.21, Al.^{c} 245.25 and 251.5, text 1003b22 – Al.^{l} 252.2, Al.^{p} 252.18, text 1004a9; – Al.^{l} 259.23, Al.^{p} 259.26, text 1004b17; – Al.^{l} 260.31 Al.^{p} 260.35, text 1004b28 – Al.^{l} 264.28–30, Al.^{p} 264.34, 265.6–8, text 1005a19–21; – Al.^{l} 273.20–21, Al.^{p} 273.23–24, text 1006a18–20; – Al.^{l} 275.22 Al.^{p} 275.26, text 1006a29–30; – Al.^{l} 320.35, Al.^{p} 321.4, text 1011a31; – Al.^{l} 336.24, Al.^{c} 336.30, text 1012a30; – Al.^{l} 362.11, Al.^{p} 362.12–13, text 1015b16. Lemmata containing *peculiar* readings that are *not disconfirmed* by the evidence in the commentary are: 15.20–21, 26.19, 30.13, 35.5, 60.11, 68.5, 73.9, 85.13, 101.11–12, 102.1, 107.14, 123.18, 133.21, 145.27, 149.14, 153.1–2, 176.17, 180.17, 183.14, 184.29, 196.29, 197.29–30, 200.33, 203.12, 203.25, 204.23, 208.26–27, 213.24, 214.20, 215.31, 225.33, 249.1, 256.19–20, 257.17, 279.15, 282.2, 292.22–23, 293.33, 296.3, 296.22, 297.27, 301.28, 329.5, 340.19, 342.21, 342.35, 350.4, 376.13.

[86] See 4.2.4.

and alignment to a *Metaphysics* text in which the error had not occurred.

A similar case is found in 264.28–30 (lemma citation of lines 1005a19–21).[87] Instead of φανερὸν δὴ ὅτι μιᾶς τε καὶ τῆς τοῦ φιλοσόφου Alexander reads φανερὸν δὴ ὅτι μιᾶς καὶ αὐτῆς. This is confirmed as a reading of ω^AL by lines 264.34 and 265.6 of the commentary. In this case, too, the lemma was not affected by the text of our transmission. The same holds for the lemma in 273.20–21 (citation of 1006a18–20): here the negation οὐ is missing. The text of the direct transmission reads the necessary and undoubtedly correct οὐ.[88] Alexander's paraphrase and comments (273.23–26; 273.34–274.2) clearly show that he read the text without the negation. Finally, the already famous[89] verb form λέγομεν (A 9, 991b3) is attested to in a lemma (106.7) as well as in Alexander's comments (106.9). All of our manuscripts have the corrupted λέγεται.

There are, however, also lemmata containing readings that disagree with the actual reading of ω^AL that have been reconstructed on the basis of Alexander's paraphrase.[90] For example, in 138.26–28, Alexander reports the textual oddity that in his text, book α begins with a ὅτι that does not fit into the syntax of the sentence (993a29–30).[91] Alexander knows of a variant reading that does not contain the ὅτι. Although Alexander's formulation makes it perfectly clear that in his own text ὅτι is the first word of book α (and it is likely that ὅτι actually represents the older reading—see 3.5.2.3), the lemma does not include the ὅτι. The lemma agrees with our direct transmission. It probably was adjusted to a version of the *Metaphysics* in which the seemingly odd ὅτι had been deleted.

In another case, 164.15, the lemma contains exactly the reading that Alexander reports in the subsequent commentary section (164.24–25) as a conjecture that earlier commentators suggested in order to emend a corrupt passage in the *Metaphysics* (α 2, 994b25–26). As indicated by Alexander's paraphrase (164.18–20), ω^AL contained the correct reading.[92] In this example as well, the incompatibility of the evidence in the lemma and the commentary leads us to conclude that the lemma has been changed in the course of the transmission.[93]

In 184.12–13 the lemma contains the text of Β 2, 996b8–10, but from the perspective of the direct transmission the quotation seems to combine the readings given in the α- and the β-versions. Alexander's comments (184.14–15), howev-

---

[87] See also 4.1.2.

[88] *Metaph.* Γ 4, 1006a18–19: ἀρχὴ δὲ πρὸς ἅπαντα τὰ τοιαῦτα οὐ τὸ ἀξιοῦν.... For a detailed discussion see 4.2.3.

[89] On this passage see Jaeger 1965 and Primavesi 2012b: 412–20.

[90] For a list see below, p. 49 n. 100.

[91] For a detailed discussion of this case see 3.5.2.3. Cf. also 3.6.

[92] See 5.1.4.

[93] See also 179.26: in the lemma the word πότερον is missing. From the subsequent paraphrase by Alexander (179.28) we know that he must have read πότερον in his text.

er, show that he read the (correct) α-version in his text.⁹⁴ The α-version reads: ἐκ μὲν οὖν τῶν πάλαι διωρισμένων τίνα χρὴ καλεῖν <u>τῶν ἐπιστημῶν</u> σοφίαν <u>ἔχει</u> λόγον ἑκάστην προσαγορεύειν.⁹⁵ The β-version, however, reads: ἐκ μὲν οὖν τῶν πάλαι διωρισμένων τίνα χρὴ καλεῖν <u>τῶν ἐπὶ</u> σοφίαν <u>οὐδαμῶς ἔχει</u> λόγον ἑκάστην προσαγορεύειν.⁹⁶ The version given in Alexander's lemma (184.12–13) can be characterized as a composite of α and β, because it reads ἐπιστημῶν like the α-version and οὐδαμῶς like the β-version: ἐκ μὲν οὖν τῶν πάλαι διωρισμένων τίνα χρὴ καλεῖν <u>τῶν ἐπιστημῶν</u> [α] σοφίαν <u>οὐδαμῶς ἔχει</u> [β] λόγον. Note that the reading in the lemma is missing the last two words ἑκάστην προσαγορεύειν. This changes the meaning slightly, but does not make the sentence more intelligible than that of the β-version.⁹⁷ Whether and how the shortening of the sentence in the lemma results from contamination by the β-version (οὐδαμῶς ἔχει instead of ἔχει) seems impossible to determine. That Alexander himself read the pure α-reading in ω^AL (i.e. without οὐδαμῶς but with the words ἑκάστην προσαγορεύειν) can be gathered from his own words (184.14–15): εἰπὼν τίνα τούτων φατέον τὴν ζητουμένην, ὅτι <u>πάσας ἐνδέχεται λέγειν δείκνυσι…</u>.⁹⁸ His words indicate that the quotation in the lemma has been changed later on.⁹⁹ Yet, of all 296 lemmata in Alexander's commentary only 28 (about 10%) display a reading that visibly disagrees with the evidence (quotation or paraphrase) that can be found in the commentary section.¹⁰⁰

---

[94] Cf. Crubellier 2009: 60 n. 17.

[95] Arist. Metaph. B 2, 996b8–10: "To judge from our previous discussion of the question which of the sciences should be called wisdom, there is reason for applying the name to each of them."

[96] The corruption of the β-version is probably due to the shortening of ἐπιστημῶν to ἐπί. The β-text, which states that it does not make sense to call every science wisdom, contradicts the logic of the subsequent argument in the Metaphysics text.

[97] It could be translated by "To judge from our previous discussion, the question as to which of the sciences should be called wisdom does not make sense." We might want to oppose this reading by asking: why does it not make sense to even ask the question about wisdom?

[98] "Having asked 'which of these should be said to be the one that is the object of inquiry?' Aristotle shows that it is possible to say that all of them are." The wording is confirmed by O and S. In the lemma, however, the Latin translation (S) follows the Latin version of the Metaphysics that Sepúlveda used, which often disagrees with the reading in ω^AL.

[99] Cf. the case in 46.5–47.1 discussed by Primavesi 2012b: 428–31. There, the α-version has contaminated the lemma.

[100] Al.ˡ 20.4 (against Al.ᵖ 21.1); Al.ˡ 46.5–6 (against Al.ᵖ 46.23–24); Al.ˡ 64.13–14 (mss.) (against Al.ᶜ 64.16–17); Al.ˡ 82.8 (against Al.ᵖ 83.18; 85.6), Al.ˡ 99.1–2 (against Al.ᶜ 99.6; 100.23; 33), Al.ˡ 111.3 (against Al.ᶜ 112.7), Al.ˡ 112.18 (against Al.ᵖ 112.4), Al.ˡ 119.13 (against Al.ᵖ 119.14–15), Al.ˡ 136.3 (against Al.ᵖ 136.4), Al.ˡ 138.24 (against Al.ᵖ 138.26), Al.ˡ 143.4 (against Al.ᵖ 143.11), Al.ˡ 148.20 (against Al.ᶜ 146.23), Al.ˡ 149.15 (against Al.ᵖ 149.25–26 et passim), Al.ˡ 150.29 (against Al.ᵖ 150.33 et passim), Al.ˡ 153.1–2 (against Al.ᵖ 155.15–16), Al.ˡ 164.15 (against Al.ᵖ 164.19), Al.ˡ 179.25–27 (against Al.ᵖ 179.28; 31), Al.ˡ 184.13 (against Al.ᵖ 184.14–15), Al.ˡ 187.14 (against Al.ᵖ 187.16–17), Al.ˡ 194.9 (against Al.ᵖ 195.4), Al.ˡ 197.29 (against Al.ᵖ 197.31), Al.ˡ 245.20 (against Al.ᶜ 251.5; Al.ᵖ 245.24–25), Al.ˡ 260.31 (against Al.ᶜ 261.15), Al.ˡ 261.17 (against Al.ᶜ 261.28, but Al.ᶜ 262.14), Al.ˡ 273.20–21 (against Al.ᵖ 273.24), Al.ˡ 275.21 (against Al.ᶜ 275.31), Al.ˡ 301.28 (against Al.ᵖ 301.32–33), Al.ˡ 315.27–28 (against Al.ᶜ 316.27–29). Cf. also appendix B.

In light of the above evidence I draw the following conclusions about the lemmata in Alexander's commentary. Alexander himself inserted lemmata in his commentary and at no time during the transmission of the commentary were all of the lemmata removed.[101] The cases in which a *peculiar* reading in the lemma agrees with the ω^{AL}-reading attested to in the commentary clearly point to the conclusion that the lemmata originally reflected the text of ω^{AL}. This means that many of the lemmata survived unscathed, while some were corrupted in the course of the transmission. Certainly, the lemmata were exposed to various influences by the neighboring transmission of the *Metaphysics*, but such influence turns out to be weaker than one might first suspect.[102] Accordingly, the lemmata do play an important role in the reconstruction of a reading from ω^{AL} and it is misguided to discount the lemma as evidence for ω^{AL}. Since the probability that a lemma contains the exact wording of Alexander's text can be much increased by additional evidence in the commentary section that confirms the wording, I will, in the present study, not draw conclusions about the relationship of ω^{AL} and the manuscript tradition of the *Metaphysics* on the basis of a reading in ω^{AL} that is attested to by a lemma alone.

3.3 THE EVIDENCE IN THE QUOTATIONS

The next type of evidence for ω^{AL} is given by quotations from the *Metaphysics* that Alexander inserts into his commentary. For this, the following questions are relevant: how do we recognize a quotation from the *Metaphysics* within the commentary? How can we distinguish a quotation from a lemma on the one hand, and from a paraphrase on the other? How reliably do Alexander's quotations testify to the wording in ω^{AL}? Were quotations exposed to secondary influence and contaminated by the transmission of the *Metaphysics* to the same degree as lemmata?

There are two different kinds of markers that indicate a quote from Alexander's *Metaphysics* copy. (i) The most important marker is the nominalization of the quoted words by the article τό.[103] The article τό marks Aristotelian phrases and

---

[101]For a different view see Primavesi 2012b: 408: "So when the lemmata of our transmission were inserted, or, for that matter, re-inserted, they were taken from a text which simply happened to be available at the time and this text need not have been particularly close to the text used centuries earlier by Alexander himself."

[102]The tradition of the *Metaphysics* text is of course by no means restricted to our directly transmitted text. This means that even if a lemma does not preserve the reading of ω^{AL} it could nevertheless preserve a reading from an otherwise lost line of the *Metaphysics* tradition.

[103]Bloch 2003: 27–31 seems to ignore this kind of linguistic designation in his analysis of the citations in Alexander's commentary on *De Sensu*. Bloch generally holds a rather pessimistic view on the usability of Alexander's quotations as evidence for Aristotle's text (29): "Therefore it cannot be established exactly when he is quoting, and this gravely diminishes the value of using Alexander's quoted variant readings as the basis of the Aristotelian text."

expressions and even whole sentences as the object of study: τὸ [quote]. Syntactically speaking, such nominalized citations are mostly objects of third person singular verbs of saying with Aristotle as the subject: τὸ δὲ [quote] εἶπεν, τὸ δὲ [quote] εἴρηκεν, ἐπήνεγκε τὸ [quote], προσέθηκε διὰ τοῦ [quote], ἐδήλωσε διὰ τοῦ προσθεῖναι τὸ [quote]. Further, explications of Aristotle's expressions are often introduced by the formula τὸ δὲ [quote] ἴσον τῷ.[104]

(ii) There are citations within Alexander's commentary that are not marked by the article but are simply introduced by a verb of saying whose subject is Aristotle: φησίν [quote], εἶπεν [quote], ἐπιφέρει [quote], ἐπήνεγκεν [quote], ἐπήνεγκεν ὅτι + [quote]. Sometimes the verb of saying is intensified by a (synonymous) participle: λέγει ὁ λέγων [quote] (e.g. 181.13), δεικνὺς ἐπήνεγκε [quote] (e.g. 247.33), ἔδειξεν ἐπενεγκών [quote] (e.g. 352.3), ἐδήλωσεν ἐπενεγκών [quote] (e.g. 352.11), ἐδήλωσεν εἰπών [quote] (e.g. 182.12–13), ἐπιφέρει εἰπών [quote] (e.g. 176.4).

This second type of quotation labeling seems ambiguous, as it can also function as an introduction to a paraphrase, which cannot be taken as faithful to the Aristotelian wording. This type of identification thus prompts the question as to how we can distinguish a quotation from a paraphrase, given that the latter is often introduced by a simple "he says."[105] The following example illustrates this. In 385.35 Alexander renders the phrase (1018b21) ἀρχὴ δὲ καὶ αὕτη τις ἁπλῶς ("and the prime mover also is a beginning absolutely") into a paraphrase that is rather close to the text: ἀρχὴ γὰρ καὶ αὕτη τις, φησίν, ἁπλῶς, τουτέστι ... .[106] The particle γάρ, which replaces Aristotle's δέ (1018b21), suggests that this rendering is not a quotation, as the signal word φησίν might lead one to think, but a paraphrase. The subsequent lines then confirm that we are dealing here with a paraphrase rather than a verbatim quotation. The γάρ is not a variant reading in Alexander's text, but Alexander's own reformulation of Aristotle's wording; Alexander simply conformed the sentence to the syntactical context of his commentary. Alexander refers again to this passage, but this time by means of a verbatim quotation, which he marks with the article τὸ (385.38): διὰ δὲ τοῦ προσθεῖναι τὸ ἀρχὴ δὲ καὶ αὕτη τις ἁπλῶς ἔοικε δηλοῦν ὅτι ... . As this citation shows, Alexander's text read the particle δέ, just like ours, and the first rendering of the passage (385.35) is a paraphrase (introduced by φησίν), but not a citation.

In many other passages it is difficult to differentiate clearly between a paraphrase and a citation (cf. 3.4). Often, a comparison with the *Metaphysics* text can help to determine whether we are dealing with a citation or a paraphrase, yet this

---

[104] Further formulas are: τὸ δὲ ἑξῆς [quote] συνάπτει τῷ [quote] (189.4–5), τοῦτο γὰρ λέγει διὰ τοῦ [quote] (298.18–19), τοῦτο γὰρ σημαίνει τὸ [quote] (207.15–16; 295.9), τὸ δὲ [quote] ἴσον ἂν σημαίνοι (193.1–2).

[105] Sometimes the syntax shows a clear differentiation between citation and paraphrase. The paraphrase is then construed as a clause subordinate to the verb of saying, whereas, in the case of the citation, the syntax of the Aristotelian clause is preserved.

[106] Hayduck even highlights these words as a citation.

means of verification is unavailable precisely in those cases in which a citation differs from our *Metaphysics* text *because* ω^{AL} reads a divergent text. In such cases we cannot determine whether we are dealing with Alexander's own reformulation or a quotation of a divergent reading in ω^{AL}, unless the commentary provides further evidence.

It can also be difficult to distinguish between a citation and a lemma. In some commentary passages we face a kind of blending of a lemma and a citation. Here, Alexander quotes a syntactically independent piece of text from the *Metaphysics* without introducing or labeling it. In some of these cases one might suspect that we are dealing with a lemma that at some point in the transmission ceased to be marked as such. The asyndetic sequel speaks in favor of its having been a lemma (cf. 3.2), e.g.,[107] in 139.19; 140.19–20; 185.1–3; 272.4–5;[108] 299.28–30;[109] 314.3–4.[110] In other instances a syntactically independent citation is so tightly embedded in the argument's train of thought that one can by no means take it as a lemma. In these cases the next sentence contains a particle: 183.20–22; 270.15–16; 325.20–21.[111]

How conclusive, then, is the evidence for ω^{AL} that is available in the quotations? What holds for lemmata holds for quotations, namely, that a quote from the *Metaphysics* in Alexander's commentary is more likely to have undergone a secondary adjustment to conform to a diverging *Metaphysics* text than a paraphrase is. Having said this, the risk of contamination seems on the whole to be lower in the case of citations than in the case of lemmata. Of about 580 quotations,[112] in only 39 cases (about 7%), as far as I can see, can a reading be shown to clearly disagree with the reading in a lemma, the text presupposed by a paraphrase, or another quotation.[113] It seems that quotations are on the whole better integrated

---

[107] All of the following examples are regarded as citations and not as lemmata in Bonitz's and Hayduck's edition.

[108] Cf. Madigan 1993: 53 with n. 302. Madigan treats this text as a lemma, too, but without using the subsequent asyndeton as evidence. Casu 2007: 823 n. 347 follows Madigan.

[109] Cf. Madigan 1993: 87 with n. 623 and Casu 2007: 834 n. 649.

[110] Cf. Madigan 1993: 105 with n. 742 and Casu 2007: 840 n. 792.

[111] I disagree with Madigan 1993: 119 n. 854, followed by Casu 2007: 843 n. 864, who takes this as a lemma.

[112] See appendix C.

[113] 982a13: Al.^{c} 10.7–8 vs. Al.^{c} 10.14–15; – 982a32–b1: Al.^{c} 13.21–23 vs. Al.^{p} 13.20 (vs. Al.^{p} 13.23); – 982b5–6: Al.^{c} 14.7 vs. Al.^{c} 14.17–18; – 984a16: Al.^{c} 31.4, 34.12–35.1 vs. Al.^{l} 28.22; – 988b22–23: Al.^{c} 64.16–17 vs. Al.^{l} 64.13–14; – 989b19–20: Al.^{c} 28.12–13 vs. Al.^{c} 70.5–6; – 990b7: Al.^{c} 96.6–7 vs. Al.^{c} 77.18, 27–28; – 990b34: Al.^{c} 91.13 vs. Al.^{c} 91.17; – 991a1–2: Al.^{c} 91.17–18, 26 vs. Al.^{c} 94.10–11; – 991a18–19: Al.^{c} 98.23–24 vs. Al.^{c} 100.32–33; – 991a19–20: Al.^{c} 99.6–7, 100.23–24, 33–34 vs. Al.^{l} 99.1; – 991b25: Al.^{c} 112.7 vs. Al.^{l} 111.3; – 993a25: Al.^{c} 137.8 vs. Al.^{c} 136.15; – 993b26–30: Al.^{c} 146.22–25 vs. Al.^{c} 148.32–149.3, 11–12; – 994b6: Al.^{c} 157.33–34 vs. Al.^{c} 159.6–7, .10–11; – 997a24: Al.^{c} 192.11, 193.1–2 vs. Al.^{c} 194.3–4; – 997a25–26: Al.^{c} 195.3–4 vs. Al.^{l} 194.9; – 998b25: Al.^{c} 205.20 vs. Al.^{c} 206.6–7; – 999b15: Al.^{c} 215.5–6 vs. Al.^{c} 218.11–13; – 999b16: Al.^{c} 215.8–9 vs. Al.^{c} 215.14; – 1001a27–28: Al.^{c} 225.8 vs. Al.^{c} 225.23–24; – 1001b23: Al.^{c} 228.24–25 vs. Al.^{p} 228.26; – 1002b24: Al.^{c} 234.22 vs. Al.^{c} 233.21–22; – 1003b20–22: Al.^{c} 251.4–5 vs. Al.^{p} 245.24–25 and Al.^{l} 245.20; – 1004a2–3: Al.^{c} 251.1–2 vs. Al.^{c} 251.6 vs. Al.^{l} 250.21; but confirmed by Al.^{c} 251.6; – 1004a18–19:

in both the syntactical and the argumentative context of the commentary and are therefore less prone to corruption.¹¹⁴

Apart from the threat of contamination and influence from the *Metaphysics* tradition, the evaluation of Alexander's quotations encounters another challenge. Alexander himself is not always careful to draw a clear distinction between a quote and a paraphrase of the Aristotelian text.¹¹⁵ Some cases show that Alexander occasionally does not refrain from shortening or slightly changing Aristotle's wording even when the passage is marked as a citation. This might be due to nonchalance¹¹⁶ or to the desire to highlight a certain word or phrase.¹¹⁷ Nonetheless, Alexander generally seems to quote the Aristotelian text far more accurately than, for example, Simplicius.¹¹⁸ This can be seen in cases such as the one discussed above, in which a clear distinction between the loose paraphrase and the accurate quotation (introduced by τό) is possible. The following two examples illustrate some of the challenges that Alexander's quotation style entails for the textual scholar of the *Metaphysics*.

In 250.21, the lemma reads Γ 2, 1004a2 (καὶ τοσαῦτα μέρη φιλοσοφίας ἔστιν ὅσαι περ αἱ οὐσίαι) in exact agreement with the directly transmitted *Metaphysics* text. In the subsequent commentary section Alexander quotes the same sentence (251.1–2): ταύτῃ γὰρ ἀκολουθεῖ τὸ καὶ τοσαῦτα μέρη φιλοσοφίας ὅσαι περ αἱ οὐσίαι. Since Alexander's purpose in quoting the sentence here is not to comment on its content, but simply to mark a section of text, he understandably omits the ἔστιν. Just a few lines later he quotes the sentence again. This time the sentence contains the ἔστιν (251.6) and this time he quotes the sentence for its content's sake and not in order to mark the beginning of a paragraph. Did Alexander therefore refer back to his *Metaphysics* exemplar to make sure he got the sentence right?

In 349.3–4 we encounter a similar situation. Here Alexander quotes Δ 2,

---

Al.ᶜ 253.34–35 vs. Al.ᶜ 254.7–8; – 1004b27–28: Al.ᶜ 261.14–15, 262.15 vs. Al.ˡ 260.31 and Al.ᵖ 260.35; – 1004b29–30: Al.ᶜ 261.27–29 vs. Al.ᶜ 262.13–14 vs. Al.ˡ 261.17–18 and Al.ᶜ 262.14; – 1005b2–3: Al.ᶜ 267.15 vs. Al.ᶜ 267.19–20 vs. Al.ˡ 266.29–31; – 1005b26–27: Al.ᶜ 270.15–16 vs. Al.ᵖ 270.17; – 1006b19–20: Al.ᶜ 280.35–36 vs. Al.ᶜ 281.36; – 1007a29: Al.ᶜ 287.4 vs. Al.ᵖ 286.29; – 1010b30: Al.ᶜ 316.27–28 vs. Al.ˡ 315.27–28; – 1012b1 Al.ᶜ 337.33 vs. Al.ᶜ 337.30; – 1013b17–18: Al.ᶜ 351.5–6 vs. Al.ᶜ 351.22; – 1014a20–22: Al. c 354.11–13 vs. Al.ᶜ 354.17; – 1018a12: Al.ᶜ 379.4–5 vs. Al.ᶜ 378.30–31; – 1018b21: Al.ᶜ 385.35, 38 vs. Al.ᵖ 385.35. For a full list of all quotations see appendix C.

¹¹⁴Cf. Barnes 1999: 37.

¹¹⁵Bloch 2003: 29: "Alexander may give all appearance of quoting literally from Aristotle, when, in fact, he has no scruples adding words."

¹¹⁶Busse 1900: 76 attributes inaccuracies in Ammonius's citations (in his commentary on *Int.*) to the habit of quoting from memory.

¹¹⁷Fazzo 2012a: 62 claims that "it is a standard use for Alexander to put the relevant words at the end of a quoted passage." The examples (only four in number) that Fazzo adduces as proof of the "standard" do not justify her conclusion.

¹¹⁸On the accuracy of quotations in Simplicius see Baltussen 2008: 27. Cf. also 42–48. See also Wildberg 1993: 193 with n. 20.

1013a27–29[119] but omits the words καὶ τὰ τούτου γένη (οἷον τοῦ διὰ πασῶν τὰ δύο πρὸς ἓν καὶ ὅλως ὁ ἀριθμός). His subsequent paraphrase (349.20–23) ensures that he has indeed read the omitted part of the sentence.[120] So we can assume either that he intended to leave out the unnecessary specification given by the omitted words or that the text of the commentary suffered a *saut du même au même* (a jump from καὶ τὰ [a27–28] to καὶ τὰ [a29]).[121] It seems impossible to decide between these two options.

These examples warn us that a reconstruction of a reading in ω^AL that is solely based on the evidence in a single quotation may stand on shaky grounds. The following numbers might be helpful for estimating how trustworthy Alexander's quotations are on the whole. There are about 95 instances where a quotation from the *Metaphysics* (with lengths ranging from one word to a full sentence) appears more than once in Alexander's commentary.[122] Among these cases of repeated quotations, there are 25 instances (about 24%) where a repeated quotation reads a (slightly) different text.[123]

Despite these caveats for establishing the text of ω^AL on the basis of quotations alone, quotations are an important factor whenever additional evidence is available either in Alexander's paraphrase or in his critical discussion of Aristotle's argument or even in another quotation. The evidence provided by a citation combined with additional pieces of evidence makes it possible to reconstruct ω^AL. As we have seen in the last two examples, Alexander's comments can confirm or correct the text presented in the quotations. Thus, the following guideline can be established for the present study: coherence between a quotation and one other type of evidence for ω^AL can be regarded as adequate proof of the quotation's authenticity.[124]

---

[119] 1013a27–29: τοῦτο δ᾽ ἐστὶν ὁ λόγος τοῦ τί ἦν εἶναι καὶ τὰ τούτου γένη (οἷον τοῦ διὰ πασῶν τὰ δύο πρὸς ἓν καὶ ὅλως ὁ ἀριθμός) καὶ τὰ μέρη τὰ ἐν τῷ λόγῳ. ("i.e. the formula of the essence, and the classes which include this (e.g. the ratio 2:1 and number in general are causes of the octave) and the parts of the formula.").

[120] See Dooley 1993: 132 n. 28.

[121] O and S confirm Hayduck's text.

[122] Cf. the list in appendix C.

[123] This list includes also minor differences, for example, between ὥστ᾽ and ὥστε: 982a13: Al.^c 10.7–8, 14–15; – 982b5–6: Al.^c 14.7, 17–18; – 989b19–20: Al.^c 28.12–13, 70.5; – 990b34: Al.^c 91.13, 17; – 991a1–2: Al.^c 91.17–18, 26, 94.10–11; – 990b7: Al.^c 96.6, 77.18, 27–28; – 991a18–19: Al.^c 98.23–24, 100.32–33; – 993a25: Al.^c 137.8, 136.15; – 993b26–30: Al.^c 146.22–25, 148.32–149.3; 149.11–12; – 994b6: Al.^c 157.33–34, 159.6–7, 159.10–11; – 997a24: Al.^c 192.11, 193.1–2, 32, 194.4; – 998b25: Al.^c 205.20, 206.6–7; – 999b15: Al.^c 215.5–6, 11–13; – 999b16: Al.^c 215.8–9, 14; – 1001a27–28: Al.^c 225.8, 23–24; – 1002b24: Al.^c 234.22, 233.21–22; – 1004a 3: Al.^c 251.6, 251.2; – 1004a18–19: Al.^c 253.34–35, 254.7–8; – 1004b29–30: Al.^c 261.27–29, 262.13–14; – 1005b2–3: Al.^c 267.15, 19–20; – 1006b19–20: Al.^c 280.35–36, 281.36; – 1012b1: Al.^c 337.30, 33; – 1013b17–18: Al.^c 351.5–6, 22; – 1014a20–22: Al.^c 354.11–13, 17; – 1018a12: Al.^c 378.30–31, 379.4–5.

[124] Busse 1900: 78 gives the following résumé for the situation in Ammonius's commentary on *Int.*: "Nur dann, wenn die Lesart durch ein Citat verbürgt wird, dürfen wir glauben, auf festem Boden zu stehen, wenngleich auch die Citate mit Vorsicht zu behandeln sind … (80–81). Wir sehen, Schreib-

3.4 THE EVIDENCE IN THE PARAPHRASE AND THE CRITICAL DISCUSSION

One can distinguish at least four modes[125] in which Alexander comments on the *Metaphysics* text.[126] First, Alexander sometimes speaks in summary mode. He often begins a commentary section with a brief review of the preceding section of the commentary and its train of thought.[127] Second, there is the mode in which Alexander reproduces the thought of a passage of the *Metaphysics* text, expanding Aristotle's concise thoughts into an extensive explanatory reformulation that sometimes includes direct quotations from the *Metaphysics*. Third, Alexander concentrates narrowly on a certain Aristotelian expression, which he quotes and then analyzes in detail.[128] In the fourth mode, Alexander subjects one of Aristotle's arguments or thoughts to critical discussion. In this mode, Alexander may refer to the opinions of other interpreters or he considers multiple solutions to a problem, sometimes without deciding in favor of any one in particular.

My definition of an Alexandrian 'paraphrase' is not restricted to one of the four modes. It includes all of those sentences or parts of sentences in which Alexander repeats a sentence or phrase from the *Metaphysics* in words that, although his own, are in close proximity to Aristotle's wording. Such paraphrases may occur in any one of the four modes, and under certain conditions it is possible to reconstruct the reading in ω^{AL} from one of these paraphrases. In addition to the paraphrase, Alexander's discussion of an Aristotelian argument or expression (see the third and fourth mode) can offer indirect access to the reading on which Alexander's analysis is based.

---

fehler, Versehen, Korrekturen finden sich hüben und drüben. Das klingt für die Verwendung des Kommentars zum Zwecke der Textkritik wenig trostreich. Doch hat dies Resultat auch eine erfreuliche Seite. Wenn der Text in den Citaten von den Lemmata so häufig abweicht, so ist das doch wohl ein Beweis dafür, dass die beiden Ueberlieferungen sich nicht gegenseitig beeinflusst, sondern selbständig fortgepflanzt haben. Das ist für die Beurteilung derjenigen Stellen, die übereinstimmend überliefert oder in überzeugender Weise durch die Konjekturalkritik in Einklang gebracht sind, von grösstem Wert. Denn wir dürfen annehmen, dass diese Bruchstücke uns in der Form erhalten sind, wie sie Ammonios in seiner Handschrift gelesen hat. Rechnen wir noch dazu, was er an Lesarten ausdrücklich im Kommentar anführt, so ist dies das ganze Material, das wir als zuverlässig ansehen können. Wir haben zwar viel unsicheres Gut preisgeben müssen, aber wir dürfen nun auch das Vertrauen hegen, dass der uns gebliebene Rest nur echtes Metall enthält."

[125] These four modes are not exhaustive nor do they exclude overlapping.

[126] There is no set order in which the various modes must appear in the commentary. Alexander's commentaries are generally known to have more formal and structural flexibility than the later, Neoplatonic commentaries. See Sharples 1990: 95, Luna 2003: 251, Fazzo 2004: 8–9 n. 26 and Kupreeva 2012: 112.

[127] These summaries are often introduced by an aorist participle, e.g. δείξας or εἰπών. See also Moraux 2001: 437–38.

[128] We sometimes encounter this mode at the beginning of a commentary passage.

The evidence available in Alexander's paraphrases seems to testify to ω^AL more reliably than lemmata or citations. It seems quite unlikely that someone would change the wording of Alexander's own paraphrase; but in those cases where Alexander quotes the works of Aristotle, as in a lemma or direct citation, changes may likely have occurred, especially if a scribe or scholar thought that Alexander quoted incorrectly. By contrast, paraphrases are Alexander's *ipsissima vox*; they are more deeply embedded in the commentary than citations and especially lemmata.[129] Why would someone want to adjust Alexander's paraphrases of the *Metaphysics* text when the very purpose of a paraphrase is to represent an author's words in slightly different terms and expressions?

Although it seems justified to regard Alexander's paraphrases as in themselves more reliable than the other types of evidence, extrapolating information on the *Metaphysics* from the paraphrases entails considerable difficulties.[130] They certainly cannot be used to reconstruct every sentence in ω^AL on which Alexander comments. Further, it is often unclear whether a divergence between Alexander's commentary and ω^αβ is due to Alexander's peculiar diction or to a genuine difference in the reading of ω^AL. A comparison of Alexander's paraphrases with our *Metaphysics* text would promptly bring to light a large number of seeming textual differences, many of which may not be due to an actual textual difference between ω^AL and ω^αβ.[131]

---

[129] However, there are cases in which the text of the paraphrase has been changed: e.g. in 46.16. The word μαλακώτερον as it occurs in Alexander's paraphrase of A 5, 987a10 is certainly not what he himself wrote. As Brandis (1836: 546 *app. crit.*) has already pointed out (see also Hayduck 1891 *app. crit.* and Primavesi 2012b: 428–31), the context shows that this cannot be the original reading of Alexander's paraphrase. The word as it stands (also attested to by O and S [*mollius*]) has to be corrected to μοναχώτερον. This case shows that a paraphrase, too, can be contaminated by the corresponding *Metaphysics* passage. However, in this case the corrupted word of the paraphrase has an exposed position because it renders precisely the term on which Alexander is commenting (cf. the direct transmission μετριώτερον α, μαλακώτερον β).

Another type of corruption in Alexander's paraphrase occurred in 229.3-4. Alexander writes ἔστι δὲ οὐκ οὐσία οἶον αἵ τε ποιότητες (ταύτας γὰρ εἶπε παθητικὰς κινήσεις). The words παθητικὰς κινήσεις render Aristotle's phrase πάθη καὶ αἱ κινήσεις in 1001b29. Did Alexander find in his text the expression παθητικὰς κινήσεις (which is nowhere attested to in Aristotle's corpus and certainly corrupt)? Alexander does not seem puzzled over the text and its meaning. As can be seen by the following lines of his commentary (229.14; cf. also 230.30) Alexander read the correct text (πάθη καὶ αἱ κινήσεις) in ω^AL. The words παθητικὰς κινήσεις must go back to a corruption that occured in the transmission of the commentary (O too reads παθητικὰς κινήσεις, S writes *motus passivos*).

[130] Barnes 1999: 39 writes concerning the textual evidence available in the paraphrase of Aspasius's commentary on *EN*: "A paraphrase is not a citation, nor is a comment. But if paraphrases and comments never display the text of the *Ethics* which lay in front of Aspasius, they will often enough imply or suggest or insinuate a certain reading. And, paradoxically enough, such insinuations are more reliable than explicit citations; for they are not liable to 'correction.' Nonetheless, it is rarely an easy matter to divine the text from the comment."

[131] This is the reason why the list of Alexander's paraphrases in appendix D does not display the numerous cases where Alexander's paraphrase 'differs' from the reading in ω^αβ.

How can we then identify paraphrases that are likely to contain valuable evidence of ω^AL? In the case of the *Metaphysics* text, the following rule of thumb is useful: in cases where the direct transmission has brought down to us two divergent, yet viable, readings (α and β), the two possible readings can be compared with Alexander's paraphrase, and the agreement between Alexander's paraphrase and one of the readings points toward the reading of Alexander's *Metaphysics* text.[132] Nevertheless, a degree of uncertainty remains when determining the reading of ω^AL by this criterion, since, as we will see in section 5.1–5.3, Alexander's comments (including his paraphrases) influenced the *Metaphysics* text of the direct transmission. Thus, agreement between a given paraphrase and α or β could possibly be due to contamination. And so the agreement between an Alexandrian paraphrase with either the α- or the β-reading is more conclusive when the nature of the textual divergence makes it unlikely that someone 'corrected' the text accordingly. It is, however, considerably less conclusive in cases where the agreement could be coincidental.[133]

The following example can illustrate these theoretical considerations. Aristotle introduces the 'coming to be' of a man from a boy as an example of 'coming to be from something' (α 2, 994a22).

Aristotle, *Metaphysics* α 2, 994a27–30

(ἀεὶ γάρ ἐστι μεταξύ, ὥσπερ τοῦ εἶναι καὶ μὴ εἶναι γένεσις, [28] οὕτω καὶ τὸ γιγνόμενον τοῦ ὄντος καὶ μὴ ὄντος· ἔστι δὲ ὁ [29] μανθάνων γιγνόμενος ἐπιστήμων, καὶ τοῦτ' ἐστὶν ὃ λέγεται, [30] ὅτι γίγνεται ἐκ μανθάνοντος ἐπιστήμων)

(for as becoming is between being and not being, so that which is becoming is always between that which is and that which is not; and the learner is a man of science in the making, and this is what is meant when we say that from a learner a man of science is being made)

28 δὲ α Ascl.^c 125.18 (δ' ν Bonitz) Bekker Christ : γὰρ β Al.^p 156.16 Ross Jaeger || 29–30 καὶ ... ἐπιστημῶν α Al.^p 156.16–18 (Ascl.^c 125.18–20) edd. : om. β : ζ in mrg.

The first difference between the α- and the β-version concerns the connective particle in line a28. The α-text reads δὲ, while the β-text reads γὰρ, and therewith expresses a causal connection. Alexander does not quote this sentence in either the lemma or his commentary, but he does paraphrase it.

Alexander, *In Metaph.* 156.14–18 Hayduck

πῶς δὲ τὰ οὕτως [15] ἔκ τινος γιγνόμενα ἐκ τοῦ γιγνομένου ἐπιτελεῖται, ἐπεδήλωσε παραθέμενος [16] τὸν μανθάνοντα. ὁ <u>γὰρ</u> μανθάνων ἐστὶν ὁ γιγνόμενος ἐπιστήμων, <u>ὃς ἐκ</u> [17] <u>τοῦ μανθάνοντος οὕτως γίγνεται</u>, ὅτι ὁ μανθάνων γίγνεται καὶ ὑπομένων

---

[132]See list in appendix D.
[133]This is especially the case when the difference consists in a particle or a spelling variant.

[18] πρόεισιν ἐπὶ τὸ ἐπίστασθαι, ἥτις ἐστὶ τελειότης τοῦ μανθάνοντος·

To show how things that come to be from something in this way come to their complete state 'from what is coming to be,' Aristotle adds the example of the learner. For the learner is a man of science in the making, who comes to be from the learner in this way because the learner is coming to be and, while remaining, progresses towards scientific knowledge, which is perfection of the learner;

Alexander's paraphrase in 156.16 stays close to the Aristotelian text: he renders Aristotle's words ἔστι <u>δὲ</u> (α) / <u>γὰρ</u> (β) ὁ μανθάνων γιγνόμενος ἐπιστήμων into ὁ <u>γὰρ</u> μανθάνων ἐστὶν ὁ γιγνόμενος ἐπιστήμων. Should we infer on the basis of this evidence that Alexander read γὰρ (β) and not δὲ (α) in ω^AL? We should not. That Alexander uses the particle γάρ in his paraphrase does not prove by any means that this is what he found in his text. There are several cases in which Alexander clearly reads δέ in his text, but interprets and paraphrases it as γάρ.[134] In the present case we simply cannot determine whether ω^AL read a δέ or a γάρ. When we take into account the possibility that, for example, the β-text might have been contaminated by Alexander's commentary[135] then one might even ask whether the γὰρ in the β-version is due to an adjustment of this passage according to Alexander's comments. But this question, too, cannot be answered on the basis of the available evidence.

Let us then have a look at the second divergence between α and β in the *Metaphysics* passage. Here the situation is different; the divergence between α and β is much more significant. In the β-version, a whole sentence, i.e. the text from καὶ to ἐπιστήμων (a29–30), is missing. The loss has likely been caused by *saut du même au même*: γιγνόμεν<u>ος ἐπιστήμων</u>, καὶ τοῦτ' ἐστὶν ὃ λέγεται, ὅτι γίγνεται ἐκ μανθάνοντ<u>ος ἐπιστήμων</u>. The sentence as it is preserved by the α-version is indispensable, because only when we read the phrase γίγνεται ἐκ μανθάνοντος ἐπιστήμων does the extensive parenthesis (ἀεὶ γάρ … μανθάνοντος ἐπιστήμων, 994a27–30) at all provide an explanation of the process ἔκ τινος. As to the question which of the two is more likely the reading of ω^AL, Alexander's commentary gives a clear answer: Alexander must have found the α-reading in his text because he paraphrases thus: ὁ γιγνόμενος ἐπιστήμων, ὃς <u>ἐκ τοῦ μανθάνοντος</u> οὕτως γίγνεται (156.16–17). In this case, then, we can safely infer simply on the basis of Alexander's paraphrase that ω^AL agrees with α and that the β-version suffered from corruption.

There is a more secure criterion available by which we can reconstruct readings of ω^AL on the basis of Alexander's paraphrase and which further does not presuppose a divergence between α and β. According to this, Alexander's paraphrase attests to a reading in ω^AL whenever this reading is confirmed by another type

---

[134] See the cases in 37.20–21; 54.11–13; 172.13–15. Cf. also 3.6.
[135] See Primavesi 2012b: 424–39 and 5.2.

of evidence in Alexander's commentary, i.e. a lemma, a citation, or Alexander's discussion of the argument. This criterion applies independently of the evidence in α and β. The following example illustrates the point:

Aristotle, *Metaphysics* Δ 4, 1015a17–19

καὶ ἡ ἀρχὴ τῆς κινή-[18]σεως τῶν φύσει ὄντων αὕτη ἐστίν, ἐνυπάρχουσά πως ἢ δυ-[19]νάμει ἢ ἐντελεχείᾳ.

And nature in this sense is the source of the movement of natural objects, being present in them somehow, either potentially or actually.

18 αὕτη **α** Al.ᶜ 360.10 edd. : ἡ αὐτὴ **β** || 19 ἐντελεχείᾳ ω^αβ Ascl.ᶜ 312.19–20 Ascl.ᵖ 312.20 edd. : ἐνεργείᾳ ω^AL Al.ᶜ 360.11 Al.ᵖ 360.12

Aristotle's term ἐντελεχείᾳ ("actually") in line 1015a19 stands for the synonymous[136] but more frequently used[137] term ἐνεργείᾳ. Alexander uses the term ἐνεργείᾳ (360.12) in his explanatory paraphrase of this passage. On the basis of this alone we cannot conclusively infer that ω^AL read ἐνεργείᾳ instead of ἐντελεχείᾳ. However, the reading ἐνεργείᾳ is confirmed by a quotation, which itself is clearly marked as such by the article (τὸ, 360.11). Taken together, these two pieces of evidence lead to the conclusion that Alexander found ἐνεργείᾳ in ω^AL.

Alexander, *In Metaph.* 360.9–12

προστίθησι δὲ ὅτι καὶ ἡ ἀρχὴ τῆς κινή-[10]σεως τῶν φύσει ὄντων αὕτη ἐστίν, ἐνυπάρχουσά πως. ἐξηγούμενος [11] δὲ τὸ πῶς, προσέθηκε τὸ δυνάμει ἢ <u>ἐνεργείᾳ</u>, δυνάμει μέν, ὡς ἐν τῷ [12] σπέρματι τῷ καταβληθέντι ἡ ψυχή, <u>ἐνεργείᾳ</u> δέ, ὅτε ἤδη ζῷόν ἐστι.[138]

He adds, "and nature in this sense is the source of the movement of natural objects, being present in them somehow." To explain the "somehow," he adds, "either potentially or actually"—potentially, as the soul is in the ejected semen; actually, [as the soul is present] whenever there is already a living thing.

12 ὅτε **A O** Bonitz : ὅταν **LF** Hayduck || ἐστι **A O** Bonitz : ἦ **LF** Hayduck

This shows that Alexander's paraphrase can provide information about the reading in ω^AL even if there is no divergence between α and β. At the same time, this example points to the crucial role Alexander's paraphrase plays in confirming a reading that is preserved in a lemma or citation.[139] Whenever a reading in a lem-

---

[136] See Bonitz 1870 s.v. ἐντελέχεια, p. 253.46–50. The terms ἐνεργείᾳ and ἐντελεχείᾳ have the same meaning when used in the formulaic dative form ("actually"). For a comparison of the two terms see Beere 2009: 218–19.

[137] In the *Metaphysics* the term ἐντελεχείᾳ occurs 21 times, the term ἐνεργείᾳ 48 times. For the whole corpus the TLG-search indicates 80 instances of ἐντελεχείᾳ compared to 171 of ἐνεργείᾳ.

[138] For the indicative with ὅτε meaning "whenever," see Kühner/Gerth II: §567.5; p. 451.

[139] Cf. Schwegler 1847a: ix: "Stimmen nun beide, die Lemmata und die Paraphrase in der Art über-

ma or a citation disagrees with one or both branches of the direct transmission, Alexander's paraphrase can confirm or disconfirm the readings as faithfully representing ω^{AL}.[140]

Therefore, my guiding rule for the present study is that we can safely reconstruct a reading in ω^{AL} whenever there is agreement between at least two of the four possible types of evidence in Alexander's commentary (i.e., lemmata, quotations, paraphrase and critical discussion), one of which is a paraphrase or critical discussion. The reading of ω^{AL} reconstructed in this way does not need to be confirmed by α, β, or ω^{αβ}.

3.5 ALEXANDER'S SOURCES FOR THE *METAPHYSICS* TEXT

Alexander writes his commentary as a philosopher who occasionally broadens his scope and includes philological aspects of Aristotle's text. Since Alexander subjects almost every sentence of the *Metaphysics* to scrupulous analysis, it comes as no surprise that he also comments on Aristotle's diction.[141] This sometimes leads him to question the validity of a given passage, or even to suggest a correction or refer to a variant reading found in another manuscript.[142] This, however, does not mean that Alexander took it upon himself to search for better readings in other manuscripts whenever he was unsatisfied with the text in front of him. Although Alexander was not indifferent to the quality of the text before him, the attitude he exhibits was nevertheless not one of a collector and collator of manuscripts (cf. 3.1). The available evidence suggests that Alexander incorporated into his commentary simply those variant readings that were noted in the margins of his own exemplar or reported in the commentaries he had read. This, however, prompts the question as to what commentaries were available to Alexander. Does Alexander give us any clues to his exegetical sources?

3.5.1 Aspasius? Others?

Nowhere in the preserved part of his commentary does Alexander mention consulting different manuscripts of the *Metaphysics* text. Although he most likely used a single exemplar (ω^{AL}), he was familiar with variant readings and even with other commentator's conjectures on how to philologically improve difficult pas-

---

ein, dass sie sich direct auf einander beziehen und sich gegenseitig bestätigen, so kann über den Text, den der Ausleger vor sich gehabt hat, kein Zweifel seyn."

[140] Cf. 3.2 and 3.3.

[141] This is especially the case when Alexander finds Aristotle unclear: e.g. in 21.30-31: ἀσαφῆ τὴν λέξιν ἐποίησε. 153.13-15; 159.6; 240.30: ἡ δὲ λέξις ἀσαφῶς ἔχει διὰ τὴν συντομίαν. Cf. Moraux 2001: 438 with n. 59. See my table B in 3.6.

[142] Cf. Moraux 2001: 429.

sages (cf. 3.6). Alexander does not indicate the sources from which he draws this information, but nearly without exception refers to them in the anonymous plural form (*some say…*). In only one instance does Alexander give the name of his source for his knowledge about a conjecture (59.6): that source is Aspasius, who preceded him by about two generations.[143] Alexander refers to this commentator in two other passages in his commentary on the *Metaphysics*, but in these instances not as a source for a *varia lectio* or conjecture. In addition to these three references to Aspasius, Alexander refers also, and just one time (166.20), to his teacher Aristotle of Mytilene. This reference does not concern the text itself, but rather the interpretation of the content. In all other instances where Alexander refers to the opinions of other scholars, he uses the anonymous pronoun τινές. The number of passages in which Alexander refers to the opinions of others is few. A great many of the few passages, however, concern the quality of the transmitted text. Are we justified in surmising that Alexander draws from other scholars especially or even primarily when philological issues are concerned? To answer this question, a closer look at the passages is needed.

I will begin by looking at those passages in which Alexander refers to his sources by name. The name of the peripatetic commentator Aspasius appears three times in the extant part of the commentary.[144] If we follow Moraux in assuming that Aspasius wrote (if even quite brief)[145] a commentary on the *Metaphysics*,[146] it is reasonable to conclude that Alexander used this commentary. One of the three comments that Alexander reports from Aspasius concerns a conjecture and therefore directly addresses the text of the *Metaphysics*. The other two involve interpretative issues, one of which, however, is closely related to a textual issue.

The first mention of Aspasius in Alexander's commentary occurs in a passage on the Pythagorean principles.[147] In 41.26–27, Alexander reports Aspasius's comments on Aristotle's statement (A 5, 986a15–18) that the Pythagoreans take numbers to be principles of things in terms of matter (ὡς ὕλη) and in terms of forming modification and states (ὡς πάθη τε καὶ ἕξεις).[148] In his commentary, Alexander offers three interpretations of the phrase ὡς πάθη τε καὶ ἕξεις ("modification and states"),[149] the second being Aspasius's understanding of the phrase. According to

---

[143] Aspasius wrote his commentary on the *Ethics* in AD 131 or slightly later (Barnes 1999: 3). Two generations separate Alexander from Aspasius, if we assume that Herminus (Moraux 1984: 361–98) was Alexander's teacher and Aspasius's student (see Simp. *In Cael*. 430.32–431.11 Heiberg). See also Moraux 1984: 361 with n. 5.

[144] For Aspasius see Moraux 1984: 226–39; Goulet 1994; Barnes 1999. For Aspasius's commentary on the *Ethics* see Alberti/Sharples 1999.

[145] Moraux 1984: 246: "einen wahrscheinlich nicht sehr umfangreichen Kommentar."

[146] Moraux 1984: 246–49. See also Luna 2003: 250.

[147] See Moraux 1984: 246–47.

[148] See Ross 1924: 147–48 and Schofield 2012: 143–46 with n. 10.

[149] Alex. *In Metaph*. 41.21–28.

Aspasius, number is matter, the even is modification, and the odd is state.[150] After this short report Alexander provides no further explication of Aspasius's position but continues forward to the third interpretation of the phrase.[151]

The second mention of Aspasius in Alexander's commentary occurs in a much-discussed passage (58.31–59.8)[152] in which Alexander reports a variant reading of the text in A 6, 988a9–11. According to Aspasius, Alexander tells us, the variant reading under discussion is a conjecture from the Middle Platonist Eudorus.[153] The passage in which Alexander refers to Eudorus's conjecture appears corrupt in the Greek manuscripts of the commentary. I follow Oliver Primavesi's reconstruction of the text, which is based on Sepúlveda's Latin translation of the commentary (58.31–59.8).[154] According to the evidence, Eudorus replaced Aristotle's wording in A 6, 988a9–11 τὰ γὰρ εἴδη τοῦ τί ἐστιν αἴτια τοῖς ἄλλοις, τοῖς δ' εἴδεσι τὸ ἕν[155] with the formulation τὰ γὰρ εἴδη τοῦ τί ἐστιν αἴτια τοῖς ἄλλοις, τοῖς δὲ εἰδόσι τὸ ἕν καὶ ἡ τοῦ εἴδους ὕλη.[156] After presenting Eudorus's alternative reading, Alexander explicates its meaning (59.2–4).[157] He then adds that he prefers the reading of his own text (which is also the one transmitted in our text).[158] Alexander further reports that Aspasius considered the first reading, i.e., the text as it appears in ω^AL and ω^αβ, to be the older one, and the second[159] a skillful conjecture

---

[150] Alex. In Metaph. 41.25–27: ἢ ὡς Ἀσπάσιος, ὁ μὲν ἀριθμὸς ὕλη, πάθος δὲ τὸ ἄρτιον, ἕξις δὲ τὸ περιττόν. / "Or, as Aspasius [explains], number is matter, the even is modification, and the odd is state."

[151] Moraux 1984: 247 understands the third interpretation as a correction of Aspasius's interpretation.

[152] See Moraux 1969, Fazzo 2012a and most recently Rashed/Auffret 2014: 65–74, with further literature.

[153] For Eudorus see Moraux 1984: 509–27 and Dillon 2000: 290–293.

[154] Primavesi presented his reconstruction of this passage in two talks in Athens and Munich delivered in 2013. For other suggested reconstructions see Moraux 1969 and Rashed/Auffret 2014: 65–74. Cf. also Fazzo 2012a.

[155] "... for the Forms are the causes of the essence of all other things, and the One [is the cause of the essence] of the Forms."

[156] "... for the Forms are the causes of the essence for the other (i.e. ordinary) people, and the One and the matter of the Form [are the causes of the essence] for those who know."

[157] I follow Primavesi's reconstruction of Alex. In Metaph. 59.2–4 and read: καὶ εἴη ἂν τὸ 'ἄλλοις' λεγόμενον ἐπὶ τοῖς οὐκ εἰδόσι τὴν Πλάτωνος δόξαν {τὴν περὶ τῶν ἀρχῶν} ὅτι τὸ ἕν καὶ ἡ ὑποκειμένη ὕλη ἀρχαὶ καὶ ὅτι τὸ ἕν καὶ τῇ ἰδέᾳ αἴτιον τοῦ τί ἐστιν. / "And 'the other (people)' should refer to those who do not know Plato's doctrine according to which the One and the underlying matter are principles, and that the One is cause of the essence for the Idea, too."

[158] Alex. In Metaph. 59.4–6: ἀμείνων μέντοι ἡ πρώτη γραφὴ ἡ δηλοῦσα ὅτι τὰ μὲν εἴδη τοῖς ἄλλοις τοῦ τί ἐστιν αἴτιον, τοῖς δὲ εἴδεσι τὸ ἕν./ "The preferable reading, however, is the first one, which makes it clear that the Forms are the cause of the essence for the other things, and the One for the Forms."

[159] I disagree with Moraux (1969: 500–501), who believes that Alexander's pronoun ἐκείνης (59.6) refers back to the (what I take to be a) conjecture that, he believes, is the older yet corrupt version of the text, while Eudorus (referred to by ταύτης in 59.7) restored the correct reading which is also given in our manuscripts.

by Eudorus.¹⁶⁰

We may reasonably assume that Alexander drew this information from Aspasius's *commentary* on the *Metaphysics*, in which the origin and the value of the alternative reading were discussed.¹⁶¹ However exactly Alexander might have had access to this information, it is striking that he mentions by name the reporter (Aspasius) of this information. He does not do this elsewhere in the extant part of his commentary on the *Metaphysics,* but he does it in his commentary on *De Sensu* (10.1-2 Wendland). There, Alexander once again references Aspasius, and again does so in an attempt to determine the correct understanding of a variant reading.¹⁶² The reason why Alexander names Aspasius as the source of a *varia lectio* at this particular passage in the *Metaphysics* commentary might simply be that Alexander discusses the significance of the readings in question in greater detail. Perhaps Aspasius (in his commentary on the passage?) had already devoted some space to the evaluation of Eudorus's conjecture. We do not know. In any case, the

---

¹⁶⁰ Alex. *In Metaph.* 59.6-8: ἱστορεῖ δὲ Ἀσπάσιος ὡς ἐκείνης μὲν ἀρχαιοτέρας οὔσης τῆς γραφῆς, μεταγραφείσης δὲ ταύτης ὕστερον ὑπὸ Εὐδώρου καὶ εὐαρμόστου. / "Aspasius relates that the former is a more ancient reading, but that it was later changed by Eudorus, and not badly so." Moraux 1969: 493-94 argued that the word <εὐ>αρμόστου (cf. Brandis's conjecture <Εὐ>αρμόστου : ἁρμοστοῦ **A**) does not refer to an otherwise unknown person named Euharmostus but is to be understood as an adjective. The new evidence in **O** confirms the reading. (Hayduck's apparatus is insufficient here.)

¹⁶¹ Cf. Moraux: 1984: 246. Fazzo 2012a: 65, however, doubts that Aspasius wrote a commentary on the *Metaphysics*. Fazzo speculates that the work in question could be a treatise on the Pythagoreans, because *one* of Alexander's three references to Aspasius is made within the context of Pythagorean doctrines. I object to this reasoning because *all* of Alexander's three references concern the *Metaphysics*, and it is for this reason that I presume that Alexander is drawing here from a commentary on the *Metaphysics*. It is extremely implausible to argue, as Fazzo does (2012a: 66), that since Aspasius's commentary on the *Ethics* extensively discusses those issues that one would expect to appear in a commentary on the *Metaphysics*, it is unlikely that such a *Metaphysics* commentary ever existed. Fazzo 2012a: 66: "At p. 4, for instance, he talks about the place of the One in the Pythagorean system. One would rather expect such a discussion in the context of a commentary on *Metaphysics* or in a treatise on metaphysical topics. Indeed one might wonder: if, in addition to the extant [*Ethics*] commentary, Aspasius undertook a commentary on *Metaphysics*, why not reserve this discussion for that other, more appropriate context?" I think the opposite conclusion is right: if there is a commentary on the *Ethics* in which Aspasius reveals himself as an expert on issues and topics related to the *Metaphysics*, we can safely assume that Aspasius worked intensively on the *Metaphysics*. What, anyway, could be more natural for a second-century AD philosopher, who works on and teaches Aristotle's *Metaphysics*, than to write a commentary on it? And Aspasius would not be the only (ancient) scholar to re-use his own work in his commentaries.

¹⁶² Alex. *In Sens.* 9.24-25 Wendland: γράφεται καὶ ἀντὶ τοῦ 'γευστικοῦ μορίου' 'τοῦ θρεπτικοῦ μορίου πάθος' [436b17-18] ... 9.29-10.3: διὸ ἄμεινον, εἰ οὕτως εἴη ἔχουσα ἡ γραφή, μὴ ἐπὶ τὴν δύναμιν τῆς ψυχῆς ἀναφέρειν τὸ θρεπτικόν, ὥς φησι δεῖν ἀκούειν τῆς λέξεως Ἀσπάσιος, ἀλλ' ἐπὶ τὸ μόριον δι' οὗ τρεφόμεθα (τούτου γὰρ ὁ χυμός), ... / "It is also written, instead of the 'part capable of taste,' 'an affection of the nutritive part'... . For this reason it would be better, if the text did say this, not to refer the nutritive to the <nutritive> power of the soul, as Aspasius says the text should be interpreted, but to the part by means of which we are nourished (for flavour <is an affection> of this <part>),..." [transl. by A. Towey]. For Aspasius's commentary on *Sens.* see Moraux 1984: 244-46.

fact that such a passage occurs only once in Alexander's *Metaphysics* commentary by no means excludes the possibility that he also relied on Aspasius in other instances where he refers to a variant reading without naming his source.

The third mention of Aspasius by name occurs in the commentary on Δ 9 (1018a12–13). Aristotle gives the following account of what *differing* (διάφορον) is: διάφορα δὲ λέγεται ὅσ' ἕτερά ἐστι τὸ αὐτό τι ὄντα, μὴ[163] μόνον ἀριθμῷ ἀλλ' ἢ εἴδει ἢ γένει ἢ ἀναλογίᾳ. Alexander understands this rule in the following way: 'We call different those things which though other are in some respect the same, <u>only not</u> in number (the same) but either in species or in genus or by analogy (the same).'[164] Alexander's explanation of μὴ μόνον as μόνον μὴ remains popular to this day.[165] Besides this, Alexander refers to the following alternative solution suggested by Aspasius:[166] 'We call different those things which though other are the same in some respect, <u>not only</u> in number (<u>different</u>) but either in species or in genus or by analogy (<u>the same</u>).'[167] Although the meaning of the sentence according to this interpretation does not differ much from Alexander's understanding of it, Aspasius's suggestion is a bold move. Alexander does not evaluate Aspasius's interpretation.

All three passages in which Alexander refers to Aspasius focus on the wording of the *Metaphysics* or even the constitution of its text. That Aspasius's commentaries included text-related issues becomes evident from several passages in Sim-

---

[163] In the α-text a καὶ precedes μὴ. This changes the meaning slightly, but does not make the meaning any more intelligible. Alexander's testimony about this aspect of the sentence is contradictory: at one point he quotes the text with καὶ (378.30), and at another point without καὶ (379.4). The καὶ, however, does not feature in his discussion of the sentence and its possible meanings.

[164] See Alex. *In Metaph*. 378.28–379.3. Alex. *In Metaph*. 378.29–32: ταῦτα ὅσα μὴ μόνον ἕτερά ἐστιν ἀλλήλων, ἀλλ' ὅσα κατά τι ἓν ὄντα ταύτῃ τὴν πρὸς ἄλληλα ἑτερότητα ἔχει. τὸ δὲ <u>καὶ μὴ</u> μόνον ἀριθμῷ προσέθηκε τῷ ταὐτό τι ὄντα ὡς ἴσον τῷ <u>μόνον μὴ</u> ἀριθμῷ ὄντα ταὐτά· τὰ γὰρ οὕτω ταὐτὰ οὐκέτι διαφέρειν δύναται. / "Those that are not only other, but that have their otherness while being in some respect the same. To the words, 'while being the same,' he adds 'and <u>not only</u> in number.' This latter is equivalent to, 'while being the same, <u>only not</u> in number,' for things that are the same in number can no longer differ."

[165] Bonitz 1849: 245 and Ross 1924: 313 defend this understanding by drawing attention to parallel passages where μὴ μόνον is used in the desired sense. See also Kirwan 1971: 151 ("Aristotle's account … is obscure") and Moraux 1984: 247–49.

[166] Alex. *In Metaph*. 379.3–5: Ἀσπάσιος δὲ ἤκουσε τοῦ [quotation of 1018a12–13] ὡς εἰρημένου ὅτι… These words make it clear that we are dealing with an interpretation by Aspasius and not with a conjecture.

[167] Alex. *In Metaph*. 379.3–8: Ἀσπάσιος δὲ ἤκουσε τοῦ [quotation of 1018a12–13] ὡς εἰρημένου ὅτι δεῖ τὰ διάφορα μὴ μόνον ἕτερα εἶναι ἀριθμῷ, ἀλλὰ καὶ κατά τι τὰ αὐτὰ ἀλλήλοις εἶναι, εἰ μέλλοι μὴ μόνον ἕτερα εἶναι ἀλλὰ καὶ διάφορα· τὰ γὰρ ἀριθμῷ ἕτερα οὐ πάντως διάφορα, ἂν μὴ καὶ κατά τι τῶν εἰρημένων τὰ αὐτὰ ᾖ. / "Aspasius, however, understood the statement, [quotation of 1018a12–13] to mean that different things must not only be other in number, but must also be the same as one another in some respect if they are to be not only other but different; for [he held] that things numerically other are not in every case different unless they are also the same in one of the ways mentioned."

plicius's *Physics* commentary.¹⁶⁸ In more than one of these passages Simplicius also mentions Alexander and his view on the reading reported by Aspasius (*In Phys.* 422.19–26 and 423.12–23; 436.13–19; 950.3–6). Might Alexander have been Simplicius's source for readings discussed by Aspasius?¹⁶⁹ Whatever the case may be, it is evident that Alexander consulted Aspasius now and then on issues of the exact meaning of Aristotle's words. Alexander does the same in the *De Sensu*-commentary.¹⁷⁰

Can we infer from this that before Alexander's day there existed a commentary tradition, in which textual problems took on a more important role? The available evidence is not strong enough to support the claim that the commentary tradition before Alexander took textual matters on the whole more seriously than Alexander did. Nevertheless, we can be sure that there *was* commentary literature preceding Alexander that also included discussions of textual issues. This is clear from Aspasius's commentary on the *Ethics* and from indirect evidence preserved by other commentators.¹⁷¹ For example, in his commentary on the *Categories* Simplicius refers several times to Boethus of Sidon,¹⁷² the late first-century AD peripatetic commentator and student of Andronicus of Rhodes, who received much praise for his outstanding commentaries.¹⁷³ Simplicius highlights the depth¹⁷⁴ of

---

¹⁶⁸Relevant passages in Simplicius's commentary are: Simp. *In Phys.* 422.19–26; 436.13–18; 714.31–715.7; 727.35–728.10; 818.27–819.3; 845.19–846.2 and 950.3–6. See Barnes 1999: 10 with n. 33 and 34. Moraux 1984: 235 states at the beginning of his treatment of Aspasius's *Physics* commentary: "Obwohl die meisten Fragmente dieses Kommentars sich auf textkritische Probleme beziehen, sind einige Spuren von Aspasios' Interpretation der Physik erwähnenswert." This does not necessarily imply that Aspasius's commentary was full of textual discussions. Rather it says that Aspasius also discussed textual issues and that those passages were transmitted in the subsequent commentary tradition. See also Moraux 1984: 238–39.

¹⁶⁹Cf. Barnes 1999: 11–12 with n. 39.

¹⁷⁰This is the only extant passage where Alexander refers to his predecessor Aspasius outside of his *Metaphysics* commentary. Other Aristotelian works on which both commentators, Aspasius and Alexander, commented, and which could lead us to further references by Alexander to Aspasius are *Cael.* and *Cat.* Unfortunately, Alexander's commentaries on these are either wholly lost or fragmentarily preserved.

¹⁷¹Moraux 1984: 238 writes about Aspasius's commentary on the *Physics*: "Die Exemplare, die etwa in den frühen Kaiserzeit umliefen, wiesen bisweilen einen viel schlechteren Text auf als unsere mittelalterlichen, auf eine sehr sorgfältig durchgeführte Translitteration zurückgehenden Manuskripte. Die durch Aspasios bezeugten Varianten lassen sich aber nicht alle durch die Nachlässigkeit der Kopisten jener Zeit erklären. Einige sind sicher keine bloßen Fehler; sie verraten den gewaltsamen und meistens nicht glücklichen Eingriff eines oder mehrerer Korrektoren. Man sieht also wie kühn und skrupellos einige Aristoteliker in den ersten zwei Jahrhunderten nach der Andronikos-Ausgabe, wenn nicht schon vorher, mit dem tradierten Text umgegangen sind."

¹⁷²See Schneider 1994. Cf. also Brandis 1833: 276 and Gercke 1897: 603–604. A more detailed discussion is offered by Moraux 1973: 143–79.

¹⁷³Moraux 1973: 147.

¹⁷⁴Simp. *In Cat.* 1.17–18 Kalbfleisch: τινὲς μέντοι καὶ βαθυτέραις περὶ αὐτὸ διανοίαις κατεχρήσαντο, ὥσπερ ὁ θαυμάσιος Βόηθος. / "Some commentators, however, also applied deeper thoughts to the

Boethus's discussions, his word-by-word exegesis,[175] and his critical reflection on the words of Aristotle.[176]

Returning to the evidence in Alexander's *Metaphysics* commentary, we see that Alexander mentions one other predecessor, who, as it happened, was also his teacher: Aristotle of Mytilene.[177] Alexander calls on his teacher not in respect to a textual issue, but instead reports his teacher's argument for the sake of showing that causes cannot be infinite in their kinds—a view held by the Stagirite (166.19–167.1). Alexander writes (166.18–20): αὐτὸς μὲν οὕτως ἐφοδεύσας ἔδειξεν ὅτι μὴ οἷόν τε ἄπειρα εἶναι τὰ αἴτια· ὁ δὲ ἡμέτερος Ἀριστοτέλης καὶ αὐτὸς ἐπιχειρῶν ἐδείκνυεν.[178]

We can therefore conclude that when composing his commentary Alexander drew on at least two earlier philosophers or commentators.[179] Of course, there may be more sources than just the two he explicitly mentions. Alexander repeatedly speaks of "some" (τινές) who favored a certain interpretation or version of the text. It is quite possible that the anonymous plural τινές refers to thinkers whose

---

work, as did the admirable Boethus" (transl. by Chase).

[175] Simp. *In Cat.* 29.28–30.3: πρὸς γὰρ ταύτην τὴν ἀπορίαν ὑπαντῶν ὁ Πορφύριος πρῶτον μέν φησιν μηδὲ ἐν πᾶσι τοῦτο γεγράφθαι τοῖς ἀντιγράφοις· μήτε γὰρ Βόηθον εἰδέναι, ὅς φησι δεικνύναι τὸν Ἀριστοτέλη τίνα ἐστὶ τὰ ὁμώνυμα λέγοντα Ὁμώνυμα λέγεται ὧν ὄνομα μόνον κοινόν, ὁ δὲ κατὰ τοὔνομα λόγος ἕτερος· καὶ ἐξηγούμενος δὲ ὁ Βόηθος καθ' ἑκάστην λέξιν τὸ τῆς οὐσίας παραλέλοιπεν ὡς οὐδὲ γεγραμμένον. / "For it is in reply to this puzzle that Porphyry says, in the first place, that this ['τῆς οὐσίας,' in *Cat.* 1a2] is not written in all the manuscripts. Boethus, he says, did not know of it, who says that Aristotle points out what homonyms are by saying: 'Those things are called homonyms of which only the name is common, but the definition in accordance with the name is different.' Although Boethus was carrying out a word-by-word exegesis, he omitted 'of the substance' as though it was not written" (transl. by Chase, slightly changed).

[176] Simp. *In Cat.* 58.27–28: ἀλλ' ὁ μὲν Βόηθος ἐνδοὺς τῇ ἀπορίᾳ μεταγράφειν ἠξίου τὴν λέξιν οὕτως· ... / "Now Boethus gave in to this problem, and suggested emending the text as follows: ..." (transl. by Chase). Gercke 1897: 603 even goes so far as to say that Boethus commented "mehr philologisch als philosophisch." The indirect evidence at our disposal, however, does not seem to warrant this conclusion. Cf. also Gottschalk 1990: 74–75.

[177] On the evidence for the teacher Aristotle of Mytilene see Moraux 1967. On his teaching see Moraux 1984: 399–425; since our passage from the *Metaphysics* commentary is not taken into account there, see also Moraux 1985. Alexander does not mention his other teachers in the *Metaphysics* commentary. That Herminus was his teacher we know from a fragment in Simplicius (Simp. *In Cael.* 430.32–431.11 Heiberg); about Sosigenes we learn from Alexander in his commentary on the *Meteorologica* (Alex. *In Meteor.* 143.12–14 Hayduck). For Herminus see 1984: 361–98; for Sosigenes see Moraux 1984: 335–60.

[178] "He [Aristotle, the Stagirite] himself proved, as he proceeded in this way, that the causes cannot be infinite; but our Aristotle [Aristotle of Mytilene] used to give a dialectical proof." Sepúlveda's Latin translation of this passage reads the additional sentence: *Caeterum potest idem alia via demonstrari ad hunc modum*. On this see Moraux 1985: 268–69.

[179] We do not know whether Aristotle of Mytilene wrote a *commentary* on the *Metaphysics*. Alexander may have drawn from private discussions. On the evidence for Aristotle of Mytilene in Syrianus, *In Metaph.* 100.3–13 Kroll (ὁ νεώτερος Ἀριστοτέλης ὁ ἐξηγητὴς τοῦ φιλοσόφου Ἀριστοτέλους) see Moraux 1984: 403–406 and Luna 2003: 250.

identity had been lost by the time of Alexander.[180]

The following list contains all of the passages in which Alexander refers to other scholars by way of an anonymous plural (expressed either by τινές, "some," or by a third person plural verb form, e.g. φασί(ν)). I distinguish between those references that are text-critical in purpose (= T) and those that are interpretative in purpose (= I). Issues concerning the text of the *Metaphysics* are not restricted to the mention of *variae lectiones* or conjectures, but also embrace discussions about the composition of the *Metaphysics* as well as the correct understanding of particular words or phrases.[181] The other category covers those passages that deal with the philosophical content of the text.

Passages in Alexander's commentary	T	I
46.23–24: ἐξηγούμενοι οἱ μὲν … οἱ δὲ[182]	X	
75.26–28: φασιν…[183]	X	X
100.25–27: ὥς τινες ἤκουσαν[184]	X	X
104.19–22: ὑπεμνηματίσαντο … [185]	X	
141.11–12: τινὲς … φασὶ[186]	X	
162.10–16: τινὲς ἤκουσαν[187]	X	X
163.6–7: τινὲς … ἤκουσαν[188]	X	X
164.22–25: τινὲς … γράφουσι καὶ ἐξηγοῦνται[189]	X	
172.20–22: τισιν ἔδοξε[190]	X	
174.25–27: τινὲς … προσγράφουσι[191]	X	
177.10: τισιν ἔδοξε[192]		X

---

[180] Cf. McNamee 1977: 92–93.

[181] Some will also fall into the category of interpretation.

[182] See below and 3.6.

[183] Alex. *In Metaph.* 75.26-28: γράφεται δὲ ἔν τισιν ἀντιγράφοις ἀντὶ τοῦ ἀδικίαν ‚ἀνικίαν'· ἀνικίαν δέ φασιν ὑπὸ τῶν Πυθαγορείων λέγεσθαι τὴν πεντάδα· / "Certain transcriptions of the text have the reading 'non-victory' (ἀνικίαν) instead of 'injustice' (ἀδικίαν). Some say that the Pythagoreans called the number 5 'non-victory' (ἀνικίαν)." We could speculate that Alexander found the opinion of the "some" on the meaning of the variant reading in the same source where he found the variant reading. See also 3.6.

[184] Alexander reports that some took Aristotle's phrase (991a20) κατ' οὐδένα τρόπον τῶν εἰωθότων / "in any of the usual senses" to mean 'in any of the senses used by those who postulate the Forms.'

[185] See below.

[186] See also 3.6.

[187] Alexander reports on how some understood the phrase πλεονάζοντα τῷ λόγῳ (994b18).

[188] Alexander reports on how some understood the phrase ἄτομα (994b21).

[189] See 3.6 and 5.1.4.

[190] Alexander refers to certain others who regarded book B as the first book of the *Metaphysics*.

[191] See 3.6 and 4.1.1.

[192] Alexander speaks of some who call the *Metaphysics* generally λογικός. In Alexander's diction, the word λογικός tends to mean "abstract, dialectical, (merely) verbal" (as it occasionally does in Aristotle, see Bonitz 1870, 432b9–11 and *Metaph.* Z 4, 1030a25; *EN* A 1, 1217b21: λέγεται λογικῶς καὶ κενῶς).

345.4-6: ὡς οἴονταί τινες ...[193]	X	
Fr. 10a Freudenthal 81.17-19:[194] Some people understood the words ... in the following way / Einige haben den Ausdruck ... in folgender Weise verstanden... / Et quidam intelligebant ... ita (Scotus)	X	X

The majority of mentions of τινές refer either primarily or partly to issues concerning Aristotle's text. In only one case (177.10) does the opinion of τινές concern the content of the *Metaphysics* in general, without regard to particular words or phrases. Given that Alexander's focus is mainly on the philosophical interpretation of the *Metaphysics*, it is striking that references to other scholars appear mostly when the accurate interpretation of one of Aristotle's words or phrases is at issue. This not only indicates that Alexander references his predecessors primarily on matters of textual criticism, but also that such criticism of Aristotle's wording was an area traditionally covered and transmitted in the commentaries. It seems probable, then, that Alexander's knowledge of textual peculiarities, conjectures, and variant readings is primarily based on this tradition.

That this is so is further suggested by cases in which Alexander notes the absence of a certain passage in other manuscripts (ἔν τισιν οὐ φέρεται) and then reports how other commentators responded. Alexander thus hints at his source. In his commentary on A 9 he says:

Alexander *In Metaph.* 104.19-22 Hayduck

991a27 Ἔσται τε πλείω παραδείγματα τοῦ αὐτοῦ ὥστε καὶ εἴδη.
[20] Αὕτη ἡ λέξις ἕως τοῦ ἔτι δόξειεν ἂν ἀδύνατον εἶναι ἔν τισιν [21] οὐ φέρεται· διὸ οὐδὲ ὑπεμνηματίσαντο αὐτήν. δείκνυσι δὲ δι' [22] αὐτῆς καὶ τοῦτο τὸ ἄτοπον ἑπόμενον τῇ περὶ τῶν ἰδεῶν δόξῃ·

> And there will be more than one model of the same thing, hence more than one Form as well.
> This text, up to, "again one would think it impossible," is not contained in certain manuscripts, and for this reason they (i.e. some commentators) did not comment on it. By it Aristotle shows this further absurd consequence of the theory about the Ideas.

20 ἂν **A O** : om. **P**ᵇ || 20-21 ἔν τισιν **A O P**ᵇ Bonitz : ἔν τισιν ἀντιγράφοις **L S** Hayduck

Alexander begins with a back reference to the passage whose beginning he quoted in the lemma. In this passage, Aristotle argues against the supposition that the Forms are paradigms. He argues that the supposition of Forms as paradigms

---

Cf. Madigan 1992: 96 n. 34.

[193] Some hold that book Δ is incomplete; Alexander argues against this opinion.

[194] On the interpretation of the expression ἐκ συνωνύμου (1070a5), put forward by some.

entails the supposition of *multiple* paradigms for the same thing.¹⁹⁵ Alexander tells his readers that this objection and the subsequent one, according to which the postulation of Forms requires that Forms be postulated for more than just the things of the sensible world¹⁹⁶ (i.e., the text up to the words ἔτι δόξειεν ἂν ἀδύνατον εἶναι in 991b1), are absent from some manuscripts.¹⁹⁷ Alexander then informs his readers that this is the reason why some commentators—who alone can be the subject of the verb ὑπομνηματίζεσθαι—did not interpret this section of the text.

These two pieces of information and their close, even causal (διὸ) connection in Alexander's commentary shed some light on Alexander's source. The way in which the reference to the missing passage and the reference to the commentator's neglect of the passage are linked in Alexander's presentation suggests that for him these are two interrelated issues. It thus may be plausible to suppose that the source of the information on the missing text is not the collation of multiple manuscripts, but the bare fact that some commentaries offer no comment on a particular text or only a short remark about the absence of the passage in the commentator's text.¹⁹⁸

It is true that much less is preserved of the commentary tradition preceding Alexander than of the tradition following him. Nonetheless, the available evidence suggests that Alexander relied on earlier commentators, who themselves relied on the work of earlier scholars.¹⁹⁹ Alexander gives us the name of one commentator whose work he was familiar with: Aspasius. It is quite probable that Aspasius's work on the *Metaphysics* was not the only source Alexander used, but it is the only

---

¹⁹⁵ *Metaph.* A 9, 991a27–29: ἔσται τε πλείω παραδείγματα τοῦ αὐτοῦ, ὥστε καὶ εἴδη, οἷον τοῦ ἀνθρώπου τὸ ζῷον καὶ τὸ δίπουν, ἅμα δὲ καὶ τὸ αὐτοάνθρωπος. / "And there will be several patterns of the same thing, and therefore several Forms, e.g. animal and two-footed and also man himself will be Forms of man."

¹⁹⁶ *Metaph.* A 9, 991a29–b1: ἔτι οὐ μόνον τῶν αἰσθητῶν παραδείγματα τὰ εἴδη ἀλλὰ καὶ αὐτῶν, οἷον τὸ γένος, ὡς γένος εἰδῶν· ὥστε τὸ αὐτὸ ἔσται παράδειγμα καὶ εἰκών. / "Again, the Forms are patterns not only of sensible things, but of themselves too, e.g. the Form of genus will be a genus of Forms; therefore the same thing will be pattern and copy." See Frede 2012: 289–92.

¹⁹⁷ Since the missing section of text is independent syntactically and semantically it is plausible to assume that someone deleted Aristotle's objections deliberately—we might not be dealing with a mechanical dropout here.

¹⁹⁸ There are other commentary passages in which Alexander connects his report of a *varia lectio* with the report of the opinion of other commentators: See 46.23–24 (on A 5, 987a9–10) and 341.30 (on Γ 8, 1012b22–31). Cf. also 58.31–59.8 (see above) and 75.26–28 (where the φασιν could, however, simply mean "it is said").

¹⁹⁹ Moraux 2001: 428: "Selbst wenn wir über diese Vorgänger wenig erfahren, dürfen wir annehmen, dass Alexander einer bereits alten Tradition der Metaphysikexegese verpflichtet war." Cf. Fazzo 2004: 6: "On the one hand, Alexander is the first Aristotelian commentator from whom we possess entire commentaries on complete works. Indeed, because of the above-mentioned tendency for works of this type to supersede one another, his commentaries almost entirely replaced the previous legacy of literature handed down by the Peripatetic school."

one we know by name. We do not know who else is meant by the anonymous plural τινές ("some"). The fact that Alexander refers to earlier scholars anonymously suggests that he, like us, did not know their names. Most likely he found the information about their criticism, textual corrections, and variant readings in the form of short notes in the margins of his manuscript or in other commentaries.

3.5.2 Did Alexander know readings from ω^{αβ}?

Some of the *variae lectiones* that Alexander mentions in his commentary are identical with the readings present in ω^{αβ}. What does this indicate? Did Alexander know readings from our *Metaphysics* text, ω^{αβ}, or an ancestor of it? Helpful for answering this question are the following analyses of four commentary passages where Alexander refers to a variant reading that agrees with the reading in ω^{αβ}. It is clear from the outset that proof of Alexander's knowledge of ω^{αβ} or an ancestor of ω^{αβ} can be reached only by showing that the reading that Alexander knows, and that is identical to the reading in ω^{αβ}, is corrupt.

3.5.2.1 Alex. *In Metaph.* 354.28–355.5 on Arist. *Metaph.* Δ 3, 1014a26–31

What we know as the fifth book of the *Metaphysics* (Δ) can be described as an encyclopedia in which Aristotle examines terms that are relevant to his inquiry in the *Metaphysics*. Chapter 3 is devoted to the term 'element' (στοιχεῖον). The beginning of the chapter reads as follows:[200]

Aristotle, *Metaphysics* Δ 3, 1014a26–31

στοιχεῖον λέγεται ἐξ οὗ σύγκειται πρώτου ἐνυπάρ-[27]χοντος **ἀδιαιρέτου τῷ εἴδει εἰς ἕτερον εἶδος**, οἷον φωνῆς [28] στοιχεῖα ἐξ ὧν σύγκειται ἡ φωνὴ καὶ εἰς ἃ διαιρεῖται [29] ἔσχατα, ἐκεῖνα δὲ μηκέτ' εἰς ἄλλας φωνὰς ἑτέρας τῷ [30] εἴδει αὐτῶν, ἀλλὰ κἂν διαιρῆται, τὰ μόρια ὁμοειδῆ, οἷον [31] ὕδατος τὸ μόριον ὕδωρ, ἀλλ' οὐ τῆς συλλαβῆς.

We call an element that which is the primary component immanent in a thing, and indivisible in kind into another kind, e.g. the elements of speech are the parts of which speech consists and into which it is ultimately divided, while *they* are no longer divided into other forms of speech different in kind from them. If they *are* divided, their parts are of the same kind, as a part of water is water, (while a part of the syllable is not [a syllable]).

27 ἀδιαιρέτου τῷ εἴδει εἰς ἕτερον εἶδος ω^{αβ} Al.^{γρ} Ar^{u} (*Scotus*) edd. : ἀδιαιρέτου τῷ εἴδει ω^{AL} : ἀδιαιρέτου εἰς τὸ αὐτὸ εἶδος Al.^{γρ} || 30 ἀλλὰ κἂν **α** : ἀλλ' ἂν καὶ **β**

An element is the primary constituent of a thing. This constituent is primary

---
[200] On this passage see Diels 1899: 23–24; for the Greek term στοιχεῖον see Burkert 1959.

because it is not further divisible in kind. Aristotle adduces the examples of the elements of human speech (φωνή),[201] which, according to his account in the *Poetics*,[202] are the letters. Were the element divided nevertheless, then it would not be divided into parts that are different in kind, but only into parts that are the same in kind (μόρια ὁμοειδῆ): for instance, water is divided only into water. Aristotle adds in brief fashion that this is not the same in the case of the syllable (ἀλλ' οὐ τῆς συλλαβῆς). This indicates that the syllable cannot qualify as the στοιχεῖον of human speech (φωνή).[203] The syllable *ba*, for instance, is divisible into the elements *b* and *a*,[204] for the actual elements of speech are letters.[205] These two examples, water and letters, point to two different types of elements: on the one hand there is water, which is divisible into homogeneous water-parts, and on the other there are letters, which are simply indivisible.[206]

In his commentary on this passage, Alexander seems to read a slightly different text, in which line 1014a27 did not contain the words εἰς ἕτερον εἶδος. This is suggested, first, by a quotation in which the three words are absent:

Alexander, *In Metaph*. 354.28–31 Hayduck

Ὅτι τὸ στοιχεῖον πολλαχῶς λέγεται, ἔδειξε, τοῦ μὲν κυρίως λεγομένου [29] στοιχείου λόγον ἀποδιδούς, ἐξ οὗ σύγκειται πρώτου ἐνυπάρχοντος, [30] ἀδιαιρέτου τῷ εἴδει· οὐ γὰρ κατὰ τὸ ποσὸν ἀδιαίρετον τὸ στοιχεῖον, [31] ἀλλὰ κατὰ τὸ εἶδός ἐστιν.

Aristotle shows that 'element' is expressed in various ways by giving the formula of element properly so called: "the primary component immanent in a thing, and indivisible in kind"; for an element is not indivisible in respect to quantity, but only in respect to kind.

28 μὲν O LF : μὴ A P^b || 30 τὸ A O : om. P^b || 30-31 ἀδιαίρετον... ἐστιν A O P^b : ἀδιαίρετον δεῖ ... εἶναι LF

---

[201] In *Metaph*. B 2, 998a23–25 Aristotle also refers to the elements of speech as example of ἐνυπάρχοντα: οἷον φωνῆς στοιχεῖα καὶ ἀρχαὶ δοκοῦσιν εἶναι ταῦτ᾽ ἐξ ὧν σύγκεινται αἱ φωναὶ πᾶσαι πρώτων, ἀλλ' οὐ τὸ κοινὸν ἡ φωνή.

[202] *Po*. 20, 1456b20–24: Τῆς δὲ λέξεως ἁπάσης τάδ᾽ ἐστὶ τὰ μέρη, <u>στοιχεῖον</u> συλλαβὴ σύνδεσμος ὄνομα ῥῆμα ἄρθρον πτῶσις λόγος. <u>στοιχεῖον μὲν οὖν ἐστι φωνὴ ἀδιαίρετος</u>, οὐ πᾶσα δὲ ἀλλ᾽ ἐξ ἧς πέφυκε συνθετὴ γίγνεσθαι φωνή· καὶ γὰρ τῶν θηρίων εἰσὶν ἀδιαίρετοι φωναί, ὧν οὐδεμίαν λέγω στοιχεῖον. / "The diction viewed as a whole is made up of the following parts: the letter, the syllable, the conjunction, the article, the noun, the verb, the case, and the sentence. The letter is an indivisible sound of a particular kind, one that may become a factor in a compound sound. Indivisible sounds are uttered by the brutes also, but no one of these is a letter in our sense of the term" (transl. by Bywater, but modified).

[203] This corresponds to what is said in the *Poetics* passage (see preceding note).

[204] Cf. *Metaph*. Z 17, 1041b11–19.

[205] See Alex. *In Metaph*. 354.35–36; Bonitz 1849: 226–27.

[206] Diels 1899: 23–24 n. 3: "Daher hätte die Definition korrekt gelautet, wenn sie beide Gattungen umfassen sollte: ἀδιαιρέτου, ἢ εἰ ἄρα, εἰς ἕτερον εἶδος (nämlich ἀδιαιρέτου)."

Alexander introduces his quotation from the *Metaphysics* as the definition of element in the proper sense (κυρίως). The quotation reads lines 1014a26–27, but without the words εἰς ἕτερον εἶδος (354.29–30). Did Alexander leave out the three last words of the sentence or were they not contained in ω^AL? The following three facts speak in favor of the latter possibility. First, Alexander does not mention the additional specification εἰς ἕτερον εἶδος when he goes on to explain what Aristotle's phrase ἀδιαιρέτου τῷ εἴδει means (354.30–31). Second, Alexander, in discussing a later section of Δ 3, comes back to the definition of στοιχεῖον given in our passage at the beginning of Δ 3 and quotes the passage, again without the words εἰς ἕτερον εἶδος.[207] Third, Alexander refers to two *variae lectiones* in the commentary passage subsequent to 354.28–31. The first of the two variant readings is identical with the reading we find in the direct transmission and contains the additional words εἰς ἕτερον εἶδος. Had that been the reading of Alexander's *Metaphysics* exemplar he could not call it a variant reading. Alexander says:

Alexander, *In Metaph.* 354.31–355.5 Hayduck

γράφεται δὲ καὶ ἀδιαιρέτου τῷ εἴδει εἰς [32] ἕτερον εἶδος καὶ
ἀδιαιρέτου εἰς τὸ αὐτὸ εἶδος. ἂν μὲν οὖν ᾖ [33] ἡ γραφὴ ἡ πρώτη, γνώριμον
τὸ λεγόμενον· ἀδιαίρετον γὰρ τὸ στοιχεῖον [34] εἰς ἕτερα καὶ διαφέροντα εἴδη.
οὔτε γὰρ τὸ πῦρ εἰς ἀνομοειδῆ διαιρεῖται [35] οὔτε τι τῶν ἄλλων τῶν ἁπλῶν· ἡ δὲ
συλλαβὴ οὐ στοιχεῖον τοῦ λόγου, [36] ἐπεὶ διαιρεῖται εἰς τὰ γράμματα ἀνομοειδῆ
ὄντα, τῷ δὲ ἀδιαίρετα εἶναι [355.1] οὐδὲ εἰς ἕτερα τῷ εἴδει διαιρεῖται. καὶ χωρὶς δὲ τοῦ
προσκεῖσθαι εἰς [2] ἕτερον εἶδος ταὐτὸν ἐσημαίνετο καὶ ὑπὸ τοῦ ἀδιαιρέτου
τῷ εἴδει· τὸ [3] γὰρ ἀδιαίρετον κατ' εἶδος οὐχ οἷόν τε εἰς ἕτερον εἶδος διαιρεθῆναι.
ἂν [4] δὲ εἰς τὸ αὐτὸ εἶδος, λέγοι ἂν εἰς στοιχεῖα. δεῖ γὰρ τὸ στοιχεῖον ἀδιαί-[5]
ρετον εἶναι εἰς στοιχεῖα· οὐ γὰρ ἂν ἔτι στοιχεῖον εἴη διαιρούμενον.

There are two variant readings of this text: (i) "indivisible in kind into another kind," and (ii) "indivisible into the same kind." (*ad i*) If the first reading is accepted, its meaning is easily understood; for the element cannot be divided into other and different kinds. For neither fire nor any of the other simple bodies is divided into parts of different kinds, but the syllable is not an element of speech because it is divided into letters that are of different kinds, whereas the result of being indivisible is that there is no division into things other in kind. Apart then from the addition of the phrase, "into another kind," [the first reading] would have the same meaning as that conveyed by "indivisible in kind," for what is indivisible in respect to kind cannot

---

[207] Alex. *In Metaph.* 356.11–14: τὸ μὲν γὰρ ὡς ὑποκείμενόν τι καὶ μέρος τι (τι Hayduck : τοῦ A O P^b) πράγματος γιγνόμενον, ὃ δηλοῦται διὰ τοῦ ὁρισμοῦ τοῦ λέγοντος στοιχεῖον (στοιχεῖον A^p.c.O S : στοιχεῖα P^b) λέγεται ἐξ οὗ σύγκειται πρῶτον ἐνυπάρχοντος (ἐνυπάρχοντος O P^b S: ἐνυπάρχοντα A) ἀδιαιρέτου τῷ εἴδει, ὃν αὐτὸς (αὐτὸς A O S : καὶ αὐτὸς P^b) ἀπέδωκεν ὡς ὄντα τοῦ κυρίως στοιχείου· / "For in one way it is understood as that which is a subject and a part of the thing, and this is the meaning that Aristotle expressed by defining element saying 'we call an element that which is the primary constituent immanent in a thing, and indivisible in kind,' a meaning that he himself gave as that of element in the primary sense."

be divided into other kinds. (*ad ii*) But if the text is read "indivisible into the same kind," it would mean, 'indivisible into elements,' for the element must be indivisible into [other] elements, for if it were to be divided it would no longer be an element.

32 ἀδιαιρέτου A^(p.c.) S : διαιρέτου A^(a.c.) O P^b || εἰς A O P^b : τῷ εἴδει εἰς LF S || 35 στοιχεῖον O LF : στοιχεῖα A P^b S || 36 δὲ A O : om. P^b || 4 εἰς τὸ αὐτὸ LF S : τὸ αὐτὸ A^(a.c.) O P^b : ἀδιαίρετον εἰς τὸ αὐτὸ A^(p.c.)|| 5 εἴη A^(p.c.) O P^b S : εἶναι A^(a.c.)

On the basis of Alexander's commentary we can reconstruct three different readings of lines 1014a26–27.[208]

ω^(AL):   ... πρώτου ἐνυπάρχοντος ἀδιαιρέτου τῷ εἴδει.
varia lectio[i] (= ω^(αβ)):   ... πρώτου ἐνυπάρχοντος ἀδιαιρέτου τῷ εἴδει εἰς ἕτερον εἶδος.
varia lectio[ii]:   ... πρώτου ἐνυπάρχοντος ἀδιαιρέτου εἰς τὸ αὐτὸ εἶδος.

The reading in ω^(αβ) (= *varia lectio*[i])[209] contains three words more than the reading in ω^(AL) (εἰς ἕτερον εἶδος / "into another kind"). These words do not contribute any thought that is not already contained in the ω^(AL)-version. Rather, they simply repeat, in slightly different terms, what is already expressed by the phrase ἀδιαιρέτου τῷ εἴδει ("indivisible in kind"). What is indivisible in kind is not divisible into another kind. Alexander describes this reading as "easily understood" (354.32–355.1; γνώριμον, 354.33), but he remarks that the three additional words do not alter the meaning of the shorter version present in his own text (355.1–3).[210]

Aristotle's idiom in the *Metaphysics* is highly economical. The repetitive nature of the ω^(αβ)-reading is therefore suspicious, and the reading in ω^(AL) seems preferable.[211] The emergence of the additional words εἰς ἕτερον εἶδος in ω^(αβ) might have been occasioned by the subsequent lines in the *Metaphysics* text, for lines 1014a29–30 stress that an element cannot be divided into another kind but only into the same kind. Concerning the indivisibility of letters Aristotle says: εἰς ἄλλας φωνὰς ἑτέρας τῷ εἴδει αὐτῶν / "into other forms of speech different in kind from them" (a29–30). Concerning water he says: ἀλλὰ κἂν διαιρῆται, τὰ μόρια ὁμοειδῆ / "If they *are* divided, their parts are of the same kind" (a30).[212] It appears that

---

[208] Diels 1899: 23–24 n. 3 incorrectly assumes that Alexander knows of (only) *two* different readings: εἰς ἕτερον εἶδος and εἰς τὸ αὐτὸ εἶδος. He seems to have overlooked the fact that Alexander's own text read the shorter version ἀδιαιρέτου τῷ εἴδει. The *Metaphysics* editors provide different information: Bekker ascribes the lack of εἰς ἕτερον εἶδος to F^b. Bonitz, Christ, and Jaeger provide the correct information and refer to all three readings. Ross gives insufficient information: he refers only to *varia lectio*[ii]. All editors put the reading of ω^(αβ) (= *varia lectio*[i]) in the text.
[209] This reading became part of the Arabic transmission also (see my apparatus).
[210] Alexander does not espouse the ω^(αβ)-reading (= *varia lectio*[i]). It therefore is unlikely that the reading in ω^(αβ) is the result of an adoption of one of Alexander's *variae lectiones*.
[211] See the cases discussed in 4.1.
[212] Diels (1899: 23 n. 3), who does not recognize that Alexander's text did not contain the words εἰς ἕτερον εἶδος, reports Alexander's statement that the omission of these words leaves the meaning unharmed, but holds on to the directly transmitted reading precisely because he sees the words confirmed by the lines a29–30.

the words εἰς ἕτερον εἶδος were added (in ω^αβ) to the definition of the element in order to adjust this definition to the subsequent exemplification. It would thereby have been overlooked that the formula ἀδιαίρετος τῷ εἴδει (1014a27) already expresses the very same idea. The explanation that εἰς ἕτερον εἶδος is a misguided later addition to the text squares well with the fact that there is no parallel passage in the Aristotelian corpus where the formula ἀδιαίρετος τῷ εἴδει is combined with the words εἰς ἕτερον εἶδος.[213] It therefore seems reasonable to conclude that the reading in ω^AL attests to the original and correct reading and that the words εἰς ἕτερον εἶδος, known to Alexander as variant reading and transmitted by ω^αβ, should be athetized.

It is less clear how the other *varia lectio* Alexander reports in his commentary and which I called *varia lectio*ⁱⁱ emerged. Was the addition εἰς τὸ αὐτὸ εἶδος meant to be an alternative to the already present addition εἰς ἕτερον εἶδος (*varia lectio*ⁱ)? *Varia lectio*ⁱⁱ does not make sense,[214] and we can only speculate whether it is an erroneously abbreviated version of a formula like διαιρετὸν οὐκ εἰς ἕτερον εἶδος, ἀλλὰ μόνον εἰς τὸ αὐτὸ εἶδος.

In any case, Alexander tries his best to extract a feasible understanding of *varia lectio*ⁱⁱ (355.3–5). He suggests that in this case εἶδος means the *kind* 'element,' which had just been defined by Aristotle. On this understanding, the definition according to the *varia lectio*ⁱⁱ would state that an element is indivisible in other *elements*. The element is the last constituent into which something can be divided. This interpretation, although ingenious, results in a contradiction between the definition and the water example, which cannot be resolved unless one denies that the water parts into which the element water is divisible are themselves elements.

In sum, Alexander knows three different readings of the passage in Δ 3, 1014a27. The reading in ω^AL is preferable to both variants; Alexander does not question this

---

[213] See *Metaph.* B 3, 999a1–4; Δ 3, 1014a31–34; Δ 6, 1016b23–24; I 1, 1052a30–34; *de An.* Γ 6, 430b14–15. See also Bonitz 1870: s.v. ἀδιαίρετος, p. 8b42–46. An only apparent parallel is given in *Cael.* Γ 3, 302a15–18: Ἔστω δὴ στοιχεῖον τῶν σωμάτων, εἰς ὃ τἆλλα σώματα διαιρεῖται, ἐνυπάρχον δυνάμει ἢ ἐνεργείᾳ (τοῦτο γὰρ ποτέρως, ἔτι ἀμφισβητήσιμον), αὐτὸ δ' ἐστὶν <u>ἀδιαίρετον εἰς ἕτερα τῷ εἴδει</u>. / "An element, we take it, is a body into which other bodies may be analyzed, present in them potentially or in actuality (which of these, is still disputable), and not itself divisible into *bodies* different in form" (transl. by Stocks, emphasis added). This cannot be taken as a parallel passage to ours because here the words εἰς ἕτερα refer to σώματα and are therefore not equivalent to our εἰς ἕτερον εἶδος. Rather, this *Cael.* passage is parallel to the first example in *Metaph.* 1014a29–30: ἀδιαίρετον εἰς ἕτερα (*sc.* σώματα) τῷ εἴδει (*Cael.* Γ 3, 302a18) corresponds to (ἀδιαίρετον) εἰς ἄλλας φωνὰς ἑτέρας τῷ εἴδει (*Metaph.* 1014a29–30).

[214] Alexander's explanation in lines 355.4–5 (δεῖ γὰρ τὸ στοιχεῖον <u>ἀδιαίρετον</u> εἶναι εἰς στοιχεῖα), whose wording is transmitted unanimously, makes it clear that the second variant of Alexander read <u>ἀδιαιρέτου</u> εἰς τὸ αὐτὸ εἶδος (cf. 354.32 and my apparatus). That the commentary manuscripts (A^a.c.O P^b) read in line 355.4 διαιρέτου instead of <u>ἀδιαιρέτου</u> shows that there occurred an early mistake in the transmission of the text, which was then corrected in A (A^p.c. ἀδιαιρέτου). That S has the correct reading points either to a later correction or to a Greek manuscript that is independent from our direct transmission (cf. 2.3).

preference, and his report of the two alternative readings appears to be nothing more than a report of other readings he happens to know, either from the margins of his own exemplar or from another commentary. One of the variant readings Alexander knows (*varia lectio*ⁱ) is identical with the reading that came down to us via direct transmission (ω^{αβ}). Did Alexander by way of this reading have access to the tradition of ω^{αβ}? The fact that this ω^{αβ}-reading is certainly corrupt shows that Alexander did indeed have access, if only by way of this reading, to the ω^{αβ}-version.[215]

3.5.2.2 Alex. Fr. 12 Freudenthal (Averroes, *Lām* 1481) on Arist. *Metaph.* Λ 3, 1070a18–19

The issues Aristotle addresses in Λ 3, although divergent, are in some sense unified by the idea of the priority of form over substance and other principles.[216] In 1070a13 Aristotle raises the question of whether the forms of composite substances exist separately from the composite substances (παρὰ τὴν συνθετὴν οὐσίαν, 1070a14).[217] Aristotle's answer for natural substances differs from that for artificial substances: a separate 'this' (τόδε τι), i.e. the form,[218] can, if at all, only exist in the case of natural substances (1070a13–19). This much established, Aristotle then makes the following statement about the theory of Forms:

Aristotle, *Metaphysics* Λ 3, 1070a18–19

διὸ δὴ οὐ κακῶς **Πλάτων ἔφη** ὅτι εἴδη ἔστιν ὁπόσα [19] φύσει, εἴπερ ἔστιν εἴδη ... .[219]

And so Plato was not far wrong when he said that there are as many Forms as there are kinds of natural things (if there are Forms at all), ...

Πλάτων ἔφη (α Michael^p 677.12–13) Ross Jaeger Fazzo vel ὁ Πλάτων ἔφη (A^b ε) Bekker Bonitz Christ, Ar.ᵘ, Al.^{γp} Fr. 12 F : οἱ τὰ εἴδη τιθέντες ἔφασαν ω^{AL} (Fr. 12 F) Themistius (8.13–14) Ar^m (Walzer 1958: 223)

Aristotle's remark is puzzling. How should we interpret his reference to Plato? Aristotle compliments Plato for positing as many Forms as there are natural things. Does this imply that Plato denies Forms for artificial substances? In Plato's dialogues, at least, Forms of artifacts are mentioned more than once (*Rep.* 596b, 597c; *Crat.* 389b-c; *Grg.* 503e),[220] and Aristotle himself speaks about Plato's theory to

---
[215] One could speculate that the variant reading Alexander knows comes from a version that only influenced ω^{αβ} in the course of the transmission. Since I do not have further evidence that could speak to this speculation, it can be dropped.
[216] Judson 2000: 110–11; 125; 131–33.
[217] Judson 2000: 131. 1070a13–20 constitutes "section 4" in Judson 2000.
[218] For this meaning of τόδε τι see *Metaph.* Z 12, 1037b26–27. See also 5.1.5.
[219] On the origin of lines 1070a18–19 in Alexander's conjecture see 5.5.
[220] See also a passage from the Seventh Letter: 342a–d. For the evaluation of this evidence see Bluck 1947: 76.

that effect in *Metaph.* A 6, 988a2-4: τὸ δ' εἶδος ἅπαξ γεννᾷ μόνον, φαίνεται δ' ἐκ μιᾶς ὕλης μία τράπεζα, ὁ δὲ τὸ εἶδος ἐπιφέρων εἷς ὢν πολλὰς ποιεῖ.[221]

The explicit exclusion of Forms of artifacts seems to be ascribed only to the early members of the Academy after Plato, especially to Xenocrates.[222] In *Metaph.* A 9, 991b6-7 Aristotle says: καὶ πολλὰ γίγνεται ἕτερα, οἷον οἰκία καὶ δακτύλιος, ὧν οὔ φαμεν εἴδη εἶναι.[223] In his commentary on A 9, Alexander refers to Aristotle's (now lost) work *On Ideas*, and in his report of the Argument (for Forms) from the Sciences he writes that "they" (i.e. the advocates of the theory of Forms) are forced to also accept Forms of artifacts, something they do not want to do.[224]

The passage of my present concern, Λ 3 (1070a18-19), and Plato's seemingly implied rejection of Forms of artifacts disturbed modern commentators of Plato's theory of Forms and Aristotle's representation of it.[225] Richard Stanley Bluck considers two different interpretations of our passage. His first interpretation amounts to understanding the pronoun ὁπόσα (1070a18) such that it does not deny that Plato believed in Forms of artifacts:

> And the remark, 'so that Plato was not wrong in saying there are χωριστὰ εἴδη of all natural objects, if there are such Forms at all' may well imply 'but of course he was wrong in saying there are Forms of artificial products.'[226]

Bluck's other interpretation leads us to Alexander's comments on the passage and the reading that was in Alexander's text. The reading of ω[AL] can be gathered from a comment made by Alexander and preserved by Averroes' commentary. Alexander's comment squares well with the *Metaphysics* text presented in Averroes' lemma (*Lām* 1481). The text of the lemma goes back to the Greek *Vorlage* upon which the Arabic version was ultimately based.[227] In the *Metaphysics* text preserved in Averroes' lemma there is no mentioning of Plato, but rather of the 'adherents of the theory of Forms.' Genequand translates: "Therefore, <u>those who postulated the Forms</u> were not wrong."[228] From this we can hypothesize the following Greek

---

[221] *Metaph.* A 6, 988a2-4. "and the form generates only once, but what we observe is that one table is made from one matter, while the man who applies the form, though he is one, makes many tables."

[222] Procl. *In Plat. Parm.* IV, 888.13-15; p. 67 Steel; See Broadie 2007: 233-34 and Krämer 2004: 107.

[223] *Metaph.* A 9, 991b6-7: "And many other things come into being (e.g. a house or a ring), of which we say there are no Forms." The "we" refers to the members of the Academy among which Aristotle counts himself. On this usage of "we" in *Metaphysics* A see Primavesi 2012b: 412-20.

[224] Alex. *In Metaph.* 79.19-80.6. 79.23-24: καὶ ὧν οὐ βούλονται ἰδέας εἶναι κατασκευάζειν ἰδέας δόξει. On this passage see Fine 1993: 81-88; concerning my question see also Broadie 2007: 233.

[225] See the discussion in Fine 1993: 82-83; Broadie 2007: 232-35; Frede 2012: 293-94.

[226] Bluck 1947: 75. See also Broadie 2007: 234: "However, the Platonic dictum which Aristotle reports here is logically compatible with admitting artefact-Ideas."

[227] In this section of book Λ Averroes used the translation by Abū Bišr Mattā for his lemmata. See Bertolacci 2005: 251 and 2.5 above.

[228] Genequand 1986: 100. Freudenthal 1885: 86.18-19: "Und aus diesem Grunde haben nicht übel gethan <u>die, welche die Ideen annehmen</u>." Scotus: *Et ideo non fecerunt male illi qui posuerunt formas.*

text: διὸ δὴ οὐ κακῶς <u>οἱ τὰ εἴδη τιθέντες ἔφασαν</u> ὅτι … .²²⁹ Also Themistius, whose paraphrase of *Metaphysics* Λ (fourth century AD) is preserved in Hebrew (which was translated into Latin by Moses Finzius) seems to have read "those who postulate the Forms" rather than "Plato."²³⁰

This reading is not only in Themistius's text and the *Vorlage* of Mattā's Arabic translation, but indeed is also found in ω^AL: Averroes remarks the following.

Genequand 1986: 100–101	Alexander says: these words refer to Plato, as is found in some manuscripts.
Fr. 12 Freudenthal (86.25–26)	Es sagt Alexander, dass er hier auf Platon hin-[26] weist, wie es sich auch in einigen Handschriften findet.
Scotus	*Dixit Alexander: Innuit in hoc Platonem.*²³¹

According to this testimony, Alexander in his commentary explains that Aristotle's remark refers to Plato. At the same time, Alexander informs his reader that there is a variant reading, which says just "Plato."²³² These two pieces of evidence allow us to draw the following two conclusions: first, Alexander's own copy of the *Metaphysics* did not read the word "Plato." What was then the reading in ω^AL? The answer that immediately suggests itself is: the reading that we can reconstruct from the lemma in Averroes' commentary and that is read also by, as his paraphrase suggests, Themistius. In all likelihood, ω^AL read the formula οἱ τὰ εἴδη τιθέντες (or an equivalent) in place of Πλάτων.²³³ Second, Alexander knew a

---

²²⁹See also Walzer 1958: 223. Freudenthal 1885: 86 n. 3 suggests: διὸ δὴ οὐ κακῶς <u>ἔφασαν οἱ τιθέμενοι τὰ εἴδη</u> ὅτι… . In the Aristotelian *corpus* there is no other instance of the construction εἴδη + middle participle τιθέμενος. There are, however, many parallel passages in which the formula οἱ τιθέμενοι (τὰς) ἰδέας appears (e.g. *Top.* B 7, 113a28; H 4, 154a19; *Metaph.* N 3, 1090a16). Since our passage is concerned with εἴδη (Andreas Lammer confirmed to me that the Arabic version implies εἴδη as the original), the middle as used with (τὰς) ἰδέας seems a dubious restoration. My proposed reconstruction οἱ τὰ εἴδη τιθέντες (εἴδη + participle of τίθημι) ἔφασαν is parallel to a passage in A 7: μάλιστα δ' οἱ τὰ εἴδη τιθέντες λέγουσιν (*Metaph.* A 7, 988a35–b1). Cf. also *Metaph.* B 2, 1002b13–14 and, not far from the passage of our concern, *Metaph.* Λ 1, 1069a35: οἱ δὲ εἰς μίαν φύσιν τιθέντες τὰ εἴδη καὶ τὰ μαθηματικά. There is only a small difference between the active and the middle form of τίθημι (see, however, LSJ s.v. τίθημι B.II: "in reference to mental action, when Med. is more freq. than Act."). Also possible is the formula οἱ λέγοντες τὰ εἴδη (cf. *Metaph.* B 2, 997b1–2; Z 16, 1040b27–28).

²³⁰Themistius, *In Metaph.* Λ 8.13–14 Landauer: *idcirco ponentes formas [abstractas] esse, formas istas rebus naturalibus tribuebant, artificialibus vero nequaquam.*

²³¹The Latin version of Averroes' commentary lacks the reference to the other manuscript reading. This should not disturb us, however. As Dag N. Hasse explained at a workshop at the Musaph in Munich (May 2012), Scotus often leaves out seemingly unnecessary comments. Cf. Freudenthal 1885: 121–23 on the Latin version of Averroes' commentary.

²³²Martin 1984: 116 n. 1 considers the possibility that the comment "as is found in some manuscripts" might be Averroes' own.

²³³Ross alone makes clear in his apparatus that Alexander's text had this reading (or Freudenthal's

variant reading that is identical with the reading preserved in all our manuscripts.

The comparison of the reading in ω^AL (οἱ τὰ εἴδη τιθέντες) with the reading in ω^αβ (Πλάτων) brings us back to Bluck's second interpretation about how to understand the somewhat puzzling remark in Λ 3, 1070a18–19.[234] The reading of ω^AL is advantageous in that the "adherents of the theory of Forms" could very well refer to the members of the Academy after and excluding Plato. Unlike Plato, these members did reject Forms of artifacts.[235]

Accepting the expression οἱ τὰ εἴδη τιθέντες as the original reading, we regard it as natural designation for the members of the Academy. At some point of the tradition of the *Metaphysics*, this expression was replaced unmindfully by the simple word "Plato." This explanation receives support by the fact that Aristotle hardly mentions Plato's *name* in the context of the theory of Forms (εἴδη).[236] Since Alexander already knows that the reading Πλάτων exists in another version of the *Metaphysics*, the substitution for οἱ τὰ εἴδη τιθέντες must have taken place before AD 200. While ω^AL, the *Metaphysics* text used by Themistius, and Mattā's *Vorlage* were not affected by this corruption, ω^αβ or one of its ancestors, was.[237]

3.5.2.3 Alex. *In Metaph.* 137.2–5; 138.24–28 on Arist. *Metaph.* α 1, 993a29–b2

The questions about the authenticity of book α ἔλαττον[238] and its status within the context of the *Metaphysics* as a whole[239] have been the subject of some mod-

---

version οἱ τὰ εἴδη τιθέμενοι). Jaeger's remark in the apparatus seems to suggest that Alexander himself proposed the reading οἱ τὰ εἴδη τιθέμενοι as substitute for "Plato." Bekker, Bonitz, and Christ do not comment on this reading. Bekker and Bonitz precede Freudenthal's work and Christ's edition (1886) was published one year after Freudenthal. The edition of *Metaphysics* book Λ by Silvia Fazzo (2012) is completely silent on this point. The most recent edition of book Λ by Alexandru (2014) offers helpful information on Alexander's testimony to this passage, yet Alexandru does not state clearly that Alexander found οἱ τὰ εἴδη τιθέντες or an equivalent in *his* text. See also Fine 1993: 289–90 n. 11 and Broadie 2007: 234 n. 11.

[234] Bluck 1947: 75.

[235] For a different interpretation see Fine 1993: 290 n. 11.

[236] There are only two passages in the corpus where Plato's name is mentioned in connection to τὰ εἴδη: *Metaph.* Z 2, 1028b19–20: οἱ δὲ πλείω καὶ μᾶλλον ὄντα ἀΐδια, ὥσπερ Πλάτων τά τε εἴδη καὶ τὰ μαθηματικὰ δύο οὐσίας… and *Ph.* Δ 2, 209b33–35: Πλάτωνι μέντοι λεκτέον, εἰ δεῖ παρεκβάντας εἰπεῖν, διὰ τί οὐκ ἐν τόπῳ τὰ εἴδη καὶ οἱ ἀριθμοί…. For Plato's name in connection with ἰδέαι see *Top.* Z 10, 148a14–17 and *Ph.* Γ 4, 203a8.

[237] Two out of three variant readings that are preserved in the fragments of Alexander's commentary on book Λ are identical with the reading in ω^αβ (Fr. 12 F on Λ 3, 1070a18; Fr. 13b F on Λ 3, 1070a20; see also 3.6).

[238] Berti 1983: 260–65 offers an overview of the question concerning α ἔλαττον's authenticity. Jaeger 1912: 114–18 regards α ἔλαττον as notes taken by Pasicles (who in a scholium in E is *seemingly* called the author of the book) from a lecture of Aristotle's. These notes then accidentally found their way into the *Metaphysics* although their content belongs to natural philosophy.

[239] Concerning the position of α ἔλαττον within the *Metaphysics* Szlezák 1983: 259 concludes: "Man

ern scholarship.[240] These issues were discussed already in antiquity, as Alexander's introduction to his commentary on the second book testifies (137.2–138.9). In his introduction Alexander addresses first the question of authorship (137.2-3) and then whether α ἔλαττον can be regarded as complete book at all (137.3-5). As Alexander informs us, the completeness of α ἔλαττον can indeed be questioned due to the book's *beginning* as well as its *brevity*. Shortly afterwards (138.26-28) it becomes clear that Alexander's mention of the beginning of the book alludes to a grammatical peculiarity in the first sentence of book α as it is presented in ω^{AL}.

Let us first look at the beginning of α ἔλαττον as it is preserved by the direct manuscript transmission (ω^{αβ}).

Aristotle, *Metaphysics* α 1, 993a29–993b2

Ἡ περὶ τῆς ἀληθείας θεωρία τῇ μὲν χαλεπὴ τῇ δὲ [30] ῥᾳδία. σημεῖον δὲ τὸ μήτ' ἀξίως μηδένα δύνασθαι τυχεῖν [993b1] αὐτῆς μήτε πάντας ἀποτυγχάνειν, ἀλλ' ἕκαστον λέγειν τι [2] περὶ τῆς φύσεως, …

The investigation of the truth is in one way hard, in another easy. An indication of this is found in the fact that no one is able to attain the truth adequately, while, on the other hand, no one fails entirely, but every one says something true about the nature of things, …

29 ἡ ω^{αβ} Al.^{γρ} Al.^{l} 138.24 Ascl.^{l} 113.3; 114.21 Ar.^{l} (*Scotus*) Bekker Bonitz Christ Ross : ὅτι ἡ ω^{AL} Jaeger ‖ 30 τυχεῖν α ζ Ascl.^{l} 114.22 Bekker Bonitz Christ Jaeger : θίγειν [sic] β Ross (θιγεῖν)

The first sentence as it appears in our manuscripts shows no unusual features. The general tone that Aristotle strikes here is common to the introductory sentences of other chapters or works.[241] So it seems that Alexander's uneasiness about the beginning of α ἔλαττον is grounded in a feature of the text that is peculiar to his version of it.

Looking at Alexander's comments on the passage brings to light that in ω^{AL} the introductory words Ἡ περὶ τῆς ἀληθείας θεωρία… were in fact preceded by the conjunction ὅτι and hence had the appearance of a subordinate clause.

Alexander, *In Metaph.* 138.24-28 Hayduck

Ἡ περὶ τῆς ἀληθείας θεωρία τῇ μὲν χαλεπὴ τῇ δὲ [25] ῥᾳδία. σημεῖον δέ.

---

wird Alpha elatton am besten dort belassen, wo es überliefert ist, aber nicht als Zeugnis der tiefen didaktischen Weisheit des Meisters, sondern als Begleitmaterial aus dem Nachlaß, das vermutlich unverändert zu unbekannter Zeit an das Ende der Rolle von A angefügt wurde, wozu außer dem editorischen Interesse sicher auch die Überschneidung mit (nicht Ergänzung zu) A 1-2 in α 1 und die Wichtigkeit des Argumentes von α 2 […] Anlaß gegeben haben." See also Jaeger 1912: 114-18.

[240]Cf. also 1; pp. 17–18.

[241]Compare the first sentence in *Metaph.* A 1 (980a21): Πάντες ἄνθρωποι τοῦ εἰδέναι ὀρέγονται φύσει or in *EN* A 1, 1094a1-2: Πᾶσα τέχνη καὶ πᾶσα μέθοδος, ὁμοίως δὲ πρᾶξίς τε καὶ προαίρεσις, ἀγαθοῦ τινὸς ἐφίεσθαι δοκεῖ.

[26] Γράφεται καὶ χωρὶς τοῦ ὅτι, ἡ περὶ τῆς ἀληθείας θεωρία· [27] καὶ μᾶλλον δοκεῖ ἐκεῖνο ἀρχὴ εἶναι, τὸ δὲ μετὰ τοῦ 'ὅτι' οὐκ ἀρχὴ ἀλλ' [28] ἑπόμενον προειρημένῳ τινί.

The investigation of the truth is in one way hard, in another easy. An indication of this...

This text is also written without ὅτι, thus: 'The investigation of the truth.' And this reading seems more clearly to be a beginning [of the book], whereas the one introduced by ὅτι is not a beginning, but a sequel to something said before it.

25 τῇ ... τῇ **A O** *Metaph.* : πῇ ... πῇ **P**[b]

Although the lemma in Alexander's commentary does not read the ὅτι and hence agrees with ω[αβ], Alexander's subsequent remarks (lines 26–27) make it undeniably clear that in his text, ω[AL], the first words of α ἔλαττον are <u>ὅτι</u> ἡ περὶ τῆς ἀληθείας θεωρία.... This is further supported by the fact that Alexander introduces a *varia lectio* from another *Metaphysics* version (γράφεται καὶ...)[242] that *differs* from his text in not reading the ὅτι.[243] The fact that ὅτι is absent in the variant reading proves that it was present in ω[AL]. And we see again that Alexander refers to a *varia lectio* that agrees with ω[αβ].

The lemma in Alexander's commentary is in tension with his own words. It is a fair assumption that the ὅτι, which originally had been written in the lemma as well, was at a later time deleted. Comparing the two versions of the beginning of book α, one must say that the sentence without the ὅτι makes more sense and thus is obviously to be preferred. It seems natural that someone got rid of the useless ὅτι in the lemma-quotation without paying attention to Alexander's indirect statement that the ὅτι was the reading of ω[AL].

As indicated above, the fact that Alexander found the conjunction ὅτι in his text squares well with his earlier remark about the conspicuous beginning of α ἔλαττον.[244] In his introduction to α ἔλαττον Alexander says:

Alexander *In Metaph.* 137.2–5 Hayduck

Τὸ ἔλαττον ἄλφα τῶν Μετὰ τὰ Φυσικὰ ἔστι μὲν Ἀριστοτέλους ὅσα [3] καὶ τῇ λέξει καὶ τῇ θεωρίᾳ τεκμήρασθαι, οὐ μὴν ὁλόκληρον ἔοικεν εἶναι, ἀλλ' ἔστιν [4] ὡς μέρος βιβλίου, <u>τεκμαιρομένοις τῇ τε ἀρχῇ</u> καὶ τῇ τοῦ βιβλίου μικρό-[5]τητι.

Book α ἔλαττον of the *Metaphysics* is the work of Aristotle so far as can be judged

---

[242] Sepúlveda translates the reference to the *varia lectio* as a request to change the text (cf. also Hayduck's apparatus): (*sic scribendum est, non ad hunc modum: Quoniam contemplatio veritatis*) *hoc enim magis videtur esse principium, quam si illud quoniam praeponas.*

[243] Bonitz 1848 notes in the apparatus incorrectly (adopted also by v. Christ 1886a): "ὅτι ἡ γρ. Al." The *varia lectio* to which Alexander refers is identical with the ω[αβ]-reading ἡ. The annotations in Ross (1924: *ad loc.* "ἡ γρ. Al.") and Jaeger (1957: *ad loc.* "γρ. καὶ χωρὶς τοῦ ὅτι Al.") are correct. Bekker (1831) and Bonitz furthermore remark that **H**[b] (= *Parisinus* 1901) contains ὅτι, **H**[b] being a manuscript of Asclepius's commentary. On the evidence in Asclepius see below.

[244] See Dooley 1992: 9 n. 3 and 11 n. 11.

from the diction and the investigation [it pursues]; but if one is to base his opinion on the evidence of its beginning and brevity, it seems to be a part of a book rather than a complete book.

---

3 τῇ λέξει καὶ τῇ θεωρίᾳ A O P^b S(*ex dicendi charactere* ipsaque disputandi ratione) LF : τῇ θεωρίᾳ Bonitz Hayduck

---

The question of α ἔλαττον's authenticity is settled promptly: both the language (λέξις) and the investigation (θεωρία) suggest Aristotle's authorship (137.2–3). Alexander's answer concerning the question whether α ἔλαττον is a complete book (ὁλόκληρον) is less assertive (137.4–5). The book's beginning (ἀρχή) and its brevity (μικρότης) speak in favor of it being part of another book (μέρος βιβλίου). While the point about α's brevity becomes clear immediately, the point about its beginning makes sense only when we know that in Alexander's text the book started with the conjunction ὅτι, which usually introduces a subordinate clause.[245] At the same time, Alexander's remark about the beginning confirms that ω^AL indeed exhibits this peculiarity.

What about the peculiar ὅτι in ω^AL? How are we supposed to make sense of a ὅτι, which does not introduce a subordinate clause? Alexander himself does not have a clue about what the ὅτι could indicate. He is sure however that this is not how a book should begin (138.27–28). By contrast, Jaeger in his edition of the *Metaphysics* even prints the ὅτι in the text.[246] His annotation in the apparatus says: *vel excerpta vel notas indicat.*[247] As Jaeger's short diagnosis indicates, the conjunc-

---

[245] The author of the *recensio altera* interpreted Alexander's remark differently. Although he also speaks about a variant reading that reads ὅτι at the beginning of book α (ἔν τισι γράφεται μετὰ τῆς τοῦ 'ὅτι' προσθήκης. Cf. Golitsis 2013a), he does not connect this information with Alexander's remark about the beginning of the book. Rather, the author of the *recensio altera* connects Alexander's concern with the beginning with what Alexander later says (137.5–7) about how well the treatment of principles in book α complements book A. So according to the *recensio altera* the reason for why book α does not seem to be a complete book is its thematic closeness to book A: ἔστι δὲ μέρος βιβλίου ἀλλ' οὐ βιβλίον ὁλόκληρον. δηλοῖ δὲ τοῦτο ἡ ἀρχὴ τοῦδε τοῦ βιβλίου, ὅτι καὶ ἐν τούτῳ περὶ ἀρχῶν ποιεῖται τὸν λόγον, καὶ οὐκ ἀπᾴδει τὰ ἐν τούτῳ λεγόμενα τῶν ἐν τῷ μείζονι A. Cf. also Asclepius (113.8–12), whence the author of the *recensio altera* might have drawn inspiration.

[246] Jaeger's note according to which a citation in Asclepius confirms the ὅτι ("Ascl.^c") is not confirmed by the evidence in Asclepius's commentary. Asclepius *paraphrases* (114.1–2): ὁ δὲ λέγει ἐν προοιμίοις τοιοῦτόν ἐστιν, ὅτι ἡ περὶ τῆς ἀληθείας θεωρία.... Here, the ὅτι functions as a conjunction and introduces a dependent statement.

[247] Jaeger further states: *quod cum scholio de Pasicle huius libelli auctore consentit*. With this remark, Jaeger links the excerpt-like character of the book to the so-called Pasicles-scholium, preserved in the *Metaphysics* manuscript E. According to Jaeger (1912: 114–18), this scholium together with Asclepius's report that some doubt the authenticity of book α (Ascl. 113.5–9) evince that the book is in fact a transcript (ὑπόμνημα) by a student of Aristotle. Yet, as Vuillemin-Diem 1983 has shown, the scholium, which does not even refer to α but rather to A, is only based on a remark by Asclepius (Ascl. 4.17–24). For a reason why the authenticity of book A could have been regarded as doubtful see Primavesi 2012b: 418–19. For my present inquiry, it matters only that the ὅτι in Alexander's text has nothing to do with the Pasicles-scholium.

tion ὅτι *in the initial position* of a section of text signals that the text is an *excerpt* taken from another context and inserted into the given place.[248] There are various examples of this use of ὅτι in Greek works.[249] Two examples will suffice: (i) The Platonist Albinus (2nd century AD) begins his Introduction to Plato's dialogues with Ὅτι τῷ μέλλοντι ἐντεύξεσθαι τοῖς Πλάτωνος διαλόγοις προσήκει πρότερον ἐπίστασθαι αὐτὸ τοῦτο, τί ποτέ ἐστιν ὁ διάλογος.[250] (ii) Proclus (5th century AD) presents his commentary on Plato's *Cratylus* as sequence of excerpts from his teacher: e.g., Ὅτι τὰς ἀρχὰς τῶν ὄντων καὶ τῆς διαλεκτικῆς νῦν παραδιδόναι βούλεται ὁ Πλάτων… .[251]

According to ω^{AL}, then, book α ἔλαττον is marked as an excerpt.[252] Following the principle *utrum in alterum* it seems by far more likely that someone deleted the seemingly useless and grammatically disturbing ὅτι than that someone arbi-

---

[248] Apart from this usage of ὅτι, there are the following uses of ὅτι when it does not function as a conjunction: ὅτι as quotation marks in introducing direct speech: Kühner/Gerth II: § 551,4; pp. 366–67 and LSJ s.v. ὅτι II; for examples, see Hdt. II, 115 and Plato, *Prt*. 318a; and ὅτι as the marking of a heading for excerpts in Stobaeus's anthology (5th century AD): for example, in Stob. I, 20, 8 (179.19–180.16 Wachsmuth); Stob. II, 11 (184.22 Wachsmuth); Stob. II, 19 (197.12 Wachsmuth).

[249] Apart from the two mentioned above, there are, for example, the Constantinian excerpts of Polybius that start with ὅτι, for instance, at Plb XX, 3 (Vol. IV, 1,10 Büttner-Wobst). For a helpful overview of the different usages of ὅτι at the beginning of a text passage see Reis 1999: 49–50. Cf. also Dickey 2007: 122.

[250] For the text see Reis 1999: 310. On the function of ὅτι see Reis 1999: 50–52.

[251] Procl. *In Cra*. VIII (3.4–5 Pasquali). On the meaning of this ὅτι see Reis 1999: 49–50 and Dickey 2007: 122.

[252] Or is there any other way of making sense of the ὅτι? The ὅτι-clause in 993a29 can certainly not be taken as causal conjunction (993a29–30), which depends on the subsequent clause introduced by σημεῖον δὲ … ("an indication is…") in 993a31. For, although in Aristotle ὅτι-clauses can depend on the phrase σημεῖον (ἐστίν) (in the sense of "it is an indication that…," e.g. *Cael*. Δ 3, 310b33; *EE* H 1, 1235a36; *GA* A 18, 725a16) and can even precede it in the sentence (*GA* Γ 5, 755b1; *HA* E 22, 553b32–554a1), such an understanding is ruled out in the present case because of σημεῖον δὲ, which clearly indicates a *new* clause (see *Cael*. Δ 4, 311b9; *HA* A 16, 497a9; *GA* E 2, 782b29; *EE* B 23, 1220a34; *EN* E 4, 1130a16. Cf. Denniston 1954: s.v. δέ II.1. Apodotic; pp. 177–81).

Another possibility might be to take the ὅτι as connecting the first sentence of book α with the last sentence of book A. Perhaps someone wanted to secure book α as genuine part of Aristotle's *Metaphysics* by tying it to the preceding book. (On other attempts to secure α's position between books A and B see 4.1.1.) The last sentence of book A reads (A 10, 993a25–27): ὅσα δὲ περὶ τῶν αὐτῶν τούτων ἀπορήσειεν ἄν τις, ἐπανέλθωμεν πάλιν· τάχα γὰρ ἂν ἐξ αὐτῶν εὐπορήσαιμέν τι πρὸς τὰς ὕστερον ἀπορίας / "But let us return to enumerate the difficulties that might be raised on these same points; for perhaps we may get some help towards our later difficulties." In this conclusion of A 10, Aristotle points towards the aporiae of book B. See Cooper 2012: 352. Connecting the first sentence of book α results in the following construction: τάχα γὰρ ἂν ἐξ αὐτῶν εὐπορήσαιμέν τι πρὸς τὰς ὕστερον ἀπορίας, ὅτι ἡ περὶ τῆς ἀληθείας θεωρία τῇ μὲν χαλεπὴ τῇ δὲ ῥᾳδία. / "For perhaps we may get some help towards our later difficulties, since the investigation of the truth is in one way hard, in another easy." This is highly problematic. Here, the ὅτι-clause gives the reason why we are to hope for help towards the later difficulties. It is very implausible that someone intended to achieve this by the addition of ὅτι to the beginning of book α.

trarily added it there. Thus, ω^AL seems in fact to represent the older reading. Yet it also stands to reason that at some point in the text-history of the *Metaphysics* someone (namely, the person who inserted the piece we know as book α into what we know as *Metaphysics*) must have added the word ὅτι at its beginning in order to signal that this is not a complete treatise but an excerpt. This person was most likely not Aristotle.[253]

In any case, Alexander, without recognizing it, finds the older reading in his own text, but is furthermore (by way of a *varia lectio*) familiar with a slightly modified, younger version of the text. This younger version of the text is what we read in ω^αβ. Thus we ask: does Alexander here have once more access to the version of ω^αβ? One should be careful with answering in the affirmative too rashly, since the elimination of ὅτι is a temptation that could have been executed in more than one manuscript independently of each other. Thus it may well be that (the tradition of) ω^αβ was not the only version of the *Metaphysics* in which ὅτι was absent. Still, it remains a strong possibility that Alexander refers to ω^αβ as the version in which the ὅτι is absent.

3.5.2.4 Alex. *In Metaph*. 169.4–11 on Arist. *Metaph*. α 3, 995a12–19

In α 3 Aristotle presents observations on methods of teaching and their relation to the relevant subject matter. The acquisition of knowledge of some subject through study or instruction is a distinct enterprise from the investigation into the correct method of the study or instruction. Further, different methods of acquisition are appropriate to different sciences:[254] mathematical accuracy[255] is inappropriate to the study of nature, because nature involves matter and matter introduces imprecision.

Aristotle, *Metaphysics* α 3, 995a12–19

διὸ δεῖ πεπαιδεῦσθαι [13] πῶς ἕκαστα ἀποδεκτέον, ὡς ἄτοπον ἅμα ζητεῖν ἐπιστήμην [14] καὶ τρόπον ἐπιστήμης· ἔστι δὲ οὐδὲ θάτερον ῥᾴδιον λαβεῖν. τὴν [15] δ' ἀκριβολογίαν τὴν μαθηματικὴν οὐκ ἐν ἅπασιν ἀπαιτη-[16]τέον, ἀλλ' ἐν τοῖς μὴ ἔχουσιν ὕλην. διόπερ οὐ φυσικὸς ὁ [17] **τρόπος**· ἅπασα γὰρ ἴσως ἡ φύσις ἔχει ὕλην. διὸ σκεπτέον [18] πρῶτον τί ἐστιν ἡ φύσις· οὕτω γὰρ καὶ περὶ τίνων ἡ φυσικὴ [19] δῆλον ἔσται.

---

[253] It seems absurd to assume that Aristotle compiled his own work and included as part of it an excerpt by himself. On the question of the composition of the *Metaphysics* see Menn 1995, Barnes 1997: 59–66, Hatzimichali 2013: 24–27; cf. 1; pp. 15–19.

[254] On the 'domain-specificity' of Aristotle's methods see Lennox 2011.

[255] Aristotle speaks about mathematical accuracy and its appropriate application also in *EN* A 1, 1094b12–27. Cf. Alex. 169.1–4; Schwegler 1847c: 112; Gigon 1983: 216–18; Szlezák 1983: 242–43. Cf. also Lennox 2011: 35–39.

Therefore one must be already trained to know how to take each subject matter, since it is absurd to seek at the same time knowledge and the method of acquiring knowledge; and neither is easy to get. The accuracy of mathematics is not to be demanded in all cases, but only in the case of things that have no matter. Therefore its **method** is not that of natural science; for presumably all nature has matter. Hence we must inquire first what nature is: for thus we shall also see what natural science treats of.

14 οὐδὲ θάτερον β (Al.ᵖ 168.25 οὐδὲ γὰρ τὸ ἕτερον) Ross Jaeger : οὐδέτερον **a** Bekker Bonitz Christ || 15 τὴν δ᾿ ἀκριβολογίαν] τὴν ἀκριβολογίαν γὰρ Al.ᶜ 169.4–5 || 17 τρόπος ωᵃᵝ Al.ʸᵖ Arⁱ edd. : λόγος ωᴬᴸ (Al.ᶜ 169.9) Arᵘ

My analysis of this section will pay close attention to the word that Aristotle uses for "method" in 995a16–17.²⁵⁶ A comparison of Aristotle's word choice as we find it in our manuscripts with the quotation in Alexander's commentary shows that ωᴬᴸ read in 995a16–17 the word ὁ λόγος in place of ὁ τρόπος.²⁵⁷ Alexander's comments confirm the words of his quotation, but also indicate that he knew of the alternative reading τρόπος.

Alexander, *In Metaph.* 169.4–11 Hayduck

τὴν ἀκριβο-[5]λογίαν γάρ, φησί, τὴν μαθηματικὴν οὐκ ἐν ἅπασιν ἀπαιτητέον, [6] ἀλλ᾿ ἐν τοῖς ἀΰλοις, ὁποῖά ἐστι τὰ ἐξ ἀφαιρέσεώς τε καὶ μαθηματικά, [7] ἴσως ἐνδεικνύμενος ἡμῖν ὅτι τοιαύτης ἀκριβολογίας χρεία καὶ πρὸς τὰ [8] παρόντα· περὶ γὰρ ἀΰλων ὁ περὶ τῶν πρώτων ἀρχῶν λόγος καὶ οὐ συνή-[9]θων. τὸ δὲ διόπερ οὐ φυσικὸς ὁ λόγος ἤτοι²⁵⁸ λέγει ὅτι ἀκριβής· προεί-[10]ρηκε γὰρ τοῦτο (γράφεται δὲ καὶ ὁ τρόπος, καὶ εἴη ἂν ὁ λόγος ὡς οὐ [11] φυσικοῦ)·

For "the minute accuracy," he says, "of mathematics is not to be demanded in all cases," but only in the case of immaterial things such as the objects of mathematics, [which are derived] from abstraction. Perhaps he is pointing out to us that precision of this sort is needed for the present inquiry too, for the treatise [λόγος] on the first principles deals with immaterial objects, not with things to which we are accustomed. He says, "Therefore the treatise is not that of natural science," because it is precise, for he said this previously. (Also the reading 'method' [τρόπος] is transmitted, and the sense of this would be that [such precision] is not characteristic of the natural philosopher.)²⁵⁹

---

²⁵⁶Cf. also 4.1.1 and 4.2.2, where I examine other aspects of this *Metaphysics* section. On the term μέθοδος in Aristotle see Lennox 2011: 28–29.

²⁵⁷The reading in ωᴬᴸ (λόγος) agrees with the Arabic translation by Ustāth; the translation by Isḥāq, however, confirms the directly transmitted reading τρόπος. See my apparatus and Walzer 1958: 223.

²⁵⁸This ἤτοι marks the first of two interpretations (see 169.11, ἤ, for the second). Since I am only covering the first, I do not translate ἤτοι.

²⁵⁹Sepúlveda apparently wanted to make clearer what Alexander says and added clarifying repetitions (*in quibusdam exemplaribus ... pro ratione ... modum*). These do not necessarily indicate that his Greek manuscripts differed from ours. However, we certainly miss the negation of *modum esse physici* in Sepúlveda's text. Perhaps it dropped out in the Latin.

8 περὶ γὰρ ἀΰλων O (περὶ ἀΰλων γὰρ e S Brandis, Bonitz, Hayduck) : περὶ ἀΰλων **A** : ἐπειδὴ περὶ ἀΰλων **P**ᵇ || 8–9 οὐ συνήθων **A O** : ἀσυνήθων **P**ᵇ || 9 ἀκριβής **A O** : ὁ ἀκριβής **P**ᵇ || 10 καὶ ὁ **LF** : ὁ **A O P**ᵇ || ὁ λόγος ὡς οὐ **A O** : λέγων ὡς ὁ λόγος οὐ **P**ᵇ || 10–11 γράφεται δὲ καὶ ὁ τρόπος, εἴη ἂν ὁ λόγος ὡς οὐ φυσικοῦ] *sed in quibusdam exemplaribus modus scribitur pro ratione, et tunc sensus esset modum esse physici* **S**

Alexander quotes from his *Metaphysics* text the statement that mathematical accuracy is not to be demanded in all areas of inquiry (995a14–16), but is appropriate only for the study of immaterial things. He then suggests understanding Aristotle's statement to mean that mathematical precision is appropriate to the inquiry undertaken in the present study, i.e., the *Metaphysics*, which he describes as ὁ περὶ τῶν πρώτων ἀρχῶν λόγος. In the subsequent lines (169.9–10), Alexander explicates Aristotle's next sentence: διόπερ οὐ φυσικὸς ὁ λόγος [ωᵃᵝ: τρόπος] (995a16–17). He offers two interpretations of it.

Alexander's first interpretation builds on the idea that Aristotle demands mathematical accuracy also for the present inquiry into the first principles, since it deals with immaterial objects, too. We can infer from this that Alexander understands the word λόγος, found in his text in place of τρόπος (995a17), as "treatise" or "theory" of the first principles: 'Therefore the *present* treatise (λόγος) is not that of natural science.' This puts metaphysics on the same level as mathematics as far as methodology is concerned. For, as Alexander says, this λόγος is ἀκριβής (169.9). Alexander supports this understanding with a reference to what Aristotle had earlier expressed (προείρηκε γὰρ τοῦτο, 169.9–10). Alexander probably has in mind *Metaphysics* A 2, 982a25–28[260] (cf. 982a13): "And the most exact (ἀκριβέσταται) of the sciences are those which deal most with first principles; for those which involve fewer principles are more exact than those which involve additional principles, e.g. arithmetic than geometry."[261]

In this first interpretation of the clause διόπερ οὐ φυσικὸς ὁ λόγος, Alexander understands Aristotle's word λόγος in just the way in which he himself used the word λόγος in the preceding sentence of his commentary, that is, as 'treatise on the first principles' (περὶ τῶν πρώτων ἀρχῶν λόγος, 169.8). It does not seem to bother Alexander that he projects his own understanding of λόγος onto Aristotle's sentence.[262] Alexander does not reconsider his interpretation even when en-

---

[260] *Metaph.* A 2, 982a25–28: ἀκριβέσταται δὲ τῶν ἐπιστημῶν αἵ μάλιστα τῶν πρώτων εἰσίν (αἱ γὰρ ἐξ ἐλαττόνων ἀκριβέστεραι τῶν ἐκ προσθέσεως λεγομένων, οἷον ἀριθμητικὴ γεωμετρίας.

[261] Just prior to our passage, at α 3, 995a8–12, Aristotle also talks about the methodological import of accuracy for teaching and investigation, but in this case there is no connection to the *Metaphysics*. For this reason it is more likely that Alexander has in mind the passage in A 2.

[262] At this point of the commentary Alexander states his understanding only briefly (169.9–10). After having introduced the second interpretation (169.11–15), he returns to the first interpretation once more saying (169.15–17): δύναται δὲ καὶ περὶ τῶν προκειμένων λόγων λέγειν, ὅτι οὔκ εἰσι φυσικοί· ἄυλα γὰρ περὶ ὧν λέγειν ἡμῖν πρόκειται, καὶ ἀκριβεστέρων λόγων ἢ κατὰ φυσικὰ δεόμενα. / "But Aristotle might also be saying this about the arguments that concern us now—that sc. they are not [the kind used in] natural [philosophy]; for it is immaterial objects about which we propose to speak, and they

countering the *varia lectio* that reads τρόπος instead of λόγος: διόπερ οὐ φυσικὸς ὁ τρόπος. Even when faced with an alternative reading, Alexander stands firm in his understanding of the sentence, and so to him τρόπος, just as λόγος, refers not to mathematics, but to the present treatise, the *Metaphysics*.[263] Thus, Alexander does not give much thought to the variant reading and the alternative it opens up, but simply transfers his understanding of λόγος (169.8) to τρόπος without further ado. He probably found the variant in the margin of his text, and for the sake of thoroughness included it in his commentary, without having any further interest in it.

Only thereafter, and hence without factoring in the variant reading, Alexander turns to his second interpretation (ἢ τὸ λεγόμενον τοιοῦτόν ἐστιν, 169.11–15) of the sentence in 995a16–17.[264] With this interpretation Alexander abandons his understanding of λόγος as 'present treatise' or 'metaphysics' and takes the sentence just as we do (although we do it on the basis of the reading τρόπος). According to this understanding, λόγος (or, for us, τρόπος) refers back to the mathematical *method* of dealing with mathematical objects. The objects of mathematics, being immaterial, admit of accuracy (ἀκριβολογία) in their treatment. The mathematical λόγος is not the same as the λόγος of natural science.

Regardless of how one rates Alexander's interpretations of the sentence in a16–17, his comments make it clear that he read λόγος in ω^AL, where ω^αβ read τρόπος. Alexander is aware of the variant reading τρόπος, but mentions it in passing and pursues none of its implications. He does not link the variant τρόπος to the expression τρόπος ἐπιστήμης in line 995a14,[265] where Aristotle seems to have introduced the word τρόπος as a relevant term for the present context.

How are the two readings τρόπος (ω^αβ) and λόγος (ω^AL) to be evaluated? The meaning of the word τρόπος ("method," "manner")[266] fits perfectly into the sentence and its context.[267] Furthermore, a few lines earlier, the term is introduced

---

require more accurate arguments than do natural objects."

[263] In line 169.10, the word λόγος means "sense," "meaning" (see Sepúlveda's *sensus*). Alexander apparently uses the word λόγος in different senses within the same passage (see Dooley 1992: 59 n. 157).

[264] Alex. *In Metaph.* 169.11–15: ἢ τὸ λεγόμενον τοιοῦτόν ἐστιν. τὰ φυσικὰ δοκεῖ πάντα σὺν ὕλῃ εἶναι, ἄυλα δὲ τὰ μαθηματικά, διὸ καὶ ἀκριβολογίαν ἐπιδέχεται, τὰ δὲ φυσικὰ οὐχ ὁμοίως. οἱ δὴ περὶ τῶν μαθηματικῶν λόγοι περὶ τῶν ἀύλων ὄντες οὔκ εἰσι φυσικοί· οὐ γὰρ τοσαύτην ἀκριβολογίαν τὰ φυσικὰ ὄντα γε σὺν ὕλῃ χωρεῖ. / "Or the statement means the following. All natural objects seem to exist with matter, but the objects of mathematics are immaterial; hence the latter also admit of precise treatment, but natural objects do not do so in the same way. Certainly the arguments about mathematical objects, since they deal with immaterial things, are not [the kind used in] natural science; for natural objects do not permit the same degree of precise statement, at least [inasmuch as] they exist with matter."

[265] That Alexander read the expression τρόπος ἐπιστήμης (995a14) in his *Metaphysics* text is confirmed by his paraphrase in 168.24–25.

[266] Cf. Bonitz 1870: s.v. τρόπος, p. 772b38–45. See, e.g. Arist. *de An.* A 1, 402a11–22.

[267] There are, however, no parallel passages in which Aristotle speaks of τρόπος φυσικὸς in the sense of "method of natural science." Cf. IA 2, 704b13: πρὸς τὴν μέθοδον τὴν φυσικήν and *Metaph.*

in the phrase τρόπος ἐπιστήμης / "the method of science," "the way of attaining knowledge" (a14), and thus is established in the passage of our concern. Nevertheless the term λόγος seems to be a viable alternative. It could, together with the reference back to the 'accuracy of mathematics' (ἀκριβολογία),[268] be understood as mathematical teaching or discipline.[269] It is also possible to take the term λόγος in the sense of 'argument' or 'argumentative method.' Aristotle uses λόγος in this sense, for instance, in *EE* A 6 (1216b35–1217a10), where he, in a way quite similar to the passage in α 3, speaks of appropriate methods of investigation. Those who lack training (ἀπαιδευσία, 1217a8; cf. πεπαιδεῦσθαι in α 3, 995a12) are unable to identify the appropriate λόγοι for each subject.[270] Assuming that Aristotle originally had written the word λόγος in 995a17, one could explain the alternative reading τρόπος as a later post-Aristotelian 'correction' that had been prompted by the occurrence of τρόπος in line 995a14. And so even if the parallel passage in *EE* does not provide conclusive evidence that λόγος is the original reading in α 3, it makes it possible to see λόγος as *lectio difficilior*. In that case Alexander's text would once more bear witness to the original reading that had been corrupted in ω^αβ.

All in all, however, the available evidence does not seem to allow for a conclusive decision in favor of λόγος as the original reading. What can be said with certainty is that here we find another instance of Alexander knowing of a variant reading (τρόπος) that is identical to the reading in our manuscript tradition. Should this reading be the correct reading of the original text (Ω), it does not necessarily follow that Alexander's knowledge of it stems from ω^αβ or one of its ancestors, for it is likely that the correct reading is found in many or even all other versions apart from ω^AL. Alexander declares the reading τρόπος to be a *varia lectio*, as indicated by the standard formula γράφεται δὲ καί (169.10), but he does not tell us anything about the variant's possible origin. By all appearances, he has told us everything he knows about it. Nevertheless if ω^AL's reading (λόγος) is the correct original (Ω) and the variant reading τρόπος, which is also in ω^αβ, is corrupt, then we have further support for the conclusion that Alexander had sporadic access to ω^αβ or one of its direct relatives via other commentaries or notes in his

---

M 1, 1076a9: ἐν μὲν τῇ μεθόδῳ τῇ τῶν φυσικῶν περὶ τῆς ὕλης. Nor are there parallel passages where he speaks of a τρόπος μαθηματικός, in the sense of "method of mathematics," which is the intended meaning of τρόπος (a17) according to our understanding of the passage.

[268] Focusing on the etymology of the term ἀκριβο-λογία in line a15, one could suppose that Aristotle refers back to it with λόγος in line a17.

[269] There is a parallel passage in which Aristotle speaks of μαθηματικοὶ λόγοι (in contrast to Σωκρατικοὶ *sc.* λόγοι) in *Rh.* Γ 16, 1416a19. A parallel case of the expression φυσικοὶ λόγοι we find in: *GC* A 2, 316a13.

[270] Arist. *EE* A 6, 1217a8–10: ἀπαιδευσία γάρ ἐστι περὶ ἕκαστον πρᾶγμα τὸ μὴ δύνασθαι κρίνειν τούς τ' οἰκείους λόγους τοῦ πράγματος καὶ τοὺς ἀλλοτρίους. / "for inability in regard to each subject to distinguish arguments appropriate to the subject from those foreign to it is lack of training" (transl. by Solomon, but modified).

Metaphysics copy.[271]

The four cases analyzed in 3.5.2.1–4 share the common feature that Alexander cites in his commentary a *varia lectio* identical to that which α and β testify to be the reading of $ω^{αβ}$. In those cases where the variant reading could also be the original one (see 3.5.2.4), we do not need to assume that Alexander knows $ω^{αβ}$ through this reading. But in those cases where the variant reading known to Alexander and preserved in $ω^{αβ}$ is a corrupt or certainly secondary version of the text (as we saw in the three cases 3.5.2.1–3),[272] it is reasonable to assume that Alexander refers indeed to $ω^{αβ}$ or one of its ancestors.[273] The only caveat is that the corrupt reading given in both $ω^{αβ}$ and in Alexander's variant is due to an error or change in the text, which, given the nature of the error or change, could have easily occurred more than once and so in different traditions of the text (cf. 3.5.2.3).[274]

In light of 3.5, it can be concluded that Alexander occasionally is acquainted with readings that differ from his own text ($ω^{AL}$) and stem from other versions of the *Metaphysics*. As shown in 3.5.1, Alexander knows about these variant readings from earlier scholars and commentators such as Aspasius or from notes in the margins or between the lines of $ω^{AL}$. Among the variant readings known to Alexander are *corrupt* readings that are identical with the reading we find in $ω^{αβ}$. On the basis of those, we are allowed to conclude that Alexander via *variae lectiones* had sporadic access to the version of $ω^{αβ}$ (3.5.2). Since Aspasius is the only commentator that Alexander refers to by name (3.5.1), it is, given the present knowledge of the issue, most economical to suppose that Alexander knew of readings from $ω^{αβ}$ precisely via Aspasius's commentary. We may assume, then, that an earlier version of $ω^{αβ}$ was among the *Metaphysics* versions used by Aspasius; let us refer to it as $ω^{ASP1}$. We may then further subsume under the siglum $ω^{ASP2-n}$ those texts from which Alexander's variant readings stem that are not identical with the reading in $ω^{αβ}$.[275] Perhaps Aspasius drew from more than one *Metaphysics* text or knew also of variant readings. The naming of $ω^{ASP1}$ and $ω^{ASP2-n}$ does not suggest, however, that Alexander necessarily knew *all* variant readings solely through Aspasius. Alexander may very well have known further commentators, whom he does not explicitly name in his commentary and whom I, for the sake of simplicity, do not include in the chart given as appendix A. Additionally, some variant

---

[271] Nothing suggests that the reading τρόπος found its way from Alexander's commentary into $ω^{αβ}$, although it is theoretically possible (cf. 5.1).

[272] The commentary passage in 174.25–27, commenting on 995a19–20, belongs also to this group (see 4.1.1).

[273] The theoretical possibility that the reading in $ω^{αβ}$ is only the result of a later implementation of the reading from Alexander's report of a *varia lectio* into $ω^{αβ}$ does not need to be further considered in those cases in which we do not have supplementary evidence that points to this explanation.

[274] Yet, on the other hand, in light of the case discussed in 3.5.2.3 one might want to argue about how likely it actually is that ὅτι (993a29) was deleted in more than one instance.

[275] See the list of variant readings known to Alexander in 3.6.

readings that Alexander reports might simply have been taken from anonymous glosses in ω^{AL}.[276]

The account I present in this chapter allows for the fact that Alexander was not familiar with all peculiarities of the ω^{αβ}-version. Primavesi assumed that since Alexander did not know that the phrase "we say in the Phaedo" (A 9, 991b3) had been changed to "it is said in the Phaedo" in ω^{αβ}, Alexander did not know the "*common text*" (which I call ω^{αβ}).[277] This assumption is compatible with my results in that I claim that Alexander's access to ω^{αβ} was limited and confined to occasional variant readings. This assumption nonetheless needs to be modified: given my findings it is incorrect to hold that Alexander did not know ω^{αβ} at all.

3.6 ALEXANDER'S DISTINCTION BETWEEN VARIANTS AND CONJECTURES

A closer look at Alexander's records of variant readings and conjectures not only sheds light on Alexander's handling of philological matters; it also gives us a clearer picture of the other *Metaphysics* versions circulating at his time as well as the scholarly debate about Aristotle's work. The present study investigates how Alexander's *Metaphysics* text as well as his commentary itself relates to the text of our directly transmitted manuscript tradition, and so it is vitally important to distinguish between Alexander's references to *variae lectiones* preserved in other manuscripts or commentaries, and his own views on how Aristotle's text could be improved. Determining whether a suspicious phrase in our text stems from a correction Alexander himself suggested depends on the evidence available in the commentary. The question is then whether Alexander himself coins the phrase or reports a variant reading.

In this section I will focus particularly on finding a criterion that allows us to clearly distinguish between a *varia lectio* on the one hand and a conjecture put forward by Alexander (or a colleague) on the other. Since this criterion will be based on the information in Alexander's commentary, the distinction it allows us to make will always be limited to Alexander's own view on the matter. In other words, the criterion will tell us when *Alexander* takes something to be a *varia lectio* of another manuscript and when he makes a suggestion of his own. Such a criterion, as I will show in the following, can be gathered from the *terminology* that Alexander applies in his reports of variant readings and conjectures. On the basis of Alexander's diction we can distinguish between two groups (list A and B):[278]

---

[276] The version(s) from which these glosses stem may be referred to as φ.
[277] See Primavesi 2012b: 414 and 423: "So it seems that he knew neither the *common text* ... ."
[278] Moraux 2001: 429–31 also assumes two groups.

A) variae lectiones[279]

Alexander		Metaphysics
36.12–13	φέρεται δὲ ἔν τισι γραφὴ τοιαύτη…	A 4, 985b12–13[280]
46.23–24	γράφεται ἔν τισιν ἀντὶ…	A 5, 987a10[281]
58.31–59.2	φέρεται ἔν τισι γραφὴ τοιαύτη…	A 6, 988a10–11[282]
59.23–27	γράφεται καὶ οὕτως…	A 6, 988a12–13[283]
75.26–27	γράφεται δὲ ἔν τισιν ἀντιγράφοις ἀντὶ…	A 8, 990a24[284]
91.5–6	γράφεται ἔν τισιν…	A 9, 990b30–31[285]
104.20–21	… ἔν τισιν οὐ φέρεται.	A 9, 991a27–b1[286]
138.26	γράφεται καὶ χωρὶς…	α 1, 993a30[287]
145.21–25	γράφεται δὲ ἔν τισιν ἀντιγράφοις…	α 1, 993b22–24[288]
169.10–11	γράφεται δὲ καὶ…	α 3, 995a16–17[289]
194.3–4	γράφεται ἔν τισιν …	B 2, 997a24[290]
251.21	γράφεται καὶ…	Γ 2, 1004a5[291]
273.34–274.1	ὡς καὶ φέρεται ἔν τισιν. / φέρεταί τις καὶ τοιαύτη γραφὴ…	Γ 4, 1006a18–21[292]
339.18–20	φέρεται δὲ καὶ οὕτως ἡ λέξις…	Γ 8, 1012b8–10[293]

---

[279] Cf. Moraux 2001: 429–30 with n. 24. Moraux does not mention the passage in 251.21–23, but includes in his list cases that I do not recognize as *variae lectiones* but rather as conjectures.

[280] Alex. refers to another version of the *Metaphysics*, in which the words καὶ ὥσπερ τῶν μαθηματικῶν follow after παθημάτων (985b12).

[281] Alex. himself reading μοναχώτερον, knows also the variant μορυχώτερον (46.23), which, as Alex. notes, scholars interpreted in two divergent ways, one of which agrees with the β-reading μαλακώτερον. The α-text reads μετριώτερον. See Primavesi 2012b: 428–31.

[282] See 3.5.1; pp. 62–63.

[283] Alex. reports a variant reading, which is in fact the correct version of the (erroneous) reading of his text (τὰ εἴδη τὰ μὲν … τὰ δὲ ἐπὶ). The correct reading is also found in β (τὰ εἴδη μὲν … τὸ δ' ἓν ἐν), and with a slight variance in α (τὰ εἴδη τὰ μὲν … τὸ δ' ἓν ἐν). Cf. Rashed/Auffret 2014: 61–65, who in addition draw on the Arabic evidence, which agrees with the variant reading of Alexander (and β).

[284] Alex. knows for the word ἀδικία (ω^{AL}, ω^{αβ}) the variant ἀνικία.

[285] Alex. cites a text in which μή (ω^{AL}, ω^{αβ}) is absent in line 990b31.

[286] Alex. states that lines 991a27–b1 are missing in some manuscripts. Cf. 3.5.1; pp. 68–69.

[287] See 3.5.2.3. Alex. presents a *varia lectio* that agrees with ω^{αβ} in not reading ὅτι in 993a30.

[288] See 5.4.2. Alex. knows for οὐκ ἀΐδιον (ω^{AL}, β) the variant οὐ τὸ αἴτιον καθ' αὑτό, which is identical to the α-reading.

[289] See 3.5.2.4. Alex. reads λόγος but also knows the variant τρόπος, the reading of the direct transmission.

[290] Alex. records the variant εἴθ' αἱ αὐταί as alternative to εἴθ' αὗται (β-reading). In his quotations, however (192.11 and 193.1–2), we find also εἴτε αὐταί, which is identical to the α-reading.

[291] Alex. knows the variant γένη ἔχον, the reading we find in β. His quotation in 251.10 shows γένη ἔχοντα to be the reading in ω^{AL}, which is identical to α.

[292] See 4.2.3. Alex. knows several variants to the text in 1006a19–20. He knows that some manuscripts read οὐ (α : οὐχὶ β), which is absent from ω^{AL}; he also knows of the variant that does not read οὐ (just as ω^{AL}) and omits the particle γὰρ (a20).

[293] Alex. quotes the variant τὸ ἀληθὲς ἢ φάναι ἢ ἀποφάναι καὶ τὸ ψεῦδός ἐστιν. The β-text reads τὸ

341.30	… οὐ φέρεται ἔν τισιν…	Γ 8, 1012b22–28[294]
348.7–8	διὸ ἔν τισι γράφεται…	Δ 1, 1013a21–23[295]
354.31–32	γράφεται δὲ καί…	Δ 3, 1014a26–27[296]
356.34–35	φέρεται δὲ ἔν τισι ἀντιγράφοις …	Δ 3, 1014b2–3[297]
417.2–3	γράφεται καί…	Δ 18, 1022a35–36[298]
439.3–5	γράφεται δὲ καὶ ἔν τισι…	Δ 30, 1025a32–33[299]
Fr. 4b Freudenthal (72.18–20)	Instead of that, another manuscript has / An Stelle dieser Worte aber findet sich in einer anderen Handschrift folgendes / *Et in alia scriptura invenitur sic…*	Λ 1, 1069a32[300]
Fr. 12 Freudenthal (86.25–26)	as is found in some manuscripts / … wie es sich auch in einigen Handschriften findet	Λ 3, 1070a18[301]
Fr. 13b Freudenthal (88.15–22)	The meaning of this passage is more clearly expressed in another manuscript / Der Sinn dieses Abschnittes liegt noch klarer in einer anderen Abschrift vor / *Et hoc invenitur manifestius in alia scriptura sic…*	Λ 3, 1070a20[302]

In Greek manuscripts and also in papyri (e.g., from the second century AD) we

---

ἀληθὲς φάναι ἢ ἀποφάναι ψεῦδός ἐστιν and **α** ἢ τὸ ἀληθὲς φάναι ἢ ἀποφάναι ψεῦδός ἐστιν.

[294] According to Alex. the passage 1012b22–28 is not transmitted in some manuscripts.

[295] See 5.4.1. Alex. knows for καλόν (**ω**^AL, **β**) the variant κακόν, which is the reading in **α**.

[296] See 3.5.2.1. Whereas **ω**^AL reads ἀδιαιρέτου τῷ εἴδει, Alex. knows also the variants ἀδιαιρέτου τῷ εἴδει εἰς ἕτερον εἶδος (**ω**^αβ) and ἀδιαιρέτου εἰς τὸ αὐτὸ εἶδος.

[297] Alex. reports the variant τῶν τριῶν μέσων, whereas in his text he reads τῶν τριῶν (356.21). The **α**-text reads ἐκ τῶν τριῶν, the **β**-text ἐκ τῶν τριῶν μέσων.

[298] Alex. knows for the word κεχωρισμένον the variant κεχρωσμένον. Cf. 1022a30–31. See also Dooley 1993: 101 n. 442.

[299] Alex. reports the words καὶ ταῦτα ἴδια αἴτια as a variant. It is not entirely clear, however, which part of the transmitted text (καὶ ταῦτα μὲν ἐνδέχεται ἀΐδια εἶναι …) should be replaced by these words.

[300] Alex. quotes as a variant reading the words ἡ δ' ἀΐδιος (1069a32), which are transmitted in **ω**^αβ but usually athetized in our editions, and which are absent in the text that Alexander prefers (as well as in the Arabic tradition in Ar^m and in Themistius's text). The text without these words is probably original. Cf. Freudenthal 1885: 44 and Walzer 1958: 224.

[301] See 3.5.2.2. Alexander's text (**ω**^AL) most likely read the words οἱ τὰ εἴδη τιθέντες ἔφασαν for which Alexander records the variant Πλάτων ἔφη that is found in our text (1070a18: Πλάτων ἔφη **α** : ὁ Πλάτων ἔφη **A**^b **ε**). Cf. Freudenthal 1885: 86 with n. 3.

[302] Alexander refers to another manuscript in which (among other differences that are difficult to determine due to the desparate state of the Arabic text) the words καὶ ἄτομος (1070a20) are absent. They are present in Alexander's text (Freudenthal 1885: 86 n. 3), but are absent in our text (Freudenthal 1885: 88: 17–23). Cf. Freudenthal 1885: 88 n. 2 ("der Text ist hier in Ar. und Hebr. unheilbar zerrüttet") and Martin 1984: 119 with n. 16.

encounter in the margins or between the lines cues whose purpose is to alert the reader to *variae lectiones*.[303] These cues are often standard formulas such as ἐν ἄλλῳ / ἔν τισι / γρ(άφεται). In Alexander's commentary we find such formulas serving exactly that purpose: φέρεται or γράφεται (+ καὶ / ἔν τισιν).[304] When Alexander mentions divergent manuscripts he speaks of them in the plural: ἔν τισι(ν) (ἀντιγράφοις) / "some (manuscripts)."[305] This plural masks an indeterminate number of manuscripts; the exact number is most likely unknown to even Alexander himself. It can safely be ruled out that Alexander uses the plural to indicate that he himself encountered the variant reading in *several* manuscripts.[306]

By contrast, the second group of remarks on textual issues includes all those commentary passages in which an emendation is suggested.[307] Here, the terminology comes in a greater variety,[308] yet the expressions Alexander uses are often similar to standard expressions that are regularly found in commentary literature and scholia.[309]

B) Conjectures and suggested emendations

Alexander		*Metaphysics*
11.4–5	ἐλλείπει...	A 2, 982a21[310]
37.20–21	εἴη δ' ἂν καταλληλότερον ἔχουσα ἡ λέξις, εἰ...	A 5, 985b26[311]

---

[303] See West 1973: 12. On papyri see McNamee 1977: 55–56 and esp. 90–96; Dover 1997: 47.

[304] Other commentators also use these or similar formulas: Simplicius repeatedly uses the formula γράφεται δὲ καὶ (οὕτως) (see *In Cael.* 460.10; 483.13–14 Heiberg; CAG VII; *In Phys.* 129.24–25; 239.28 Diels; CAG IX). See also Michael of Ephesus (for instance, *In Metaph.* 446.29; 537.15; 541.26 Hayduck). The formula γράφεται + ἔν τισι, which Alexander uses several times, can be found in commentators and grammarians such as Aristonicus of Alexandria, *De signis Iliadis*, Z 240, p. 121 Friedländer; H 5, p. 126 Friedländer; Θ 213, p. 143 Friedländer; Anonymous, *In EN* 195,1 Heylbut; and Michael of Ephesus, *In SE* 18.22; 110.7 Wallies.

[305] Cf. Fazzo 2012a: 62 n. 35.

[306] See 3.1 and McNamee 1977: 92–93 on ἔν τισι and τινές in papyri.

[307] There are some passages (not included in the list) in which Alexander seems to propose an alternative formulation of Aristotle's words simply for the purpose of commenting on it. See, e.g., that Alexander's oft-used formula "[*Metaphysics* text] ... ἴσον τῷ [Alexander's reformulation]" introduces, we can say confidently, an explanation of the text rather than an emendation. Cf. 21.21–25, 44.9–10, 59.16–19, 93.10, 109.30–110.2, 141.29–30, 206.9–11 (see 5.1.1), 316.27–29, 321.1–3; on the formula δύναται x (καὶ) ἀντὶ τοῦ y εἰρηκέναι see 5.1.2.

[308] In Moraux's (2001: 430 n. 25) list of Alexander's conjectures the following are absent: 11.3–5, 70.7–9, 164.22–25, 172.13–15, 174.25–27, 185.21–25, 244.31–32, 264.17–18, 270.12–17, 330.1–3. Moraux includes in his list 54.11–13, where Alexander points out that δέ should be taken as γάρ (τὸ δὲ ἄπειρον εἶπεν ἀντὶ τοῦ τὸ γὰρ ἄπειρον, 54.11–12). This is not about changing the text; it is about understanding what Aristotle means to say.

[309] See the overview given by Dickey 2007: 150–66.

[310] See 4.2.1. Alex. wants to add τὰ πάντα to the text.

[311] Alex. suggests reading γάρ instead of δέ.

46.20–23	... λείποι ἄν...	A 5, 987a11–13[312]
68.3–4	ἄμεινον γεγράφθαι...	A 8, 989a26[313]
70.7–8	ἐλλείπει...	A 8, 989b20–21[314]
114.22;	ἂν ᾖ ... γεγραμμένον... δοκεῖ δέ μοι	A 9, 992a2–3[315]
116.25–27	ἡ λέξις μὴ οὕτως ἔχειν...	
141.11–13	ὃ φεύγοντές τινες ἐνηλλάχθαι φασὶ τὴν λέξιν καὶ εἶναι τὸ κατάλληλον αὐτῆς...	α 1, 993b4–7[316]
141.19–21	καὶ εἴη ἂν ὑπερβατὸν ἐν τῇ λέξει τοιοῦτον...	
141.24–26	δύναται καὶ λείπειν τῇ λέξει τὸ...	
164.22–25	τὸ αὐτὸ δὲ σημαίνοι ἂν καὶ εἰ εἴη γεγραμμένον... τινὲς δὲ ... γράφουσι, καὶ ἐξηγοῦνται τὴν λέξιν ...	α 2, 994b25–26[317]
167.11–12	εἰ δὲ εἴη γεγραμμένον... λείποι ἄν...	α 3, 995a1[318]
172.13–15	εἴη δὲ ἂν καταλληλότερον, εἰ ... εἴη γεγραμμένον	B 1, 995a27[319]
174.25–27	τινὲς ... προσγράφουσι...	α 3, 995a19–20[320]
185.22–24	τὸ κατάλληλόν ἐστι τῆς λέξεως·...	B 2, 996b18–20[321]
186.31–33	ἄμεινον γεγράφθαι ...	B 2, 996b24–25[322]
193.32–33	λείπει γὰρ τοῦτο τῇ λέξει ... ἐνδεῖ γὰρ ...	B 2, 997a24[323]

[312] Alexander's cautious formulation makes it difficult to decide whether he really wants to have τινὲς ἐξ αὐτῶν added to the text. Such an addition would narrow down the number of philosophers Aristotle refers to. Cf. Dooley 1989: 73 n. 152.

[313] See 5.3.4. Alex. proposes to read ἀλόγως instead of εὐλόγως.

[314] Alex. indicates that the word ἀκόλουθα is missing from the text (70.8). Jaeger, inspired by Alexander's suggestion, adds ἀκολουθεῖ to line 989b13 of the *Metaphysics*.

[315] Moraux (2001: 430 n. 24) is unsure about how to classify this passage, but opts for a *varia lectio*. Alexander extensively discusses the two readings ἀδιάφοροι and διάφοροι (ω^AL and ω^αβ), in the end (116.25–27) supporting διάφοροι (cf. the lemma in 114.20). Al. does not indicate that the alternative reading ἀδιάφοροι is a *varia lectio* (*pace* Primavesi 2012c). Rather, it looks as though Alex. engages an emendation proposed by a predecessor. The rejected reading was incorporated into the Arabic version of the text (see Crubellier 2012: 315).

[316] Moraux 2001: 430 n. 24 only mentions the last of a total of three suggestions. Alex. discusses in detail the sentence in 993b4–7 and declares it contradictory. First, he refers to a conjecture made by others (τινες ... φασί), then he offers two suggestions that are probably his own: first he suggests a transposition (141.19–21), second the addition of πάντας ἅμα (141.24–29). See also below.

[317] See 5.1.4. Alex. conjectures ἐν κινουμένῳ (ω^αβ) for κινουμένῳ.

[318] See 4.1.1. Alex. wants to delete ἔτι (ω^AL).

[319] Alex. conjectures γὰρ for δέ. Cf. Laks 2009: 39.

[320] See 3.5.2.4 and 4.1.1. Alex. reports that "some" added a sentence at the end of book α.

[321] Alex. suggests transposing an epexegetic relative clause. It is questionable whether he proposes an emendation to the text here or rather just explicates it.

[322] After having gone through several interpretations of the transmitted text (186.9–31), Alex. suggests adding to the sentence the negation οὐκ (see also pp. 257–58 n. 282).

[323] Alex. states that one should add (in thought only?) the words μιᾶς ἔσται θεωρεῖν τε καὶ δεικνύναι from the previous sentence. It is questionable whether this is meant to be an emendation to the text (as in Moraux). It seems more likely that Alexander wants to draw attention to Aristotle's elliptical expression.

224.18–19	τὸ αὐτὸ δὲ σημαίνει, κἂν ᾖ γεγραμμένον ...	B 4, 1001a19–20[324]
233.26	ἄμεινον δὲ γεγράφθαι...	B 6, 1002b24[325]
244.31–32	τὸ μὲν κατάλληλον τῆς λέξεώς ἐστιν·...	Γ 2, 1003b19[326]
251.2–5	ἦν ἂν σαφέστερα τὰ λεγόμενα, εἰ ἦν ἡ λέξις αὕτη κειμένη πρὸ ...	Γ 2, 1003b22–1004a4[327]
264.17–18	λείπει δὲ τῇ λέξει ...	Γ 2, 1005a14–15[328]
267.14–21	δοκεῖ δέ μοι αὕτη ἡ λέξις... τὴν τάξιν ἔχειν	Γ 3, 1005b2–11[329]
270.15–17	... ἐκ περισσοῦ δοκεῖ προσκεῖσθαι...	Γ 3, 1005b26–27[330]
273.34–36	καταλληλότερον δὲ ἡ λέξις ἔχοι ἄν, εἰ ... εἴη γεγραμμένον	Γ 4, 1006a20[331]
285.34–36	λείπει γάρ...	Γ 4, 1007a20–23[332]
288.9–11	δύναται γεγράφθαι...	Γ 4, 1007a33–34[333]
321.1	ἐνδεῖ δὲ τῇ λέξει ...	Γ 6, 1011a28–31[334]
330.1–3	λείποντος τῇ λέξει ...	Γ 7, 1011b23–1012a1[335]
349.5–6	ἐλλείποι δ' ἄν...	Δ 2, 1013a27–9[336]

[324] Alex. reformulates the sentence to make its meaning clearer. Moraux speaks of a *varia lectio*, but it might rather be the case that Alex.'s λέγει in 224.19 is paraphrasing Aristotle's θήσεται (cf. Madigan 1992: 176 n. 393).

[325] Alex. suggests reading ἀλλ' εἴδει in place of καὶ εἴδει in 1002b24. Ross follows this suggestion (cf. Ross 1924: 250). Jaeger athetizes καὶ εἴδει.

[326] In this case, too, it is not clear whether Alex. reorganizes Aristotle's sentence in his own paraphrase or whether he proposes an actual emendation to the text.

[327] Alex. suggests transposing the passage 1003b22(εἰ δή)–1004a2(ἐναντίων) after 1004a9(μαθήμασιν) so that the sentence καὶ ... οὐσίαι (1004a2–3) follows directly upon the sentence διὸ ... εἰδῶν (1003b21–22). Jaeger puts the text in 1003b22–1004a2 into double-brackets, marking it as a later addition by Aristotle. Madigan (1993: 149 n. 115) seems to interpret Alex. as suggesting connecting (at least in thought) just the sentence καὶ ... οὐσίαι (1004a2–3) so that is follows after διὸ ... εἰδῶν (1003b21–22).

[328] It is unclear whether Alex. merely wants to draw attention to an elliptical expression or proposes an emendation (cf. 193.30–33).

[329] Alex. suggests (δοκεῖ δέ μοι) transposing 1005b2–5 behind 1005b5–8.

[330] According to Alex., δὲ in 1005b26 is to be discarded. It seems that Alex. wants to have the sentence follow asyndetically, perhaps to mark it as the reason for the aforesaid. The text of the commentary is problematic. Madigan (1993: 50 with n. 287) thinks that Alexander wants to replace δὲ with γάρ. Sepúlveda's translation of this commentary passage (*dictio enim plena est...*) suggests rather that Alexander proposes deleting δέ without substitution.

[331] See 4.2.3.

[332] See 5.2.2. Alex. proposes adding καί.

[333] Alex. conjectures to read in 1007a34 καθ' οὗ instead of the transmitted καθόλου. The editors Bonitz (see 1842: 116), Christ, Ross, and Jaeger adopted Alexander's suggestion. Cassin/Narcy and Hecquet-Devienne do not follow his suggestion.

[334] Alex. wants to heal a lacuna in Aristotle's text (1011a28–31) by adding ῥᾳδία ἡ ἀπάντησις. For other suggestions on the text see Jaeger 1917: 513–16; Ross 1924: 281–82; Cassin/Narcy 1989: 249–51.

[335] See 4.3.3. Alex. suggests supplementing ἔχει.

[336] Alex. misses the word δηλωτικός, which would clarify the connection between ὁ λόγος and τί ἦν

357.24	εἴη δ' ἂν τὸ εἰρημένον σαφές, εἰ εἴη προσκείμενον...	Δ 4, 1014b18–19[337]
368.7–15	δοκεῖ δὲ ἡ λέξις αὕτη ἐλλιπῶς εἰρῆσθαι. ... εἴη δ' ἂν σαφὴς ἡ λέξις... εἰ εἴη γεγραμμένον...	Δ 6, 1016b11[338]
433.15–16	ἔνεστι καὶ...ἀναγνῶναι ..., ἵνα ᾖ τὸ λεγόμενον	Δ 29, 1024b26–28[339]
Fr. 12 Freudenthal (87.10–11)	Es ist möglich, den Sinn dieser Stelle einfacher zu gewinnen, wenn wir die Worte umkehren... / it would be easier to understand what he means if the word ... was transposed from its place ... / Et erit manifestior iste sermo si mutaverint hanc particulam ...	Λ 3, 1070a18–19[340]

The passages of list B share the feature that Alexander either reports a conjecture made by others or discusses an emendation proposed by himself. Alexander's diction in discussing these alternative manners of expression varies greatly depending to the kind of problem and the suggested solution—there is no fixed set of formulae such as γράφεται or φέρεται. When Alexander regards a word or expression as missing he indicates this by (ἐλ)λείπει or similar expressions (11.5, 70.8, 264.17, 285.35; cf. ἐλλείποι ἂν in 349.5; λείποντος in 330.2; ἐλλιπῶς εἰρῆσθαι in 368.7).[341] There are various possibilities of expressing criticism of a transmitted phrase or sentences: e.g., δοκεῖ δέ μοι αὕτη ἡ λέξις... τὴν τάξιν ἔχειν (267.14–21) or ἐνηλλάχθαι φασὶ τὴν λέξιν καὶ εἶναι τὸ κατάλληλον αὐτῆς... (141.11–12).

We see that the verb γράφειν is not confined to the description of *variae lectiones* but appears also among passages discussing conjectures or emendations. The mode in which the verb appears nevertheless differs. When Alexander considers or suggests an emendation for the text he uses phrases such as: ἄμεινον γεγράφθαι... (68.3; 186.31; 233.26); εἰ (δὲ) (...) εἴη γεγραμμένον... (164.22–23; 167.11; 172.13–14; 368.14); ἂν ᾖ ... γεγραμμένον... (114.22; 224.19). We often find here conditional clauses and a tone considerably more cautious than that found

---

εἶναι: the λόγος is the ὁρισμός of the essence. Whether Alexander here really wants to change the text is unclear. Cf. Dooley 1993: 17 with n. 29.

[337] Alex. suggests adding ἥτις in order to make the meaning clearer (σαφές).

[338] Alex. suggests changing ἐπεὶ δ' ἔστι into ἔτι ἔστι. He (correctly) misses a main clause. This conjecture has been incorporated into the γ-branch of the *Metaphysics* text, and the editors Bonitz, Christ, Ross and Jaeger follow Alexander's suggestion.

[339] Alex. discusses the possibility of reading ἢ ψευδής for ᾖ ψευδής.

[340] See 5.1.5. Alex. suggests transposing a part of the sentence and reading ὅτι εἴδη ἔστιν ὁπόσα φύσει, εἴπερ ἔστιν εἴδη (ω^αβ) instead of ὅτι εἴπερ ἔστιν εἴδη, ἔστιν ὁπόσα φύσει.

[341] On the use of (ἐλ)λείπει in scholia and commentaries see Dickey 2007: 119.

in his reports of a variant reading.³⁴² This cautiousness is programmatic and leads to the reason why Alexander's tone differs between the description of a *varia lectio* and that of a conjecture: many of the supposed suggestions for a reformulation of the Aristotelian text are not so much meant as a call for textual intervention but rather intended as clarification and explication of what Aristotle actually meant.³⁴³

The distinction between group A (*varia lectio*) and B (conjecture) has so far been drawn on the basis of Alexander's diction. That Alexander's terminology indeed provides a trustworthy criterion is confirmed by the following passage, which is indeed the exception that proves the rule. In this single case, Alexander treats the very same alternative reading first as a conjecture and then as a *varia lectio*.

Alexander, *In Metaph*. 273.34–36 Hayduck (*ad* 1006a18–21)

καταλληλότερον δὲ ἡ λέξις ἔχοι ἄν, εἰ [35] ἀντὶ τοῦ τοῦτο μὲν γὰρ τάχα ἄν τις ὑπολάβοι τὸ ἐξ ἀρχῆς αἰτεῖν [36] εἴη γεγραμμένον τὸ τοῦτο μὲν τάχα ἄν τις, ὡς καὶ φέρεται ἔν τισιν.

The sentence would be more consistent if, instead of "*for* this one might perhaps take to be begging the question," it read "this one might perhaps ...," as it is found in some witnesses.

In this passage Alexander wants to eliminate an inconsistency he encounters in a passage of his *Metaphysics* text (ω^AL). We know on the basis of the evidence in ω^αβ that this inconsistency had been caused in ω^AL by a dropout of the negation οὐ (see 4.2.3). Alexander discusses possible ways of removing the incongruity in the *Metaphysics* text. His first suggestion is to delete the particle γάρ so that the sentence no longer appears to be an explanation of what was said before.

Given Alexander's presentation of this solution, one expects it to be his own conjecture. In the sentence introducing his suggestion, the verb in the apodosis stands in the potential optative, followed by an optative + εἰ in the protasis (καταλληλότερον δὲ ἡ λέξις ἔχοι ἄν, εἰ ... εἴη γεγραμμένον):³⁴⁴ "the sentence would be more consistent if it read... ." Yet, this diagnosis hits only half of the truth. For Alexander continues saying that the reading without γάρ is actually found in some manuscripts: καὶ φέρεται ἔν τισιν. Here Alexander uses one of his characteristic formulas and thus marks the deletion of γάρ as a *varia lectio*.

In this exceptional passage Alexander first presents a solution as his own emendation and then refers to the evidence in another manuscript, where the proposed

---

³⁴² Moraux 2001: 430: "In solchen Fällen schlägt er Änderungen, Streichungen, Ergänzungen oder Transpositionen vor. Er tut es aber meistens mit vorsichtigen Formeln wie etwa: 'Mir scheint, es wäre besser, passender, sinnvoller, so und so zu schreiben.'"

³⁴³ Again, this is in line with the terminological evidence in scholia and other commentaries.

³⁴⁴ The optative in the protasis expresses an imaginary state of affairs ("bloße Vorstellung") or arbitrary assumption ("etwas willkürlich Angenommenes"): Kühner/Gerth II, § 576; p. 477.

emendation already occurs as part of the text (ὡς καὶ φέρεται). Why does Alexander discuss the deletion of the γὰρ in such a curious way? In other words, why does he present it as both a conjecture and a *varia lectio*? Two related motivations come to mind. On the one hand, presenting the deletion of γὰρ as his own idea allows him to show off that he came up with a textual solution that is actually found in other manuscripts. On the other hand, letting his reader know that his suggestion is in fact a *varia lectio* lends credence to the validity of his suggestion. Alexander would not want to miss the opportunity to use the evidence that favors his case. These two entangled motivations entail an important consequence: whenever Alexander speaks in favor of an alternative reading without adducing evidence from another manuscript, we may safely assume that such evidence was not available to him.

Thus, this case of a twofold characterization as a conjecture *and* alternative reading does not falsify my thesis that Alexander's diction allows us to draw a clear line between what he takes to be a *varia lectio* and what he poses as a conjecture. Quite to the contrary, this passage confirms the validity of the rule. For, this passage shows precisely that a phrase like 'it would be better if the text read x' does *not* introduce or even include the possibility of a *varia lectio*. Moreover, Alexander distinguishes clearly between two types of sources, which in this case, are in agreement: there is, first, his own creativity and, second, the evidence of another version of the *Metaphysics*. As this commentary passage shows, the two sources are markedly distinct and Alexander does not merge them together in order to make his statement shorter. Rather, he can easily combine them precisely because they are clearly differentiated in his thoughts and in his words.

The differentiation between a *varia lectio* and a conjecture pertains only as far as Alexander's knowledge about the status of the alternative reading extends. What Alexander describes as a *varia lectio* most likely was marked as such already in his source, i.e., in other commentaries or marginal notes in his *Metaphysics* copy. At the same time, such a variant reading can still have emerged from a prior scholar's conjecture. Yet, those readings that Alexander ascribes to conjecture are surely either attributable to his own interpretation of the transmitted text or to someone else's.

While Alexander's diction allows for a clear distinction between *varia lectio* and conjecture,[345] there is no such clear differentiation between Alexander's own emendations and those proposed by others. Alexander rarely mentions other commentators or exegetes by name (see 3.5.1). It is only on those rare occasions when he is discussing a textual problem in some detail that he distinguishes between his own proposed solution and that of his predecessors. See, for instance,

---

[345] It is not always clear whether Alexander's suggestion for improving a formulation in the *Metaphysics* is actually meant as a call for textual intervention or is rather to be taken as a commentator's clarifying explication of Aristotle's words. For examples, see above n. 307.

the passage in 141.6-8 (commenting on α 1, 993b4-7). Alexander observes that Aristotle's statement appears inconsistent (ἔχει δέ τι ἡ λέξις ἀκατάλληλον…) with what was said before. Alexander then reports a solution that has been proposed by others (τινές): "In an attempt to avoid [this seeming inconsistency], certain [interpreters] say that the text has been reversed, and that the appropriate reading is…" (ὃ φεύγοντές τινες ἐνηλλάχθαι φασὶ τὴν λέξιν καὶ εἶναι τὸ κατάλληλον αὐτῆς…, 141.11-12). Subsequent to this Alexander presents further proposals for solving the problem; evidently these are his own. The caution with which he proposes this interpretation indicates so much: "Perhaps, however, it is better to understand the text …" (μήποτε δὲ ἄμεινον τῆς λέξεως ἀκούειν …, 141.14). Further on he refers back to this explanation, his own: "in the way we have just explained" (ὡς προειρήκαμεν, 141.14-15). He continues with suggestions for transposition (καὶ εἴη ἂν ὑπερβατὸν ἐν τῇ λέξει τοιοῦτον, 141.19-20) or addition (δύναται καὶ λείπειν τῇ λέξει τὸ, 141.24-25) of words or phrases. These also seem to be Alexander's own, as he does not mention "some" others anymore.[346]

In many cases, however, it is difficult to determine whether Alexander presents his own conjecture or just reports an emendation he found in the literature.[347] One might ask whether we are allowed to assume that whenever Alexander does not mention τινές we are dealing with his own suggestion on the text. This question will be addressed in those cases where we want to find out whether Alexander's comments influenced the transmission of the *Metaphysics*.

---

[346] See also the comparison between earlier conjectures and Alexander's own in 164.15-165.5 (see 5.1.4).

[347] Since Aristotle's usage of the particle δέ is one of Alexander's pet issues, we may assume that most of the proposed textual changes that involve the particle δέ go back to Alexander himself. At several places, he wants to either delete δέ or substitute it by γάρ. In 172.14-15 (comments on 995a27), Alexander wants to interpret δέ in the sense of a γάρ (cf. Denniston 1954, s.v. δέ I.C.1(i), pp. 169-70) or even replace it: εἴη δὲ ἂν καταλληλότερον, εἰ ἀντὶ τοῦ ἔστι δέ εἴη γεγραμμένον ἔστι γάρ. / "It would be more consistent with this if, instead of 'but it is,' he had written 'for it is.'" Further passages in which Alexander critically engages with Aristotle's usage of δέ are: 54.11-12, 270.12-17, and 295.29-32 (see 5.2.4). See also Laks 2009: 39 n. 43.

CHAPTER 4

Alexander's Text (ω^{AL}) and the Direct Transmission ($\omega^{\alpha\beta}$)

Alexander used one copy of the *Metaphysics* (ω^{AL}) when composing his commentary. Now and then he shows knowledge of variant readings, obtained either from marginal notes in his own copy or from other, earlier commentaries (see 3.5). His commentary can be dated to about AD 200; this date thus gives us the *terminus ante quem* for ω^{AL}. In order to determine how Alexander's copy of the *Metaphysics* relates to the text of the direct transmission ($\omega^{\alpha\beta}$) I will examine both versions for "indicative errors" (*errores significativi*). In doing so, I will follow the rules of textual criticism that were most succinctly set out by Paul Maas.[1] The rules for establishing relationships between textual witnesses by means of "separative errors" and "conjunctive errors" (*Trenn-* and *Bindefehler*) hold also for reconstructed witnesses.[2] In the following, I will deal solely with reconstructed versions of the text: α and β are reconstructed from our manuscripts, $\omega^{\alpha\beta}$ from the agreement of α and β. These versions can be reconstructed in their entirety, that is, for the complete text of the *Metaphysics*. By contrast, only parts of the versions ω^{AL} and (to a lower degree) ω^{ASP1} and ω^{ASP2-n} (see 3.5) can be reconstructed from Alexander's commentary.

4.1 SEPARATIVE ERRORS IN $\omega^{\alpha\beta}$ AGAINST ω^{AL}

I will first examine separative errors in $\omega^{\alpha\beta}$ that are not shared by ω^{AL}. The possibility that the text in all our manuscripts is corrupt while Alexander's commentary alone bears witness to the correct reading (or a reading prior to the corruption) has already been considered, however fleetingly and hesitantly, by Brandis in 1823.[3] Since then, and especially since Bonitz's *Metaphysics* edition in 1848, the

---

[1] See 1; pp. 12–14.
[2] Maas 1958: 3–4.
[3] See for example Brandis 1823 *app. crit. ad* α 1, 993b1 (= 35.21 Brandis): *fort leg.* μήτε πάντως cf. *Alex.*

number of readings of Aristotle's text that are based on Alexander's testimony has increased steadily.[4] Among these readings, however, are corrections whose basis in Alexander is doubtful.[5] The following three examples are cases in which Alexander clearly and justifiably reads the correct text while our *Metaphysics* text has suffered corruption:[6]

(i) A 9, 991b3-4: In the context of his critique of the Platonic theory of Forms in A 9, Aristotle speaks on several occasions in the first person plural ("we"…). In doing this, he regards himself as a member of the Academy.[7] Whereas the α-version (agreeing with Alexander's testimony) preserves these verb forms in several cases in A 9, the β-version's verbs have been "corrected" to third person plural forms ("they…"). In one passage of the text, *both* versions, α and β, and hence ω^{αβ}, have the third person form "it is said," whereas Alexander's commentary alone bears witness to what is the correct reading "we say," which was read in ω^{AL}: A 9,

---

Bonitz 1848 takes up Brandis's suggestion and puts πάντως in his text.

[4] Bonitz follows Alexander's authority (64.27 Hayduck), for example, in A 8, 988b26 and deletes the words καὶ φθορᾶς.

[5] For example, it is problematic to replace ἐκτοπωτέρως (989b30 ω^{αβ}) by ἐκτοπωτέροις on the basis of Alexander's paraphrase (71.14), as done by Bonitz (followed by Christ, Ross, and Jaeger), or to add καὶ λίθος to the text (1008a25-26) on the basis of Alexander's amplifying paraphrasis (290.29-31; 295.25-26), as done by Jaeger.

[6] The same holds for the cases analyzed in 3.5.2.1-3. Further passages to be considered are: A 3, 983b32: τῶν ποιητῶν ω^{αβ} : non reddit Al.^{P} 25.18; – A 8, 989b20-21: νῦν φαινομένοις μᾶλλον ω^{αβ} : φαινομένοις μᾶλλον Al.^{c} 28.12-13 Al.^{c} 70.7; – B 2, 996a24: αὐτοῦ ω^{αβ} : αὐτοῦ Al.^{c} 182.6, 13 Al.^{P} 182.5; – Δ 27, 1024a27: ἔχῃ ω^{αβ} : ἔχει Al.^{P} 428.1; – A 5, 987a10: μαλακώτερον β, μετριώτερον α : μοναχώτερον Al. 46.23-24; – Γ 4, 1007a6: ἐστὶ ω^{αβ} : ἔσται Al.^{P} 283.29; – Δ 2, 1013b6: εἶναι ω^{αβ} : εἶναι καθ' αὐτὸ καὶ Al.^{P} 350.22 Al.^{c} 350.25; – Δ 6, 1016a34: τί ἦν εἶναι : non reddit Al. 366.11-13, 15-16; – Δ 7, 1017a35: ἀσύμμετρος ω^{αβ} : σύμμετρος Al.^{P} 372.6-9; – Δ 30, 1025a15: δὲ β : om. α : ὡς Al.^{P} 437.21; – Γ 2, 1003b28-29: τὸ ἔστιν ὁ ἄνθρωπος καὶ ἄνθρωπος καὶ εἷς ἄνθρωπος α : τὸ εἷς ἔστιν ἄνθρωπος καὶ ἔστιν ἄνθρωπος β : τὸ ἔστιν ὁ ἄνθρωπος ἄνθρωπος καὶ ἔστιν ἄνθρωπος Al. (I am grateful to Stephen Menn for this example); – Λ 1, 1069a32: ἡ δ' ἀΐδιος ω^{αβ} Al.^{yp} : om. Al. Fr. 4b Freudenthal (72.18-25) (see also 3.6).

There is further a group of passages given in ω^{αβ}, about which Alexander is completely silent in his commentary and which therefore are likely to have been absent from ω^{AL}: A 3, 984a22-25 (λέγω δ' οἷον οὔτε τὸ ξύλον … τῆς μεταβολῆς αἴτιον) : non reddit Al. 29.5-8; – A 4, 985a13-17 (ἀλλ' οἷον ἐν ταῖς μάχαις οἱ ἀγύμναστοι … λέγουσιν) : non reddit Al. 34.3-6; – B 1, 995b29-31 (οἷον πότερον ζῷον … καθ' ἕκαστον) : non reddit Al. 177.26-178.2; – Γ 2, 1004b28-29 (οἷον στάσις τοῦ ἑνὸς κίνησις δὲ τοῦ πλήθους) : non reddit Al. 260.30-261.16; – Γ 4, 1008a19-20 (οἷον ὅτι λευκὸν καὶ πάλιν ὅτι οὐ λευκόν) : non reddit Al. 294.34-295.9; – Δ 2, 1013b26-27 (καὶ τέλος τῶν ἄλλων ἐθέλει εἶναι) : non reddit Al.^{c} 352.4; – Δ 6, 1016a27: (οἷον ἵππος ἄνθρωπος κύων ἕν τι ὅτι πάντα ζῷα) : non reddit Al. 364.40-365.7; – Δ 6, 1016b28-31 (καὶ ἀντιστρέψαντι δὴ τὸ μὲν διχῇ διαιρετὸν … στιγμή.) : non reddit Al. 368.34-37; – Δ 27, 1024a15-16 (εἰ κύλιξ κολοβός, ἔτι εἶναι κύλικα) : non reddit Al. 426.33-427.7 (cf. 1024a 24 and Al. 427.32-36). All these passages share characteristic features of later additions: they offer examples or slightly repetitive explications of something already said. And so it is tempting to regard them as later additions in ω^{αβ}, of which ω^{AL} was free. Yet, it is always risky to make an argument *e silentio* and in this case one could counterargue that Alexander has disregarded them in his commentary precisely because these passages contain examples and the like.

[7] See Jaeger 1965 and Primavesi 2012b: 412-20.

991b3–4: λέγομεν Al.¹ 106.7 Al.ᶜ 106.9 Ascl.¹ 90.19 : λέγεται ω^{αβ} Ar^n.

(ii) A 9, 993a5: Alexander (together with the Arabic transmission) reads in his text the correct example: the syllable ζα and its components σ, δ, and α. The reading in ω^{αβ} has been corrupted in a way that is etymologically plausible:[8] ζα Al.^P 132.14–133.4 C (ex ξα) Ar^n (Walzer 1958: 224) : ξα ζ : σμα ω^{αβ} ‖ δ Al.^P 132.17 ζ Ar^n (Walzer 1958: 224) : μ ω^{αβ}.

(iii) Γ 4, 1008b11–12: While our manuscripts all read πεφυκότων (ω^{αβ}), Alexander's paraphrase reveals that he read φυτῶν or γε φυτῶν:[9] πεφυκότων ω^{αβ} Bekker Cassin/Narcy Hecquet-Devienne : ex Al.^P 298.31, 299.2, 7 γε φυτῶν Ross, φυτῶν Bonitz Christ Jaeger.

In the following I will analyze in detail four cases in which we can determine separative errors in ω^{αβ} that ω^{AL} does not share. In these cases Alexander's testimony can correct ω^{αβ}. In two (4.1.2 and 4.1.3) of the following four cases the evidence in Alexander's commentary has not yet been recognized as leading to the more authentic reading of the *Metaphysics* text. In the other two cases (4.1.1 and 4.1.4) Bonitz has already taken the reading presented in Alexander's commentary as the correct one. Nevertheless, it is worth looking at these two passages more closely, as they contain crucial evidence for our present purpose. The first passage to be discussed (4.1.1) shows once more (cf. 3.5.2) that Alexander had sporadic access to a predecessor of ω^{αβ} and that the error we find in ω^{αβ} can thus be dated to a time before AD 200—a fact especially relevant for the present section.

4.1.1 Alex. *In Metaph*. 174.5–6; 25–27 on Arist. *Metaph*. α 3, 995a12–20

In the third, and last, chapter of book α ἔλαττον, Aristotle observes that different sciences apply different methods of investigation to their respective subject matters. In some cases mathematical accuracy is to be employed, while in others the authority of a poet is demanded (995a6–8). According to both the α- and the β-version, the last lines of book α ἔλαττον read the following text.

Aristotle, *Metaphysics*, α 3, 995a12–20

διὸ δεῖ πεπαιδεῦσθαι [13] πῶς ἕκαστα ἀποδεκτέον, ὡς ἄτοπον ἅμα ζητεῖν ἐπιστήμην [14] καὶ τρόπον ἐπιστήμης· ἔστι δὲ οὐδὲ θάτερον ῥᾴδιον λαβεῖν. τὴν [15] δ' ἀκριβολογίαν τὴν μαθηματικὴν οὐκ ἐν ἅπασιν ἀπαιτη-[16]τέον, ἀλλ' ἐν τοῖς μὴ ἔχουσιν ὕλην. διόπερ οὐ φυσικὸς ὁ [17] τρόπος· ἅπασα γὰρ ἴσως ἡ φύσις ἔχει ὕλην. διὸ σκεπτέον [18] πρῶτον τί ἐστιν ἡ φύσις· οὕτω γὰρ καὶ περὶ τίνων ἡ φυσικὴ [19] δῆλον ἔσται [καὶ εἰ μιᾶς ἐπιστήμης ἢ πλειόνων τὰ αἴτια καὶ [20] τὰς ἀρχὰς θεωρῆσαί ἐστιν].

---

[8] See Ross 1924: 210–11 and Crubellier 2012: 331 n. 83.
[9] See Bonitz 1847: 88–89.

Therefore one must be already trained to know how to take each subject matter, since it is absurd to seek at the same time knowledge and the method of acquiring knowledge; and neither is easy to get. The minute accuracy of mathematics is not to be demanded in all cases, but only in the case of things that have no matter. Therefore its method is not that of natural science; for presumably all nature has matter. Hence we must inquire first what nature is: for thus we shall also see what natural science treats of [and whether it belongs to one science or to more to investigate the causes and the principles of things.]

14 οὐδὲ θάτερον β, cf. Al.ᵖ 168.25 οὐδὲ γὰρ τὸ ἕτερον, Ross Jaeger : οὐδέτερον **a** Bekker Bonitz Christ || 15 τὴν δ' ἀκριβολογίαν] τὴν ἀκριβολογίαν γὰρ Al.ᶜ 169.4-5 || 17 τρόπος ω^{αβ} Al.^{γρ} edd. : λόγος Al.ᶜ 169.9 || ἅπασα γὰρ ἴσως] ἴσως ἅπασα γὰρ ? Al.ᶜ 169.17-18 || 18 ἡ φύσις β Al.ᶜ 169.20 et 137.15 edd. : φύσις **a** || τίνων β Al.ᶜ 169.21 et 137.16 Bonitz Ross Jaeger : τίνος **a** Bekker Christ || 19-20 καὶ ... ἐστιν ω^{αβ} Arᵘ (*Scotus*) Bekker Schwegler : om. ω^{AL} del. Bonitz Christ Ross Jaeger

Different subject matters require different methods. One ought to know these methods *before* one starts to treat a subject, because it is impossible to learn about a subject while at the same time learning about the proper method of dealing with the subject. In order to know the proper method, however, one has to have some familiarity with the subject matter to be dealt with. One has to know what kind of subject matter nature is, in order to see what method of study would be appropriate to it. Knowing that nature includes matter, which excludes presicion, one knows also that mathematics, which is precise, is inappropriate to it.

In all of our manuscripts (ω^{αβ}), these thoughts are followed by a sentence seemingly unfitting with them.[10] This sentence states that the natural scientist also has to determine whether it belongs to one science or to more than one to investigate the causes and the principles of things (καὶ εἰ μιᾶς ἐπιστήμης ἢ πλειόνων τὰ αἴτια καὶ τὰς ἀρχὰς θεωρῆσαί ἐστιν, a19–20). This last sentence of book α ἔλαττον is excised in the *Metaphysics* editions from Bonitz's 1848 edition onwards.[11]

The information we find in Alexander's commentary strongly speaks in favor of this deletion. There are two passages in Alexander's commentary that indicate that the additional sentence at the end of book α ἔλαττον was not found in ω^{AL}. The first passage is part of Alexander's introduction to book α ἔλαττον. There he discusses the correct position of book α within the treatise of the *Metaphysics* and even whether it belongs to the *Metaphysics* at all (137.1–138.9). He compares the endings of book A and α, in order to find out which of the two offers a proper transition to book B.[12] After quoting and discussing the last sentence of book A

---

[10] Here I concentrate on the wording in lines 995a19–20. For a consideration of the textual situation of lines a16–17 see 3.5.2.4.

[11] Schwegler 1847, who sporadically corrects Bekker's text according to the evidence in Alexander, does not mark lines a19–20 as an interpolation.

[12] Book A, Alexander concludes, offers the better transition to book B. See *In Metaph.* 138.2-6: διὸ δόξει τῷ μείζονι A τὸ B μᾶλλον ἀκολουθεῖν· συνῳδὸς γὰρ ἡ τούτου ἀρχὴ τῷ ἐκείνου τέλει. ἐκεῖ τε γὰρ προέθετο περὶ ὧν ἀπορήσειεν ἄν τις περὶ τῶν εἰς τὰς ἀρχὰς καὶ τὴν εὕρεσιν αὐτῶν συντεινόντων εἰπεῖν, ἔν τε τῷ B φαίνεται τοῦτο ποιῶν. / "Hence it would seem that B rather than α ἔλαττον follows

(137.8–9),[13] Alexander quotes the concluding sentence (ἐπαύσατο, 137.15) of book α. He says: διὸ σκεπτέον πρῶτον τί ἐστιν ἡ φύσις· οὕτω γὰρ καὶ περὶ τίνων ἡ φυσικὴ δῆλον ἔσται.[14] We see that this sentence is not the last sentence of book α that our direct transmission testifies to. This proves that the interpolation at the end of book α was not contained in ω^AL.[15]

The second passage in Alexander's commentary that indicates that the last additional sentence of book α was not part of the ω^AL we find in his comments on the final words of book α. There, Alexander does not mention the content of the suspicious last sentence of ω^αβ (169.19–170.11). Since Alexander's commentary offers a thorough analysis of the *Metaphysics* text and comments, as far as possible, on every sentence, his total silence about this additional sentence suggests that ω^AL did not contain the sentence at the end of book α.

Alexander's commentary does more than just provide evidence that the additional sentence of book α was not contained in ω^AL. Alexander himself speaks quite explicitly about the origin of the interpolation. In his commentary on the first announcement of the first aporia in B 1, Alexander comes back to the closing clause of book α. In B 1, 995b4–6, we read again, though in slightly different words, the question whether the investigation of the causes belongs to one or more sciences (πότερον μιᾶς ἢ πολλῶν ἐπιστημῶν θεωρῆσαι τὰς αἰτίας).[16] Aristotle intro-

---

A, since the beginning of B is consistent with the conclusion of A; for in the latter Aristotle promises to deal with the difficulties that should be raised about matters relevant to the discovery of the causes, and this is obviously what he does in B."

[13] *Metaph*. A 10, 993a24–27: ὅσα δὲ περὶ τῶν αὐτῶν τούτων ἀπορήσειεν ἄν τις, ἐπανέλθωμεν πάλιν· τάχα γὰρ ἂν ἐξ αὐτῶν εὐπορήσαιμέν τι πρὸς τὰς ὕστερον ἀπορίας. / "But let us return to enumerate the difficulties that might be raised on these same points; for perhaps we may get some help towards our later difficulties." For an analysis of the concluding sentence of A 10 see Laks 2009: 27–34 and Cooper 2012: 351–54. Cf. also Jaeger 1912: 17–19.
Alexander commented on the last sentence of book A already in his commentary on book A (136.8–17). There, Alexander by no means draws the conclusion that book B is the direct sequel to book A. Rather, he regards the announced aporiae as treated in both α and B. Cf. Dooley 1992: 10–11 n. 7 and Cooper 2012: 352 n. 34.

[14] "Hence we must inquire first what nature is: for thus we shall also see what natural science treats of." It is reasonable to take the actual last sentence of book α (995a17–19) as an introduction to an investigation into natural science (or the *Physics*). This is how Jaeger 1912: 114–18 takes it. Cf. also Ross 1924: 213 and 221; Gigon 1983: 218–19; Szlezák 1983: 241–45. Also Alexander asks critically whether this sentence fits the *Metaphysics* at all. He offers two possible interpretations of the sentence (169.19–170.4): Either (i) the sentence introduces the study of nature; book α therefore does not belong to the *Metaphysics*, but is an introduction to theoretical philosophy in general. Or (ii) the sentence fits just right and it simply describes the task of another discipline, which is to be distinguished from the task of the *Metaphysics*.

[15] One could speculate about a possible motivation for the addition of the sentence in ω^αβ. Someone might have intended to make book α ἔλαττον end in such a way that it offers a better transition to book B than the actual last sentence does.

[16] Cf. the description and discussion of the first aporia in B 2, 996a18–20: πότερον μιᾶς ἢ πλειόνων ἐστὶν ἐπιστημῶν θεωρῆσαι πάντα τὰ γένη τῶν αἰτίων.

duces this aporia in the following way (995b4–5): ἔστι δ' ἀπορία πρώτη μὲν περὶ ὧν ἐν τοῖς πεφροιμιασμένοις διηπορήσαμεν. / "The first problem concerns the subject which we discussed in our prefatory remarks."[17] Alexander comments that a misunderstanding of this introductory sentence led to the addition of a sentence at the end of book α.

> Alexander, *In Metaph.* 174.5–6; 25–27 Hayduck
>
> 995b4 Ἔστι δ' ἀπορία πρώτη μὲν περὶ ὧν ἐν τοῖς πεφροι-[6]μιασμένοις διηπορήσαμεν.
> ... τινὲς μέντοι διὰ τὸ νῦν εἰρημένον ὑπ' αὐτοῦ ἐπὶ τέλει τοῦ τῶν [26] Α ἐλάττονος αὐτὴν ταύτην τὴν ἀπορίαν προσγράφουσι, κατ' οὐδένα λόγον ἐκεῖ [27] κειμένην.
>
> The first problem concerns the subject which we discussed in our prefatory remarks.
> ... [*After a discussion about the different interpretations of this clause Alexander adds the following:*] Some, however, on account of the statement Aristotle has just made, insert this very aporia at the end of α ἔλαττον, although it is positioned there without good reason.
>
> ---
> 26 αὐτὴν ταύτην τὴν ἀπορίαν **A O P**ᵇ **S**(*hanc ipsam dubitationem*) : αὐτὴν ταύτην ἀπορίαν **L** : ταύτην τὴν ἀπορίαν Hayduck Bonitz

Alexander says that "some" misinterpreted Aristotle's reference back to his "prefatory remarks" (B 1, 995b4–5). These interpreters related Aristotle's reference not to the subject of the aporia, namely the causes, which indeed were treated in book A.[18] Instead, they understood the reference to bear on an earlier treatment of this very aporia.[19] Alexander thus explains the additional sentence at the end of book α as an interpolation that some interpreters intended as a correction of the Aristotelian text (174.25–26: τινὲς ... προσγράφουσι, cf. 3.6). The intention was simply to create a referent to which Aristotle's back-reference in B 1 (understood as a reference to a treatment of the aporia) could relate.

Alexander does not explicitly quote the text of the interpolation. Nevertheless, we are allowed to assume that Alexander speaks about the interpolation we find in our text ω^{αβ}, because Alexander precisely describes its content as well as its place in the text (25–26: ἐπὶ τέλει τοῦ τῶν Α ἐλάττονος). As a result, we can conclude that the false addition at the end of book α was part of a textual tradition before AD 200.[20] It is reasonable to assume that our text ω^{αβ}, the *terminus ante quem*

---

[17] See also Laks 2009: 28–29, who argues that we find in book A not only the subject matter, that is, the causes, but also the question regarding to which science these belong. He points to A 2, 982b7–10: ἐξ ἁπάντων οὖν τῶν εἰρημένων ἐπὶ τὴν αὐτὴν ἐπιστήμην πίπτει τὸ ζητούμενον ὄνομα· δεῖ γὰρ ταύτην τῶν πρώτων ἀρχῶν καὶ αἰτίων εἶναι θεωρητικήν.

[18] See Alex. *In Metaph.* 174.7–25.

[19] See Ross 1924: 224.

[20] Asclepius's exemplar (early sixth century AD) also read the addition. Commenting on the in-

of which is the end of the fourth century and which shares this mistake, derives from (or was influenced by) this tradition.[21] By contrast, ω^AL (a text from around AD 200) does not share this mistake. Since we are dealing here with a separative error in ω^αβ that did not occur in ω^AL, we can infer that ω^AL is independent of the tradition of ω^αβ.

4.1.2 Alex. *In Metaph.* 264.28–35; 265.6–9 on Arist. *Metaph.* Γ 3, 1005a19–23

At the beginning of the third chapter of book Γ, Aristotle repeats the question he raised as second aporia in book B (B 1, 995b6–10 and B 2, 996b26–997a15) and then answers it. *Do the principles of demonstration belong to the same science as the principles of substance?* Before we look at how Aristotle answers this question, we should have a look at how he formulates it. In B 1, Aristotle gives the following statement of aporia 2:

Aristotle, *Metaphysics* B 1, 995b6–8

καὶ πό-[7]τερον τὰς τῆς οὐσίας ἀρχὰς τὰς πρώτας ἐστὶ τῆς ἐπιστήμης [8] ἰδεῖν μόνον ἢ καὶ περὶ τῶν ἀρχῶν ἐξ ὧν δεικνύουσι ἅπαντες …

… and, whether this science should survey only the first principles of substance, or also the principles on which all men base their proofs …

The formulation given in B 2 (where the proper discussion of the aporia occurs), is by comparison somewhat extended. Aristotle says:

Aristotle, *Metaphysics* B 2, 996b26–27; 31–33

ἀλλὰ μὴν καὶ περὶ τῶν ἀποδεικτικῶν ἀρχῶν, πότερον [27] μιᾶς ἐστὶν ἐπιστήμης ἢ πλειόνων, ἀμφισβητήσιμόν ἐστιν … πότερον μία τούτων ἐπιστήμη καὶ τῆς οὐσίας ἢ [32] ἑτέρα, κἂν εἰ μὴ μία, ποτέραν χρὴ προσαγορεύειν τὴν ζη-[33]τουμένην νῦν.

But, regarding the starting-points of demonstration also, it is a disputable question whether they are the object of one science or of more. … the question is whether the same science deals with them as with substance, or a different science, and if it is not one science, which of the two must be identified with that which we now seek.

---

troduction of the first aporia in B 1 (995b4) Asclepius writes (Ascl. *In Metaph.* 140.22–27): ἔστιν οὖν, φησίν, ἀπορία πρώτη μὲν περὶ ὧν ἐν τοῖς πεφροιμιασμένοις διηπορήσαμεν· ἠπόρησε γὰρ πρὸς τῷ τέλει τοῦ ἐλάττονος Α, ἡνίκα ἔλεγεν "εἰ μιᾶς ἐπιστήμης [ἐστὶν] ἢ πλειόνων τὰ αἴτια καὶ τὰς ἀρχὰς θεωρῆσαί ἐστι." τοῦτο οὖν καὶ ἐνταῦθα λέγει, ὅτι πότερον μιᾶς ἐπιστήμης ἢ πολλῶν ἐστι τὸ θεωρῆσαι τὰς αἰτίας. / "'The first problem concerns,' as he says, 'the subject which we discussed in our prefatory remarks.' For, he posed this aporia at the end of α ἔλαττον when he said 'whether it belongs to one science or to more to investigate the causes.'" Scotus's translation suggests that the Arabic text contained the additional sentence as well.

[21] Concerning Alexander's acquaintance with the version of ω^αβ see 3.5.2.

Here Aristotle introduces the aporia by way of the more general question whether the axioms are the object of one or more sciences,[22] and from there focuses on the question whether the axioms belong to the same science as substance.

Let us then look at the beginning of Γ 3. According to our manuscript tradition (ω^{αβ}) the text reads as follows:

Aristotle, *Metaphysics* Γ 3, 1005a19–23

λεκτέον δὲ πότερον μιᾶς ἢ ἑτέρας ἐπιστήμης περί τε [20] τῶν ἐν τοῖς μαθήμασι καλουμένων ἀξιωμάτων καὶ περὶ [21] τῆς οὐσίας. φανερὸν δὴ ὅτι μιᾶς **τε καὶ τῆς τοῦ φιλοσόφου** [22] καὶ ἡ περὶ τούτων ἐστὶ σκέψις· ἅπασι γὰρ ὑπάρχει τοῖς [23] οὖσιν ἀλλ᾽ οὐ γένει τινὶ χωρὶς ἰδίᾳ τῶν ἄλλων.

We must state whether it belongs to one or to different sciences to inquire into those things which are called in mathematics axioms, and into substance. It is evident, then, that also the inquiry into these belongs to one science, **and that science is the philosopher's**; for these axioms hold good for everything that is, and not for some special genus apart from others.

21 τε καὶ τῆς τοῦ φιλοσόφου ω^{αβ} Ar^u (*Scotus*) edd. : καὶ τῆς αὐτῆς ω^{AL} (Al.^{l} 264.30, Al.^{p} 264.34–35, 265.6–9) || 22 ἐστὶ σκέψις α edd. : ἐπίσκεψις β

Axioms belong to the same science as substance, and that science is philosophy. Aristotle's reasoning is as follows: in Γ 2 he made clear that the science of οὐσία is philosophy,[23] and here in Γ 3 he points out that the axioms concern all that is, and the science of the axioms belongs to the science of substance. When we turn to the closing paragraph of aporia 2 in B 2, 997a12–15, we see Aristotle hint at this affirmative answer still to be given in Γ 3. In B 2 he says: "The axioms are most universal and are principles of all things. And if it is not the business of the philosopher, to whom else will it belong to inquire into what is true and what is untrue about them?"[24]

[22] For the status of this question see Madigan 1999: 40 and Crubellier 2009: 63–64.

[23] Aristotle stated it one time indirectly (Γ 2, 1003b16–19): δῆλον οὖν ὅτι καὶ τὰ ὄντα μιᾶς θεωρῆσαι ᾗ ὄντα. πανταχοῦ δὲ κυρίως τοῦ πρώτου ἡ ἐπιστήμη, καὶ ἐξ οὗ τὰ ἄλλα ἤρτηται, καὶ δι᾽ ὃ λέγονται. εἰ οὖν τοῦτ᾽ ἐστὶν ἡ οὐσία, τῶν οὐσιῶν ἂν δέοι τὰς ἀρχὰς καὶ τὰς αἰτίας ἔχειν τὸν φιλόσοφον ("It is clear then that it is the work of one science also to study all things that are, *qua* being. But everywhere science deals chiefly with that which is primary, and on which the other things depend, and in virtue of which they get their names. If, then, this is substance, it is of substances that the philosopher must grasp the principles and the causes") and one time directly (Γ 2, 1004a31–1004b1): φανερὸν οὖν [ὅπερ ἐν ταῖς ἀπορίαις ἐλέχθη] ὅτι μιᾶς περὶ τούτων καὶ τῆς οὐσίας ἐστὶ λόγον ἔχειν (τοῦτο δ᾽ ἦν ἓν τῶν ἐν τοῖς ἀπορήμασιν), καὶ ἔστι τοῦ φιλοσόφου περὶ πάντων δύνασθαι θεωρεῖν ("It is evident then that it belongs to one science to be able to give an account of these concepts as well as of substance. This was one of the questions in our book of problems. And it is the function of the philosopher to be able to investigate all things").

[24] B 2, 997a12–15: καθόλου γὰρ μάλιστα καὶ πάντων ἀρχαὶ τὰ ἀξιώματά ἐστιν, εἴ τ᾽ ἐστὶ μὴ τοῦ φιλοσόφου, τίνος ἔσται περὶ αὐτῶν ἄλλου τὸ θεωρῆσαι τὸ ἀληθὲς καὶ ψεῦδος; Cf. Crubellier 2009: 69–70.

The text of Γ 3, 1005a21–22 in our manuscript tradition is different from that in Alexander's exemplar. According to Alexander's text, Aristotle does not say that the axioms and the substance are studied by one science, namely the philosopher's, but just that they are studied by *one and the same science*:

Alexander, *In Metaph.* 264.28-35; 265.6-9 Hayduck

Λεκτέον δὲ πότερον μιᾶς ἢ ἑτέρας ἐπιστήμης περί τε [29] τῶν ἐν τοῖς μαθήμασι καλουμένων ἀξιωμάτων καὶ περὶ τῆς [30] οὐσίας. φανερὸν δὴ ὅτι <u>μιᾶς καὶ τῆς αὐτῆς</u>.[25]
[31] Τῶν ἐν τῷ δευτέρῳ[26] κειμένων ἀποριῶν μέμνηται νῦν· ἔστι δὲ αὕτη, [32] πότερον τὰς τῆς οὐσίας ἀρχὰς τὰς πρώτας ἐστὶ τῆς ἐπιστήμης τῆς προκει-[33]μένης ἰδεῖν μόνον, ἢ καὶ περὶ τῶν ἀρχῶν ἐξ ὧν δεικνύουσι πάντες, ἃ [34] ἀξιώματά εἰσιν· περὶ ὧν ζητεῖ νῦν εἰ <u>τῆς αὐτῆς ἐπιστήμης</u> ἐστὶ περί τε [35] τῆς οὐσίας θεωρεῖν καὶ περὶ ἐκείνων. … [264.35–265.6] λέγει δὲ <u>μιᾶς καὶ τῆς αὐτῆς</u> εἶναι ἐπιστήμης τήν τε περὶ [7] οὐσίας τε καὶ τοῦ ὄντος θεωρίαν καὶ τὴν περὶ τῶν ἀξιωμάτων· αὕτη δέ [8] ἐστιν ἡ πρώτη φιλοσοφία· καὶ ὅτι <u>τῆς αὐτῆς</u> ἐστι δείκνυσι διὰ τοῦ πᾶσιν [9] αὐτὰ τοῖς οὖσιν ὑπάρχειν, ἀλλ' οὐκ ἀφωρισμένῳ τινὶ τοῦ ὄντος γένει.

We must state whether it belongs to one or to different sciences to inquire into those things which are called in mathematics axioms, and into substance. It is evident, then, that it belongs <u>to one and the same science</u>.
He now mentions the aporiae posited in the second book. This aporia is, whether it belongs to the proposed science to take in only the primary principles of substance, or to take in the principles which all use to prove things, i.e. the axioms, as well. Concerning these he now inquires whether it belongs <u>to the same science</u> to consider both substance and the axioms. … [*Explanation about what is meant by axioms*] He says that the consideration of substance and being and the consideration of the axioms belong <u>to one and the same science</u>. This science is first philosophy. That they belong <u>to the same science</u> he shows by way of the fact that the axioms belong to all beings, not to some determinate genus of being.

---

28 ἢ ἑτέρας ἐπιστήμης O P^b F (cf. *Metaph.*) : ἐπιστήμης A : ἐπιστήμης ἢ ἑτέρας L Hayduck || 29 ἐν A O : om. P^b || 30 καὶ τῆς αὐτῆς A O S : τε καὶ τῆς φιλοσόφου P^b || 31 δευτέρῳ Hayduck : β´

---

[25] Sepúlveda's Latin translation (§ 21, f. p.iv.v) of the lemma agrees with **A** and **O** in reading a text that differs from our *Metaphysics* text, but which will turn out to be the reading that Alexander found in ω^AL. By contrast, the commentary manuscript **P^b** reads the lemma-text in agreement with our direct *Metaphysics* tradition. The reading attested to by **A**, **O**, and **S** is the *lectio difficilior* and, as will be seen below, is confirmed by Alexander's paraphrase (unanimously attested to by **A**, **O**, and **P^b**). It is likely that the lemma in **P^b** was later adjusted to the *Metaphysics* text.

[26] Alexander counts B as the second book, since he regards α as something like an appendix to book A. See his introduction to his commentary on book α: 137.2–9: ὡς μέρος βιβλίου … δόξει καὶ οὐκ ἀπᾴδειν τοῦτο τοῦ μείζονος A, ἀλλ' ἕπεσθαι ἐκείνῳ and 138.2–6: διὸ δόξει τῷ μείζονι A τὸ B μᾶλλον ἀκολουθεῖν. Further, Alexander's back reference in 184.14–16 suggests that, according to his understanding, book B directly follows upon book A. At the end of book A, however, Alexander does speak of α ἔλαττον as if it were the next book (136.12–17). See also Dooley 1992: 10–11 n. 7.

A O : βῆτα P^b || 32 ἐστὶ add. A : om. O P^b S || 33 μόνον add. A (cf. Metaph. 995b8) : om. O P^b S || 7 τε LF : om. A O P^b || 8 τοῦ O LF : τὸ A P^b

The text in the lemma[27] indicates that Alexander's reading differs from that which we find in ω^αβ: In the lemma we read (264.30) μιᾶς καὶ τῆς αὐτῆς / "one and the same [science]"[28] instead of the words μιᾶς τε καὶ τῆς τοῦ φιλοσόφου / "one [science], and [the science] of the philosopher."[29] That this reading is in fact the reading of ω^AL is confirmed by two passages in Alexander's paraphrase (264.34 and 265.6).

In lines 265.6-8, Alexander paraphrases Aristotle's words in 1005a21-22, as indicated by the word λέγει / "he [Aristotle] says": λέγει δὲ μιᾶς καὶ τῆς αὐτῆς εἶναι ἐπιστήμης. / "He says that [the considerations ...] belong to one and the same science." Alexander then explains what is meant by "one and the same science." In lines 265.7-8 he says: αὕτη δέ ἐστιν ἡ πρώτη φιλοσοφία. / "This science is first philosophy." There is nothing to suggest that this short remark is a paraphrase of Aristotle's text. The expression ἡ πρώτη φιλοσοφία is quite different from καὶ τῆς τοῦ φιλοσόφου, which we find in ω^αβ.[30] The words αὕτη δέ ἐστιν ἡ πρώτη φιλοσοφία are rather one of Alexander's typical explanatory additions,[31] wherein Alexander spells out what is only implicit in Aristotle.

After this paraphrase and clarification, Alexander goes on to analyze Aristotle's argument (1005a22-23) as to why it belongs to one and the same science to inquire into substance and the axioms: 265.8-9: ὅτι τῆς αὐτῆς ἐστι δείκνυσι... / "That they belong to the same science he shows." This time the δείκνυσι makes clear that Alexander reports Aristotle's thought. Although we do not find the phrase μιᾶς καὶ τῆς αὐτῆς in its entirety, Alexander says τῆς αὐτῆς, which works to confirm the reconstruction of the reading of ω^AL once more. From these statements we can infer that the reading of Alexander's copy is identical with the reading given in the lemma and that this reading differs from ω^αβ accordingly.

This difference between ω^AL and our text has not been noted by any of the editors of the *Metaphysics*.[32] Let us, then, compare the two versions with each other:

---

[27] This holds for the text in the commentary manuscripts **A**, **O**, and **S** (see p. 108 n. 25).

[28] That Alexander's lemma does not read the last part of the sentence in 1005a22 should not disturb us too much. The omitted words (καὶ ἡ περὶ τούτων ἐστὶ σκέψις) are not necessary to understand the sentence. Alexander's paraphrase in 265.6-7 indicates that he read them in his text.

[29] The Greek *Vorlage* on which Usṭāth's Arabic translation is based also contained the ω^αβ-reading. Scotus writes: *Manifestum est igitur quod consideratio de istis est unius scientie, scilicet scientie philosophi.*

[30] As for the expression ἡ πρώτη φιλοσοφία, Alexander might have drawn inspiration from lines 1005a33-b2, where Aristotle mentions πρώτη σοφία.

[31] Alexander regularly puts such explications in his report of Aristotle's word. He introduces them by the formula demonstrative pronoun + δέ / γάρ ἐστιν: e.g. 63.3-5: λέγει μὲν περὶ τῆς κατὰ τὸ τέλος αἰτίας (αὕτη γάρ ἐστι τἀγαθόν τε καὶ τὸ ὡς τέλος αἴτιον τοῖς οὖσι), δείκνυσι δὲ ὅτι οὐδεὶς οἰκείως τῶν πρὸ αὐτοῦ περὶ ταύτης τῆς αἰτίας ἐποιήσατο τὸν λόγον. See also 143.14-15; 181.36-37; 389.1-3.

[32] Even the newest editions of *Metaphysics* Γ (Cassin/Narcy 1989 and Hecquet-Devienne 2008) do

ω^αβ   φανερὸν δὴ ὅτι μιᾶς **τε καὶ τῆς τοῦ φιλοσόφου** καὶ ἡ περὶ τούτων ἐστὶ σκέψις·

ω^AL   φανερὸν δὴ ὅτι μιᾶς **καὶ τῆς αὐτῆς** καὶ ἡ περὶ τούτων ἐστὶ σκέψις·

It seems as though either version could be right. As for the expression given in ω^AL, Aristotle uses the expression "one and the same" (εἷς/μία/ἕν + form of ὁ αὐτός) often and in various contexts.[33] As for the version given in ω^αβ, we observe that although there is no exact parallel in other parts of Aristotle's works, the connection of μιᾶς and τῆς τοῦ φιλοσόφου by the words τε καί fulfills the idiomatic requirement according to which the use of τε καί demands that the two connected terms are closely related.[34] That a numeric adjective (μιᾶς) and an attributive genitive (τῆς τοῦ φιλοσόφου) are connected by τε καί is no objection against the phrase, as there are some other cases in Aristotle where words from different classes are connected by τε καί.[35]

Both the reading of ω^AL and the reading of ω^αβ can claim authenticity, and due to the context they mean more or less the same thing. The *one* science which inquires into substance and into the axioms is philosophy. In order to settle between these readings it is necessary to address the question whether Aristotle explicitly used the term "philosopher" here (as given in the ω^αβ); or whether he restricted his answer solely to the question he posed in line a19 (πότερον μιᾶς ἢ ἑτέρας ἐπιστήμης) and stated simply that the same science studies substance and the axioms. When we examine the immediate context of lines 1005a21–22, we see that the subsequent γάρ-clause fits both viable readings. The fact that the axioms hold good for everything connects them with the οὐσία named in line a21 just as well as it connects them, if we follow the ω^αβ-reading, with the philosopher, for philosophy was said in Γ 2[36] to be the science of being *qua* being.

Let us then look at the broader context of our passage. In the opening of Γ 3, Aristotle states and then answers the second aporia: both substance and axioms belong to one science (which ω^αβ declares immediately to be philosophy)

---

not mention it. This is especially surprising in the case of Hecquet-Devienne 2008, who says that she intensively examined Alexander's commentary (39–53). Madigan 1993: 153 n. 228 alone notes it in his translation of Alexander's commentary.

[33]The following passages are from the *Metaphysics*: A 9, 991a5; B 1, 995b9; B 4, 999a28; Γ 4, 1007a5; Δ 6, 1016a31; Z 4, 1029b22; Z 6, 1031b19; Z 14, 1039a28; H 6, 1045b18–19. In *Metaphysics* B (and K) we find the phrase "one and the same science" (cf. B 2, 997a22–25; see K 3, 1061a18–20: τὰ ἐναντία πάντα τῆς αὐτῆς καὶ μιᾶς ἐπιστήμης θεωρῆσαι; 1061b1–3: μίαν πάντων καὶ τὴν αὐτὴν τίθεμεν ἐπιστήμην τὴν γεωμετρικήν). Cf. also Bonitz 1870: s.v. εἷς, μία, ἕν; p. 223.

[34]Kühner/Gerth II: § 522, p. 249: "τε … καί (…) drücken aus, dass das erstere und das durch καί hinzugefügte Glied in einer innigen oder notwendigen Verbindung mit einander stehen."

[35]A comparison with other uses of τε καί in the *Metaphysics* shows that the two connected terms are of the same word class: ἑτέρας τε καὶ ἐναντίας 985a31; γῆ τε καὶ ἀέρι 985b2; σχῆμά τε καὶ τάξιν 985b14. However there are also a few uses where the two terms are of different word classes: ἀρχαῖόν τε καὶ πάντες ὡμολόγησαν 984a33; ἀρχή τε καὶ μᾶλλον 995b31.

[36]Cf. Γ 1; Γ 2, 1003b16–19; Γ 2, 1004a31–1004b2.

(1005a19–22). Aristotle gives the following explanation for this. Axioms hold for everything that is (1005a22–23), and all scientists use them for their proofs within their respective fields (1005a23–27). Nevertheless, the investigation of the axioms falls under the science that investigates being *qua* being (τοῦ περὶ τὸ ὂν ᾗ ὂν γνωρίζοντος καὶ περὶ τούτων ἐστὶν ἡ θεωρία, 1005a28–29). The axioms do *not* fall under a special science, despite the claim of some natural philosophers that they inquired into the whole of nature (1005a29–33). Since first *sophia* precedes physics, the axioms belong to this *sophia* (1005b4–5). Aristotle then concludes and transitions over to the next paragraph by stating that it is the *philosopher* who inquires into the axioms:[37]

Aristotle, *Metaphysics* Γ 3, 1005b5–8

ὅτι μὲν [6] οὖν τοῦ φιλοσόφου, καὶ τοῦ περὶ πάσης τῆς οὐσίας θεωροῦντος [7] ᾗ πέφυκεν, καὶ περὶ τῶν συλλογιστικῶν ἀρχῶν ἐστὶν ἐπι-[8]σκέψασθαι, δῆλον.

Evidently then the philosopher, who is studying the nature of all substance, must inquire also into the principles of deduction.

Having considered the whole of the first paragraph of Γ 3, what can we infer about the reading at its beginning? Did Aristotle state already at the beginning that the one science of both substance and the axioms is philosophy? Aristotle presents his thought in such a way as to suggest that the identification of the philosopher as the one responsible for the study of both, substance and the axioms, comes as a conclusion to the argument. For that reason it is unlikely that he declared the philosopher as such already in lines 1005a21–22. These considerations point to the authenticity of the reading in ω^AL.[38]

Is a means of settling this difficulty conclusively available to us? Does a comparison of Aristotle's answers to other aporiae in the first chapters of book Γ offer further support for the ω^AL-reading? In Γ 2, 1004a31–b1, we read the following answer to the fourth aporia, which concerns the question whether it belongs to one science to investigate the substance and the *per se* attributes:[39]

Aristotle, *Metaphysics* Γ 2, 1004a31–1004b1 (on aporia 4)

φανερὸν [32] οὖν [ὅπερ ἐν ταῖς ἀπορίαις ἐλέχθη] ὅτι μιᾶς περὶ τού-[33]των καὶ τῆς οὐσίας ἐστὶ λόγον ἔχειν (τοῦτο δ' ἦν ἓν [34] τῶν ἐν τοῖς ἀπορήμασιν), καὶ ἔστι τοῦ

---

[37] Cf. Crubellier 2009: 70–72.

[38] It is, however, not outright impossible to understand the given distinction between the philosopher and the physicist to be a supplementary elucidation of an answer that was already given at the beginning of the paragraph. *This* consideration would then speak in favor of the reading of ω^αβ. The formula ὅτι μὲν οὖν … δῆλον does not unambiguously stipulate whether the conclusion results in a new thought (e.g. *de An.* B 3, 415a12–13) or a summary of something that has been said before (e.g. *EE* B 10, 1226a17–18; *Pol.* Θ 1, 1337a22–34).

[39] Cf. 4.3.1.3.

φιλοσόφου περὶ πάν-[1004b1]των δύνασθαι θεωρεῖν.

It is evident then that it belongs to one science to be able to give an account of these [the attributes] as well as of substance. This was one of the questions in our book of problems. And it is the function of the philosopher to be able to investigate all things.

Does this passage, with its mention of the philosopher, speak in favor of the ω^{αβ}-reading in Γ 3, 1005a21–22? On the basis of the formulation of the fourth aporia in B 1 we are inclined to answer in the negative. In the summary of the fourth aporia in B 1 we see that Aristotle not only asked whether the *per se* attributes and "the same" and "the other" and such belong to the science of substance; he also asked explicitly *what* science investigates these attributes: τίνος ἐστὶ θεωρῆσαι περὶ πάντων; / "whose business is it to inquire into all these?" (995b24–25).[40] This question is answered in Γ 2, 1004a31–1004b1: it is the philosopher's task to investigate all these attributes.

By contrast, in B 1 and B 2, among the questions introducing the second aporia, we do not find a question to which the ω^{αβ}-reading in Γ 3 would be the appropriate answer. Therefore, the comparison of our passage in Γ 3 (1005a21) to the answer given to aporia 4 in Γ 2 (1004a31–1004b1) does not at all confirm the reading in ω^{αβ}. Rather, it is possible that the answer to aporia 4 given in Γ 2 was taken as a model, according to which the original reading of our Γ 3 passage (preserved in ω^{AL}) was expanded in ω^{αβ} through the interpolation of the words "and the philosopher's."

So what remains to be done is to ascertain, according to the principle of *utrum in alterum*, which of the two readings can be more easily explained as emerging from the other. An accidental scribal error can safely be ruled out. We deal here with a deliberate intervention in which one of the two readings was changed into the other. Which of the following two is more likely, that someone changed μιᾶς τε καὶ τῆς τοῦ φιλοσόφου to μιᾶς καὶ τῆς αὐτῆς, or changed the latter to the former? The expression μιᾶς καὶ τῆς αὐτῆς (ω^{AL}) is idiomatic in Aristotle, but in this particular usage the words καὶ τῆς αὐτῆς do not supply any additional meaning to the word μιᾶς, which alone suffices to express that the science of substance and the science of the axioms is *one*. By contrast, in ω^{αβ}, in place of the words καὶ τῆς αὐτῆς we find the words τε καὶ τῆς τοῦ φιλοσόφου, which indeed supply the phrase with additional content. This additional content is found again at the end of the passage (see discussion above). The phrase τε καὶ τῆς τοῦ φιλοσόφου therefore would not have been excised for being incorrect, and it is strange to suppose that the phrase was cut in anticipation of a redundancy of content and meaning. It is even more peculiar to suppose that this phrase would be cut and replaced with καὶ τῆς αὐτῆς, which provides no new information at all, which is indeed more rhetorical than anything. What is more probable, then, is that someone was

---

[40] Cf. the discussion in B 2, 997a32–33: τίς ἔσται ἡ θεωροῦσα περὶ τὴν οὐσίαν τὰ συμβεβηκότα;

dissatisfied with the information provided by the expression μιᾶς καὶ τῆς αὐτῆς and accordingly deleted αὐτῆς and added the words τε and τοῦ φιλοσόφου, thus making clear at the outset that this science is philosophy.[41]

Might the corrector have been inspired by Alexander's explanatory remark that the *one* science that investigates both substance and the axioms is "first philosophy" (αὕτη δέ ἐστιν ἡ πρώτη φιλοσοφία, 265.7–8)? We do not have evidence to answer in the affirmative. What speaks in favor of answering in the negative is, first, that the formulation in Alexander's paraphrase differs from the reading in ω^{αβ}, and, second, that the insight expressed in Alexander's remark could be gained by any alert reader of the Aristotelian text, as its content is confirmed by the lines 1005b5–8 of this text.

These arguments considered, I conclude that Alexander's commentary provides evidence that questions the authenticity of the ω^{αβ}-reading in line 1005a19. From Alexander's lemma and his paraphrase we can extract the reading of ω^{AL}. This reading seems to be older than the ω^{αβ}-reading, which appears to be a later correction. If this is the case, we have another example of a separative error in ω^{αβ} of which ω^{AL} is free, and therewith yet another *Metaphysics* passage that can be corrected through Alexander's testimony.

4.1.3 Alex. *In Metaph.* 220.1–4 on Arist. *Metaph.* B 4, 1000a26–32

In B 4, 1000a5–1001a3 Aristotle discusses as the tenth aporia (cf. B 2, 996a2–4) the question whether the principles of perishable things are the same as the principles of imperishable things.[42] If they are the same, Aristotle asks, how is it possible that some things are perishable and others imperishable (1000a7–8)? For Aristotle, the explanation given by Hesiod or other theologians of the difference between eternal divinities and mortal humans is not satisfying. According to myth, the consumption of nectar and ambrosia is decisive for divine status, and so imperishability (1000a9–24). Even the answer given by Empedocles, whom one would readily expect to speak in a more self-consistent way (ὁμολογουμένως αὑτῷ, 1000a25), is unconvincing (1000a24–b20).

Empedocles says that Strife is the cause of destruction, but Aristotle points out that it must also be a cause of generation (for all things are generated through the destruction of the One). At the same time, Empedocles says that Love is the cause of generation, but Aristotle points out that it must also be a cause of destruction (for all things are destroyed in order to generate the One). Finally, Empedocles makes all things perishable, and so his principles cannot explain the existence of imperishable things.[43]

---

[41] Alternatively, the words καὶ τῆς τοῦ φιλοσόφου could have their origin in a marginal note.

[42] Cf. Madigan 1999: 97–107; Wildberg 2009: 159–74.

[43] Whether or not Aristotle's reproach is justified is another question. Empedocles might have agreed with what Aristotle says, for, according to Empedocles' theory, everything but the principles

In explicating Empedocles' mistake, Aristotle concentrates in 1000a26–1000b11 on the principle of Strife (νεῖκος). He criticizes the notion that Strife not only is the cause of destruction but also of genesis (1000a26–29). Aristotle illustrates his point of critique by quoting four verses from Empedocles' *Physics* (lines 269–72a Primavesi,[44] quoted in 1000a29–32). My following remarks concern the words in 1000a28–29, which immediately precede Aristotle's quotation of Empedocles. These words are transmitted differently in ω^{αβ} and ω^{AL}. In order to determine which of the two versions constitutes the better or more authentic text we first have to acquaint ourselves with the transmission of the verse quotation in 1000a29–32. Only after we have found firm ground regarding the wording and context of Empedocles' verse in the *Metaphysics* will we be able to judge between Alexander's testimony and our transmission.

The *Metaphysics* passage according to the direct transmission runs as follows:

Aristotle, *Metaphysics* B 4, 1000a26–32

τίθησι μὲν γὰρ ἀρχήν τινα αἰτίαν [27] τῆς φθορᾶς τὸ νεῖκος, δόξειε δ' ἂν οὐθὲν ἧττον καὶ τοῦτο [28] γεννᾶν **ἔξω τοῦ ἑνός**· ἅπαντα **γὰρ** ἐκ τούτου τἆλλά ἐστι [29] πλὴν ὁ θεός. λέγει γοῦν "ἐξ ὧν πάνθ' ὅσα τ' ἦν ὅσα τ' [30] ἐσθ' ὅσα τ' ἔσται ὀπίσσω, δένδρεά τ' ἐβλάστησε καὶ ἀνέ-[31]ρες ἠδὲ γυναῖκες, θῆρές τ' οἰωνοί τε καὶ ὑδατοθρέμμονες [32] ἰχθῦς, καί τε θεοὶ δολιχαίωνες."

for he maintains that strife is a principle that causes perishing, but none the less, this [Strife], too, would seem to produce **except the One; for** from this [Strife?] come all other things excepting God. At least he says: "From which comes all that was and that is and that will be hereafter: Trees sprang forth, and men and women, and beasts and birds and fish abiding in water, and gods who live for many ages …"

27 δόξειε δ' **α** Al.^{1} 220.1 edd. : ὡς δόξειεν **β** || οὐθὲν **α** Bekker Bonitz Ross Jaeger (οὐδὲν Al.^{1} 220.1 Christ) : οὐδὲν δὲ **β** || 28–29 γεννᾶν ἔξω τοῦ ἑνός· ἅπαντα γὰρ ω^{αβ} Ar.^{u} (*Scotus*) edd. : γεννᾶν. ἐκ γὰρ τοῦ ἑνός ἅπαντα καὶ ω^{AL} (Al.^{1} 220.1 Al.^{c} 220.2–3 Al.^{p} 219.29–34) || 29 ὁ θεός (Al.^{p} 219.33)] θεός Al.^{c} 220.2–3 || 29 πάνθ' edd. : πάντα **α β** || ὅσα τ' ἐσθ' I^b Bekker Bonitz Christ Ross : ὅσα τ' ἐστὶν β Al.^{c} 220,5 (ὅσα τ' ἔστιν Jaeger) : om. **α** || 30 ὀπίσσω **α** Bekker Bonitz Christ Ross : om. β Jaeger || 31 ἠδὲ **α** edd. : τ' ἠδὲ β

I will first look at the verse quotation. Following Martin/Primavesi 1999, the verses that Aristotle quotes in 1000a29–32 can be identified as lines 269–272a from the first book of Empedocles' *Physics*. The *Strasbourg Papyrus* enables us to verify Aristotle and to understand the quotation in its original context. I will briefly comment on these two aspects.[45]

Before the discovery of the *Strasbourg Papyrus* the quotation in Aristotle had been compared with an almost identical quotation preserved by the Neoplatonic

---

(i.e. the four elements and Love and Strive) is perishable (cf. 1000b17–20).

[44] My quotations from Empedocles' *Physica* follow Mansfeld/Primavesi 2012.

[45] For a detailed elucidation see Primavesi 1998.

commentator Simplicius (B 21.9-12 DK) and accordingly judged faulty or inaccurate.[46] The context of the *Metaphysics* passage clearly suggests that the quoted verse illustrates the effect of Strife (1000a26-28). But the parallel citation in Simplicius makes it clear that the verses describe *Love's* agency. The accuracy of Aristotle's quotation was vindicated when Primavesi 1998 demonstrated that the newly discovered verses reveal that Aristotle is not in fact quoting the same lines as Simplicius. It is true that lines 270-72 Primavesi (= *Metaph.* B 4, 1000a29-32) and lines 318-20 Primavesi (= Simp. *In Phys.* 159.22-24 = B 21.10-12 DK) are identical. However, the preceding lines (317 in Simplicius and 269 in Aristotle) differ in the following way:

Simp. *In Phys.* 159.21 = B 21.9 DK
ἐκ τούτων γὰρ πάνθ' ὅσα τ' ἦν ὅσα τ' ἔστι καὶ ἔσται   *Physics* I, 317

Arist. *Metaph.* B 4, 1000a29-30[47]
ἐξ ὧν πάνθ' ὅσα τ' ἦν ὅσα τ' ἐσθ' ὅσα τ' ἔσται ὀπίσσω   *Physics* I, 269

P.Strasb. a(i) 8
[ἐξ ὧν πάνθ' ὅσα τ' ἦν ὅσα τ' ἐσθ' ὅ]σα τ' ἔσσετ' ὀπίσσω   *Physics* I, 269

The papyrus clearly shows that the verse in Aristotle is not the verse we find in Simplicius. What is more, Aristotle's quotation can help to restore the first half of the verse, which is missing from the papyrus. The papyrus verse ends with ὀπίσσω, and so we know for metrical reasons that it has to begin with ἐξ ὧν (as it does in Aristotle) and not with ἐκ τούτων γάρ (as it is in Simplicius).[48] As pointed out above, the preceding verse in Simplicius deals with Love's agency (Simp. *In Phys.* 159.20a = 316a Primavesi: σὺν δ' ἔβη ἐν Φιλότητι), yet the context of Aristotle's verse treats the effect of Strife. Since the papyrus gives us the endings of the line (268) that precedes those quoted by Aristotle, it verifies that Aristotle quotes from a passage that deals with Strife.[49]

Empedocles, *Physics* I, 265-72

265   ₁ἀλλ' αὔτ' ἐστιν ταῦτα, δι' ἀλλήλων₁ γε θέοντα
266   ₁γίγνεται ἄλλοτε ἄλλα καὶ ἠνεκὲ₁ς αἰὲν ὁμοῖα.
267   [– ‿‿ – ‿‿ – ‿ συνερχό]μεθ' εἰς ἕνα κόσμον,

[46] Primavesi 1998: 29-30.

[47] A few words on the transmission of this verse within the *Metaphysics* text (see my apparatus): In α the verse is lacking the words ὅσα τ' ἐσθ'. This makes the verse metrically impossible and was most probably caused by *saut du même au même* (ὅσα τ' ἦν <u>ὅσα τ' ἐσθ' ὅσα τ'</u> ἔσται ὀπίσσω). The β-text preserves the right reading (as does Alexander in 220.5), but lacks the ending ὀπίσσω.

[48] See Primavesi 1998: 34-35. As Primavesi points out, the only (metrically relevant) difference between the verse in Aristotle and the papyrus, namely the difference between ἔσται and ἔσσετ', should be taken as "sekundäre Normalisierung" in the Aristotelian text.

[49] The papyrus also shows clearly that Simplicius quotes from a passage of Empedocles' poem quite removed from the passage Aristotle quotes.

268 [– ⏑⏑ – ⏑⏑ – διέφυ πλέ]ον' ἐξ ἑνὸς εἶναι,
269 ⌊ἐξ ὧν πάντ' ὅσα τ' ἦν ὅσα τ' ἔσθ' ὅ⌋σα τ' ἔσσετ' ὀπίσσω·
270 ⌊δένδρεά τ' ἐβλάστησε καὶ ἀνέρες⌋ ἠδὲ γυναῖκες,
271 ⌊θ⌋ῆρές τ' οἰωνοί ⌊τε καὶ⌋ ὑδατοθρ⌊έμμονες ἰχθῦς⌋
272 ⌊κ⌋αί τε θεοὶ δολιχα⌊ίων⌋ες τιμῇσι[ι φέριστοι.]

265 Rather, just these things are: as they run through each other
266 they become different things at different times, yet these are throughout always similar.
267 ... [under love] we [the elements] come together to a single ordered world ...
268 [under strife] we grow apart from each other to become many out of one,
269 —out of which come all beings that were and that are and that will be hereafter
270 trees sprang forth and men and women
271 and beasts and birds and fish abiding in water,
272 and gods who live for many ages and are preeminent in their honors.

Now I turn to Aristotle's train of thought in our *Metaphysics* passage. I will first look at the directly transmitted text ($\omega^{\alpha\beta}$) and then at the evidence in Alexander (ω^{AL}). Aristotle says that Strife, although it is the principle of destruction, none the less brings about all things (τὸ νεῖκος, δόξειε δ' ἂν οὐθὲν ἧττον καὶ τοῦτο γεννᾶν ..., a27–28). The One alone is not a product of Strife (... ἔξω τοῦ ἑνός). Strife brings about everything except the god (*Sphairos*), that is, the world as a state of complete unity (ἅπαντα γὰρ ἐκ τούτου τἆλλά ἐστι πλὴν ὁ θεός, a28–29).

The fragments of Empedocles' *Physics* tell us about a *cosmic cycle*.[50] The reign of Strife lasts for 6000 years.[51] It begins with a cosmic state of complete unity in the spherical god *Sphairos* and it ends with a state of complete separation of the four elements. During this reign, mortal beings and the world as we know it come into existence.[52] The state of complete separation of the four homogeneous masses lasts for 4000 years. After 2000 years of complete separation Love starts to gain strength, and her reign begins 2000 years thereafter. During her rule, which also lasts for 6000 years, the elements gradually unite and heterogeneous combinations come about. Again a zoogony of mortal beings takes place. Love's reign ends with the Sphairos, the complete unification of everything, which lasts for 4000 years.[53]

A passage in *De caelo* Γ 2, 301a14–20 (= DK 31 A 42)[54] shows Aristotle believed

---

[50] For Empedocles' theory see Primavesi 2013: 694–721.
[51] For a reconstruction of the cycle's timetable see Rashed 2001b and Primavesi 2006.
[52] Primavesi 2013: 704–707 and 709–13.
[53] See Primavesi 2013: 705.
[54] *Cael.* Γ 2, 301a14–20: ἐκ διεστώτων δὲ καὶ κινουμένων οὐκ εὔλογον ποιεῖν τὴν γένεσιν. διὸ καὶ Ἐμπεδοκλῆς παραλείπει τὴν ἐπὶ τῆς Φιλότητος· οὐ γὰρ ἂν ἠδύνατο συστῆσαι τὸν οὐρανὸν ἐκ κεχωρισμένων μὲν κατασκευάζων, σύγκρισιν δὲ ποιῶν διὰ τὴν Φιλότητα· ἐκ διακεκριμένων γὰρ συνέστηκεν ὁ κόσμος τῶν στοιχείων· ὥστ' ἀναγκαῖον γίνεσθαι ἐξ ἑνὸς καὶ συγκεκριμένου. / "But it

that Empedocles' cosmic cycle implies that the world as we know it can come about only during the rule of Strife. This understanding forms the background of our *Metaphysics* passage. Aristotle says here that Strife, by dividing up *Sphairos*, brings about all things (i.e. the world), with the exception of *Sphairos*, which is the end product of Love's rule (1000a27–29).

In Alexander's commentary on this passage (220.1–10 and 219.29–34) we find evidence of ω^{AL} in a lemma (220.1), a citation (220.2–3) and a paraphrase (already in 219.29–34). The paraphrase confirms the ω^{AL}-reading that we can reconstruct from the lemma and citation. Hayduck did not recognize that Alexander here quotes verbatim from ω^{AL}.[55] His view might have been distorted by the fact that Alexander's quote differs from our text in wording and punctuation. Hayduck did not print the words spaced out, as is his practice when indicating quotations in Alexander's commentary, but instead placed it in single quotation marks.[56] That we indeed are dealing with a *verbatim* quotation from Alexander's *Metaphysics* text is indicated by his use of the definite article τῷ, by means of which he nominalizes the Aristotelian phrase.[57] In the following excerpt of the commentary text the words that I take as verbatim evidence of ω^{AL} (220.2–3) appear spaced out.

Alexander, *In Metaph.* 220.1–4 Hayduck

Δόξειε δ' ἂν οὐδὲν ἧττον καὶ τοῦτο γεννᾶν.
[2] τουτέστι γεννητικὸν εἶναι καὶ ποιητικόν. τ ῷ δ ὲ ἐ κ γ ὰ ρ τ ο ῦ ἑ ν ὸ ς [3] ἅ π α ν τ α κ α ὶ ἐ κ τ ο ύ τ ο υ τ ὰ ἄ λ λ α ἐ σ τ ὶ π λ ὴ ν θ ε ό ς δεῖ προστιθέναι 'τὰ γιγνό-[4]μενα ὑπὸ τοῦ νείκους.'

But none the less this [strife], too, would seem to produce.
That is, [it would appear to be] generative and productive.[58] To 'For from the One

is unreasonable to start generation from an original state in which bodies are separated and in movement. Hence Empedocles begins after the process ruled by Love; for he could not have constructed the heaven by building it up out of bodies in separation, making them to combine by the power of Love, since our world has its constituent elements in separation, and therefore presupposes a previous state of unity and combination" (transl. by J. L. Stocks)

[55] *Metaphysics* editors since Bonitz recognize the divergence of Alexander's report from our text, but apart from Jaeger they have not spoken of a citation in Alexander. That this citation might even lead us to *another* version of the *Metaphysics* text has not yet been considered. Bonitz and Christ in app. crit.: Al. fort. ἐκ γὰρ τοῦ ἑνὸς ἅπαντα καὶ … . Ross writes "ἐκ γὰρ τοῦ ἑνὸς ἅπαντα καὶ Al.," the abbreviation "Al." standing for Alexander's paraphrase or own formulation. Jaeger alone cites the words as "Al^c" but adds *varias lectiones miscet*. Does Jaeger believe that Alexander himself blends different readings?

[56] It is not entirely clear what Hayduck wants to illustrate with these single quotation marks. Madigan 1992: 168 n. 357 suggests that Hayduck took the words to be a quote from Empedocles. Bonitz 1847 does not mark the words in question at all.

[57] See 3.3.

[58] In later authors τὸ ποιητικόν denotes the *efficient cause*. LSJ s.v. ποιητικός, cf. Plotinus VI, 7, 20,8. In Alexander ἡ ποιητικὴ αἰτία means *efficient cause*, e.g. 22.7–8. Cf. 32.1–9. See 5.2.5.

come all things, and from this the other things, except God' we must add the words
'the things that come to be under the influence of Strife.'

1 δόξειε δ' **P**ᵇ : δόξειεν **A O** ‖ 2 τῷ **O P**ᵇ **LF** : τὸ **A** ‖ 2–3 ἐκ γὰρ τοῦ ἑνὸς ἅπαντα καὶ ἐκ τούτου τὰ ἄλλα ἐστὶ πλὴν θεός] *cuncta nanque caetera ex hac ipsa sunt* **S**⁵⁹ ‖ 2 γὰρ **A O** : om. **P**ᵇ ‖ 3 προστιθέναι **A O** : προστεθῆναι **P**ᵇ

In the lemma the Aristotelian text appears abbreviated. Instead of the sentence δόξειε δ' ἂν οὐθὲν ἧττον καὶ τοῦτο γεννᾶν ἔξω τοῦ ἑνός, as given in ωᵃᵝ (a27–28), Alexander's lemma reads δόξειε δ' ἂν οὐθὲν ἧττον καὶ τοῦτο γεννᾶν. The subsequent (220.2–3) quotation of lines a28–29 contains words that we do not find in the directly transmitted version of the text, but which are a perfect sequel to the seemingly abbreviated text of the lemma. When we connect the words of the lemma with those in the quotation the difference between ωᴬᴸ and ωᵃᵝ is plain to see.⁶⁰

1000a27–29 according to ωᵃᵝ

δόξειε δ' ἂν οὐθὲν ἧττον καὶ τοῦτο γεννᾶν ἔξω τοῦ ἑνός· ἅπαντα γὰρ ἐκ τούτου τἆλλά ἐστι πλὴν ὁ θεός.

1000a27–29 according to ωᴬᴸ

δόξειε δ' ἂν οὐθὲν ἧττον καὶ τοῦτο γεννᾶν. ἐκ γὰρ τοῦ ἑνὸς ἅπαντα καὶ ἐκ τούτου τὰ ἄλλα ἐστὶ πλὴν θεός.

In ωᴬᴸ the first sentence ends with the word γεννᾶν. Instead of ἔξω ("except")⁶¹ we read ἐκ ("out of"). Since the second sentence starts already with ἐκ, γάρ follows as particle. In ωᵃᵝ the second sentence begins with ἅπαντα followed by γάρ. At the same spot in ωᴬᴸ we read καί. Furthermore, in ωᴬᴸ the noun θεός is not preceded by the article. The following translation illustrates the differences between the two versions:

1000a27–29 according to ωᵃᵝ

But none the less, this (Strife), too, would seem to produce except the One. For from this (Strife) come all other things excepting God.

⁵⁹ In Sepúlveda's Latin translation of Alexander's commentary, the quotation from the *Metaphysics* (in 220.2–3) agrees with Sepúlveda's Latin version of the *Metaphysics*, but differs from the reading that we find in the Greek manuscripts of the commentary. On the reliability of lemmata and quotations in Sepúlveda see 2.3.

⁶⁰ This reconstruction of ωᴬᴸ gains support from Alexander's proposed addition of τὰ γιγνόμενα ὑπὸ τοῦ νείκους (220.3–4). Alexander's proposal indicates that in *his* text ἐκ τούτου in a28 refers back to "the One" (τοῦ ἑνός) and that Strife was mentioned only in the previous sentence.

⁶¹ The adverb ἔξω here means "except" (LSJ s.v. ἔξω III.); it is parallel to the expression πλὴν ὁ θεός in the subsequent sentence. The adverb ἔξω can indeed mean with *verbs of motion* "out of" (LSJ s.v. ἔξω I.1.), as in "to go out of the house." The verb γεννᾶν "to generate" does not describe this kind of motion. It is not the case that generation takes place "out of the elements" in a *local sense*. Aristotle does use the combination γεννᾶν ἔξω in *HA* A 1, 487a21, but the context clearly shows that this ἔξω means ἔξω τοῦ ὑγροῦ / "outside of water." Cf. Bonitz 1870: s.v. ἔξω; p. 262b49–263a29.

1000a27–29 according to ω^AL

> But none the less, this (Strife), too, would seem to produce. For from the One come all things and from this (the One) come all other things excepting God.

This reconstruction of the wording in ω^AL is confirmed by Alexander's paraphrase in 219.29–34. There, Alexander writes: ᾗ δὲ λέγει καὶ τὸ νεῖκος αὐτὸ τοῦτο <u>γεννᾶν</u> … (ἐκ γὰρ τοῦ ἑνός, ὃν θεὸν ἐκεῖνος καὶ σφαῖρον λέγει …).[62]

Both the ω^αβ- and ω^AL-versions are grammatically possible.[63] According to the reading in ω^αβ, Aristotle understands Strife as a generating principle (1000a26–29) by pointing out that Strife, despite being the principle of destruction, generates nonetheless. Only the One has to be excluded (ἔξω τοῦ ἑνός) from the list of things generated by Strife. Aristotle further argues that everything other than God (πλὴν ὁ θεός) comes from this (ἐκ τούτου), i.e., Strife. This version's first sentence is striking in that it is somewhat unconnected to the expression ἔξω τοῦ ἑνός. Also striking is this very expression, for Aristotle in the next sentence repeats the very same idea, but this time with the expression πλὴν ὁ θεός.[64] When we further remind ourselves that the verse quotation functions to characterize Strife, it seems odd that Strife does not appear in Empedocles' own words, but only in the words with which Aristotle introduces the quote.[65] Why would Aristotle adduce a quotation that does not provide clear, explicit evidence for his contention?

According to the reading in ω^AL lines 1000a26–32 read as follows:

Aristotle, *Metaphysics* B 4, 1000a26–32 according to ω^AL

τίθησι μὲν γὰρ ἀρχήν τινα αἰτίαν [27] τῆς φθορᾶς τὸ νεῖκος, δόξειε δ' ἂν οὐθὲν ἧττον καὶ τοῦτο [28] **γεννᾶν. ἐκ γὰρ** τοῦ ἑνὸς **ἅπαντα καὶ** ἐκ τούτου τἄλλά ἐστι [29] πλὴν θεός. λέγει γοῦν …

> For he maintains that strife is a principle that causes perishing, but none the less, this [Strife], too, would seem to produce. For from the One come all things and from this [the One] come all other things excepting God. At least he says: …

The statement made in this version of the text differs slightly from the statement

---

[62] *In Metaph.* 219.29–34: "But insofar as he says that this very Strife also begets … for it is out of the One (which he calls God and Sphairos)… ."

[63] For the following I am much indebted to Oliver Primavesi. I further thank Peter Adamson and Christof Rapp for discussing this passage with me.

[64] Such a repetition makes sense only when taken as parallelism, in which Aristotle first speaks in the language of (Empedoclean) *Physics* (ἔξω τοῦ ἑνός), then uses a compatible expression from the (Empedoclean) mythical story (πλὴν ὁ θεός). On the interaction of physics and myth in Empedocles' philosophy see Primavesi 2013: 713–21.

[65] The words ἐξ ὧν with which the citation begins (1000a29) refer to the four elements out of which everything comes to be. See the context in the papyrus (verse 265–69) and Primavesi 2008: 47–57. In the Arabic transmission we find the words ἐξ ὧν / "out of which" replaced by *ex lite* (Scotus) / "out of Strife." This discrepancy can be explained as an attempt to bring in the missing Strife.

in ω^{αβ}. It seems advantageous that the sentence δόξειε δ' ἂν οὐθὲν ἧττον καὶ τοῦτο … (1000a27) ends with γεννᾶν: there is no appended ἔξω τοῦ ἑνός. The fact that Sphairos does not result from the rule of Strife is made sufficiently clear in the subsequent sentence (τὰ ἄλλα … πλὴν θεός). Yet this text too has striking features: Aristotle mentions the One unexpectedly, and Aristotle does not mention Strife, even though Strife is the pivotal element.[66] Having read ἐκ γὰρ τοῦ ἑνός it is impossible to take what follows, ἐκ τούτου, to mean "out of Strife," as was naturally done in the ω^{αβ}-version. In the ω^{AL}-version the καὶ introduces an explication of the ἅπαντα. It means: "For from the One come all things, that is to say (καί),[67] from this come all other things excepting God." Still, why does Aristotle explain the generative power of Strife by pointing out that everything comes out of the One, but never again mentions Strife?[68] An answer to this question can be found in Empedocles as well as in Aristotle. I will first look at the answer given by Empedocles.

The *Strasbourg Papyrus* preserves, albeit fragmentarily, the verse in Empedocles' poem that preceded the verses quoted by Aristotle (see above). This verse reads: [– ⏑⏑ – ⏑⏑ – διέφυ πλέ]ον' ἐξ ἑνὸς εἶναι (*P.Strasb.* a(i) 7 = 268).[69] Whereas verse 267 describes the unification that takes place under the influence of Love (συνερχόμεθ' εἰς ἕνα κόσμον, *P.Strasb.* a(i) 6 = 267),[70] verse 268 describes the influence of Strife.[71] Under Strife, Many (i.e., the four elements), come *out of the One*.[72] When the words ἐξ ἑνὸς εἶναι precede the verse ἐξ ὧν πάνθ' ὅσα τ' ἦν ὅσα τ' ἐσθ' ὅσα τ' ἔσται ὀπίσσω, quoted by Aristotle, then the reading alone attested in ω^{AL} becomes quite plausible. In this context, Aristotle's words ἐκ γὰρ τοῦ ἑνὸς ἅπαντα (1000a28) reveal themselves to be a close paraphrase of the Empedoclean verse, which precedes the verse Aristotle quotes and which describes the generative power of Strife and its effect of separating all things out of the One. By means of this close paraphrase (ἐκ … ἑνὸς) Aristotle brings in Strife, whose presence we had expected to see in the verse quoted, indirectly and at that in the words of Empedocles himself.

Let us now look at how Aristotle provides an answer to our question. As evident in the *De caelo* passage (Γ 2, 301a14–20 = DK 31 A 42) quoted above (pp. 115

---

[66] In the sentence that follows after the verse quotation (1000a33–b1) Strife is the implicit subject of the verb ἐνῆν (1000b1).

[67] See LSJ s.v. καί A.I.2. "to add a limiting or defining expression." Bonitz 1870: s.v. καί, 357b13–20.

[68] As we saw above in the ω^{αβ}-version Strife is preserved, if nevertheless in Aristotle's own peculiar words, by the expression ἐκ τούτου.

[69] See Martin/Primavesi 1999: 179–83. Martin/Primavesi suggest, on the evidence of a fragment in Lysias, the following reconstruction (182): [ἐν δ' ῎Εχθρηι γε πάλιν διέφυ πλέ]ον' ἐξ ἑνὸς εἶναι.

[70] Concerning the "we" in συνερχόμεθ' see Primavesi 2008: 47–57.

[71] Empedocles *Physics*, 267–68: "[under love] we [the elements] come together to a single ordered world … [under strife] we grow apart from each other to become many out of one."

[72] Primavesi 2008: 12: "die Herrschaft der *Mehreren* (d.h. der vier chemisch rein voneinander getrennten, zu homogenen Massen verbundenen Elemente)."

n. 54), Aristotle holds that Empedocles is forced to contend that our world comes about only under the rule of Strife, since (for Aristotle) our world could not have come about by the unification of separate elements.[73] Thus Aristotle says: ὥστ' ἀναγκαῖον γίνεσθαι ἐξ ἑνὸς καὶ συγκεκριμένου ("[our world] therefore necessarily comes out of a state of unity and combination"). We see that from Aristotle's point of view generation under the rule of Strife is equivalent to generation "out of the One." Thus the expression "out of the One" implies that we are under the rule of Strife, under which everything, the whole world, is generated. It is therefore quite plausible for Aristotle to characterize the work of Strife without explicit mention of Strife but simply with the formula "generation out of the One."[74]

What then about the redundancy of the two phrases ἐκ … τοῦ ἑνὸς and ἐκ τούτου given in line 1000a28 of ω^AL? It could be understood in the following way: Aristotle first speaks in the words of *Empedocles*, verse 268, (ἐκ γὰρ τοῦ ἑνὸς…) and then states the matter in his own words (… ἐκ τούτου τἆλλά ἐστι πλὴν θεός). Everything coming out of the One means (καί as limiting) everything apart from God. However, this twofold statement does not just express the same thought in two idiosyncratic ways. Rather, Aristotle ascribes to Empedocles a thought (namely, that all things come out of the One) that Aristotle needs to clarify, especially because it serves the purpose of his argument. Aristotle intends to show that Strife "generates no less than Love" (τὸ νεῖκος … οὐθὲν ἧττον καὶ τοῦτο γεννᾶν a27–28). When Aristotle therefore starts with the notion that everything comes to be out of the One (ἐκ γὰρ τοῦ ἑνὸς ἅπαντα), which is to say that everything is generated by Strife, he seems already to presuppose his own understanding of the cosmic cycle, according to which the world can *only* be generated during the rule of Strife (*De caelo* Γ 2, 301a14–20 = DK 31 A 42). But since *Sphairos* evidently is not generated by Strife, Aristotle has to exempt the god from his rule and rephrase his "everything comes out of the One (i.e. from Strife)" to "everything apart from *Sphairos* comes out of the One (i.e. from Strife)" (ἐκ τούτου τἆλλά ἐστι πλὴν θεός, a28–29).

To conclude: the reading preserved only in ω^AL is confirmed by the newly discovered *Strasbourg Papyrus*. The papyrus thus shows that the ω^AL-reading is preferable to the reading of our direct transmission (ω^αβ), whose oddities are in fact eliminated when we follow the ω^AL-text. The question then is, how did it happen that the reading preserved in ω^AL deteriorated into the reading preserved in ω^αβ? We can only speculate. A reader who was unfamiliar with the Empedoclean context and who did not understand Aristotle's words ἐκ γὰρ τοῦ ἑνὸς as a quasi-citation of Empedocles' verse could have wondered why Aristotle so suddenly and

---

[73] See Primavesi 2013: 698–99.

[74] Alexander wants to secure this meaning of the passage by proposing the addition (δεῖ προστιθέναι, 220.3) τὰ γιγνόμενα ὑπὸ τοῦ νείκους. Those things that come out of the One are the products of the work of Strife.

seemingly inexplicably spoke of the One out of which everything comes about. In order to understand the subsequent ἐκ τούτου (a28) as a reference to Strife, which is supposed to generate things, the preceding ἐκ γὰρ τοῦ ἑνός had to be removed and then integrated into the preceding sentence as ἔξω τοῦ ἑνός. Line a29, where πλὴν ὁ θεός expresses the same idea in different words, probably served as the model for this integration.

4.1.4 Alex. *In Metaph.* 204.23-31 on Arist. *Metaph.* B 3, 998b14-19

In B 3, 998b17-18 the α- and the β-version offer different, but equally unsatisfactory readings. As Bonitz has pointed out, it seems that Alexander's paraphrase alone offers the correct reading.[75] Let us take a closer look. We are in the third book of the *Metaphysics* at the beginning of the seventh aporia (B 1, 995b29-31; B 3, 998b14-999a23).[76] The following passage contains the conditional clause in line 998b17 (εἰ μὲν ... ἀρχαί) as restored on the basis of Alexander's commentary and read by all editors since Bonitz.

Aristotle, *Metaphysics* B 3, 998b14-19

πρὸς δὲ τούτοις εἰ καὶ ὅτι μάλιστα ἀρχαὶ τὰ γένη εἰσί, [15] πότερα δεῖ νομίζειν τὰ πρῶτα τῶν γενῶν ἀρχὰς ἢ τὰ [16] ἔσχατα κατηγορούμενα ἐπὶ τῶν ἀτόμων; καὶ γὰρ τοῦτο ἔχει [17] ἀμφισβήτησιν. εἰ μὲν γὰρ **ἀεὶ** τὰ καθόλου μᾶλλον **ἀρχαί**, [18] φανερὸν ὅτι τὰ ἀνωτάτω τῶν γενῶν· ταῦτα γὰρ λέγεται [19] κατὰ πάντων.

Besides this, even if the genera are in the highest degree principles, should one regard the first of the genera as principles, or those which are predicated directly of the individuals? This also admits of dispute. For if the universal is **always** more of a principle, evidently the uppermost of the genera are the principles; for these are predicated of all things.

15 πότερα α Al.¹ 204.24 : πότερον β Al.ᵖ 204.26 edd. ‖ 17 ἀεὶ ω^AL (Al.ᵖ 204.29) Bonitz Christ Ross Jaeger : δεῖ β : ὅτι α Ascl.¹ 177.10 Ascl.ᵖ 177.11 Bekker ‖ ἀρχαί α Al.ᵖ 204.29 edd. : ἀρχάς β

In the seventh aporia[77] Aristotle asks: if the genera are the principles (a presupposition taken from the sixth aporia), is it the first and most remote genera or the lowest and most proximate genera that are the principles of things (998b14-16)? If it is true that the more universal is always (ἀεὶ) more of a principle (ἀρχαί), then the uppermost and most universal of the genera are principles (b17-18). According to Alexander's paraphrase, he must have read the above text. Before looking in more detail at Alexander's paraphrase, I will evaluate the text as it is transmitted through our manuscripts.

[75] Bonitz 1848 *ad loc.*: ἀεὶ *scripsi cum Alex.*
[76] For the seventh aporia see Madigan 1999: 68-80 and Berti 2009. See also 5.1.1.
[77] According to Berti 2009: 119-20 the seventh aporia is to be identified as a special case of the sixth aporia. Schwegler 1847c: 131 already treats this passage as part of the sixth aporia.

The conditional clause as transmitted by α and β entails the following difficulties from a syntactical point of view. According to the α-version, it reads:[78]

α-text: Aristotle, *Metaphysics* B 3, 998b17–18

εἰ μὲν γὰρ **ὅτι** τὰ καθόλου μᾶλλον **ἀρχαί**, [18] φανερὸν ὅτι τὰ ἀνωτάτω τῶν γενῶν·

For if it is the case that the universal is more of a principle, evidently the uppermost of the genera are the principles;

In line b17 the α-text reads ὅτι instead of ἀεί. The predicative nominal ἀρχαί stands in the nominative case as it does in Alexander's paraphrase. Even if the α-reading seems syntactically less problematic than the β-reading (which we turn to below),[79] the ὅτι and the construction subsequent to it are difficult to integrate into the rest of the sentence. The protasis (εἰ μὲν γάρ ...) contains a dependent clause which is introduced by ὅτι. The protasis itself, however, is either highly elliptical or not a clause at all. It just says: εἰ μὲν γάρ, (ὅτι...) / "For if, (that)." We do not find anywhere else in Aristotle a phrasing such as this. In Plotinus, however, we can find this sort of phrase, where it has the sense of "if it is the case that...."[80] So we are dealing here in the α-text with an un-Aristotelian, but nevertheless grammatically possible idiom.

In the β-version, we find the following text:

β-text: Aristotle, *Metaphysics* B 3, 998b17–18

εἰ μὲν γὰρ **δεῖ** τὰ καθόλου μᾶλλον ἀρχάς, [18] φανερὸν ὅτι τὰ ἀνωτάτω τῶν γενῶν·

For if it is necessary that the universal ... more of a principle, evidently the uppermost of the genera are the principles;

The β-text deviates in two respects from the text attested to by Alexander's paraphrase. Instead of ἀεί there is δεῖ and instead of ἀρχαί in the nominative case there is ἀρχάς in the accusative case. These deviations are connected to each other: since δεῖ takes an accusative with infinitive construction, the subject τὰ καθόλου and the complementary predicative (ἀρχάς) are in the accusative case. The accusative form ἀρχάς seems to be a later correction that aims at making sense of the δεῖ. Yet, adjusting ἀρχαί to ἀρχάς does not solve the problem that the new construction (subsequent to δεῖ) lacks an infinitive.[81] In light of this, the nominative case of

---

[78] Scotus's Latin translation of the Arabic version of the *Metaphysics* does not reveal what the Greek *Vorlage* read. Scotus writes: *Quoniam si universalia sunt magis prima quam alia, manifestum est quod principia sunt genera altissima*. This seems to be closer to the α-version than to the β-version or Alexander's text.

[79] Bekker and Schwegler read α. Asclepius also had the α-reading in his text (177.10–12).

[80] For the expression εἰ μὲν γάρ, ὅτι... in Plotinus see e.g. II 9.9,66 and VI 3.21,30.

[81] The copula ἐστί / εἰσί can be naturally left out in an independent nominal sentence. In our case, however, we are dealing with an accusative with infinitive construction.

ἀρχαί given in α and in Alexander's paraphrase appears to be the older reading. Alexander's paraphrase of the passage reads as follows:

Alexander, *In Metaph.* 204.23-24; 29-31 Hayduck

998b14 Πρὸς δὲ τούτοις εἰ καὶ ὅτι μάλιστα ἀρχαὶ τὰ γένη [24] εἰσί, πότερα δεῖ νομίζειν τὰ πρῶτα τῶν γενῶν ἀρχάς.
... [29] εἰ μὲν γὰρ <u>ἀεὶ</u> τὰ καθόλου τῶν μὴ ὁμοίως καθόλου μᾶλλον <u>ἀρχαί</u>, διὰ [30] τὸ τὴν ἀρχὴν κεῖσθαι τὸ καθόλου εἶναι ἀρχήν, τὰ ἀνωτάτω ἂν γένη καὶ [31] τὰ κοινότατα εἶεν ἀρχαί·

Besides this, even if the genera are in the highest degree principles, should one regard the first of the genera as principles...?
... [*summary of aporia in Alexander's words*] For if the universal is <u>always</u> more of a principle than things which are not in a like manner universal—on account of its being laid down at the outset that the universal is a principle—then the highest and most common genera would be principles.

23 καὶ ὅτι P^b S : ἔτι καὶ A O || γένη A O S : γένη τῶν ὄντων P^b || 24 πότερα A P^b S : πρότερα O || 30 τὸ A O S : τῷ P^b || ἀνωτάτω P^b : ἀνώτατα O : ἀνω^ττ’ A

Alexander formulates his paraphrase of line 998b17 such that it stays close to Aristotle's words. He merely adds an object of comparison and says τῶν μὴ ὁμοίως καθόλου ("than things which are not in a like manner universal"). Apart from these words, he seems simply to copy the words of line 998b17, as far as we can judge on the basis of α and β. Note that Alexander's paraphrase is identical with the α-text except for one word: ἀεὶ (204.29: εἰ μὲν γὰρ <u>ἀεὶ</u> τὰ καθόλου [...] μᾶλλον ἀρχαί).

As discussed in 3.4 (pp. 57-60), we can reconstruct a reading in ω^AL on the basis of Alexander's paraphrase alone (i.e., without the need of further evidence in a lemma or quotation) when α and β differ significantly and one of the two agrees with the reading attested to by Alexander's paraphrase. In the present case, the confirmation Alexander's paraphrase receives from either α or β is only indirect. The reading in α (ὅτι) and β (δεῖ) *both* differ from what Alexander's paraphrase suggests. Nevertheless, the genesis of these two incorrect readings can best be explained as having originated in the reading that we find in Alexander's paraphrase.[82]

Taking ἀεί ... ἀρχαί (ω^AL) as the correct reading, the β-version (δεῖ) can be seen as the result of a rather common mistake in majuscule script.[83] The visual difference between ΛΕΙ and ΔΕΙ is slight.[84] This scribal error seems to already have occurred in the ω^αβ-text, as can be inferred from the fact that both versions

---

[82]Cf. the case in 4.3.3.
[83]Cf. e.g. *MA* 1, 698a16: ἀεὶ β : δεῖ α.
[84]Cf. v. Christ 1886a: VI-VII and 1 above.

α and β exhibit different strategies for dealing with the problematic δεῖ.⁸⁵ In the β-text, ἀρχαί became ἀρχάς in order to adapt the predicative nominal to the new (incomplete) infinitive construction. In the α-text, ἀρχαί remained unchanged, but the δεῖ (ω^{αβ}) was changed to ὅτι, which rendered the sentence grammatically acceptable to the time at which the correction was likely made (cf. Plotinus's use of the idiom εἰ μὲν γάρ, ὅτι…, mentioned above), but unidiomatic to Aristotle's time. ω^{AL}, however, preserved the original reading unscathed.

On the basis of the four cases analyzed here (4.1.1–4.1.4) it can be concluded that ω^{αβ} contains signs of corruption and errors that are not shared by ω^{AL}. These separative errors in ω^{αβ} rule out the possibility that ω^{AL} is a copy of ω^{αβ}.

4.2 SEPARATIVE ERRORS IN ω^{AL} AGAINST ω^{αβ}

I have analyzed separative errors in ω^{αβ} that are not shared by ω^{AL}. I now turn to the investigation of separative errors in ω^{AL} against ω^{αβ}. Such errors show ω^{αβ} to be independent of ω^{AL} in the sense that ω^{αβ} is not a descendent of ω^{AL}.

4.2.1 Alex. In Metaph. 11.3–6 on Arist. Metaph. A 2, 982a19–25

After having introduced σοφία as science (ἐπιστήμη) concerned with causes and principles (A 1, 982a1–3), Aristotle continues his characterization of this science at the beginning of A 2 by reviewing generally accepted views or presupposed assumptions about the wise person (σοφός) (982a4–8).⁸⁶ Aristotle introduces the first view as follows:

Aristotle, *Metaphysics* A 2, 982a8–10

ὑπολαμβάνομεν δὴ πρῶτον μὲν ἐπίστασθαι πάντα τὸν [9] σοφὸν ὡς ἐνδέχεται, μὴ καθ' ἕκαστον ἔχοντα ἐπιστήμην [10] αὐτῶν·

We suppose first, then, that the wise person knows all things, as far as possible, although he has not knowledge of each of them individually;

---

8 πάντα β Al.^P 9.29–32 10.1–2, cf. Ascl.^P 15.30 Bekker Bonitz Christ Ross Jaeger : μάλιστα πάντα α Primavesi⁸⁷

⁸⁵It is not impossible but very unlikely that an error occurred in both versions α and β independently of each other at precisely the same point in the text.

⁸⁶For an analysis of chapter A 2 of the *Metaphysics* see Broadie 2012: 43–67. For Aristotle's procedure of beginning with an analysis of widely held assumptions see Broadie 2012: 55.

⁸⁷I follow the β-text in 982a8 (*pace* Primavesi 2012) and read πάντα instead of μάλιστα πάντα for the following reasons: first, to say that the wise person is supposed to know all things (object of knowledge) to the highest degree, μάλιστα, (degree of knowledge) does not square well with the immediately following restriction "as far as possible" (ὡς ἐνδέχεται). This seems to rule the superlative out. Second, the specification μάλιστα becomes relevant only at the *later* passage, in which Aristotle spells out the first assumption more precisely (982a21, see below). There, the μάλιστα is part of his account of the

In 982a10–19 Aristotle enumerates other opinions about the wise, and at 982a19–21 Aristotle declares his list complete. Thereafter he examines closely these views and their implications.[88] To the first opinion Aristotle says:

Aristotle, *Metaphysics* A 2, 982a19–23

τὰς μὲν οὖν [20] ὑπολήψεις τοιαύτας καὶ τοσαύτας ἔχομεν περὶ τῆς σοφίας [21] καὶ τῶν σοφῶν· τούτων δὲ τὸ μὲν **πάντα** ἐπίστασθαι τῷ μά-[22]λιστα ἔχοντι τὴν καθόλου ἐπιστήμην ἀναγκαῖον ὑπάρχειν [23] (οὗτος γὰρ οἶδέ πως ἅπαντα τὰ ὑποκείμενα)

Such and so many are the assumptions, then, which we hold about wisdom and the wise. Now of these characteristics, that of knowing **all things** must belong to the person who has in the highest degree universal knowledge; for this person knows in a sense all the subordinate objects;

20 καὶ τοσαύτας **α** ζ Ascl.ᵖ 16.19 edd. : om. β ‖ τῆς **α** Ascl.ᵖ 16.20 edd. : om. β ‖ 21 πάντα **α** Ascl.ᶜ 16.21 edd. : ἅπαντα β : om. ω^AL (Al.¹ 11.3), τὰ πάντα ci. Al. 11.5 ‖ 22 τὴν καθόλου ω^αβ Al.ᵖ 11.6–7 Ascl.ᶜ 16.21–22 edd. : τὴν κατὰ πάντων Al.¹ 11.3–4 ‖ 23 πως **α** ζ : πῶς ἔχει β : om. Ascl.ᶜ 16.25

The word πάντα in line 982a21, which both versions attest to (the difference of πάντα [α] and ἅπαντα [β] being irrelevant for the present purpose), is in two ways anchored in the context. First, Aristotle has already at the beginning of this chapter (982a8–9) introduced knowledge of all things as the first generally accepted view we have about the wise. Second, the sentence taken by itself and without its context would not make good sense without πάντα. It is not at all a satisfactory characterization of the wise person, who has universal knowledge to the highest degree, simply to say that he knows. Knows what? Some particular thing, knows generally? The mere ability to know (ἐπίστασθαι) is too general a characteristic to describe a person who knows in a special way, namely, who has the ability to know the universal.[89]

The version of ω^AL, however, did not contain the word πάντα (a21), as Alexander's lemma and his comments on the passage indicate.

Alexander, *In Metaph.* 11.3–6 Hayduck

982a21 Τούτων δὲ <u>τὸ μὲν ἐπίστασθαι</u> τῷ μάλιστα ἔχοντι τὴν [4] κατὰ πάντων ἐπιστήμην.
[5] Ἐλλείπει τῷ ἐπίστασθαι τὸ 'τὰ πάντα'· τὸ γὰρ πάντα ἐπίστασθαι τῷ [6] μάλιστα ἔχοντι τὴν καθόλου ἐπιστήμην ὑπάρχει· τοῦτο γὰρ ἦν τὸ κείμενον.

---

wise person's πάντα ἐπίστασθαι: the person who knows all things has universal knowledge *to the highest degree*. The word μάλιστα in 982a8 of the **α**-text seems to be an overcorrection aimed at aligning the text to the only seeming parallel phrase in 982a21.

[88] Broadie 2012: 54: "Aristotle's responses to the assumptions."

[89] Furthermore, the parenthetical explication given in 982a23 (οὗτος γὰρ οἶδέ πως ἅπαντα τὰ ὑποκείμενα) takes up the word πάντα and thereby presupposes it.

Now of these characteristics, that of knowing must belong to the person who has in the highest degree knowledge concerning everything; [In this text] the words 'all things' are omitted before 'knowing,' for to know all things belongs to the man who possesses universal knowledge in the highest degree; for this was the assumption.

---

5 τὸ γὰρ **A P**ᵇ : τῷ γὰρ **O**

In the *Metaphysics* text presented in the lemma πάντα is absent.[90] Alexander begins his comments by diagnosing the absence of the words τὰ πάντα from his text as a mistake, and so suggests supplementing them. Alexander demonstrates that the absence of τὰ πάντα is a mistake by recalling Aristotle's earlier exposition of the first view: τοῦτο γὰρ ἦν τὸ κείμενον (11.6). There, Aristotle made clear that knowing all things belongs to the wise. Since nothing in Alexander's words suggests that he knew the reading τὰ πάντα from another manuscript,[91] we can ascribe Alexander's recognition of its absence to his thorough reading of Aristotle's text.[92]

Alexander's supplement is warranted, but it does not exactly coincide with the wording in ω^{αβ}.[93] There we read πάντα without article. The reading in ω^{αβ} more closely agrees with the parallel passage at the beginning of A 2 (ἐπίστασθαι πάντα, 982a8),[94] and so it should be preferred. Given that the reading suggested by Alexander does not exactly match the reading transmitted in ω^{αβ}, there is no need to speculate that at an earlier stage πάντα had been missing also in ω^{αβ}, but was later added to the text at Alexander's suggestion. Therefore we conclude that we are dealing here with a separative error in ω^{AL} of which ω^{αβ} is free. This demonstrates that ω^{αβ} does not derive directly from ω^{AL}.

4.2.2 Alex. In Metaph. 167.7–14 on Arist. Metaph. α 3, 994b32–995a3

In α 3 Aristotle comments on how teaching methods relate to the subject matter being taught. He starts off with the following considerations on pedagogy.

---

[90] Interestingly, the lemma diverges in yet another way from our text. It reads τὴν κατὰ πάντων ἐπιστήμην instead of τὴν καθόλου ἐπιστήμην (982a22). That this is not what Alexander read in ω^{AL}, but a later corruption of the lemma, is made clear by Alexander's paraphrase: his words at 11.6 show clearly that he read τὴν καθόλου ἐπιστήμην.

[91] Dooley 1989: 29 n. 52 seems to understand Alexander in that way.

[92] The *recensio altera* (**L**) reads ἅπαντα (11 *app.*) in the lemma but adopts Alexander's remark that πάντα should be supplemented. Asclepius (16.21–22) cites the sentence in the **α**-version (πάντα), but does not further comment on it. A scholium in the *Metaphysics* manuscript **E** (see Brandis 1836) mentions two different versions of the text: ἔν τισι τῶν ἀντιγράφων λείπει τὸ πάντα, ὡς ἀπὸ κοινοῦ λαμβανόμενον, 527a12–13. This might go back to Alexander's comment on the passage. That the absence of πάντα should be understood as an ἀπὸ κοινοῦ construction is an attempt to make sense of the reading without the πάντα.

[93] Ross 1924: clxii is imprecise when he says: "om. Al., who desiderates πάντα."

[94] τὰ πάντα means "the whole" (LSJ s.v. πᾶς B.II), whereas πάντα means "all things, everything."

Aristotle, *Metaphysics* α 3, 994b32–995a3

Αἱ δ' ἀκροάσεις κατὰ τὰ ἔθη συμβαίνουσιν· ὡς γὰρ [995a1] εἰώθαμεν οὕτως **ἀξιοῦμεν λέγεσθαι**, καὶ τὰ παρὰ ταῦτα οὐχ [2] ὅμοια φαίνεται ἀλλὰ διὰ τὴν ἀσυνήθειαν ἀγνωστότερα καὶ [3] ξενικώτερα· τὸ γὰρ σύνηθες γνωριμώτερον.

The effect which lectures produce on a hearer depends on the hearer's habits; for we expect the lecturing style we are accustomed to, and that which is different from this seems not in keeping but somewhat unintelligible and foreign because it is not customary. For the customary is more intelligible.

1 λέγεσθαι **ω**^αβ Ascl.^P 134.32–33 Ar^i (*Scotus*) edd. : ἔτι τὸ λέγεσθαι **ω**^AL (Al. 167.11 Al.^c 167.10; 12) || 3 γνωριμώτερον **α** Bekker Bonitz Christ : γνώριμον **β** Ross Jaeger

Aristotle's remark that learning is easier when we are accustomed to the manner of teaching is transmitted in our text (**ω**^αβ) without any grammatical oddities. Alexander's text (**ω**^AL), however, differs from **ω**^αβ in that ἔτι τὸ appears in front of λέγεσθαι (995a1). According to Alexander's comments on the text, the ἔτι is superfluous.

Alexander, *In Metaph.* 167.7–14 Hayduck

αἱ γὰρ ἀκροάσεις γίγνονται κατὰ τὴν ἰδίαν τῶν ἐθῶν οἰκειό-[8]τητα· οἷς γὰρ συνειθίσμεθα, τούτοις ἀξιοῦμεν καὶ τὰ λεγόμενα συμφωνεῖν, [9] τὰ δὲ παρὰ τὰ συνήθη ἡμῖν τῶν λεγομένων ἀγνωστότερα φαίνεται τῷ [10] ξενικὰ εἶναι. ὃ εἶπε διὰ τοῦ ἔτι τὸ λέγεσθαι καὶ τὰ παρὰ ταῦτα οὐχ [11] ὅμοια φαίνεται. δοκεῖ δὲ τὸ ἔτι περιττὸν εἶναι. εἰ δὲ εἴη γεγραμμένον, [12] λείποι ἂν τῷ ἔτι τὸ λέγεσθαι τὸ 'ἄλλως'· τὸ γὰρ ἄλλως λέγεσθαί τινα καὶ ὡς [13] μὴ εἰθίσμεθα καὶ ἄλλα, ἀλλὰ μὴ ὧν εἰθίσμεθα ἀκούειν, ὃ εἶπε διὰ τοῦ καὶ παρὰ ταῦτα,^95 ἀγνωστότερα ποιεῖ [14] τὰ λεγόμενα.

For the effect produced by lectures is determined by the habits of the individual [auditors], for we demand that the [lecturer's] words agree with the things to which we have become accustomed, and if [he] says anything beyond what is familiar to us we think it somewhat unintelligible because it is foreign to us. This he [Aristotle] expresses by the words "Moreover, the lecturing that is different from this seems not in keeping." The word "moreover" [ἔτι] seems superfluous, but if it is to be written it would require [the addition of] "in a different manner" [ἄλλως] to "moreover, the lecturing"; for lecturing in a different manner and not as we are accustomed to and about other things than we are accustomed to hear—this last he expresses by the words 'and beyond what [is familiar to us]'—makes what is said somewhat unintelligible.

8 συνειθίσμεθα **A O** : συνεθίσμεθα **P**^b || 9 συνήθη **A O** : συνήθως **P**^b || ἡμῖν **O P**^b : ἡμῶν **A** || τῷ **O P**^b : διὰ τὸ **LF** : τὸ **A** || 10–14 ὃ ... τὰ λεγόμενα] om. **LF** || 10 ξενικὰ Hayduck : ξενὰ codd. || 12 τῷ ἔτι **O P**^b : τῷ ἔστι **A** || λέγεσθαι τὸ ἄλλως **P**^b : λέγεσθαι ἄλλως **A O** || τὸ γὰρ **A P**^b : τῷ γὰρ **O** || 13 εἰθίσμεθα καὶ ἄλλα, ἀλλὰ μὴ ὧν εἰθίσμεθα **O A P**^b (cf. *aliter dici quem consuevimus,*

---

[95] Hayduck did not mark these words as a quotation from Aristotle's text.

et diversa ab iis, qu(a)e audire solemus videntur S) : εἰθίσμεθα Bonitz Hayduck || τοῦ καὶ **A O** : τοῦ καὶ τὰ **P**ᵇ || ποιεῖ **P**ᵇ : ποιεῖν **A O**

In lines 167.7-10, Alexander paraphrases lines 994b32–995a3 of the *Metaphysics*. In the line that follows, 167.10-11, Alexander quotes lines 994a1–2 as they appear in his copy of the *Metaphysics*. In this quotation we see that ω^AL read in line 995a1 ἔτι τὸ λέγεσθαι instead of simply λέγεσθαι. At first glance, it looks as though Alexander quoted only the middle part of the sentence in 994b32–995a2, leaving out the beginning as well as the end. That Alexander indeed found the words ἔτι τὸ in his exemplar is confirmed by his subsequent remark about their superfluity. By contrast, in ω^αβ, the infinitive λέγεσθαι is preceded by and syntactically connected to the word sequence ὡς γὰρ εἰώθαμεν οὕτως ἀξιοῦμεν / "for we expect it (i.e. the lecturing) to be as we are accustomed to" (994b32–995a1).[96] The infinitive λέγεσθαι is the (accusative) object of the verb ἀξιοῦμεν.[97] Yet, when the infinitive λέγεσθαι is preceded by ἔτι τὸ, as it is in ω^AL, the syntax of the sentence, and hence its punctuation, changes. Thus, it makes sense that Alexander quotes ἔτι τὸ λέγεσθαι … (167.10-11), taking it as an independent sentence; it is a term in a dependent clause only from the perspective of ω^αβ.

On the basis of Alexander's quotation and subsequent remark we can reconstruct the following wording for lines 994b32–995a2 in ω^AL.

Reconstruction of lines 994b32–995a2 according to ω^AL

ὡς γὰρ [995a1] εἰώθαμεν οὕτως ἀξιοῦμεν. **ἔτι τὸ** λέγεσθαι καὶ τὰ παρὰ ταῦτα οὐχ [2] ὅμοια φαίνεται ἀλλὰ …

For we expect it [the lectures] to be as we are accustomed to. **Moreover**, the lecturing and that which is different from the customary seems not in keeping but …

The phrase that Alexander quotes (ἔτι τὸ λέγεσθαι…) indicates clearly that according to ω^AL the words ἔτι τό introduce a new sentence. That the words ἔτι τὸ function as the beginning of the sentence is consistent with Aristotle's typical use of ἔτι. In most cases ("usitatissimum," Bonitz 1870: s.v. ἔτι; p. 291a13), ἔτι (δέ) stands at the beginning of a clause.[98] Accordingly, when a new sentence begins in

---

[96] Both Asclepius (Ascl. *In Metaph*. 134.32-33) and the Arabic tradition (Scotus: *Dicimus enim illud quod assueti sumus audire et…*) testify to this syntactical connection between "to expect" and "lecturing" in the Aristotelian sentence.

[97] Aristotle usually uses the verb ἀξιόω in the sense of "to expect" with the infinitive (LSJ s.v. ἀξιόω III; Bonitz 1870: s.v. ἀξιοῦν, p. 70.20-29; Goodwin 1867: §92; p. 189). A few lines later we find another instance of the same construction: ἀξιοῦσιν ἐπάγεσθαι (995a8). Cf. the parallel construction in *SE* 6, 168b32 (ἀξιοῦμεν εἶναι ταὐτά) and *Top*. H 3, 153a37 (ἀξιοῦμεν κατηγορεῖσθαι).

[98] At the beginning of a clause, the adverb ἔτι can stand without a particle, as, for example, in the context of our passage in *Metaph*. α, 994b20, but also in *Metaph*. Δ 5, 1015a33; Δ 7, 1017a31; 1017a35; *Ph*. Γ 8, 208a11; Δ 4, 211a10; *Pol*. 1252b34; *EN* 1132b30; 1147a10; *Cael*. 275b25; 301a4. Cf. Bonitz 1870: s.v. ἔτι; p. 291a13-15.

995a1 (ἔτι), the preceding ἀξιοῦμεν ("we expect") has to be taken in an absolute sense. In the new subsequent sentence, then, τὸ λέγεσθαι must be the subject[99] and syntactically equal to τὰ παρὰ ταῦτα.

Yet, even in this position, the infinitive λέγεσθαι, "lecturing," does not make much sense. Why should an *unspecified* lecturing "not be in keeping" (οὐχ ὅμοια)? In addition, the function that ἔτι usually has at the beginning of a sentence, namely as a means of introducing an additional argument,[100] is not present in our passage. Here, ἔτι adds not an additional point, but an example that illustrates the point just made. Yet, if one wants to read the ἔτι not as beginning a new sentence, as Alexander does, but rather as introducing a climatic apposition[101]—a function that ἔτι rarely serves in Aristotle—then the difficulty arises that there is no preceding enumeration in this passage to which the thought "on top of that the lecturing" could be added as a culmination.

Alexander, too, recognizes the syntactical difficulties presented by the sentence in ω^AL and thus declares the ἔτι to be superfluous (167.11).[102] The deletion of the ἔτι has the effect of merging the two sentences into one, thereby making τὸ λέγεσθαι the object of ἀξιοῦμεν. Besides the deletion of ἔτι, Alexander proposes another solution (167.11-12): According to his second suggestion, the sentence could be corrected by leaving the ἔτι in the text and adding the adverb ἄλλως to the infinitive λέγεσθαι. The addition of ἄλλως would lend the sentence a sufficiently coherent meaning: "Moreover, lecturing *in a foreign manner* and that which is different from what we are accustomed to seems not in keeping but...."[103]

Since the reading in ω^αβ (i.e. without the ἔτι τὸ) does not cause any problems, it is preferable to the reading in ω^AL even when implementing Alexander's emendations (deletion of ἔτι or addition of ἄλλως). How could the reading in ω^AL have emerged? Perhaps someone erroneously took the sequence ὡς γὰρ εἰώθαμεν οὕτως ἀξιοῦμεν ("for we expect it to be as we are accustomed to") as a complete sentence, and then rendered the subsequent text as a new sentence and introduced it with the word ἔτι. As λέγεσθαι is the subject of the new sentence, the article τό needs to be supplied as well. In any case, the words ἔτι τὸ, which are preserved only in ω^AL, appear to be an erroneous addition to the text.

---

[99] In 167.7-10, Alexander's paraphrase already hints at the fact that in his copy of the *Metaphysics* the verb λέγεσθαι was separated from the preceding (οὕτως ἀξιοῦμεν) and pulled into the subsequent sentence. Alexander paraphrases: τὰ δὲ παρὰ τὰ συνήθη ἡμῖν τῶν λεγομένων ἀγνωστότερα φαίνεται.

[100] Bonitz 1870: s.v. ἔτι; p. 291a13-15.

[101] Bonitz 1870: s.v. ἔτι; p. 291a6-11.

[102] The author of the *recensio altera* does not adopt Alexander's textual remarks (see apparatus above). Most likely his text, just as ω^αβ, did not contain the ἔτι τὸ either. Nor does Asclepius (134.30-35) comment on textual issues here; his paraphrase confirms our transmission of the text.

[103] At this point in the commentary (167.13), I was able to expand, on the basis of the evidence in **O** (which is confirmed by **S**), the text edited by Bonitz and Hayduck and include an additional dependent clause. The new collations of Golitsis have shown that this actually is the reading of **A**, **O** and **P^b**. It seems that Hayduck adopted this error from Bonitz.

The absence of the two disruptive words in ω^{αβ} likely points to a tradition independent from ω^{AL} (or even its ancestor) in which the erroneous addition of ἔτι τό never occurred. In the given case, however, we might still ask whether the tradition of text ω^{αβ} might have contained the words ἔτι τό, too, but then was corrected in line with Alexander's comments on the passage. That the words ἔτι τό were present in an ancestor of ω^{αβ} and then deleted cannot entirely be ruled out. It is extremely unlikely, however, that this hypothesized deletion of ἔτι τό in ω^{αβ} occurred at the prompt of Alexander's commentary, for Alexander suggests discarding only the ἔτι, not the τό. Consequently, although Alexander's emendation comes close to the reading given in ω^{αβ}, it still differs in respect to the article τό in front of λέγεσθαι.[104] Therefore, it is justified to regard ἔτι τό as a peculiar error in ω^{AL} that never occurred in ω^{αβ} and hence as a further separative error in ω^{AL} against ω^{αβ}.

4.2.3 Alex. In Metaph. 273.20–26; 34–274.2 on Arist. Metaph. Γ 4, 1006a18–24

At the beginning of Γ 4, Aristotle engages with the deniers of the principle of non-contradiction. Although the principle's validity cannot be positively proved, the absurdity of its denial can be demonstrated negatively.[105] The first step in Aristotle's strategy for engaging with the opponents is not to assert that they have to admit that something is or is not (εἶναί τι … ἢ μὴ εἶναι)—for this would already imply acceptance of the principle—but rather to get them to admit that there is something (σημαίνειν … τι) that is of any significance at all to them and to others.[106]

Aristotle, *Metaphysics* Γ 4, 1006a18–24

ἀρχὴ [19] δὲ πρὸς ἅπαντα τὰ τοιαῦτα **οὐ** τὸ ἀξιοῦν ἢ εἶναί τι λέγειν [20] ἢ μὴ εἶναι (τοῦτο μὲν γὰρ τάχ' ἄν τις ὑπολάβοι τὸ ἐξ [21] ἀρχῆς αἰτεῖν), ἀλλὰ σημαίνειν γέ τι καὶ αὑτῷ καὶ ἄλλῳ· [22] τοῦτο γὰρ ἀνάγκη, εἴπερ λέγοι τι. εἰ γὰρ μή, οὐκ ἂν [23] εἴη τῷ τοιούτῳ λόγος, οὔθ' αὑτῷ πρὸς αὑτὸν οὔτε πρὸς [24] ἄλλον.

The starting-point for all such arguments[107] is **not** the demand that our opponent

---

[104] Ross 1924: clxii points to this passage and states: "But if the manuscript reading were due to Alexander's note the MSS. would have to read τὸ λέγεσθαι."

[105] *Metaph.* Γ 4, 1006a11–18.

[106] According to Rapp 1993: 531–41 σημαίνειν should be taken in the sense it has in *De Int.* and *Poetics*. Therefore, the opponent's utterance has to contain at least one ὄνομα (i.e. a noun, cf. *Po.* 1457a10; *Int.* 16a19) that signifies something. See *Int.* 2, 16a16–18: σημεῖον δ' ἐστὶ τοῦδε· καὶ γὰρ ὁ τραγέλαφος σημαίνει μέν τι, οὔπω δὲ ἀληθὲς ἢ ψεῦδος, ἐὰν μὴ τὸ εἶναι ἢ μὴ εἶναι προστεθῇ ἢ ἁπλῶς ἢ κατὰ χρόνον. / "A sign of this is that even 'goat-stag' signifies something but not, as yet, anything true or false—unless 'is' or 'is not' is added (either simply or with reference to time)" (transl. by Ackrill). Cf. also Kirwan 1971: 91–92; Flannery 2003: 117–18.

[107] Aristotle calls the kind of demonstration that shows it to be impossible to deny this axiom

shall say that something either is or is not (for this one might perhaps take to be begging the question), but that he shall signify something which is significant both for himself and for another; for this is necessary, if he really is to say anything. For, if he does not signify anything, such a man will not make any statement, neither he himself to himself nor to another person.

19 οὐ **a** Al.ʸᵖ (273.37-274.1) Bekker Bonitz Ross Jaeger Cassin/Narcy Hecquet-Devienne : οὐχὶ β Christ : om. ω^AL (Al.¹ 273.20-21 Al.ᵖ 273.23-24) Ar^u (*Scotus*) ‖ ἢ εἶναί τι λέγειν] εἶναί τι λέγειν Al.¹ 273.20-21 : λέγειν τι εἶναι Al.ʸᵖ 274.1 ‖ γὰρ] om. Al.ʸᵖ ‖ 21 τι **a** edd. : om. β ‖ αὐτῷ EˢIᵇ Bekker Bonitz Ross Jaeger : ἑαυτῷ β Al.ᵖ 274.3 Christ : αὐτῷ **a**

Applying this strategy to the example Aristotle uses in the passage quoted below (1006a31-b11), the section says that when the opponent expresses "human being" and signifies something determinate by it (e.g. a rational animal) the opponent cannot at the same time signify "non human being." This condition, however, does not imply that a human being is or is not.

Comparing lines 1006a18-20 with the text used by Alexander (ω^AL) shows that the negation οὐ was absent from ω^AL.[108] This much is clear from the lemma:

Alexander, *In Metaph*. 273.20-21 Hayduck

Ἀρχὴ δὲ πρὸς ἅπαντα <u>τὰ τοιαῦτα τὸ ἀξιοῦν</u> εἶναί τι [21] λέγειν ἢ μὴ εἶναι.

The starting-point for all such arguments <u>is the demand</u> that our opponent shall say that something either is or is not.

According to the lemma, lines 1006a18-20 of ω^AL read "The starting-point for all such arguments *is* the demand that..." instead of "The starting-point for all such arguments is *not* the demand that...."[109] Without the negation οὐ Aristotle's statement is turned on its head. Yet, this was the reading in Alexander's copy of the *Metaphysics*, as is confirmed by his paraphrase:

Alexander, *In Metaph*. 273.22-26 Hayduck

Τοῦ ἐλεγκτικοῦ συλλογισμοῦ καὶ τοῦ πρὸς ἄλλον γινομένου περὶ τῆς [23] τοῦ προκειμένου ἀξιώματος δείξεως ἀρχήν φησιν εἶναι <u>τὸ ἀξιοῦν</u> τὸν προσ-[24]

(1006a11-18) ἔλεγχος (a18), "negative proof" or "refutation," and not ἀπόδειξις (proof), which is only given when one begins with positive assumptions—and such is not the case when we are dealing with a first axiom. Cf. Rapp 1993: 521-24.

[108] In addition to the negation οὐ, the ἢ before εἶναι is also missing in ω^AL. The omission of ἢ, however, which seems to have occurred only in the lemma quotation (cf. the paraphrase in 273.24; 25; 31), does not change the meaning of the sentence.

[109] In Sepúlveda's Latin translation the *lemma* agrees, as it usually does, with the Latin *Metaphysics* version that precedes the commentary sections and therefore reads a *non* (f. q.ii.v.). Yet, Sepúlveda's translation of Alexander's *paraphrase* agrees with the evidence in the Greek manuscripts of the commentary (*principium esse ait, petere ab adversario, ut dicat esse quidpiam...*), and thus confirms that Alexander himself did not find the οὐ in his text.

διαλεγόμενον ἢ εἶναί τι λέγειν ἢ μὴ εἶναι, τουτέστιν ἐρωτητέον αὐτὸν εἰ [25] μὴ δοκεῖ αὐτῷ πᾶν ἢ εἶναι τοῦτο ὃ λέγεται ἢ μὴ εἶναι, οἷον ἄνθρωπον [26] ἢ εἶναι ἄνθρωπον ἢ μὴ εἶναι, ὁμοίως ἵππον, κύνα, τὰ ἄλλα.

In the syllogism of refutation, carried on in reply to someone else about the proposed axiom, he says that <u>the starting-point of proof is to insist</u> that the respondent say that something is the case or that it is not the case. That is, one should ask him whether it seems to him that everything either is that which it is said to be or not. For example, a human: whether it seems that it either is human or not; and likewise a horse, a dog, and the rest.

22 περὶ τῆς **O** LF : περὶ **A** : τῆς περὶ **P**[b] || 23 ἀρχήν **A O** : ἀρχή **P**[b] || 24 ἐρωτητέον **A O S** : ἐρωτᾶν **P**[b] || 25 ἢ εἶναι **A O** : εἶναι **P**[b] **S**

Since there is no οὐ in the text in front of him, Alexander interprets Aristotle as saying that the opponent must be made to affirm or deny that something is a human being (ἢ εἶναι ἄνθρωπον ἢ μὴ εἶναι). According to this interpretation, Aristotle demands a much greater concession from his opponent than the verb σημαίνειν τι (1006a21) indicates. And yet this is not what troubles Alexander about the reading in ω[AL]. What troubles Alexander is the logical inconsistency between the sentence (without οὐ) in 1006a18–20 and Aristotle's subsequent explanation of it (γάρ, a20).

In lines a20–21 (ω[αβ]), Aristotle explains why the starting-point *cannot* consist in the opponent's assertion that something is or is not: *for* this (τοῦτο μὲν γάρ..., a20) could be regarded as begging the question. Aristotle could not have used this reasoning to justify (γάρ) his strategy had he said, as it is preserved in ω[AL], that the starting point is to demand that the opponent affirm or deny an assertion. Rather than simply declare the text of ω[AL] to be inconsistent, Alexander tries to maintain the logic of Aristotle's argument by means of the following interpretation (273.26–32): Aristotle's explanation (a20–21) that the demand (a18–20) would beg the question is meant to show that the strategy (as given in ω[AL]) is inappropriate. Therefore, according to Alexander (273.33–34), Aristotle abandons this strategy and, in lines 1006a21–22, presents a more appropriate one.

Since Alexander recognizes that this interpretation does not eliminate the dissonance that the particle γάρ creates between the two sentences, he suggests emending the text by discarding the γάρ. For, without the particle γάρ the statement that the demand would beg the question could more easily be taken as Aristotle's own correction of his previous sentence.[110]

Alexander, *In Metaph.* 273.34–36 Hayduck

καταλληλότερον δὲ ἡ λέξις ἔχοι ἄν, εἰ [35] ἀντὶ τοῦ τοῦτο <u>μὲν γὰρ τάχα</u> ἄν τις ὑπολάβοι τὸ ἐξ ἀρχῆς αἰτεῖν [36] εἴη γεγραμμένον τὸ τοῦτο <u>μὲν τάχα</u> ἄν τις, ὡς καὶ φέρεται ἔν τισιν.

[110] Madigan 1993: 157 n. 327 remains too vague in his analysis of Alexander's intention. He seems to be unaware that Alexander's emendation was prompted by the missing οὐ in ω[AL].

The sentence would be more consistent if, instead of "for this one might perhaps take to be begging the question," it read "this one might perhaps ...," as it is found in some witnesses.

36 τις **A O S** : τις ὑπολάβοι τὸ ἐξ ἀρχῆς αἰτεῖν **P^b**

Alexander informs us that the correction he suggests is actually found in other manuscripts.[111] This other textual tradition apparently agrees with ω^AL in not reading the οὐ.[112] The deletion of γάρ may well have been a reaction to the absence of the οὐ. (By contrast, in the ancestor of our tradition, ω^αβ, the οὐ seems not to have dropped out in the first place.) The conjunctive error shared by ω^AL and the exemplar(s) Alexander refers to must have occurred only *after* their common ancestor split from the ancestor of ω^αβ. This split of ω^AL and ω^αβ must therefore have occurred sometime before AD 200, since Alexander already knows of a version that shares with his own text the loss of the οὐ, but differs from it through the deletion of the γάρ.

Is there any evidence that suggests that the ancestor of ω^αβ first shared the error of not reading the οὐ, but had the οὐ inserted into it at some later time? In order best to answer this question, let us clarify the picture with the following piece of information: Alexander knows of yet another *Metaphysics* version, one in which there is no logical inconsistency between the two Aristotelian sentences.[113]

Alexander, *In Metaph.* 273.37–274.2 Hayduck

φέρεταί τις καὶ τοιαύτη γραφή ἀρχὴ δὲ πρὸς ἅπαντα τὰ τοιαῦτα οὐ [274.1] τὸ ἀξιοῦν λέγειν τι εἶναι ἢ μὴ εἶναι. καὶ ἔστι γνωριμώτερον τὸ λε-[2]γόμενον οὕτως.

A reading is also found as follows: "the starting-point for all such arguments, is not the demand that our opponent shall say that something either is or is not." And this way the meaning makes more sense.

1 λέγειν τι εἶναι **A** : λέγειν τι εἰ **O** : εἶναί τι λέγειν **P^b S** (*Metaph.*)

The other variant reading which Alexander knows and judges favorably of agrees with the text preserved in ω^αβ. It contains the negation οὐ in line a19, thus allowing the γάρ-sentence to follow consistently. Since this version shows that in Alexander's time there existed a tradition that preserved the original οὐ, the same very well may have held for the tradition of ω^αβ, too.

To sum up: ω^AL, the text containing Alexander's first cited variant reading

---

[111] Cf. 3.6, p. 90.

[112] This indicates that the dropout of the οὐ was not confined to text ω^AL. The negation is also absent from the Arabic translation by Ustāth (as Scotus's translation shows). Since the succeeding sentence in the Latin version of the Arabic text is far removed from the Greek original it is impossible to say whether or not the Greek *Vorlage* contained the γάρ.

[113] On this commentary passage see also Flannery 2003: 124–25.

(273.34–36), and the Greek *Vorlage* of the Arabic tradition (see apparatus) all share the error of the missing οὐ. ω^{αβ} (or rather an ancestor of it), however, was free of this error, and none of the evidence I have encountered suggests that the οὐ in ω^{αβ} is the result of a later correction. Thus, ω^{αβ} is independent of ω^{AL}, and, additionally, it is quite possible that the version Alexander knows as having read the οὐ is just ω^{αβ}, as in the cases discussed in 3.5.2.

4.2.4 Alex. *In Metaph.* 228.29–229.1 on Arist. *Metaph.* B 5, 1001b26–28

Aristotle introduces his treatment of the twelfth aporia (B 5) with the following words:[114]

> Aristotle, *Metaphysics* B 5, 1001b26–28
>
> Τούτων δ' ἐχομένη ἀπορία πότερον οἱ ἀριθμοὶ καὶ [27] τὰ σώματα **καὶ τὰ ἐπίπεδα** καὶ αἱ στιγμαὶ οὐσίαι τινές [28] εἰσιν ἢ οὔ.
>
> A question connected with these is whether numbers and bodies **and planes** and points are substances or not.
>
> ---
> 27 καὶ τὰ ἐπίπεδα ω^{αβ} Ascl.^{l} 208.24 edd. : om. ω^{AL} (Al.^{l} 228.30 Al.^{p} 228.32–229.1) : *et superficies et linee* Ar.^{u} (*Scotus*) ex Al.^{p} 229.1?

The listed items, whose status as substance is under dispute, consist of numbers,[115] bodies, planes, and points, that is, of mathematical terms and geometrical figures. Both Alexander's lemma and his paraphrase indicate that in ω^{AL} the list does not include the planes (καὶ τὰ ἐπίπεδα).

> Alexander, *In Metaph.* 228.29–229.1 Hayduck
>
> Τούτων δ' ἐχομένη ἀπορία, πότερον οἱ ἀριθμοὶ καὶ [30] <u>τὰ σώματα καὶ αἱ στιγμαὶ</u> οὐσίαι τινές εἰσιν ἢ οὔ.[116]

---

[114]In the summary of the aporiae in B 1, the twelfth aporia is described thus (996a12–15): πρὸς δὲ τούτοις πότερον οἱ ἀριθμοὶ καὶ τὰ μήκη καὶ τὰ σχήματα καὶ αἱ στιγμαὶ οὐσίαι τινές εἰσιν ἢ οὔ, κἂν εἰ οὐσίαι πότερον κεχωρισμέναι τῶν αἰσθητῶν ἢ ἐν τούτοις. / "Further, whether numbers and lines and figures and points are a kind of substance or not, and if they are substances whether they are separate from sensible things or present in them." Mueller 2009: 191 points to the differences between the two versions of the aporia. On the twelfth aporia as a whole see Mueller 2009: 189–209.

[115]Aristotle neither mentions numbers in the first description of this aporia in B 1 nor do they play any role in the further treatment of the aporia (cf. 1002a12).

[116]Again the testimony in the lemma in S (n.iii.v.) is questionable (cf. 2.3). The reading in the lemma is based on the Latin *Metaphysics* version that precedes the commentary section in Sepúlveda's commentary and which agrees with the directly transmitted version of the text (i.e. it includes the words καὶ τὰ ἐπίπεδα or *plana*). By contrast, Alexander's paraphrase in the Latin version agrees with the commentary text transmitted in the Greek manuscripts and does not include "planes" (*et superficies videlicet et linea*).

[31] Δείξας δι' ὧν ἠπόρησεν, ὅτι οἱ ἀριθμοὶ οὐκ οὐσίαι ἔσονται, ἐφεξῆς [32] ἀπορεῖ τοῖς προειρημένοις, πότερον οἱ ἀριθμοὶ καὶ τὰ σώματα καὶ <u>αἱ στιγ-[229.1]μαί</u>, δηλονότι καὶ ἐπιφάνειαι καὶ γραμμαί, οὐσίαι εἰσὶν ἢ οὔ·

A question connected with these is whether numbers and <u>bodies and points</u> are substances or not.
Having shown, through the aporiae he raised, that numbers will not be substances, Aristotle next raises—in addition to the aporiae already discussed—the aporia, whether numbers and bodies and points, that is, surfaces and lines as well, are substance or not.

---

30 καὶ αἱ στιγμαὶ **A O** : καὶ αἱ γραμμαὶ καὶ τὰ ἐπίπεδα καὶ αἱ στιγμαὶ **P**ᵇ || 32 τοῖς προειρημένοις **A O P**ᵇ **S**(*praeterea*) Bonitz : τοῖς προηπορημένοις **LF** Ascl. Hayduck[117]

The lemma quotes lines 1001b26–28 without the words καὶ τὰ ἐπίπεδα.[118] Looking at the subsequent commentary text, we find confirmation that this is in fact the reading of ω^AL. For in his paraphrase, Alexander repeats the same list, saying οἱ ἀριθμοὶ καὶ τὰ σώματα καὶ αἱ στιγμαί (228.32–229.1), which confirms the wording of the lemma (οἱ ἀριθμοὶ καὶ τὰ σώματα καὶ αἱ στιγμαί). Thus the words καὶ τὰ ἐπίπεδα ("and planes") were missing from his text. Following the paraphrase, Alexander expands on the list by naming those terms that are clearly implied, but not explicitly stated in the list: καὶ ἐπιφάνειαι καὶ γραμμαί ("surfaces and lines"). His expansion, introduced by δηλονότι, "that is to say," would not make sense had καὶ τὰ ἐπίπεδα ("and planes") preceeded αἱ στιγμαὶ in his *Metaphysics* text.

The absence of the words καὶ τὰ ἐπίπεδα from ω^AL could be explained by a scribe's jumping from the second καὶ directly to the third, leaving out τὰ ἐπίπεδα in the process. Or is it instead possible that Alexander's text preserves Aristotle's original wording, which spoke only of the two extremes bodies and points, rather than of the tripartite hierarchy of bodies, planes and points? Should that prove to be the case, the words καὶ τὰ ἐπίπεδα in ω^αβ are to be taken as a later addition.[119]

Before examining the lists of mathematical terms in the remainder of the aporia one could ask whether it is possible that an ancestor of ω^αβ shared Alexander's reading (i.e., καὶ τὰ ἐπίπεδα was absent), but was later at the provocation of Alexander's comments supplemented and expanded to result in what we see in α and β. This seems unlikely, however, for the readings in α and β do not exactly coincide with what Alexander suggests is implied in Aristotle's words. Alexander asserts that ἐπιφάνειαι καὶ γραμμαί ("surfaces and lines") are implied,[120] but in our

---

[117] Hayduck's information on the reading in **A** (προειρημένοις) is incorrect.

[118] I follow the reading of **A** and **O**. In **P**ᵇ something interesting happened: apparently someone added τὰ ἐπίπεδα (most likely following the *Metaphysics* text) *and* καὶ αἱ γραμμαί (which is not even found in the *Metaphysics* text). Alexander's commentary remarks clearly indicate, however, that these additions were not part of his original lemma (see my comments below).

[119] Bonitz 1848 *app. crit.*: *omissa esse testatur Alex, fort. recte.*

[120] It is most interesting to see that the Arabic tradition apparently adopted Alexander's comments into the text. Scotus writes: *Et istam questionem consequitur alia difficilis, et est utrum numeri et corpo-*

text, we find only καὶ τὰ ἐπίπεδα ("surfaces").[121]

In order, then, to decide whether the reading in ω^AL is original and correct, it seems best to look for other passages in the twelfth aporia in which Aristotle lists mathematical entities. In all of the following five parallel passages (i–v) Aristotle *never* mentions only the two extremes of the hierarchy of geometrical terms, i.e. bodies and points. In other words, the text of ω^AL diverges from Aristotle's idiom.

Aristotle, *Metaphysics* (twelfth aporia)

i   [1002a4] ἀλλὰ μὴν τό γε σῶμα ἧττον οὐσία τῆς ἐπιφανείας, [5] καὶ αὕτη τῆς γραμμῆς, καὶ αὕτη τῆς μονάδος καὶ τῆς [6] στιγμῆς· But, on the other hand, a body is surely less of a substance than a surface, and a surface less than a line, and a line less than a unit and a point.

ii   [1002a15] ἀλλὰ μὴν εἰ τοῦτο μὲν ὁμολογεῖται, ὅτι μᾶλλον οὐσία τὰ [16] μήκη τῶν σωμάτων καὶ αἱ στιγμαί ... But if this is admitted, that lines and points are substance more than bodies ...

iii   [1002a23] οὐκ ἄρα οὐδ' ἐπιφάνεια (...), ὁ δ' [25] αὐτὸς λόγος καὶ ἐπὶ γραμμῆς καὶ ἐπὶ στιγμῆς καὶ μονάδος... therefore the surface is not in it either; ... And the same account applies to the line and to the point and the unit. ...

iv   [1002a32] τὰς δὲ στιγμὰς καὶ τὰς γραμμὰς καὶ τὰς [33] ἐπιφανείας οὐκ ἐνδέχεται οὔτε γίγνεσθαι οὔτε φθείρεσθαι... but points and lines and surfaces cannot be in process of becoming nor of perishing ...

v   [1002b8] ὁμοίως δὲ δῆλον ὅτι ἔχει καὶ περὶ [9] τὰς στιγμὰς καὶ τὰς γραμμὰς καὶ τὰ ἐπίπεδα· And evidently the same is true of points and lines and planes;

These parallel passages show that, while the way in which the mathematical terms are enumerated may vary, in none of these cases does the list include only the two extremes, bodies and points.[122] Perhaps it was on account of this idiom that Alexander felt the need to extend the short list given in ω^AL by the terms γραμμή and ἐπιφάνεια.

These parallel passages also show that Aristotle more frequently used the term ἐπιφάνεια ("surface") than the equivalent ἐπίπεδα ("surface, plane"). Yet, the phrase καὶ τὰ ἐπίπεδα, which is transmitted by ω^αβ and which most likely dropped out of ω^AL, is by no means unusual. In the concluding sentence of the

---

ra *et superficies et linee* et puncta sunt substantie aut non.

[121]This is what Ross 1924: clxii briefly notes. My emphasis on the divergence between the terms used relates to a difference in language not in content. As we will see, Aristotle himself uses the terms ἐπίπεδα and ἐπιφάνειαι interchangeably. Cf. Mueller 2009: 189.

[122]This conclusion receives further support from parallel passages in the rest of the *Metaphysics*. See e.g. *Metaph.* Z 2, 1028b16–18: δοκεῖ δέ τισι τὰ τοῦ σώματος πέρατα, οἷον ἐπιφάνεια καὶ γραμμὴ καὶ στιγμὴ καὶ μονάς, εἶναι οὐσίαι, καὶ μᾶλλον ἢ τὸ σῶμα καὶ τὸ στερεόν. Cf. also K 2, 1060b12–16; M 2, 1076b5–7; 1077a34–35 and *Ph.* B 2, 193b23–25.

aporia (1002b9, see v above), we see that ἐπίπεδα is used in a sense equivalent to ἐπιφάνεια.[123] Thus, the ω^{αβ}-reading of the introductory sentence (οἱ ἀριθμοὶ καὶ τὰ σώματα καὶ τὰ ἐπίπεδα καὶ αἱ στιγμαὶ οὐσίαι, 1001b26–27) should not arouse any suspicion, but should rather be understood as the *lectio difficilior* and as preferable to Alexander's own proposed augmentation of the sentence (καὶ ἐπιφάνειαι καὶ γραμμαί).

In sum, the most plausible conclusion is that in this case, as in the case before, we are dealing with an omission in ω^{AL} that did not occur in the tradition leading to ω^{αβ}.

In the four cases I have analyzed (4.2.1–4.2.4), ω^{AL} is found to contain errors that do not appear in ω^{αβ}.[124] On the basis of these discovered errors it is possible to conclude that ω^{αβ} is not a copy of ω^{AL}. This is a new result. With this, the assumption held by Primavesi 2012b that the text of the *Metaphysics* presupposed by Alexander's commentary and the "original text" (which Primavesi takes to be the text established by Andronicus[125]) are identical is shown to be incorrect. Since ω^{AL} contains separative errors against ω^{αβ} we have to assume that the versions Ω and ω^{AL} are not identical, and that ω^{αβ} is independent of ω^{AL}.[126]

Following the basic rules of Maas's *Textual Criticism*, the fact that both versions ω^{AL} and ω^{αβ} contain separative errors against each other allows us to conclude that the two versions are independent witnesses to the *Metaphysics* text. Neither is the direct source of the other.[127] Alexander's commentary can be used to date the text ω^{AL} to roughly AD 200, but the present knowledge of ω^{αβ} permits us to say only that it was produced before AD 400 (see 1; pp. 4–5).[128]

One could raise the following objection to the claim that ω^{αβ} is independent from ω^{AL}. If we assume that ω^{αβ} is younger than ω^{AL}, there is the theoretical possibility that the correct reading given in ω^{αβ} is only the result of a correction of a reading that had been previously shared with ω^{AL}.[129] In those cases where it was

---

[123] Alexander also confirms this use of ἐπίπεδα. In his commentary on the description of the twelfth aporia in B 1 (996a12–15), Alexander, whose text here agrees with ours, explains Aristotle's term τὰ σχήματα in the following way (180.26–27): τὰ σχήματα (λέγοι δ' ἂν τὰ ἐπίπεδα, τρίγωνον, τετράγωνον, κύκλον, τὰ τοιαῦτα)... / "shapes (he would mean plane figures: triangle, square, circle, and the like)."

[124] Cf. also the following erroneous reading in ω^{AL}: A 6, 988a11–12: τὰ εἴδη τὰ μὲν ... τὰ δὲ ἐπὶ ω^{AL} : τὰ εἴδη μὲν ... τὸ δ' ἓν ἐν β : τὰ εἴδη τὰ μὲν ... τὸ δ' ἓν ἐν α (β preserves the correct reading, which most likely was given in ω^{αβ}, and was slightly mutilated in α) (cf. 3.6).

[125] Primavesi 2012b: 457: "the 'original text' as edited in the first century BC and used by Alexander c. AD 200."

[126] This does not mean that ω^{AL} and ω^{αβ} could not have a conjunctive error. Such an error could be given in A 6, 987b22: τὰ εἴδη εἶναι τοὺς ἀριθμοὺς ω^{αβ} ω^{AL} (Al.^c 53.5–6, Al.^p 53.9–11), where τοὺς ἀριθμοὺς seems to be a gloss that had found its way into the text. This possible error can either be attributed to Ω itself or to a copy of Ω, from which ω^{AL} and ω^{αβ} both descend.

[127] See Maas 1958: 42–43. Cf. Erbse 1979: 549–52.

[128] A secure *terminus post quem* will be determined in section 5.1.

[129] Maas 1958: 42: "We can prove that a witness (B) is independent of another witness (A) by finding

appropriate to raise this question (see above pp. 133–34 and 135–36) I concluded that there were no indications that could lend credence to this objection; accordingly I did not pursue this possibility.[130] Given that it is highly unlikely that in the two centuries between AD 200 and 400 all of the errors preserved in ω^{AL} had been corrected in $\omega^{\alpha\beta}$, we can conclude that the two versions ω^{AL} and $\omega^{\alpha\beta}$ are independent from one another.

What we may assume, then, is that in the roughly 250 years that lie between the edition produced in the first century BC and the date of ω^{AL}, several different versions of the *Metaphysics* were circulating. At some point between the first century BC and the second century AD the text of ω^{AL} emerged, which contained errors that were not part of the first-century-BC edition. At some point before the end of the fourth century AD, $\omega^{\alpha\beta}$ was produced as a copy of a version that was not identical with ω^{AL}. This version may be referred to as ω^{ASP1}, since, as was shown by the four case studies in 3.5.2 and is now further corroborated by the cases in 4.1.1 and 4.2.3, Alexander had indirect access (presumably via other commentators such as Aspasius) to $\omega^{\alpha\beta}$ or one of its ancestors. This version (ω^{ASP1}) contained errors that were neither present in the first-century-BC edition nor in ω^{AL}. Our direct transmission descends from this copy.[131]

4.3 ω^{AL} AS CRITERION FOR PRIORITY IN CASES OF DIVERGENCE BETWEEN α AND β

The fact that $\omega^{\alpha\beta}$ and ω^{AL} (second century AD) are two independent witnesses to the *Metaphysics* text lends considerable strength to the evidence available in Alexander's commentary. The transmission of the *Metaphysics* text brought down to us two divergent versions, whose readings often compete. The divergences between the two versions are either due to mistakes or to intentional changes that occurred in one (or perhaps both) of the two versions.[132] Since versions α and β often offer different, yet viable readings, a third witness that is independent from the two and even from their ancestor $\omega^{\alpha\beta}$ is most welcome. Such a third witness could assist us in identifying the older reading in those instances where α and β differ from one another. Alexander's commentary, despite providing only re-

---

in A as against B an error so constituted that our knowledge of the state of conjectural criticism in the period between A and B enables us to feel confident that it cannot have been removed by conjecture during that period." Cf. also Erbse 1979: 550 and Pöhlmann 2003b: 140. Cf. also 1, p. 14.

[130] Apart from this specific kind of contamination (Cf. Maas 1960: 8–9; Pöhlmann 2003b: 143–49) there is, of course, the general uncertainty that unavoidably attends the determination of manuscripts, which is due to no other fact than that in most cases we do not know all participants in the transmission process of a given text.

[131] From this perspective, it would be an astonishing coincidence if $\omega^{\alpha\beta}$ had been a direct copy of ω^{AL}.

[132] Cf. 1. On the character of the two versions see Frede/Patzig 1988: 13–17 and Primavesi 2012b.

stricted access to ω^AL, functions as just such a witness.¹³³ Therefore, the agreement of ω^AL with one of the two readings in α or β is a crucial criterion for identifying the older reading that had been given in ω^αβ.

This fact can be illustrated with the following seven exemplary cases (4.3.1–4.3.3). In analyzing these cases, I intend to show that, given that ω^AL and ω^αβ are two independent versions of the *Metaphysics* (4.1–4.2), the agreement of ω^AL with α or the agreement of ω^AL with β can and should be regarded as weighty evidence in favor of the agreed upon reading. Nevertheless for each of the seven cases (and for various reasons) the reading that is shared between ω^AL and α or β could have been or indeed has been questioned as being the correct or preferable reading. My analysis therefore aims to accomplish two things: (i) to find out which of the two divergent readings in α and β is confirmed by the evidence in Alexander's commentary to be the reading in ω^αβ and (ii) to show that this is the preferable and most likely correct reading.

Regarding the second aim, the following should be noted. Readings that are shown to be those of ω^αβ through the agreement of ω^AL and either α or β are not necessarily the correct and preferable reading of Aristotle's text. It remains possible that a reading shared between ω^AL and either α or β (and therewith ω^αβ) is wrong, while the diverging reading in α or β alone is the correct reading. Such a scenario can be explained in the following ways: an error in ω^AL that is shared with either α or β against the other could point to an old error shared by ω^AL and ω^αβ that survived in α or β and was later *corrected* in the divergent version.¹³⁴ Alternatively, a corruption shared by ω^AL with α or β respectively could theoretically be due to contaminations by an otherwise unknown version that occurred in ω^AL and α or β independently.¹³⁵ We do not expect this to happen very often, however. A *considerable number* of such *conjunctive errors* (ω^AL + α or ω^AL + β) against the

---

¹³³ See Ross 1924: clxi and especially Primavesi 2012b: 409–10. Cf. also 1.

¹³⁴ The possibility of later corrections in α or β has further implications for the assessment of the agreement of ω^AL with either α or β. The agreement does not necessarily bring us to the *original* reading of the first century BC edition. It only brings us to ω^αβ, which, of course, could have been corrupt. Therefore, the editor of the *Metaphysics* is not in all cases necessitated to prefer the ω^αβ-reading.

¹³⁵ Rashed/Auffret 2014 recently drew attention to seven passages in *Metaphysics* A in which it seems as if Alexander's text shares readings with α against β that could, according to the argument proposed by Rashed/Auffret, be seen as the result of an editorial redaction (executed by the middle Platonist Eudorus of Alexandria) and so could (according to my parameters) be regarded as "conjunctive errors" of α and ω^AL against β. Yet in two of the seven passages (981a10–12 and 986a15–21) I think it is questionable whether Alexander's text actually had the α-reading. In three of the remaining five passages, it is quite possibly that α and ω^AL share the correct readings, while a scribal error occurred in β (in 985b23–26 due to *saut du même au même*, 988b24–26, 989b6–9). If there are still sporadic additions (of possible Eudorian origin) found in α, only some of which are known to Alexander (cf. Rashed/Auffret 2014: 60; 82–3), but the split of ω^αβ into α and β happened at a time *after* AD 200 (see 5.1), then these additions are likely to be attributed to contaminations by a *Metaphysics* version of possibly Eudorian origin that occurred in α and ω^AL independently, at different times, and to a different extent (cf. also the passages discussed in 5.4).

correct reading in the divergent version (α or β) would suggest dependencies between two of them against the other and thus could undermine the view that ω^AL and α and β (ω^αβ) are independent witnesses. Therefore, it will not be sufficient in the following case studies to only work out that ω^AL agrees with either α or β, but also that the agreed reading is preferable.

4.3.1 Separative errors in α against β + ω^AL

4.3.1.1 Alex. *In Metaph.* 299.5–9 on Arist. *Metaph.* Γ 4, 1008b12–19

In the fourth chapter of book Γ Aristotle demonstrates the validity of the principle of non-contradiction by pointing to the difficulties that result from its denial. In 1008b7–8 he says that when every statement is simultaneously true and false it is impossible to make any statement at all. Even if the statement in question is just an opinion rather than an assertion (μηθὲν ὑπολαμβάνει, b10), it nevertheless would result in opining something and at the same time not opining it (ὁμοίως οἴεται καὶ οὐκ οἴεται, b10–11). In a world such as this, where the principle does not obtain, a human is effectively a plant (τί ἂν διαφερόντως ἔχοι τῶν γε φυτῶν).[136] Aristotle continues the argument in the following way:

> Aristotle, *Metaphysics* Γ 4, 1008b12–19
>
> ὅθεν καὶ μάλιστα φανερόν ἐστιν ὅτι οὐδεὶς οὕτω διά-[13]κειται οὔτε τῶν ἄλλων οὔτε τῶν λεγόντων τὸν λόγον τοῦτον. [14] διὰ τί γὰρ βαδίζει Μέγαράδε ἀλλ' οὐχ ἡσυχάζει, οἰόμε-[15]νος **βαδίζειν δεῖν**; οὐδ' εὐθέως ἔωθεν πορεύεται εἰς φρέαρ ἢ εἰς [16] φάραγγα, ἐὰν τύχῃ, ἀλλὰ φαίνεται εὐλαβούμενος, ὡς οὐχ [17] ὁμοίως οἰόμενος μὴ ἀγαθὸν εἶναι τὸ ἐμπεσεῖν καὶ ἀγαθόν; [18] δῆλον ἄρα ὅτι τὸ μὲν βέλτιον ὑπολαμβάνει τὸ δ' οὐ βέλ-[19]τιον.
>
> Thus, then, it is in the highest degree evident that neither any one of those who maintain this view nor any one else is really in this position. For why does a man walk to Megara rather than stay at home when he thinks **he ought to walk**? Why does he not walk in the morning straight into a well or over a precipice, if one happens to be in his way, but evidently guards himself against this, not thinking that falling in is alike good and not good? Evidently he judges the one thing to be better and the other worse.

---
15 βαδίζειν δεῖν β ω^AL (Al.^P 299.7–9) Ar^u (*Scotus*) Ross Jaeger : βαδίζειν α Bekker Bonitz Christ Cassin/Narcy Hecquet-Devienne ‖ ἢ εἰς β Al.^P 299.10 edd. : ἢ α

At this point in the argument against the denial of the principle of non-contradiction Aristotle's focus is on human actions. Although it may appear possible to

---

[136] For a reconstruction of the text in 1008b11–12 made on the basis of Alexander's commentary (πεφυκότων ω^αβ Bekker : ex Al.^P 298.31; 299.2, 7 γε φυτῶν Ross, φυτῶν Bonitz Christ Jaeger) see Bonitz 1847: 88–89.

deny the principle, it is not possible in practice. The denial flies in the face of the fundamentals of human behavior (1008b12–14). Aristotle illustrates this with two examples. These show human beings do certain things and avoid others, because they assert that it is better to do so and they do not simultaneously believe that it is not better to do so. As explained in *De anima*, Γ 9–10 and even more so in *De motu animalium* (6, 700b15–b29), motion in human beings is caused by two capacities, desire (ὄρεξις) and reason (νοῦς).[137] Movement is preceded by choice (προαίρεσις), which consists in recognizing the object of desire as the goal of the action.[138] In such a decision-making process it is impossible that the desired object be thought to be good and not good at the same time. If that were the case, no choice would be made and no movement would follow.

The *second* of the two examples (1008b15–17) that Aristotle adduces in this passage is highly intuitive and, what is more, α and β do not display a decisive difference in the transmission of the text. This example is visceral: we take care not to fall into a well or precipice because we think doing so would be damaging and bad for us. By contrast, the *first* example (1008b14–15) appears to be less obvious, and on top of that is transmitted differently in the α- and the β-versions. According to the α-version, Aristotle states in the first example that the opponents of the principle, who say that it is possible to opine something and simultaneously not opine it, cannot explain why someone would walk to Megara (βαδίζει Μέγαράδε) rather than just stay at home and think that one is walking (οἰόμενος βαδίζειν). According to the β-version, whose reading is confirmed here by the Arabic transmission of the text,[139] Aristotle challenges the opponents by asking why someone would walk to Megara rather than just stay at home when one thinks he ought to walk (οἰόμενος βαδίζειν δεῖν). The β-version thus poses a pressing question for the deniers of the principle that the α-version does not. According to the β-version, it makes no difference to the deniers of the principle whether one thinks one ought to go or not.

Bekker, Bonitz, and Christ followed the α-reading (οἰόμενος βαδίζειν). It must be noted, however, that these editors do not even mention the β-reading (οἰόμενος βαδίζειν δεῖν) in their apparatus. This absence strongly suggests that they did not know that this reading exists.[140] Cassin/Narcy also follow the α-reading, but they clearly knew of and consciously rejected the alternative β-reading.[141] Hecquet-Devienne 2008 follows their decision and reads the α-text, too. By contrast,

---

[137] *MA* 6, 700b17–b19: ὁρῶμεν δὲ τὰ κινοῦντα τὸ ζῷον διάνοιαν καὶ φαντασίαν καὶ προαίρεσιν καὶ βούλησιν καὶ ἐπιθυμίαν. ταῦτα δὲ πάντα ἀνάγεται εἰς νοῦν καὶ ὄρεξιν. Cf. *de An.* Γ 10, 433a13–14: ἄμφω ἄρα ταῦτα κινητικὰ κατὰ τόπον, νοῦς καὶ ὄρεξις, νοῦς δὲ ὁ ἕνεκά του λογιζόμενος καὶ ὁ πρακτικός.

[138] *MA* 6, 700b25–b28: διὸ τὸ τοιοῦτόν ἐστιν τῶν ἀγαθῶν τὸ κινοῦν, ἀλλ' οὐ πᾶν τὸ καλόν· ᾗ γὰρ ἕνεκα τούτου ἄλλο καὶ ᾗ τέλος ἐστὶν τῶν ἄλλου τινὸς ἕνεκα ὄντων, ταύτῃ κινεῖ.

[139] Scotus: *quia opinatur quia ambulandum est*.

[140] See Ross 1924: 272: "*I have restored the reading of* A^b ..." (emphasis added).

[141] See their commentary in Cassin/Narcy 1989: 227–29.

Ross and Jaeger opted for the β-reading.[142] From a paleographical point of view it seems entirely justified to take the β-reading as the original from which the α-version emerged due to *homoioteleuton*. The corruption from ΒΑΔΙΖΕΙΝΔΕΙΝ to ΒΑΔΙΖΕΙΝ seems a much more plausible explanation than that the word δεῖν was added in the course of the transmission.

Alexander's paraphrase of the *Metaphysics* passage as well as his explication of it indicates that his text contained the reading οἰόμενος βαδίζειν δεῖν (β).[143]

Alexander, *In Metaph.* 299.5–9 Hayduck

εἵπετο γὰρ τοῖς οὕτω διακειμένοις καὶ ταύ-[6]την τὴν ὑπόληψιν περὶ τῶν πραγμάτων ἔχουσι τὸ μηδὲ πράσσειν τι, ἀλλ' [7] ὡς φυτοῖς ζῆν. τί γὰρ μᾶλλον περιπατητέον τῷ οἰομένῳ <u>δεῖν</u> περιπατεῖν [8] τοῦ μὴ περιπατητέον, εἰ ἴσαι εἰσὶν αἱ περὶ ἑκάστου ὑπολήψεις, καὶ ἴσον [9] τὸ οἴεσθαι <u>δεῖν</u> ἢ μὴ οἴεσθαι <u>δεῖν</u> ἢ οἴεσθαι μὴ <u>δεῖν</u>;

For people in such a condition, i.e. having this supposition concerning things, it would follow that they would not even do anything but rather would live like plants. For why should the person who thinks he must go for a walk go for a walk rather than the person who thinks that he does not need to go for a walk, if the suppositions concerning each thing are equivalent, i.e. if thinking that one must go for a walk is equivalent to not thinking that one must go for a walk, or to thinking that it is not necessary to go for a walk?

6 περὶ O LF S : om. A P^b ‖ μηδὲ A P^b : μὴ O ‖ 7–8 τῷ ... περιπατητέον O S LF : om A P^b ‖ 9 ἢ pr. A O : τῷ P^b ‖ μὴ ... δεῖν A P^b : om. O

According to Alexander's explication of Aristotle's argument, the first example illustrates that the walking would not happen if the thought that one ought to go and the thought that one does not need to go were identical. Alexander's testimony confirms the reading we find in β (βαδίζειν δεῖν), and so it is more likely that the β-text here preserves the reading that was in ω^{αβ}. The α-reading (βαδίζειν), then, as we already suspected, is the result of a corruption in which the δεῖν dropped out.

Nevertheless, Cassin/Narcy (1989: 227–29) defend the α-reading resolutely.[144]

---

[142]Ross 1924: 272: "I have restored the reading of A^b and Alexander, βαδίζειν δεῖν. The point is, as the corresponding instance of the precipice shows, not that a man cannot think both that he is walking to Megara and that he is not, but that he cannot think both that he ought to walk to Megara and that he ought not." Kirwan 1971 translates Ross's text.

[143]Asclepius seems to have found the α-reading in his *Metaphysics* text (Ascl.^{p} 272.24–25). Syrianus's paraphrase suggests that he read β (Syr.^{p} 73.10–12 Kroll), although he seems not to have understood what is meant by the walk to Megara (see below).

[144]Hecquet-Devienne 2008 follows them. Cassin/Narcy 1989: 111 note that they generally prefer the agreement of E and J (i.e. α) against A^b (one main source for the β-version). It appears to me that Cassin/Narcy overdo this preference. In so doing, they seem to ignore the most basic fact that even if the α-version on the whole offers a more reliable text, it must also contain mistakes, however few, which is a natural and unavoidable result of the transmission process. For a general characterization

According to Cassin/Narcy, Aristotle introduces with the example of the walk to Megara a new argument, but Alexander's explanation of the passage and the reading that is given in the β-text both amount to an unacceptable attempt to make continuous a train of thought that is, according to Cassin/Narcy, in fact not.[145] The purpose of the example of the walk to Megara was not to illustrate the "apraxie"[146] that results for the deniers of the principle. Rather, Aristotle wants to point out the absurdity that confront the deniers of the principle: they think that walking and not walking are indistinguishable and yet they conduct themselves in daily tasks (as, for example, in walking to Megara) as if walking and not walking were distinct.[147]

Cassin and Narcy's argument for keeping the α-reading, however, does not hold water, because in the *Metaphysics* passage there is no mention of someone who thinks simultaneously of walking and not walking. Rather, Aristotle asks *why* (διὰ τί) someone would walk to Megara, when one thinks that one is walking (α) or ought to walk (β). The weakness of the α-version is that Aristotle's question loses its rhetorical force, since it poses an unanswerable question. There could be many reasons why someone would walk to Megara when he thinks that he is walking. A pressing connection between thinking and walking is given only when the thinking consists in a call to action. This is only the case in the β-reading.[148]

The preference shown by some editors for the α-reading seems to rest on a misinterpretation of the overall meaning of the walk to Megara. Whereas the avoidance of a well or precipice is easily comprehensible one could ask why in the world one should think that one ought to go to Megara. A passage from the beginning of Plato's *Phaedrus* offers help. In this passage Socrates reaffirms his desire to listen to Lysias's speech, which Phaedrus carries with him, thus:

Plato, *Phaedrus* 227d2–5

ἔγωγ' οὖν οὕτως ἐπιτεθύμηκα ἀκοῦσαι, ὥστ' ἐὰν βαδίζων ποιῇ τὸν περίπατον Μέγαράδε καὶ κατὰ Ἡρόδικον προσβὰς τῷ τείχει πάλιν ἀπίῃς, οὐ μή σου ἀπολειφθῶ.

---

of the two families α and β see 1.

[145] Cassin/Narcy 1989: 228: "Alexandre masque la rupture qui s'opère à cet endroit dans l'argumentation d'Aristote."

[146] Cassin/Narcy 1989: 228.

[147] Cassin/Narcy 1989: 228–29: "Le nouveauté de cet argument peut être résumée de la façon suivante: jusque-là, Aristote a montré que, si l'on s'en tient au discours de l'adversaire, on doit conclure qu'il ne dit rien et ne diffère pas d'une plante. Si, à l'inverse, argumente-t-il maintenant, on s'en tient à l'ordinaire de ses actions, on constate qu'il ne diffère pas des autres hommes. Mais cela, à soi seul, implique qu'il n'agit pas conformément à son discours, qu'il ne le soutient pas."

[148] Further, the phrase οἰόμενος ... δεῖν, which we read in the β-text, is a common Greek idiom that means "to believe that something is right to do," "to decide to do." Cf. the cases in *EE* B 10, 1226a5–6, *EE* Θ 15, 1249a14–15, *EN* Δ 6, 1126b13–14, *EN* H 10, 1152a5–6 and Plato *Smp.* 173a2 (see Dover 1980: 78, who translates "'thinking it to be necessary,' i.e. 'choosing,' 'preferring.'"), *Phd.* 83b5.

However, I'm so eager to hear it [the speech] that I vow I won't leave you even if you extend your walk as far as Megara, up to the walls and back again as recommended by Herodicus.[149]

Socrates here recalls Herodicus of Megara, a famous doctor and physical therapist (παιδοτρίβης) from the fifth century BC, who hailed from Megara and later lived in Selymbria.[150] To him is attributed a strict regimen of exercises, which apparently mostly consisted in extensive walks.[151] When Socrates here imagines a walk from Athens to the walls of Megara and back again he seems to be exaggerating his point—the distance between Athens and Megara is about 40 kilometers. Hermias of Alexandria, the Neoplatonic commentator on the *Phaedrus*, assumes that Herodicus's recommended exercise was to start a moderate distance from the city walls and then walk there and back repeatedly.[152]

The example of a "walk to Megara" is an instance of Aristotle's practice of citing walking as a paradigmatic example of the means by which we reach the goal of health. In the canonical chapter on the four-cause theory, *Physics* II 3, health functions as the final cause of walking.[153]

Aristotle, *Physics* B 3, 194b32–35 (= *Metaphysics* Δ 2, 1013a32–35)

ἔτι ὡς τὸ τέλος· τοῦτο δ' ἐστὶν [33] τὸ οὗ ἕνεκα, οἷον τοῦ περιπατεῖν ἡ ὑγίεια· διὰ τί γὰρ περι-[34]πατεῖ; φαμέν ἵνα ὑγιαίνῃ, καὶ εἰπόντες οὕτως οἰόμεθα ἀπο-[35]δεδωκέναι τὸ αἴτιον.

Again, in the sense of end. This is that for the sake of which a thing is done, e.g. health is the cause of walking about. ('Why is he walking about?' We say: 'to be healthy,' and, having said that, we think we have assigned the cause.)

Even more generally, walking functions as a standard example of illustrating how actions are performed for the sake of attaining some good. In a passage in Plato's *Gorgias* walking and standing still are introduced as means to attain a good. In his

---

[149] Translation by Hackforth, slightly changed.

[150] Touwaide 1998: 468. Plato mentions Herodicus also in *Prot.* 316e1 and *Rep.* III 406a7–b8.

[151] Heitsch 1997: 72 says in his commentary on the *Phaedrus*: "auf Grund eigener Erfahrungen überzeugt, dass Krankheiten ihre Ursachen in falscher Lebensführung haben, entwickelte er Anweisungen, die allerdings so aufwendig waren, dass dem, der sie befolgen wollte, für andere Tätigkeiten keine Zeit mehr blieb."

[152] Hermias, *In Platonis Phaedrum scholia* 24.25–30 Couvreur: Ὁ δὲ Ἡρόδικος ὁ Σηλυμβριανὸς ἰατρὸς ἦν καὶ τὰ γυμνάσια ἔξω τείχους ἐποιεῖτο, ἀρχόμενος ἀπό τινος διαστήματος οὐ μακροῦ ἀλλὰ συμμέτρου ἄχρι τοῦ τείχους, καὶ ἀναστρέφων, καὶ τοῦτο πολλάκις ποιῶν ἐγυμνάζετο. "Ὅπερ οὖν ὁ Ἡρόδικος ἐποίει ἔξω τοῦ τείχους, ἐὰν σὺ τοῦτο ἄχρι Μεγάρων πολλάκις ποιῇς, οὐ μή σου ἀπολειφθῶ." / "Herodicus of Selymbria was a physician and he did his exercises outside of the city walls by starting from a certain distance which was not too far but reasonably close to the walls and then going back and forth. His exercise consisted in doing this repeatedly. 'So, even if you do repeatedly and up to Megara what Herodicus did outside of the walls I won't leave you' (227d5)."

[153] Cf. also *Metaph.* α 2, 994a9.

conversation with the young Polus, Socrates distinguishes between the means and the end of an action and determines that the end of our actions is always the good.

Plato, *Gorgias* 468b1-4

Τὸ ἀγαθὸν ἄρα διώκοντες [2] καὶ <u>βαδίζομεν</u> ὅταν βαδίζωμεν, <u>οἰόμενοι βέλτιον εἶναι</u>, καὶ [3] τὸ ἐναντίον ἕσταμεν ὅταν ἑστῶμεν, τοῦ αὐτοῦ ἕνεκα, τοῦ [4] ἀγαθοῦ· ἢ οὔ;

Then it is in pursuit of the good that we both walk when we walk, thinking it is better, and on the other hand stand still when we stand still, for the sake of the same thing, the good. Isn't that so?[154]

Returning to our *Metaphysics* passage and the walk to Megara we find two aspects combined: the general aspect of walking as goal-oriented, and the more specific aspect of walking to Megara as health-oriented. Aristotle's question amounts, then, to this: why would one walk to Megara rather than stay at home when one *thinks* one *ought to* walk? And one ought to walk to Megara because, so say the physicians of Aristotle's time, it is good for one's health. But in order to actually walk one has both to believe in the effects of walking therapy and think that one ought to go, for otherwise, one would not walk at all. In any case, it is impossible to believe at the same time that one should and should not go.

Further, Aristotle's talk about the human being who walks to Megara and does not stay at home presupposes Aristotle's theory of human motion, which he develops in *De anima* Γ 9–10 and especially *De motu animalium*.

Aristotle, *De motu animalium* 7, 701a13-15

οἷον ὅταν νοήσῃ ὅτι παντὶ βαδιστέον ἀνθρώπῳ, αὐτὸς [14] δὲ ἄνθρωπος, βαδίζει εὐθέως, ἂν δ᾽ ὅτι οὐδενὶ βαδιστέον νῦν [15] ἀνθρώπῳ, αὐτὸς δ᾽ ἄνθρωπος, εὐθὺς ἠρεμεῖ·

For example, whenever someone thinks that every man should take walks, and that he is a man, at once he takes a walk. Or if he thinks that no man should take a walk now, and that he is a man, at once he remains at rest.[155]

According to Aristotle an action is the conclusion of a practical syllogism (701a22–23: ὅτι μὲν οὖν ἡ πρᾶξις τὸ συμπέρασμα).[156] Its two premises concern the good and the possible:[157] A man thinks that it is good to walk and he is able to walk, therefore he walks.

This structure is presupposed in our *Metaphysics* passage and shows the ab-

---

[154] Translation by T. Irwin.

[155] The text of this passage of *MA* follows the edition currently prepared by Oliver Primavesi. The translation is by Nussbaum. (There are no editorial changes in this passage that affect the translation.)

[156] See *MA*, 7, 701a10-20. On the practical syllogism see Rapp/Brüllmann 2008.

[157] *MA*, 7, 701a23-25: αἱ δὲ προτάσεις αἱ ποιητικαὶ διὰ δύο εἰδῶν γίνονται, διά τε τοῦ ἀγαθοῦ καὶ διὰ τοῦ δυνατοῦ.

surdity of denying the principle of non-contradiction. Prerequisite to an act, for example, of walking is the belief that one ought to walk (for walking is for the sake of health, for example). The possibility of thinking simultaneously that one ought to walk and does not need to walk (for the sake of health) is thereby excluded. For if this were possible, the syllogism could not be completed and so no action would take place. Or, were it possible, it could result in dangerous and life-threatening actions, as Aristotle illustrates in his second example (1008b15–17). The example of the walk to Megara and the example of the avoidance of the well and the precipice do differ, but not in the way Cassin/Narcy suppose. In the former, Aristotle shows that it is necessary to firmly believe that something is good in order *to act* in a certain way, and in the latter, Aristotle shows, that it is necessary to firmly believe that something is good in order *to avoid* a certain hindrance.

4.3.1.2 Alex. *In Metaph.* 419.25–420.3 on Arist. *Metaph.* Δ 22, 1022b32–36

The 22nd chapter of book Δ treats of privation (στέρησις). First, Aristotle distinguishes between four types of privation: we speak of privation (i) when a given thing does not possess what can be naturally possessed by some other thing, e.g., plants lack eyes (1022b22–24); (ii) when a given thing lacks what it or its genus naturally possess, a blind man or mole lacks vision (b24–27); (iii) when a given thing lacks an attribute at a time at which it would naturally possess it, e.g., blindness at an age at which one would naturally have sight (b27–31); (iv) when a given thing has lost an attribute by violence (b31–32).

In the lines below Aristotle determines that the spectrum of the term privation coincides with that of the alpha privative (b32–33). The following three cases aim at this result.[158]

Aristotle, *Metaphysics* Δ 22, 1022b32–36

καὶ ὁσαχῶς δὲ αἱ ἀπὸ τοῦ α ἀποφάσεις λέγον-[33]ται, τοσαυταχῶς καὶ αἱ στερήσεις λέγονται· (i) ἄνισον μὲν [34] γὰρ τῷ μὴ ἔχειν ἰσότητα πεφυκὸς λέγεται, (ii) ἀόρατον δὲ [35] καὶ τῷ ὅλως μὴ ἔχειν χρῶμα [**καὶ τῷ φαύλως**], (iii) καὶ ἄπουν [36] καὶ τῷ μὴ ἔχειν ὅλως πόδας καὶ τῷ φαύλους.

There are just as many kinds of privations as there are of words with negative prefixes; (i) for a thing is called *unequal [an-ison]* because it has not equality though it would naturally have it, (ii) but *invisible [a-horaton]* also[159] when it has no color at

---

[158] For the present I confine myself to these three cases. These three share a common feature: under discussion is natural possession or natural lack of possession. Later on Aristotle discusses other applications of the alpha privative, which are introduced by ἔτι (b36); the ἔτι marks these others as their own distinct group. I will come back to this group below.

[159] The καί ("also") indicates that the criterion of invisibility given here holds *in addition* or *alternatively* to the criterion given in type (i): not having something although one would naturally have it.

all,¹⁶⁰ (iii) and footless [*a-poun*] either because it has no feet at all or because it has imperfect feet.

---

32 αἱ ἀπὸ τοῦ α ἀποφάσεις] αἱ ἀποφάσεις αἱ ἀπὸ τοῦ ἄλφα Al.ᶜ 419.23 || 34 τῷ **EV**ᵈ edd. : τὸ γ β || ἰσότητα **α** Al.ᵖ 419.29 edd. : om. β || 35 τῷ **α** Al.ᶜ 419.32 edd. : τὸ β || ὅλως μὴ **α** Al.ᶜ 419.32 edd. : μὴ ὅλως β || καὶ τῷ φαύλως **α** Ascl.ᵖ 348.9-10 Bekker Christ Ross Jaeger, *secl.* Bonitz : om. β ω^AL (Al. 419.30-420.3) Ar^u (*Scotus*) || 36 τῷ bis **α** edd. : τὸ bis β || φαύλους **α** Al.ᵖ 420.2 edd. : φαύλως β

In the following I will focus on lines b34-35 in particular. The α-text reads ἀόρατον δὲ καὶ τῷ ὅλως μὴ ἔχειν χρῶμα <u>καὶ τῷ φαύλως</u> ("something is called *invisible* either because it has no color at all or because it has a poor color"). From a syntactical perspective, the words καὶ τῷ φαύλως seem to fit perfectly into the immediate context, as they seem to be anticipated by the preceding καὶ (in καὶ τῷ ὅλως, 1022b35). As for the content, the last three words καὶ τῷ φαύλως (b35) are suspicious if we assume that in Greek, just as in English or German, something that is, however poorly, *visible* cannot be called *in*visible.¹⁶¹

The words καὶ τῷ φαύλως are missing from the β-version. Might they have fallen out due to *saut du même au même*? Since another καὶ (in καὶ ἄπουν, b35) immediately follows καὶ τῷ φαύλως a scribe might have jumped accidentally from <u>καὶ</u> τῷ φαύλως to <u>καὶ</u> ἄπουν. From this perspective the β-reading appears inferior. The Arabic version (translation by Ustāth), however, confirms the β-text.¹⁶² Most of the modern commentators settle for the α-version: Bekker, Christ,¹⁶³ Jaeger and Ross read the α-text;¹⁶⁴ Bonitz alone followed the β-version and athetized the words καὶ τῷ φαύλως.

The picture of the α-reading as the superior reading, however, is blurred by two stains and Alexander's commentary calls attention to both of them. The first stain is a weakness in the logical consistency of Aristotle's trifurcation of the alpha privative. The second shows up when we compare it with the reading in ω^AL. We have access to ω^AL through Alexander's citation as well as paraphrase (419.22-420.3):

I come back to this καὶ below.

¹⁶⁰For the phrase ὅλως μὴ ἔχειν see *Metaph.* I 5, 1055b4-5: ἡ δὲ στέρησις ἀντίφασίς τίς ἐστιν· ἢ γὰρ τὸ ἀδύνατον ὅλως ἔχειν, ἢ ὃ ἂν πεφυκὸς ἔχειν μὴ ἔχῃ, ἐστέρηται ἢ ὅλως ἢ πὼς ἀφορισθέν ("Privation is a kind of contradiction; for what suffers privation, either in general or in some determinate way, is either that which is quite incapable of having some attribute or that which, being of such a nature as to have it, has it not."). Cf. also Θ 1 1046a31-33: ἡ δὲ στέρησις λέγεται πολλαχῶς· καὶ γὰρ τὸ μὴ ἔχον καὶ τὸ πεφυκὸς ἂν μὴ ἔχῃ, ἢ ὅλως ἢ ὅτε πέφυκεν.

¹⁶¹The LSJ does not offer any instances of ἀόρατον where the meaning is "poorly visible." But as I will discuss below, Aristotle seems to use ἀόρατον in the sense of "poorly visible" in *de An.* B 10, 422a28.

¹⁶²This can be inferred from Scotus's translation: *et dicitur "non visibile" quod non habet colorem omnino, et dicitur "non habens pedem"*... .

¹⁶³See, however, Christ 1853: 21.

¹⁶⁴Kirwan 1971: 57 follows this reading in his translation.

Alexander, *In Metaph.* 419.25–420.3 Hayduck

ὁσαχῶς δή, φησίν, αἱ διὰ τοῦ α ἀναιρέσεις τε [26] καὶ ἀποφάσεις τινῶν γίνονται, τοσαυταχῶς λέγεσθαι καὶ τὰς στερήσεις· [27] στερήσεως γὰρ ἡ διὰ τοῦ α ἀπόφασις δηλωτική. καὶ παρατίθεται καὶ διὰ [28] τῶν παραδειγμάτων δείκνυσιν αὐτῶν τὴν διαφοράν. ἄνισον μὲν γὰρ τὸ [29] πεφυκὸς ἰσότητα ἔχειν καὶ μὴ ἔχον λέγεται, ὥστε ἡ στέρησις καὶ τοῦτο [30] σημαίνει· ἀόρατον δὲ λέγεται μὲν καὶ τὸ πεφυκὸς ὁρᾶσθαι καὶ μὴ ὁρώ-[31]μενον,[165] ὃ ἐδήλωσε διὰ τοῦ τῷ προειρημένῳ τῷ ἐπὶ τοῦ ἀνίσου προσθεῖναι [32] καὶ τῷ ὅλως μὴ ἔχειν χρῶμα· λέγεται γὰρ καὶ τὸ τὴν ἀρχὴν μήτε [33] πεφυκὸς ὁρᾶσθαι μήτε ἔχον χρῶμα, ὡς ὅταν λέγωμεν τὴν φωνὴν ἀόρατον [34] εἶναι· ὥστε καὶ κατὰ τούτου ἡ στέρησις τοῦ σημαινομένου, ὅ ἐστι τὸ ἀδύ-[420.1]νατον. πάλιν ἄπουν λέγεται τό τε μηδὲ ὅλως πεφυκὸς πόδας ἔχειν, ὡς [2] τὰ ἑρπετά, καὶ τὸ φαύλους ἔχον, ὥστε καὶ κατὰ φαύλως τι ἔχοντος ἡ στέ-[3]ρησις λέγεται.

He is saying, then, that there are as many privations as there are denials through the letter *alpha*, i.e. negations, of certain attributes, for the negation expressed by the letter *alpha* indicates a privation; and he adds examples to show the difference among them. [i] For a thing is called 'unequal' if it could naturally have equality and does not have it, so that privation also signifies this fact; [ii] but a thing is 'invisible' also if it could naturally be seen but is unseen. He makes this point by adding, to his previous statement about unequal, the words, "also when it has no color at all," for a thing is called 'invisible' also if it is by its nature simply incapable of being seen and has no color, as when we say that sound is invisible; so that there is privation in this sense too, that namely of the impossible. [iii] Again a thing is called 'footless' either if it is by nature completely incapable of having feet, as are reptiles, or if it has defective feet, so that the term 'privation' is also used in reference to what has some attribute in an imperfect way.

---

27 στερήσεως P[b] S : στέρησιν A : στέρησις O ‖ 30–31 post καὶ μὴ ὁρώμενον addendum καὶ τὸ ὅλως μὴ πεφυκὸς ὁρᾶσθαι censet Bonitz ‖ 31 τῷ προειρημένῳ τῷ e S ci. Bonitz (*quod significavit post ea quae de inaequali dixerat, adijciens*) : ὡς προειρημένου τοῦ A O P[b] S : προειρημένου LF ‖ 32 τὸ A P[b] S : τῷ O ‖ 420.1 πόδας S : πόδα A O P[b] ‖ 2 φαύλους LF Hayduck : φαύλως A O P[b] S

Alexander goes through Aristotle's representative examples of the three types of privations expressed by the alpha privative, paying especial attention to the second case, which is also our concern. Alexander emphasizes that the three examples represent three *different* (αὐτῶν τὴν διαφοράν) types (419.27–28). We will see that the second example as it is transmitted by the α-version breaks this rule: reading the words καὶ τῷ φαύλως in line b35 annihilates the difference between

---

[165] It is not necessary to follow Bonitz (1847: 385, 19 *app. crit.*), followed by Dooley (1993: 105 and 173 n. 472) and Borgia (2007: 1118 n. 798), and supplement the commentary text here with the words καὶ τὸ ὅλως μὴ πεφυκὸς ὁρᾶσθαι. Alexander adopts Aristotle's καί (1022b35) in the sense of "also" and defines the term invisible in the same way as the previous term unequal. That "invisible" can be applied also to those things which are in general incapable of being seen, Alexander shows only in the subsequent lines (31 et seqq.) of his commentary.

the second and the third example.

Alexander's treatment of the first example (ἄνισον / "unequal") is brief; most of his attention is devoted to the second (ἀόρατον / "invisible"). The example unequal and the invisible share a characteristic: unequal is that which can be equal, but is not (419.28–30), and invisible is that which can be seen but is not (419.30–31). Distinctive to the second example is that "invisible" are also those things that have no color at all (419.31–32). Alexander illustrates this through the example of the voice, which is invisible inasmuch as it cannot be seen (419.32–34).[166]

Does Alexander's *Metaphysics* text coincide here with the α- or the β-version? According to the α-version (ἀόρατον δὲ καὶ τῷ ὅλως μὴ ἔχειν χρῶμα καὶ τῷ φαύλως, 1022b34–35), things called invisible either have no color at all or have a poor color. The β-version, on the other hand, reads ἀόρατον δὲ καὶ τῷ ὅλως μὴ ἔχειν χρῶμα and says—however tersely—what Alexander says in more detail, namely that invisible means *also* (i.e., in addition to the characteristic that it shares with unequal) that something is naturally invisible. By means of the word καὶ (b35), which in the β-version does not mean "either" (which it would if followed by another καὶ) but "also," Aristotle expresses the thought that something is invisible when it is naturally visible but not in the given case (this feature is shared by the example of inequality), *and* when it is naturally invisible.[167] Alexander therefore read the β-version.

What about the internal consistency of the *Metaphysics* passage, which, as I announced above, is warranted only by the reading in ω[AL] and β? Looking at the third example (ἄπουν, b35–36; Alex. 420.1–3), we see that Aristotle calls "footless" what does not have feet at all (τῷ μὴ ἔχειν ὅλως) as well as what has poor feet (τῷ φαύλους). When we follow the α-reading of the second example and read the addition καὶ τῷ φαύλως, the problem results that the third example illustrates exactly the same type of privation as the second example, for on the α-reading both cases are about something that either altogether lacks the attribute or has it in a poor way. Such repetition does violence to the economy of Aristotle's examples. Following the β-reading, Aristotle's examples are economical in that each of the three represents a type of privation that differs from the other two types, though they may overlap in one aspect or another:

---

[166] Perhaps Alexander was thinking of the parallel passage in *MA* 4, 699b17–b21, which I cite below. Aristotle also points to the voice as an example of something that is impossible to be seen in *Metaph.* K 1066a35–b1 and *Ph.* Γ 4, 204a4. Cf. also *Ph.* E 2, 226b11 and *Metaph.* Δ 15, 1021a25–26.

[167] Bonitz, who alone athetized the words καὶ τῷ φαύλως as a later addition, seems not to have been bothered by the καὶ that precedes in line b35. He neither mentions it in his commentary (1848) nor renders it in his translation (revised by Seidl, 1989: "... dem Unsichtbaren, weil es Farbe überhaupt nicht besitzt ...").

Types of privation	characteristic naturally possessed but not in given case	characteristic naturally not possessed	characteristic possessed but poorly
ἄνισον	x		
ἀόρατον	x	x	
ἄπουν		x	x

Even if one follows my explication of the passage and accepts that the economy of the examples is an important criterion for interpretation, one could still challenge my interpretation by taking the α-reading to express that the invisible differs from the other two examples in that it showcases all three characteristics. This challenge would require an understanding of the α-reading that diverges from the accepted understanding. The καὶ τῷ ... καὶ τῷ (1022b35) could then not be taken in the sense of "both ... and" (or "either ... or"), but in such a way that the first καί (b35) means "also" and the second καί (b35 according to α) means "and." Yet, it is highly questionable whether this interpretation is grammatically reasonable given the clear parallel structure of the two καί (καὶ τῷ ... καὶ τῷ).[168] Therefore, I am inclined to recognize καὶ τῷ φαύλως (b35) as a later addition to the α-text, which was, as is shown by ω^AL and β, not present in ω^αβ. Since this addition, when taken according to the common understanding of the phrase καὶ ... καί, evokes an uneconomic equation between two different examples,[169] it is reasonable to conclude that we are dealing here with a later addition to the text.

There are two parallel passages in *De motu animalium* and *De anima*, in which Aristotle speaks about the invisible (ἀόρατον). The *De motu* passage confirms the correctness of the reading in ω^AL and β, and most likely ω^αβ, in our passage. The *De anima* passage seems, at first glance, to speak in favor of the α-reading. On closer inspection, however, this assertion reveals itself to be unwarranted. Indeed it is not unthinkable to ask whether the addition in the α-text was modeled after the passage in *De anima*.

First to the passage in *De motu*: Aristotle argues that those things that are ἀόρατος or ἀδύνατον ὁραθῆναι can be invisible in two different ways. The naturally invisible voice is different from the men in the moon who, though naturally visible, are invisible to us. This difference covers exactly the two meanings of

---

[168] Kühner/Gerth II, §522, p. 249 and §523, pp. 252–56 do not discuss this case of καὶ ... καί. Bonitz 1870: s.v. καί; p. 357b31–34 does not mention it either. Denniston 1954 is silent about it in the section "corresponsive καὶ ... καί"; s. v. καί III., pp. 323–25," but he mentions the rare case that both καί in καὶ ... καί are to be taken adverbially in the sense of "also" (p. 324). In section II., p. 293, however, Denniston refers to one case in which καί is repeated while only the first καί means "also" and the second καί ("and") *has another reference*. In this case (X. HG 4.8.5,10–11), the two references differ from each other also in terms of their cases. In our passage of the *Metaphysics* the two references that are connected by καὶ ... καί are clearly parallel to each other.

[169] Also the further examples that Aristotle adduces for the meaning of the alpha privative in Δ 22 (1022b36–1023a7) represent a *different* case each.

ἀόρατος that are given in our *Metaphysics* passage according to the reading in β and ω^AL.

Aristotle, *De motu animalium* 4, 699b17–b21[170]

ἐπεὶ δὲ τὸ [18] ἀδύνατον λέγεται πλεοναχῶς (οὐ γὰρ ὡσαύτως τήν τε φω-[19]νὴν ἀδύνατον εἶναί φαμεν ὁραθῆναι καὶ τοὺς ἐπὶ τῆς σελήνης [20] ὑφ' ἡμῶν. τὸ μὲν γὰρ ἐξ ἀνάγκης, τὸ δὲ πεφυκὸς ὁρᾶ-[21]σθαι οὐκ ὀφθήσεται),

Now 'impossible' has several senses: for when we say it is impossible to see a sound and for us to see the men in the moon, we use two different senses of the word. The former is invisible of necessity; the later, though of such a nature as to be visible, will not actually be seen.

Whereas this passage from *De motu* speaks clearly in favor of the reconstructed ω^αβ-reading in the *Metaphysics* passage, the *De anima* passage appears at first sight to support the α-reading understood according to the special interpretation of καὶ ... καὶ.[171]

Aristotle, *De anima* B 10, 422a20–31

ὥσπερ δὲ καὶ ἡ ὄψις ἐστὶ τοῦ τε ὁρατοῦ καὶ τοῦ ἀοράτου (τὸ [21] γὰρ σκότος ἀόρατον, κρίνει δὲ καὶ τοῦτο ἡ ὄψις), ἔτι τε τοῦ [22] λίαν λαμπροῦ (καὶ γὰρ τοῦτο ἀόρατον, ἄλλον δὲ τρόπον τοῦ [23] σκότους), ὁμοίως δὲ καὶ ἡ ἀκοὴ ψόφου τε καὶ σιγῆς, ὧν [24] τὸ μὲν ἀκουστὸν τὸ δ' οὐκ ἀκουστόν, καὶ μεγάλου ψόφου [25] καθάπερ ἡ ὄψις τοῦ λαμπροῦ (ὥσπερ γὰρ ὁ μικρὸς ψόφος [26] ἀνήκουστος, τρόπον τινὰ καὶ ὁ μέγας τε καὶ ὁ βίαιος), ἀόρα-[27]<u>τον δὲ τὸ μὲν ὅλως λέγεται, ὥσπερ καὶ ἐπ' ἄλλων τὸ [28] ἀδύνατον, τὸ δ' ἐὰν πεφυκὸς μὴ ἔχῃ ἢ φαύλως, ὥσπερ [29] τὸ ἄπουν καὶ τὸ ἀπύρηνον</u>—οὕτω δὴ καὶ ἡ γεῦσις τοῦ γευστοῦ [30] τε καὶ ἀγεύστου, τοῦτο δὲ τὸ μικρὸν ἢ φαῦλον ἔχον χυμὸν [31] ἢ φθαρτικὸν τῆς γεύσεως.

Just as sight apprehends both what is visible and what is invisible (for darkness is invisible and yet is discriminated by sight); and also what is over-brilliant (for this is also invisible, but in another way than darkness), and as hearing apprehends both sound and silence, of which the one is audible and the other inaudible, and also loud sound as sight does what is bright (for as a faint sound is inaudible, so in a sense is a loud or violent sound); <u>and as one thing is called invisible absolutely (as in other cases of impossibility), another if it is adapted by nature to have the property but does not have it or has it only in a very low degree, as when we say that something is footless or pitless</u>—so too taste has as its object both what can be tasted and the tasteless—the latter in the sense of what has little flavor or a bad flavor or one destructive of taste.

---

[170] The text of this passage of *MA* follows the edition currently prepared by Oliver Primavesi. The translation is by Nussbaum. (The editorial changes for this passage do not affect the translation.)

[171] The text of this passage of the *De anima* follows the edition by Förster 1912; the translation is that of Smith, with modification.

The similarity between the formulation in lines 422a26–29 and in the α-version of our *Metaphysics* passage might incline us to accept the α-reading (and *nolens volens* the unusual meaning of καὶ … καὶ) as correct. In this passage of *De anima* (422a26–29), the invisible is described as either that which naturally lacks the characteristic of being seen or that which has the characteristic but is either invisible at the given moment or barely visible (ἢ φαύλως).

However, this understanding of ἀόρατον corresponds only *seemingly* to our *Metaphysics* passage in the α-version. In *De anima* B 10 Aristotle compares and contrasts the different sense organs. In doing this, Aristotle is not interested in a nuanced analysis of the word ἀόρατον and its various meaning. When Aristotle discusses the sense of sight here in *De anima*, he regards as equal (ὥσπερ, 422a28) three terms that he neatly distinguishes in the *Metaphysics* passage. These three terms are ἀόρατον, ἄπουν and ἀπύρηνον. While these three terms are treated in the *De anima* passage as *equal* representatives of the group characterized *by nature having the property, but not in the given case*, Aristotle in our *Metaphysics* passage introduces them as three clearly *distinct* cases of the alpha privative. There—according to the β-reading—ἀόρατον (1022b34–35) refers to what is either naturally or in the given case invisible, ἄπουν (1022b35–36) (in contrast to *de An.* 422a28–29!) to what is naturally footless or poorly footed, ἀπύρηνον (1022b36–1023a2) (in contrast to *de An.* 422a28–29!) to what has a small or bad stone.

The three terms have quite a different purpose in the *Metaphysics* passage than they do in the *De anima* passage, and accordingly they have different meanings in both passages. The meanings of ἄπουν and ἀπύρηνον in the *De anima* passage are incompatible with the nuanced distinctions made between them in the *Metaphysics* passage.[172] Therefore it is unreasonable to expect that the precise meanings of the third term, ἀόρατον, in both passages conform, and thereby also unreasonable to prefer the α-reading in the *Metaphysics* passage in order to maintain that conformity. And so it is more natural to follow the β-reading in the *Metaphysics* passage.

All in all the evidence speaks to the fact that line 1022b35 of the α-text is a later addition,[173] which was not part of $ω^{αβ}$.[174] The addition might have resulted from

---

[172] By contrast, the term ἀόρατον in the *Metaphysics* passage conforms not only to the meaning of ἀόρατον in the *De motu* passage, but also to its function so far as it illustrates an ἀδύνατον. In the *De anima* passage, by contrast, this aspect of ἀόρατον is of marginal importance.

[173] In case one wants to nevertheless defend the α-reading in our *Metaphysics* passage, disregarding the evidence in *De anima* and *De motu*, and accepting without qualm the special usage of καὶ … καὶ, then one has to assume that the α-version was secondarily corrected, whereas the β-text and $ω^{AL}$ preserve the reading of $ω^{αβ}$. (It is extremely unlikely that β had deleted the words καὶ τῷ φαύλως in accordance with the evidence in $ω^{AL}$). Such later correction of α is theoretically possible. Yet, in the given case the evidence strongly suggests that the words καὶ τῷ φαύλως in α are a later addition.

[174] The addition must have emerged before Asclepius (early sixth century AD) wrote his commentary, for his paraphrase (348.9–10) suggests that he had the α-addition in his text.

a scribal error caused by an anticipation of what follows in line 1022b36 (καὶ τῷ φαύλως). Such a visual error could have been prompted by the words μὴ ἔχειν occurring in b35 and again in b36. Given that the passages in the *Metaphysics* and the *De anima* appear to be quite close in content, one might also wonder whether a reader or scribe of the *Metaphysics* passage, who also knew the *De anima* passage but failed to recognize the fine yet substantive difference between the two passages, expected the two passages to express the same thought, and so added to the α-text of the *Metaphysics* the words καὶ τῷ φαύλως in accordance with the *De anima* passage. Such an explanation, however, might demand too much from our hypothetical reader. Whatever the case may be, it can be conluded that Alexander's testimony and its agreement with β allow for the reconstruction of the reading in ω^{αβ}, which is also the preferable reading.

4.3.1.3 Alex. *In Metaph.* 257.7–16 on Arist. *Metaph.* Γ 2, 1004a31–b3

The fourth aporia, as raised in book B (B 1, 995b18–27 and B 2, 997a25–34), asks whether the science that Aristotle seeks to delineate in B studies only substances, or also their *per se* attributes and predicates, for instance, those such as "same," "other," "like," and "unlike." In Γ 2 Aristotle argues that what is said πρὸς ἕν, that is, with reference to one common term,[175] belongs to one science.[176] Therefore, although "being" is said in many ways, everything that is belongs to *one* science, because "being" is said ultimately in reference to one term,[177] substance. This one science, philosophy, also studies the attributes of substances. This further specification of philosophy seems to be a reply to the fourth aporia.[178] Aristotle says:

Aristotle, *Metaphysics* Γ 2, 1004a31–1004b3

φανερὸν [32] οὖν [**ὅπερ ἐν ταῖς ἀπορίαις ἐλέχθη**] ὅτι μιᾶς περὶ τού-[33]των καὶ τῆς οὐσίας ἐστὶ λόγον ἔχειν (τοῦτο δ' ἦν ἓν [34] τῶν ἐν τοῖς ἀπορήμασιν), καὶ ἔστι τοῦ φιλοσόφου περὶ πάν-[1004b1]των δύνασθαι θεωρεῖν. εἰ γὰρ μὴ τοῦ φιλοσόφου, τίς ἔσται [2] ὁ ἐπισκεψόμενος εἰ ταὐτὸ Σωκράτης καὶ Σωκράτης καθή-[3]μενος, ...

It is evident then that it belongs to *one* science to be able to give an account of these concepts as well as of substance. This was one of the questions in our book of problems. And it is the function of the philosopher to be able to investigate all things. For if it is not the function of the philosopher, who is it who will inquire whether Socrates and Socrates seated are the same thing, ...

---

32 ὅπερ ... ἐλέχθη α Bekker Bonitz Christ Cassin/Narcy Hecquet-Devienne, *secl.* Ross Jaeger : om. ω^{AL} Ar^u β V^k <E>γρ || 33–34 τοῦτο ... ἀπορήμασιν secl. Hecquet-Devienne || 1004b1 ἔσται] ἐστιν Al.^{1} 257.17

[175] Aristotle's example is health (ὥσπερ καὶ τὸ ὑγιεινὸν ἅπαν πρὸς ὑγίειαν, a34–35): 1003a34–b4.
[176] *Metaph.* Γ 2, 1003a33–b16. See my analysis of Γ 2, 1005a2–8 in 5.2.3.
[177] *Metaph.* Γ 2, 1003a33–34: Τὸ δὲ ὂν λέγεται μὲν πολλαχῶς, ἀλλὰ πρὸς ἓν καὶ μίαν τινὰ φύσιν.
[178] Madigan 1999: 50 raises doubt as to whether this passage in Γ 2 really gives an answer to aporia 4.

Lines 1004a32–33 are so formulated as to present the answer to the aporia raised in book B concerning the appropriate subject matter and the unity of the sought-for science. Aristotle's specification reads: μιᾶς [*sc.* ἐπιστήμης] περὶ τούτων καὶ τῆς οὐσίας ἐστὶ λόγον ἔχειν / "it belongs to *one* science to give an account of these concepts as well as of substance" (cf. B 2, 995b18–20; 997a25–26). That this statement indeed refers back to the aporia in book B is made explicit by the following remark (a33–34): τοῦτο δ᾽ ἦν ἓν τῶν ἐν τοῖς ἀπορήμασιν. / "this was one of the questions in our book of problems." This reference to book B is transmitted in α and β unanimously and therefore attests the reading of ω^αβ. The α-version is distinctive in that it contains an *additional* back reference to book B. This additional reference, which expresses the exact same idea (a32), appears as a relative clause one line before the unanimously transmitted one. It reads: ὅπερ ἐν ταῖς ἀπορίαις ἐλέχθη / "what was said in the books of problems."

Bekker and Christ were not bothered by the double reference to book B and so follow the α-reading.[179] Bonitz, too, reads the α-text, though hesitantly. He remarks that a twice-occurring back reference in the same sentence to the same passage goes against Aristotle's habit.[180] Jaeger makes the case that it is impossible to keep both references in the text.[181]

Given that the reference in line a32 is attested only by the α-version, it is rather implausible to assume that this is the original reference of ω^αβ, which was deleted in the β-version, and that the reference in line a33–34, which is attested by α and β (and hence ω^αβ), is a secondary addition. Although this explanation of the given textual situation is not impossible, it is much more plausible to assume that the α-version alone obtained a later addition.[182] The Arabic tradition supports the assumption of a later addition restricted to the α-version: Ustāth does not translate the α-reference.[183] Might someone have added the hint to book B at the beginning of the argument, before recognizing that there was already a reference in the text?

---

[179] Cassin/Narcy 1989 faithfully follow the α-text.

[180] Bonitz 1849: 181: *Ceterum quum praeter consuetudinem Aristotelis esse videatur, quod bis in eodem enunciato superioris disputationis lectores commonefacit,* ὅπερ—ἐλέχθη *et* τοῦτο—ἀπορήμασιν, *non negligendum est quod A^b et mg E priora illa verba omittunt.*

[181] Jaeger 1971: 491. In his edition (1957), Jaeger deletes the additional reference in α. According to Jaeger 1917: 491 both references, taken individually, are viable. Jaeger mentions this case also in his *praefatio* (1957: xiv): *patet etiam hoc loco Π varias lectiones in fonte suo invenisse et contaminasse, A^b aut unam tantum legisse aut alteram reiecisse, quod minus probabile est.* Hecquet-Devienne does not like the duplication either, but she decides for the α-reference (1004a32) and deletes the reference that is unanimously transmitted by α and β in 1004a33. See Hecquet-Devienne 2008: 114–15 n. 10.

[182] The important α-manuscript E contains a marginal gloss pointing to the absence of the α-reference in other manuscripts: ἔν τισι λείπει τὸ ὅπερ ἐν ταῖς ἀπορίαις ἐλέχθη. Bekker 1831 already reports this marginal gloss in his apparatus. See also Walzer 1958: 224.

[183] Walzer 1958: 224. This is confirmed by Scotus's translation: <u>Manifestum est igitur quod</u> oportet scire ista et declarare definitionem eorum et definitionem substantie. <u>Et ista questio est una earum de quibus perscrutati fuimus in capitulo questionum.</u>

It is also possible that a gloss in the margin, referencing a parallel passage, was later incorporated into the text. In any case, the direct textual evidence suggests that the α-reference is a later addition to the text and should be deleted.

Is there indirect textual evidence that supports this claim? In addressing this question, I turn to Alexander's commentary on the passage.

Alexander, *In Metaph.* 257.7–16 Hayduck

δείξας δὲ πῶς ἀφ' ἑνός ἐστι καὶ πρὸς ἓν ταῦτα λεγόμενα, ἐπιφέρει τὸ τῆς [8] αὐτῆς εἶναι περί τε τούτων ἃ ἀπὸ τῆς οὐσίας τὸ εἶναι ἔχει καὶ περὶ τῆς [9] οὐσίας ἔχειν ἐπιστήμην. ἢν δὲ περὶ τὴν οὐσίαν ἡ πραγματεία τῷ φιλο-[10]σόφῳ, καὶ περὶ τούτων ἄρα τοῦ φιλοσόφου διαλαμβάνειν. εἰπὼν δὲ ταῦτα [11] καὶ δείξας ὅτι τοῦ φιλοσόφου τὸ περὶ πάντων τῶν ὄντων γνῶσιν ἔχειν, [12] ἐπισημαίνεται ὅτι διὰ τῶν δεδειγμένων λύεται τῶν ἀποριῶν μία τῶν ἀπο-[13]ρηθεισῶν ἐν τῷ δευτέρῳ,[184] περὶ ἧς ἠπόρει, πότερον μιᾶς ἢ πλειόνων ἐστὶν [14] ἐπιστημῶν τὸ θεωρῆσαι περί τε τῆς οὐσίας καὶ τῶν τῇ οὐσίᾳ συμβεβη-[15]κότων, καὶ τῶν οἷς εἶπε χρῆσθαι τὴν διαλεκτικὴν κατὰ τὸ ἔνδοξον, ἃ ἦν [16] ἐναντία, ταὐτὸν ἕτερον καὶ τὰ ἄλλα ὡς φθάνομεν εἰρηκότες.

Having shown how these things are said by derivation from one thing and with reference to one thing, he adds that it belongs to the same science to have scientific knowledge both of substance and of these things which possess their being from substance. But the study concerning substance belongs to the philosopher, and so it belongs to the philosopher to deal with these things as well. Having said these things, and having shown that it belongs to the philosopher to have knowledge of all beings, he indicates in addition that <u>one</u> of the aporiae raised in the second book is solved by way of what has been shown: the aporia which he raised, whether it belongs to one science or to several sciences to consider substance and the accidents of substance and also the items which he said dialectic uses, on the basis of accepted opinion, namely, the contraries, sameness, otherness, and the other which we have mentioned earlier.

---

8 περί τε A[p.c.] O LF S Ascl. : παρά τε A[a.c.] ‖ ἔχει O A[p.c.] P[b] L Ascl. : ἔχειν A[a.c.] ‖ 9 τὴν οὐσίαν A O L Ascl. : οὐσίαν P[b] : τῆς οὐσίας Bonitz Hayduck ‖ 12 διὰ A O S : ἐκ P[b] ‖ 13 δευτέρῳ A O : βῆτα P[b] ‖ 14 τῆς A P[b] : τὰς O ‖ 16 ὡς A O P[b] L S(*ut*): ἃ Bonitz Hayduck

Alexander mentions one reference to the book of aporiae (257.12–15). In light of Alexander's commenting practice, we can assume that Alexander would have said something about the curious doubling of the reference to book B had he read both in his text. Given Alexander's silence we can infer that Alexander's exemplar contained only one reference. So far, however, it is not immediately clear which of the two possible references Alexander found in ω[AL].

Alexander's paraphrase gives us the following picture of his text: Aristotle *first* declares that it belongs to one science to investigate being, substances and their

---

[184] Alexander calls book B "second book" also in 264.31: τῶν ἐν τῷ δευτέρῳ κειμένων ἀποριῶν μέμνηται νῦν. This squares with his assertion in 137.2–9 that α ἔλαττον is a sort of appendix to book A. See also 3.5.2.3.

attributes, and that this science is philosophy, and *then* refers to book B. Accordingly, Alexander's text appears not to have contained the α-reference, for this reference occurs in the *Metaphysics* text *before* the answer to the aporia is given. It is true, Alexander's comments do not exactly reproduce the unanimously attested reference and its place in Aristotle's thought, but his comments are more in line with this reference than with the α-reference.[185]

In addition to this, there is a very clear sign that Alexander read the reference given in ω^{αβ} and not the one given in α alone. Alexander's paraphrase corresponds in one crucial detail with the formulation of the reference in ω^{αβ}. Alexander writes (257.12–13): ἐπισημαίνεται ὅτι διὰ τῶν δεδειγμένων λύεται τῶν ἀποριῶν μία τῶν ἀπορηθεισῶν ἐν τῷ δευτέρῳ... ("he indicates in addition that one of the aporiae raised in the second book is solved by way of what has been shown"). Precisely speaking, Alexander's comments concern not the book of problems but *one* (μία) of the aporiae discussed there. The α-reference speaks of the book of problems, but the reference in ω^{αβ} focuses on one (εἷς) of the problems discussed there: τοῦτο δ' ἦν ἓν τῶν ἐν τοῖς ἀπορήμασιν ("this was one of the questions in our book of problems"). The correspondence in respect to this small detail strongly suggests that Alexander found in ω^{AL} exactly the reference of ω^{αβ}.[186] This evidence in Alexander contributes a great deal to the argument that the reference in a33–34 is the original one.

What remains to be done is to consider the wording of each of the references as well as their distinctive features. Of special importance is the name of the book of aporiae: ἐν τοῖς ἀπορήμασιν (a34, in ω^{αβ}) and ἐν ταῖς ἀπορίαις (a32, in α). When Aristotle speaks about aporiae in a technical sense, that is, as the set of problems treated in book B of the *Metaphysics*, he says ἐν τοῖς (δι)απορήμασιν.[187] We find this phrase also in the reference attested by ω^{αβ} (ἐν τοῖς ἀπορήμασιν). In stark contrast, there is not one other passage in the corpus where Aristotle refers to book B by the phrase ἐν ταῖς ἀπορίαις.[188] This fact makes the additional α-reference (ἐν ταῖς ἀπορίαις, 1004a32) suspicious yet again.[189]

---

[185] The reference transmitted by α and β precedes the specification of the philosopher as the appropriate scientist, whereas in Alexander's paraphrase the reference comes only after it. See the aorist participles: εἰπὼν δὲ ταῦτα καὶ δείξας ὅτι τοῦ φιλοσόφου ... ἐπισημαίνεται. In any case, we should not be too strict about demanding an exact reproduction of Aristotle's argumentative steps in Alexander's commentary, and Alexander does occasionally alter the position of an argument's steps (cf. e.g. the case analyzed in 5.3.3).

[186] Bonitz remains hesitant about the evidence in Alexander. In his apparatus he writes of the reference in 1004a32: *om. fort. Alex.* Ross adopts this in his apparatus.

[187] *APr.* 93b20 (ἐν τοῖς διαπορήμασιν), *Metaph.* I 2, 1053b10 (ἐν τοῖς διαπορήμασιν); M 2, 1076b1 (ἐν τοῖς διαπορήμασιν), M 2, 1077a1 (ἐν τοῖς ἀπορήμασιν), M 9, 1086b16 (ἐν τοῖς διαπορήμασιν).

[188] The exact words ἐν (ταῖς) ἀπορίαις appear in an Aristotelian fragment (209.9–10 Rose = A. Gellius *N.A.* XX, 4.3–4), but they do not refer to the third book of the *Metaphysics*, but to precarious living conditions.

[189] This makes it extremely unlikely that the α-reference is authentic and was deleted in ω^{AL} and in β.

The internal evidence in the *Metaphysics* passage as transmitted through the direct transmission taken together with the evidence in Alexander's commentary thus compels the conclusion that the reference in a32, transmitted solely by the α-version, is an inauthentic interpolation into the text that is to be excised from our text.

In the three cases analyzed in 4.3.1 we encountered, first (4.3.1.1), the loss of a word in the α-text, which did not occur in β and ω^{AL}; second and third (4.3.1.2–3), the addition of a phrase in the α-text, which did not occur in β and ω^{AL}.[190] In each case, the agreement of ω^{AL} with the β-version leads us to the correct reading that was given in ω^{αβ}.

4.3.2 Separative errors in β against α + ω^{AL}

4.3.2.1 Alex. *In Metaph.* 292.13–16 on Arist. *Metaph.* Γ 4, 1007b29–1008a2

In his discussion of the principle of non-contradiction in Γ 4 Aristotle examines the absurdities that result from denying the principle. From 1007b18 onwards, Aristotle shows the absurdity that follows from Protagoras's relativism, which states that anything can be affirmed or denied of anything because all assertions are only opinions.[191] The upshot of Protagoras's position is that everything must blur into one:[192] trireme is a human being is a wall (ἔσται γὰρ τὸ αὐτὸ καὶ τριήρης καὶ τοῖχος καὶ ἄνθρωπος, 1007b20–21).

In order to illustrate the consequences of this position Aristotle adduces the following example: if someone opines (δοκεῖ, b23) that a certain human being is not a trireme, then the human being is not a trireme (according to Protagoras: opinion = assertion), but since everything can equally well be affirmed or denied, the human being is a trireme after all (1007b23–25).[193] Therefore nothing can be true and the adherents of Protagoras's position deal with the indeterminate and with non-being rather than being.[194] Aristotle then goes on to flesh out this refutation with some logical rigor:

---

[190] Cf. the "α-supplements" Primavesi 2012b: 439–56 collected from the first book of the *Metaphysics*.

[191] Aristotle discusses Protagoras's phenomenalistic position extensively in Γ 5 and 6.

[192] 1007b18–20: ἔτι εἰ ἀληθεῖς αἱ ἀντιφάσεις ἅμα κατὰ τοῦ αὐτοῦ πᾶσαι, δῆλον ὡς ἅπαντα ἔσται ἕν. For an analysis of this argument see Kirwan 1971: 102–103.

[193] 1007b23–25: εἰ γάρ τῳ δοκεῖ μὴ εἶναι τριήρης ὁ ἄνθρωπος, δῆλον ὅτι οὐκ ἔστι τριήρης· ὥστε καὶ ἔστιν, εἴπερ ἡ ἀντίφασις ἀληθής.

[194] Kirwan 1971: 103 takes this statement of Aristotle to be the starting point of an argument of its own: since the opponents take "being" in the sense of "potential being," Aristotle can disprove their position.

Aristotle, *Metaphysics* Γ 4, 1007b29–1008a2

ἀλλὰ μὴν λεκτέον γ᾿ αὐτοῖς κατὰ [30] παντὸς <παντὸς>[195] τὴν κατάφασιν ἢ τὴν ἀπόφασιν· ἄτοπον γὰρ [31] εἰ ἑκάστῳ ἡ μὲν αὑτοῦ ἀπόφασις ὑπάρξει, ἡ δ᾿ ἑτέρου ὃ μὴ [32] ὑπάρχει αὐτῷ οὐχ ὑπάρξει· λέγω δ᾿ οἷον εἰ ἀληθὲς εἰπεῖν τὸν [33] ἄνθρωπον ὅτι οὐκ ἄνθρωπος, δῆλον ὅτι καὶ **οὐ** [34] **τριήρης**. εἰ μὲν οὖν ἡ κατάφασις, ἀνάγκη καὶ τὴν ἀπόφασιν· [35] εἰ δὲ μὴ ὑπάρχει ἡ κατάφασις, ἥ γε ἀπόφασις ὑπάρξει [1008a1] μᾶλλον ἢ ἡ αὑτοῦ. εἰ οὖν κἀκείνη ὑπάρχει, ὑπάρξει καὶ ἡ [2] τῆς τριήρους· εἰ δ᾿ αὕτη, καὶ ἡ κατάφασις.

But they must predicate of every subject every attribute and the negation of it indifferently. For it is absurd if of every subject its own negation is to be predicable, while the negation of something else which cannot be predicated of it is not predicable of it; for instance, if it is true to say of a man that he is not a man, evidently it is also true to say that he is not a trireme. If, then, the affirmative can be predicated, the negative must be predicable too; and if the affirmative is not predicable, the negative, at least, will be more predicable than the negative of the subject itself. If, then, even the latter negative is predicable, the negative of 'trireme' will also be predicable; and if this is predicable, the affirmative will be so too.

30 παντὸς <παντὸς> ci. ex Al.ᴾ 292.5-6 (sed Al.ˡ 292.1-2) Bonitz Christ Ross Jaeger Hecquet-Devienne || 32-33 τὸν ἄνθρωπον **α** edd. : τὸ ἄνθρωπος **β** || 33 οὐ τριήρης **α** ωᴬᴸ (Al.ᴾ 292.15-16) Ascl.ᵖ 268.9-10 Arᵘ (*Scotus*) Bekker Bonitz Christ Cassin/Narcy Hecquet-Devienne : ἢ τριήρης ἢ οὐ τριήρης **β** Ross Jaeger || 1008a1 ἢ ἡ **α** Al.ᴾ 292.13 Bonitz Christ Ross Jaeger Cassin/Narcy Hecquet-Devienne : ἢ **Vᵈ** (ἡ Bekker) : om. **β**

According to the relativistic position, one may predicate of every subject every predicable and at the same time every negation (b29–30). If it is possible to predicate the negation of what a thing is then it is all the more possible to predicate the negation of what the thing is not (b30–32).[196] Aristotle illustrates this with an example (οἷον εἰ…, b32): If a human being is also not a human being, then it is all the more correct that a human being is not a trireme (b33–34 according to **α** and Arᵘ [*Scotus*]).

In lines 1007b34–1008a2 Aristotle presents his argument showing the absurdity

---

[195]Since Bonitz (see Bonitz 1849: 194-95) and based on Alexander's paraphrase (292.3-6) editors (except for Cassin/Narcy: 1989: 215-16) have conjectured an additional παντὸς: Δείξας ἑπόμενον τῷ τὴν ἀντίφασιν συναληθεύειν λέγοντι τὸ οὗ ἡ ἀπόφασις ἐπί τινος ἀληθής, ἐπ᾿ ἐκείνου καὶ τὴν κατάφασιν αὐτοῦ ἐκείνου ἀληθῆ γίγνεσθαι, νῦν δείκνυσιν ὅτι ἀνάγκη αὐτοῖς λέγειν ἐπὶ παντὸς πᾶσαν ἀντίφασιν κατηγορεῖσθαι. / "Having shown that for one who says that contradictories are both true, it follows that, where the negation is true in the case of a certain thing, the affirmation of that [predicate] also turns out to be true in that case, he now shows that it is necessary for them to say that in every case every pair of contradictories is predicated." Because the text of the lemma (292.1-2) matches that of our transmission (that is, there is no second παντὸς) we have to assume that this lemma was corrupted in the course of the tradition. Or is it possible that Alexander, having found only one παντὸς in ωᴬᴸ, expanded the text on his own in his reformulation (ἐπὶ παντὸς πᾶσαν ἀντίφασιν)? Cf. also Jaeger 1923: 258 and Hecquet-Devienne 2008: 137 n. 23.

[196]Cf. Kirwan 1971: 103.

of the opponent's position.[197] If the affirmative (κατάφασις, b34) is predicable ("a human being is a human being"), then, the negation is also predicable (ἀπόφασιν, b34; "a human being is not a human being"). But if the affirmative is not predicable—for a human being is in fact not a trireme—then the negation ("a human being is not a trireme"), as shown in b30–32, is more predicable than the negation of what it in fact is ("a human being is not a human being"). Therefore (1008a1), if the negation of what something is (a human being) is predicable, then *a fortiori* the negative of what something is not (a trireme) is predicable. If this negation ("a human being is not a trireme") can be predicated (1008a2), then the affirmation ("a human being is a trireme") can also be predicated.

Understanding the argument in this way, the passage quoted above is self-consistent. This construal, however, passes over a textual difficulty presented by lines b33–34 of the β-version. Unlike in the α-version[198] we do not find there the words "if a man is not a man, evidently he is also not a trireme" (δῆλον ὅτι καὶ οὐ τριήρης), but instead "if a man is not a man, evidently he is *either a trireme or* not a trireme" (δῆλον ὅτι καὶ ἢ τριήρης ἢ οὐ τριήρης). In terms of content, what speaks against the β-version is that the alternative "either a trireme or not a trireme" appears out of place at this stage of the argument. In other words, the alternative appears too early. The proper place of the alternative is rather the conclusion; for only at the end of the argument does it become clear that according to the opponent's view a human being is a trireme. That the alternative "either a trireme or not a trireme" comes too early is also made clear by the context in lines b30–34: The alternative does not fit into Aristotle's *a fortiori* argument, stating that, for example, the negation non-trireme is more predicable of a human being than non-human. That a human being is therefore also a trireme does not matter at this point of the argument. Furthermore, the formulation of an alternative as in ἢ τριήρης ἢ οὐ τριήρης ("either a trireme or not a trireme") does not match with the result that is achieved in the subsequent lines. There it says that a human being is *both* a trireme *and* not a trireme. The ἢ ... ἢ is therefore misleading.[199]

The testimony in Alexander's commentary shows that ω[AL] agrees with the α-reading.

Alexander, *In Metaph.* 292.13–16 Hayduck

μᾶλλον γὰρ ἡ ἄλλου ἀπόφασις ἀληθὴς κατά τινος ἢ ἡ αὐτοῦ· εἰ γὰρ ἀλη-[14]θὲς κατὰ τοῦ ἀνθρώπου τὸ ὅτι οὐκ ἔστιν ἄνθρωπος, πολὺ εὐλογώτερον [15] ἐπ' αὐτοῦ

---

[197] Cf. Cassin/Narcy 1989: 216–17 and Hecquet-Devienne 2008: 137 n. 23

[198] This reading is further confirmed by the text Asclepius used for his commentary and the Arabic tradition of the *Metaphysics*.

[199] To the objection that the ἢ ... ἢ should be taken as inclusive, I answer that had the author of ἢ τριήρης ἢ wanted to express an inclusive meaning (ἢ ... ἢ / "either ... or" in the sense of "both ... and") he certainly would have used a single ἢ.

ἀληθὲς λέγειν τό τε οὐκ ἔστιν ἵππος καὶ τὸ οὐκ ἔστι τοῖχος [16] καὶ τὸ οὐκ ἔστι τριήρης, καὶ τὰ ἄλλα ὅσα μή ἐστιν.

For the negation of something else is more true of a thing than the negation of the thing itself. For example, if it is true of a human that he is not human, it is much more reasonable to say, in the case of the human, 'he is not a horse' and 'he is not a wall' and 'he is not a trireme' and not the other things which he is not.

Like Aristotle, Alexander first introduces the *a fortiori* argument as a general rule (292.13) and then illustrates it with an example. The scope of Alexander's argument exceeds that of Aristotle only in the terms used to signify the negatives that may be predicated of a human being (not a horse, nor a wall, nor a trireme).[200] The agreement of α and ω^{AL} confirms the suspicion that this is the older reading and was also in ω^{αβ}, while the β-version suffered a later interpolation of the words ἢ τριήρης ἢ.[201]

Consequently, it is all the more surprising that Ross, followed by Jaeger,[202] puts the β-reading in the text. Ross does not justify his decision.[203] His diagnosis that the α-version contains a corrupted text, which lost some words due to *homoioteleuton*,[204] is mistaken and betrays his hasty judgment on this issue. If a scribe had jumped from one similar word to the next (καὶ ἢ τριήρης ἢ οὐ τριήρης) he would have written out *one* of the similar words (ἢ), which we would then find in the α-version. This, however, is not the case.

A possible explanation of how the β-reading came about and why some editors preferred it could be that a misunderstanding arose about which affirmation is meant by ἡ κατάφασις in line b34 of Aristotle's text.[205] Whether κατάφασις (b34) means the affirmation of being a human being or of being a trireme is only made clear by the following sentence in b35–1008a1. Since there, in line b35, κατάφασις and ἀπόφασις must refer to being a trireme the κατάφασις (and ἀπόφασις) of the previous sentence (b34) must, in order to avoid an exact reiteration, refer to being a human being.[206] Yet if one assumes wrongly that κατάφασις and ἀπόφασις in line b34 refer to being a trireme, then it is natural to wish for such an affirmation in the preceding line. The addition of ἢ τριήρης ἢ may well be the result of this wish.

---

[200] Cf. the Aristotelian examples in 1007b20–21.

[201] This does not violate my rule, which states that a reading of ω^{AL} is to be reconstructed on the basis of two different types of evidence in Alexander's commentary (one of which is a paraphrase or discussion) (see 3.4). The two types of evidence in this section are (i) Alexander's paraphrase in 292.16 (καὶ τὸ οὐκ ἔστι τριήρης) and (ii) his discussion of the argument as a whole in 292.13–16. Alexander's presentation of the argument makes it clear that he did not read ἢ τριήρης ἢ in his text.

[202] Kirwan 1971 bases his translation on Jaeger's text. Cf. his comments in Kirwan 1971: 103 ("(h)").

[203] Ross 1924: 271: "The logic of the passage requires A^b's reading ἢ τριήρης ἢ οὐ τριήρης." It seems that Ross, as Jaeger puts it in his *app. crit.*, had in view the subsequent sentence in b34.

[204] Ross 1924: 271.

[205] Cf. Cassin/Narcy 1989: 216.

[206] See Cassin/Narcy 1989: 217.

4.3.2.2 Alex. *In Metaph.* 182.32–38 on Arist. *Metaph.* B 2, 996a29–996b1

As his first aporia Aristotle poses the question whether one, single science investigates all the kinds of causes (B 1, 995b4–6; B 2, 996a18–996b26). The second objection against the thesis that all causes are studied by one science is that not all things are subject to all the kinds of causes. A final cause, for example, is not operative among unchanging objects (996a21–29).[207] Aristotle continues in the following way:

Aristotle, *Metaphysics* B 2, 996a29–996b1

> διὸ καὶ ἐν τοῖς μαθήμασιν οὐθὲν δείκνυται διὰ [30] ταύτης τῆς αἰτίας, οὐδ' ἔστιν ἀπόδειξις οὐδεμία διότι βέλτιον [31] ἢ χεῖρον, ἀλλ' οὐδὲ τὸ παράπαν μέμνηται οὐθεὶς οὐθενὸς τῶν [32] τοιούτων, ὥστε διὰ ταῦτα τῶν σοφιστῶν τινὲς οἷον Ἀρίστιππος [33] προεπηλάκιζεν αὐτάς· ἐν μὲν γὰρ ταῖς ἄλλαις τέχναις, [34] καὶ ταῖς βαναύσοις, οἷον ἐν τεκτονικῇ καὶ σκυτικῇ, διότι [35] βέλτιον ἢ χεῖρον λέγεσθαι πάντα, τὰς δὲ μαθηματικὰς [996b1] οὐθένα ποιεῖσθαι λόγον περὶ ἀγαθῶν καὶ **κακῶν**.

> This is why in mathematics nothing is proved by means of this kind of cause, nor is there any demonstration of this kind—'because it is better, or worse'; indeed no one even mentions anything of the kind. And so for this reason some of the Sophists, e.g. Aristippus, ridiculed mathematics; for in the arts, even in handicrafts, e.g. in carpentry and cobbling, the reason always given is 'because it is better, or worse,' but the mathematical sciences take no account of goods and **evils**.

34 βαναύσοις **α** edd. : βαναύσοις αὐταῖς **β** || b1 κακῶν **α ζ** Al.ᵖ 182.38 Ascl.ᵖ 153.3–5 Ar.ᵘ (*Scotus*) edd. : καλῶν **β**

The final cause has no place in mathematics, for in mathematics there is no good for the sake of which something is done. This is why some of the Sophists (Aristotle mentions Aristippus as representative) disdained mathematics. Even the lowly handicrafts aim at the better and avoid the worse, but in mathematics criteria like good or bad are no issue at all. This, according to the **α**-text, is what Aristotle reports as Aristippus's disdain of mathematics. By contrast, the **β**-version reads in line 996b1 not λόγον περὶ ἀγαθῶν καὶ κακῶν ("account of goods and evils") but περὶ ἀγαθῶν καὶ καλῶν ("account of goods and *beauty*"). According to the **β**-text the Sophists disdain mathematics because it does not aim towards goods *and beauty*.

From a grammatical point of view both versions are viable. The **α**-reading, preferred by all editors, receives confirmation from lines 996a30–31, where it is stated that mathematics does not care about the better or worse. On the other hand, defenders of the **β**-reading could point to the fact that what is at issue in the broader context is the final cause, which is the good and the beautiful and not the good and

---
[207] Cf. Madigan 1999: 34–36; Crubellier 2009: 53

the bad. What did Alexander read in ω^AL?

Alexander, *In Metaph.* 182.32–38 Hayduck

καὶ Ἀριστίππου μνημονεύει, ὃς [33] καὶ αὐτὸς ὁμοίως ἄλλοις τισὶ τῶν σοφιστῶν ἔλεγε τὰς μαθηματικὰς ἐπι-[34]στήμας, ὡς καὶ τῶν εὐτελεστάτων[208] τεχνῶν καταδεεστέρας· ἐκείνων μὲν γὰρ [35] ἑκάστης εἶναί τι τέλος καὶ ἀγαθὸν προσκείμενον καὶ μεμνῆσθαι αὐτὰς ἐν [36] τοῖς γιγνομένοις ὑπ' αὐτῶν τοῦ ὅτι βέλτιον γὰρ οὕτως, τὰς δὲ μαθηματι-[37]κὰς μηδὲν ἔχειν αἴτιον τοιοῦτον μηδὲ ποιεῖσθαί τινα λόγον περὶ ἀγαθῶν [38] καὶ <u>κακῶν</u>.

And he (Aristotle) mentions Aristippus who, like some other sophists, spoke of the mathematical sciences as deficient even relative to the simplest crafts; for each of these crafts has an end and a good proposed to it, and in what takes place under their influence they attend to the argument 'because it is better that way,' while the mathematical sciences have no such cause, nor do they take any account of goods and <u>evils</u>.

32 καὶ **O LF** : καὶ ὅτι **A** : καὶ ἔτι **P^b** ‖ 33 ἔλεγε **A O** : προεπηλάκιζε **P^b** ‖ 34 εὐτελεστάτων **O** : εὐτελεστέρων **A P^b** ‖ 35 ἑκάστης **A O** : ἑκάστη **P^b** ‖ 35–36 καὶ ἀγαθὸν ... τοῖς γιγνομένοις **LF S** : in lac. om. καὶ ἀγαθὸν ... τοῖς γιγνο) **A O** : ὑποτίθεται γὰρ **P^b** ‖ 36 τοῦ **A O** : τὸ **P^b** ‖ γὰρ **A O** : om. **P^b S**

Alexander's paraphrase reveals that in his copy of the *Metaphysics* the reading was identical to that of the α-text. This paraphrase is the only type of evidence available in Alexander's commentary for these lines of the *Metaphysics* and so we should be cautious when inferring what Alexander read in his text. Nevertheless the claim that ω^AL read the α-reading may be justified when we compare Alexander's paraphrase with the two divergent readings in α and β (cf. 3.4, pp. 57–59). On the assumption that the agreement between the paraphrase and the α-text point to a textual agreement between α and ω^AL, the β-reading appears as a later modification of the *Metaphysics* text.[209]

Did the β-reading emerge from a simple scribal error? A misreading of ΚΑΚΩΝ as ΚΑΛΩΝ is entirely plausible.[210] Or did someone not understand that the α-reading is not a confused description of the final cause, but rather a description of the criteria according to which actions (the final cause of which is the

---

[208] The form εὐτελεστάτων is the superlative (or, when following **A** and **P^b**, the form εὐτελεστέρων is the comparative) of the adjective εὐτελής, which means "cheap," "easy," "mean." It does *not* mean what Madigan 1992, followed by Lai 2007, suggests it means when he translates "most complete and perfect" (Lai: "più perfette"). This mistranslation breaks the logic of the argument: that mathematics is deficient in comparison to "the most perfect art" is obvious, simply because all other arts are deficient in comparison to the most perfect art.

[209] The α-reading was also in the copy of Asclepius (153.3–5) as well as in the Greek *Vorlage* of the Arabic tradition. Scotus translates: *et artifices istius artis non perscrutantur omnino de bonis et malis*.

[210] Such misreading could have happened in either direction. If we take the α-reading as original, then one could suppose that a scribe wrote the common hendiadys ἀγαθῶν καὶ καλῶν (καλὸς καὶ ἀγαθός, or with crasis καλοκἀγαθός) instead of the polar expression ἀγαθῶν καὶ κακῶν.

good) are performed? These are both viable options, but in this case there is an even better explanation to hand. *Metaph.* M 3, 1078a31–34 offers a *seemingly* parallel passage, which at first glance gives the impression that the β-reading in our passages is preferable. At second glance, however, the passage reveals itself to be a candidate for the model used in modifying the β-version in our passage in book B.

In the following passage from book M Aristotle refers once again to the disdainful attitude that some have towards mathematics.[211]

Aristotle, *Metaphysics* M 3, 1078a31–34

ἐπεὶ δὲ τὸ ἀγαθὸν καὶ τὸ καλὸν ἕτερον (τὸ [32] μὲν γὰρ ἀεὶ ἐν πράξει, τὸ δὲ καλὸν καὶ ἐν τοῖς ἀκινήτοις), [33] οἱ φάσκοντες οὐδὲν λέγειν τὰς μαθηματικὰς ἐπιστήμας περὶ [34] καλοῦ ἢ ἀγαθοῦ ψεύδονται.

Now since the good and the beautiful are different (for the former always implies conduct as its subject, while the beautiful is found also in motionless things), those who assert that the mathematical sciences say nothing of the beautiful or the good are in error.

Here in M 3, just as before in the β-version of B 2, we hear about the view that the good and beautiful does not play any role in mathematics. Aristippus is nowhere mentioned, but we may assume that Aristotle is alluding here to the group that he explicitly referred to as Sophists in B 2 (τῶν σοφιστῶν τινὲς).

Does this parallel passage authenticate the β-reading? Hardly. For while the words in M 3 seem at first glance to echo those of B 2, the contexts of these two passages clearly differ widely. This divergence in context prohibits the equation of the mention of the good and the beautiful in M 3 and the statement in B 2. In the context of M 3 Aristotle discusses the good and beautiful in terms of their status in mathematics, granting a place in mathematics to the beautiful (order and symmetry are forms of the beautiful, 1078a36–1078b1),[212] but not to the good. The opponents to this view, discussed in the text cited above, say that mathematics has nothing to do with either the good or the *beautiful*. The topic of the passage in B 2, by contrast, is the lack of a final cause in mathematics and especially the resulting irrelevance of questions of *good* or *bad*.

Moreover, the immediate context of B 2[213] confirms the very phrase "the good *and bad*" in line 996b1: In the preceding lines we read both in Aristotle's own words and in his report of the Sophists' opinion that mathematics does not include demonstrations involving the criterion of better or worse (ἀπόδειξις …διότι

---

[211]Cf. Crubellier 2009: 53. On the passage in M 3 see Annas 1976: 151–52.

[212]M 3, 1078a36–b1: τοῦ δὲ καλοῦ μέγιστα εἴδη τάξις καὶ συμμετρία καὶ τὸ ὡρισμένον, ἃ μάλιστα δεικνύουσιν αἱ μαθηματικαὶ ἐπιστῆμαι. / "The main forms of the beautiful are order, symmetry, and definiteness, which are what the mathematical branches of knowledge demonstrate to the highest degree" (transl. by Annas).

[213]*Metaph.* B 2, 996a29–35.

βέλτιον ἢ χεῖρον, 996a30-31 and διότι βέλτιον ἢ χεῖρον λέγεσθαι, 996a34-35). The concluding remark in 996b1, stating that this science takes no account of goods and evils, accords perfectly. Therefore, the α-reading (περὶ ἀγαθῶν καὶ κακῶν), which is supported by the evidence in Alexander's paraphrase, is preferable to the β-reading (περὶ ἀγαθῶν καὶ καλῶν).

The following can be said about the possible origin of the β-reading. As Primavesi showed concerning A 9 and the more or less identical passages in M 4-5, the β-version in A 9 contains traces of contamination with the text in M 4-5.[214] The most peculiar signs of the alignment of the β-text in A 9 to the wording in M 4-5 is the correction of the original first person plural forms, which Aristotle uses to include himself among the members of the Academy, to verbal forms in the third person, which Aristotle employs in M 4-5 to speak more objectively about the Academy's teachings. Our case in book B can be compared to this. The comparison of the passage in M 3 could have prompted the β-version's adjustment of line B 2, 996b1 (κακῶν to καλῶν). As in the case of the "we"-corrections in the β-version of A 9, the evidence in Alexander's commentary, providing a third witness to α and β, can help us to declare justifiably the α-reading the original reading of $ω^{αβ}$.

4.3.2.3 Alex. In Metaph. 303.23-29 on Arist. Metaph. Γ 5, 1009a22-28

In Γ 5, 1009a26, the β-version contains an explanatory addition that is lacking in the α-version. In the fifth chapter of book Γ Aristotle critically engages those who deny the principle of non-contradiction on the basis of the relativistic phenomenalism of Protagoras. Aristotle diagnoses those whose denial depends on a flimsy, easily unveiled misconception as easily curable (εὐΐατος, 1009a19). Those who deny the principle of non-contradiction for the sake of argument are more difficult to treat (1009a17-18). The former group has to be persuaded (οἱ μὲν γὰρ πειθοῦς δέονται ...), while the latter defeated (... οἱ δὲ βίας). Aristotle describes the misguided, sensualistic presupposition of the former group in the following way:[215]

> Aristotle, *Metaphysics* Γ 5, 1009a22-28
>
> ἐλήλυθε δὲ τοῖς δια-[23]ποροῦσιν αὕτη ἡ δόξα ἐκ τῶν αἰσθητῶν, ἡ μὲν τοῦ ἅμα [24] τὰς ἀντιφάσεις καὶ τἀναντία ὑπάρχειν ὁρῶσιν ἐκ τοῦ αὐτοῦ [25] γιγνόμενα τἀναντία· εἰ οὖν μὴ ἐνδέχεται γίγνεσθαι τὸ μὴ [26] ὄν, προϋπῆρχεν ὁμοίως τὸ πρᾶγμα **ἄμφω ὄν**, ὥσπερ καὶ [27] Ἀναξαγόρας μεμῖχθαι πᾶν ἐν παντί φησι καὶ Δημόκρι-[28]τος·
>
> Those who really feel the difficulties have been led to this opinion by observation of the sensible world. They think that contradictions or contraries are true at the same

---

[214] Primavesi 2012b: 412-14.
[215] Cf. Bonitz 1849: 200; Kirwan 1971: 107; Cassin/Narcy 1989: 231.

time, because they see contraries coming into existence out of the same thing. If, then, that which is not cannot come to be, the thing must have existed before **as both contraries** alike, as Anaxagoras says all is mixed in all, and Democritus too;

24 ὑπάρξειν **α** Cassin/Narcy Hecquet-Devienne : ὑπάρχειν **β** Bekker Bonitz Christ Ross Jaeger || 25 γίνεσθαι **β** Christ Ross Jaeger, cf. Al.ᵖ 303.28 : γενέσθαι **α** Bekker Bonitz Cassin/Narcy Hecquet-Devienne || 26 ἄμφω ὄν **α** Al.ᵖ 303.27–28 Ar.ᵘ (*Scotus*) edd. : ἄμφω ὄν, τούτεστιν ὄν καὶ μὴ ὄν **β** (fort. Ascl. 275.17)

Those who deny the principle because they are confused are led into this position by the following mistake: they take the supposedly valid rule that no being comes out of non-being and combine it with their observation of the sensible world, where contraries appear to come out of the same thing. From this they infer that both contraries were already present in the thing. Aristotle steps out of this confusion (1009a30–36) by holding to the position that the rule stating that nothing comes out of non-being calls for a crucial differentiation of what is meant by being and non-being. Aristotle is hereby led to distinguish between what is potential and what is actual.

In line 1009a26 of the β-text, the words ἄμφω ὄν ("as both") are followed by τούτεστιν ὄν καὶ μὴ ὄν ("i.e. being and non-being"). This specification articulates what Aristotle means by the word "both" (ἄμφω): The opponents' opinion is based on the assumption that the thing (πρᾶγμα) already contains both being and non-being, when in fact it only has the capacity to be and not to be. The β-words τούτεστιν ὄν καὶ μὴ ὄν, which present no problems in terms of content or grammar, look like a later addition that was put into the text (perhaps by first having been put into its margins) in order to explicate what Aristotle means to say.[216] Or, are the words τούτεστιν ὄν καὶ μὴ ὄν original and did they drop in the α-text due to a *saut du même au même* (ἄμφω <u>ὄν</u> τούτεστιν ὄν καὶ μὴ <u>ὄν</u>)?

We should have a look at Alexander's commentary and determine what his comments reveal about the text in his copy. Did ω^AL contain the additional β-words?

Alexander, *In Metaph*. 303.23–29 Hayduck

καὶ πρῶτα μὲν λέγει [24] ὑπὸ τίνος παρεκρούσθησαν οἱ ἐπὶ παντὸς τὴν ἀντίφασιν συναληθεύειν λέ-[25]γοντες· ὁρῶντες γὰρ ἐκ τοῦ αὐτοῦ γινόμενα τὰ ἐναντία, προειληφότες δὲ [26] καὶ ὅτι ἀδύνατον γίνεσθαί τι ὅλως ἐκ τοῦ μὴ ὄντος (κοινὴ γὰρ αὕτη ἡ δόξα [27] τῶν περὶ φύσεώς ἐστί τι ἀποφηναμένων), ὑπέλαβον <u>ἀμφότερα τὰ</u> [28] <u>ἐναντία</u> τὸ πρᾶγμα εἶναι. οὐ γὰρ ἂν ἄλλως ἐξ αὐτοῦ δύνασθαι αὐτὰ γίνε-[29]σθαι, <u>εἰ μὴ προϋπάρχοντα ἐν αὐτῷ εἴη</u>.

And first he tells under what influence those who say that in every case contradictories are both true have been misled: seeing that contraries come to be from the same thing, and having assumed in advance that it is impossible for something to come

[216] See Jaeger *app. ad loc.*

to be altogether from non-being (this view is common to those who made any statement concerning nature), they supposed that the object was both the contraries; for they thought there was no other way in which the contraries could come to be from the object, than if they were preexistent within it.

23 πρῶτα **A O** : πρῶτον **P^b** ‖ 26 γίνεσθαί **S** Bonitz Hayduck : γενέσθαι **A O P^b** : τὸ μὴ ὂν γενέσθαι **LF** ‖ 27 ἐστί τί ἀποφηναμένων **A O** : τι αποφηναμένων ἐστιν **P^b** ‖ 28 τὸ πρᾶγμα **A O S** : πράγματα **P^b** ‖ 28–29 γίνεσθαι **S** Bonitz Hayduck : γενέσθαι **A O P^b** ‖ 29 αὐτῷ **C M° R** : αὐτ~ **A** : αὐτῇ **O S** : αὐτοῖς **P^b**

Although Alexander's comments provide us here with only one type of evidence about the reading in ω^AL, namely, an explanatory paraphrase, we may conclude from *two* formulations within his paraphrase that he did not read the words τούτεστιν ὂν καὶ μὴ ὄν in his text. Alexander presents Aristotle's argument in a slightly expanded way, using two sentences to render one of Aristotle's. In lines 303.26–28 Alexander amplifies the content of 1009a25–26 with an account of the origin of this common opinion (κοινὴ γὰρ αὕτη ἡ δόξα... 303.26). For τὸ πρᾶγμα ἄμφω ὄν Alexander writes ἀμφότερα τὰ ἐναντία τὸ πρᾶγμα εἶναι (27–28). Although one might call this an amplified and hence altered version of the Aristotelian original, it is evident that Alexander does not include anything that suggests he read τούτεστιν ὂν καὶ μὴ ὄν in his text.[217] Furthermore, in the sentence thereafter, 303.28–29, Alexander is still covering the same Aristotelian line (1009a25–26), saying that they could not understand how a thing could be and then not be unless both contraries were preexistent in it. Also in this sentence, Alexander does not explicitly state that the contraries that are preexistent in the thing are being and non-being (εἰ μὴ προϋπάρχοντα ἐν αὐτῷ εἴη, 303.29). And so also this sentence points to the conclusion that he did not find the β-addition in ω^AL.

Therefore, we may conclude on the basis of both direct and indirect evidence in Alexander that the words τούτεστιν ὂν καὶ μὴ ὄν, transmitted by the β-version only, are a secondary addition to the β-text that was not contained in ω^αβ.[218]

---

[217] The Greek *Vorlage* of the Arabic translations, as Scotus's translation confirms, also did not contain the additional words.

[218] Asclepius's commentary on this passage (275.14–17) invites a far-reaching suspicion: εἰ οὖν μὴ ἐνδέχεται γενέσθαι τὸ μὴ ὄν, προϋπῆρχεν ὁμοίως τὸ πρᾶγμα ἄμφω ὄν, καὶ λευκὸν καὶ οὐ λευκόν· ὥστε ἅμα τὰ ἐναντία, καὶ ὂν καὶ οὐκ ὄν/ "If, then, that which is not cannot come to be, the thing must have existed before as both contraries alike, both white and non-white, so that both contraries exist simultaneously, both being and non-being." In 15–16 (εἰ ... ἄμφω ὄν), Asclepius stays very close to the Aristotelian text. Then (16, καὶ λευκὸν καὶ οὐ λευκόν) he illustrates the thought in terms of the color "white" ("both white and non-white"); Asclepius uses this example multiple times in the context of this passage. The explication that follows (16, ὥστε ἅμα <u>τὰ ἐναντία</u>) then seems to derive, as is often the case, from Alexander's commentary on the passage (ἀμφότερα <u>τὰ ἐναντία</u>, see above). What thereafter follows, seems to be Asclepius's own contribution to the thought. With this he makes clear what is meant by Alexander's τὰ ἐναντία. These contraries are καὶ ὂν καὶ οὐκ ὄν (16) / "being and non-being." If this assessment of Asclepius's commentary is correct then one has to admit that the formulation in Asclepius comes very close to the wording of the β-addition (τούτεστιν ὂν καὶ μὴ ὄν). Does the

ALEXANDER'S TEXT AND THE DIRECT TRANSMISSION 167

In all three cases we could reconstruct the reading in ω^αβ on the basis of the reading in α and its confirmation by the evidence available in Alexander's commentary. In the first (4.3.2.1) and the third case (4.3.2.3), β contains a later supplement. In the second case (4.3.2.2), the β-text suffered a slight change of one letter, which might have been caused by a comparison with an only seemingly parallel passage in book M of the *Metaphysics*.

4.3.3 Reconstruction of an ω^αβ-reading from ω^AL and two differently corrupted readings in α and β: Alex. *In Metaph.* 329.33–330.8 on Arist. *Metaph.* Γ 7, 1011b35–1012a1

In the cases discussed so far in 4.3 the reading in ω^AL coincided *either* with α *or* with β. The reading in ω^αβ was then reconstructed on the basis of the agreement of ω^AL with one of our two versions. In the following case the situation is a bit more complex: here, the reconstruction of the reading in ω^αβ is built on a comparison of the two *differently corrupted* readings in α and β with the testimony in Alexander's commentary.

In Γ 4–6 Aristotle defends the validity of the principle of non-contradiction by examining and disputing possible arguments against the principle. In Γ 7 Aristotle turns to the principle of excluded middle. According to this principle there can be no intermediate between contradictory terms (οὐδὲ μεταξὺ ἀντιφάσεως ἐνδέχεται εἶναι οὐθέν, 1011b23–24).[219] Aristotle argues for the validity of this principle by first defining truth and falsehood (1011b25–27): "To say of what is that it is not, or of what is not that it is, is false; while to say of what is that it is, and of what is not that it is not, is true." So, to say that something is (or that something is not) is either true or false.[220] Truth and falsity defined (1011b29–1012a1), Aristotle reduces to absurdity the supposition of an intermediate (μεταξὺ …) between being F and not being F (… τῆς ἀντιφάσεως) by a two-part argument.[221] The two parts correspond to the two types of intermediates, both ruled out as possible intermediates between being F and not being F. The text of the argument reads as follows:

Aristotle, *Metaphysics* Γ 7, 1011b29–1012a1

ἔτι [30] ἤτοι μεταξὺ ἔσται τῆς ἀντιφάσεως ὥσπερ τὸ φαιὸν [31] μέλανος καὶ λευκοῦ, ἢ ὡς τὸ μηδέτερον ἀνθρώπου καὶ ἵππου. [32] εἰ μὲν οὖν οὕτως, οὐκ ἂν μεταβάλλοι (ἐκ μὴ ἀγαθοῦ γὰρ [33] εἰς ἀγαθὸν μεταβάλλει ἢ ἐκ τούτου εἰς μὴ ἀγαθόν), νῦν [34]

β-addition go back to Asclepius's commentary? Or is it rather that Asclepius worked with a β-copy—despite the ample evidence that his *Metaphysics* text shows strong affinities with the α-version? (On Asclepius's relationship to the α-version see Kotwick 2015.)

[219] Cf. also *Metaph.* I 5, 1056a22–b2; I 7, 1057a18–b34.
[220] For an analysis of this argument see Kirwan 1971: 117–18. See also Ross 1924: 285: "The argument thus has value only *ad hominem*."
[221] See Kirwan 1971: 118–19.

δ' ἀεὶ φαίνεται (οὐ γὰρ ἔστι μεταβολὴ ἀλλ' ἢ εἰς τὰ ἀντι-[35]κείμενα καὶ μεταξύ)· εἰ δ' ἔστι μεταξύ, καὶ οὕτως **εἴη ἄν** [1012a1] **τις** εἰς λευκὸν οὐκ ἐκ μὴ λευκοῦ γένεσις, νῦν δ' οὐχ ὁρᾶται.

Again, either the intermediate between the contradictories will be so in the way in which grey is between black and white, or as that which is neither man nor horse is between man and horse. If it were thus [i.e. of the latter kind], it could not change, for change is from not-good to good, or from that to not-good; but in fact it evidently always does, for there is no change except to opposites and to their intermediate. But if there is an intermediate, in this way too **there would be** some sort of [process of] coming to be white which was not from not-white; but as it is, this is never seen.[222]

30 μεταξὺ **α** Al.¹ 329.5[O] Ascl.¹ 294.8 Ross Jaeger Cassin/Narcy Hecquet-Devienne : τὸ μεταξὺ **β** Al.¹ 329.5[Hayduck] Bekker Bonitz Christ || ἔσται τῆς **α** edd. : ἐστι τῆς **β** : ἐστιν Al.¹ 329.5 || 34 δ' ἀεὶ **α** Al.ᴾ 329.18 Al.ᶜ 329.26 edd. : δὲ **β** || 35 μεταξύ] τὰ μεταξύ Jaeger coni. ex Al.ᴾ 329.21 et Ascl. 294.20 || 35–1012a1 εἴη ἄν τις **α** Ascl.ᶜ 294.23–24 Arᵘ (*Scotus*) edd. : ἢ ἡ ἀντίφασις **β** : ἡ ἀντίφασις, εἴη ἄν τις ωᴬᴸ (Al.ᶜ 330.1–2 Al.ᴾ 330.7–8) : ἡ ἀντίφασις ἔχει, εἴη ἄν τις ci. Alex 330.2 || 1 οὐκ ἐκ μὴ ωᵃᵝ : ἐκ οὐ μὴ Ascl.ᶜ 294.24

Aristotle's train of thought seems to be the following: the goal of the argument is to show the absurdity of an intermediate between the contradictories being F and not being F. Did such an intermediate exist, then it would have to be either (i) as grey is between the two contraries black and white[223] (1011b30–31) or (ii) as (e.g.) a stone[224] is between a horse and a human being (1011b31).[225] The second option is easily ruled out, since change from horse into human is impossible.[226] But, as Aristotle adds, it is obvious that there is change between intermediates and opposites (1011b32–34), and so the fate of intermediates between contradictories hangs on the first option. Aristotle rules this option out with the following consideration. If there was an intermediate between contradictories (being F and not being F) such as grey is between black and white, then there would have to exist a not non-F (as this intermediate can neither be F nor not be F). This, however, amounts to saying that white could come from not non-white, which is absurd (1011b35–1012a1). For it is clear that white comes solely from non-white, that is, black or every shade of grey.[227]

The sentence in lines 1011b35–1012a1 as it is transmitted by the α-version is

[222] The process of coming to be white out of white is invisible. The assumption of such a process is absurd.

[223] For grey as intermediate between black and white see *Cat.* 10, 12a2–11; 12a17–20 and *Metaph.* Δ 10, 1018a20–25.

[224] I take over the example of the stone from Asclepius's commentary: *In Metaph.* 294.14 Hayduck.

[225] Cf. *Cat.* 5, 3b24–27 and also *Metaph.* Ι 5, 1056a30–b2, where Aristotle points out that in contrast to good and bad there is no intermediate between a shoe and a hand.

[226] Cf. *Metaph.* Λ 1, 1069a36–b7.

[227] Ross 1924: 285: "There is of course transition to white from grey, which is not *simpliciter* not-white. But the transition is from grey *qua* not-white; it is the specks of black in the grey that change to white."

comprehensible and grammatically correct. The situation is different in the β-version. Here, we read instead of the words καὶ οὕτως **εἴη ἄν τις** εἰς λευκὸν οὐκ ἐκ μὴ λευκοῦ γένεσις ("in this way too there would be some sort of [process of] coming to be white which was not from not-white") the words καὶ οὕτως **ἢ ἡ ἀντίφασις** εἰς λευκὸν οὐκ ἐκ μὴ λευκοῦ γένεσις ("in this way too [is?] indeed the contradiction a coming to be white which was not from not-white"). Two things are odd about the β-reading. The first problematic point is the particle ἦ. Affirmative ἦ stands usually at the beginning of a sentence and is in prose almost always confined to dialogue.[228] Aristotle does not at all use the affirmative ἦ by itself,[229] and only rarely in the combination ἦ που and ἦ μήν.[230] Second, the apodosis (introduced by καὶ οὕτως) does not have a verb. This results in the nominal construction, "the contradiction (ἀντίφασις) *is* a coming to be (γένεσις) white which was not from not-white." For these reasons, the α-reading is preferable, and indeed the editors of the *Metaphysics* always have followed it.[231] I note in passing that all editors place a comma after μεταξύ so that the apodosis begins with the words καὶ οὕτως.[232]

It is striking that despite the difference in meaning the two readings in α and β are typographically quite similar, visually and aurally. Does the word ἀντίφασις go back to a misreading, fostered by the context, of ἄν τις? Was εἴη misspelled as ἢ ἡ due to an iotacism? When did the corruption occur? The tendency to iotacism (ι-sound for η, ει) begins already in Hellenistic times.[233] Alexander can help us in this matter.

Alexander, *In Metaph.* 329.33–330.8 Hayduck

λείπεται ἄρα ὡς ἀντικειμένων εἶναι αὐτῶν [34] μεταξύ τι, εἰ ἔστιν, ὡς τοῦ λευκοῦ καὶ μέλανος τὸ φαιόν, ὃ καὶ κυρίως [35] ἐστὶ μεταξύ. διὸ καὶ οὕτως εἶπεν εἰ δὲ ἔστι

---

[228] Denniston 1954, s.v. ἦ; pp. 279–80. Cf. Denniston 1954: 280: "[ἦ] Affirmative, mostly with adjectives and adverbs. This is mainly a verse idiom, and is hardly found at all in oratory, except for ἦ μήν, and the common use of ἦ που in *a fortiori* argument."

[229] My statement is based on a TLG-search.

[230] There is no lemma for ἦ in Bonitz's index. Interestingly though, Denniston 1954: 281 points to some exceptional cases in which ἦ does not stand at the beginning of the sentence. One of the exceptions is ἦ "at the opening of an apodosis." It therefore seems possible that the ἦ in our passage does not, as one might suppose, go back to an erroneous dittography of the article ἡ, but to a later attempt to mark ἡ ἀντίφασις *as the beginning of the apodosis*. I will come back to the question of where exactly in this sentence the apodosis starts.

[231] The α-version is attested also by Asclepius, *In Metaph.* 294.23–25: διό φησιν εἰ δὲ ἔστι μεταξύ, καὶ οὕτως εἴη ἄν τις (τι mss.) εἰς λευκὸν ἐξ οὐ μὴ λευκοῦ γένεσις and the Arabic tradition (Scotus): *Et si medium fuert secundum hanc dispositionem, tunc erit aliquid quod transmutatur in album non ex albo*. In Scotus's version, καὶ οὕτως / *secundum hanc dispositionem* is still part of the protasis. See previous note and below.

[232] This avoids a reading like "but if there is an intermediate also in this way, there would be...." For this question cf. Cassin/Narcy 1989: 262 and below.

[233] Adrados 2002: 187.

μεταξύ, τουτέστιν εἰ δὲ [36] ἔστι κυρίως αὐτῶν μεταξύ, καὶ οὕτως ἔχει ἡ ἀντίφασις ὡς ἀντικεῖσθαί τε καὶ [330.1] μεταξύ τι ἔχειν αὐτῆς (τοῦτο γὰρ σημαίνει τὸ καὶ οὕτως ἡ ἀντίφασις, [2] λείποντος τῇ λέξει τοῦ 'ἔχει,' ἢ καὶ οὕτως ἡ ἀντίφασις, τουτέστι [3] καὶ οὕτως ἐστὶ τὸ εἶναι τῇ ἀντιφάσει), συμβήσεται, φησί, τὴν γένεσιν [4] καὶ τὴν μεταβολὴν τὴν εἰς λευκὸν γίγνεσθαι ἐξ οὐχὶ οὐ λευκοῦ, ἀλλ' οὐκ [5] ἐκ τοῦ οὐ λευκοῦ· τοῦτο δέ, ἐπεὶ διὰ τῶν μεταξὺ γίνεται εἰς τὰ ἄκρα ἡ [6] μεταβολή. τὸ δὴ ἐκ τοῦ μεταξὺ τοῦ τε λευκοῦ καὶ τοῦ οὐ λευκοῦ μετα-[7]βάλλον εἰς λευκὸν ἐξ οὐχὶ οὐ λευκοῦ μεταβάλλοι ἂν εἰς λευκόν, καὶ γίνοιτο [8] ἄν τι λευκὸν ἐξ οὐχὶ οὐ λευκοῦ.

So it remains[234] that there is an intermediate between the contradictories, if at all, as grey, which is intermediate in the proper sense, is [intermediate] between white and black. This is why he spoke thus: "but if there is an intermediate," that is, but if there is something intermediate between them in the proper sense, and the contradiction is such that [the contradictories] are opposed and have some intermediate between them (this is what the expression "and the contradiction thus"—the "it is" [ἔχει] is lacking to the phrase—signifies; or "and the contradiction thus" [signifies as follows], that is, "such is the being of the contradiction") [if there is an intermediate] then it will occur, he says, that coming-to-be, i.e. change, into white will take place from not-non-white, not from non-white; this, because change proceeds to extremes by way of intermediates. That, then, which changes from the intermediate between white and non-white into white, would change from not-non-white into white, and something would come to be white from not-non-white.

33–34 εἶναι αὐτῶν μεταξύ τι O L S : εἶναι αὐτῶν μεταξύ τις A : ὄντων αὐτῶν εἶναί τι μεταξύ P^b || 34 εἰ ἔστιν O LF : ἐστιν A : om. P^b A^p.c. || καὶ A O : om. P^b S || 36 ὡς O P^b LF S : καὶ A || 2–3 ἡ ἀντίφασις ... οὕτως A P^b S : om. O || 4–5 οὐκ P^b LF S : om. A O || οὐ λευκοῦ] λευκοῦ (albo) legit S || 7–8 μεταβάλλοι ... λευκοῦ O LF S : om. A : ἔσται P^b

The evidence for ω^AL that is available in this commentary passage has to be extracted from Alexander's words. At first glance, lines 329.35–330.3 suggest that Alexander had the β-reading (καὶ οὕτως ἢ ἡ ἀντίφασις) in his text or a slightly altered version of it, namely, without the un-Aristotelian ἤ: καὶ οὕτως ἡ ἀντίφασις.[235] For this is the wording which Alexander quotes two times (330.1–2), and at which he targets his suggested corrections (cf. λείποντος τῇ λέξει τοῦ ἔχει, 330.2).

However, a closer look at Alexander's conjecture prompts the question whether Alexander could have thought to improve his text through the addition of the word ἔχει if his text was (apart from the ἤ) identical to the β-version. After examining Alexander's explanation it becomes clear that adding ἔχει in the sense of "it

---

[234] After having ruled out the second of the two named types of intermediates.

[235] Schwegler, Ross, Jaeger, Cassin/Narcy, Hecquet-Devienne and (the translators) Madigan 1993: 177 n. 891 (with hint to the other reading) and Casu 2007: 844 n. 903 supposed that such was the reading Alexander found in his text. Ross 1924: 285 thinks that in the subsequent part of the sentence Alexander read γένεσις with an article: εἰς λευκὸν οὐκ ἐκ μὴ λευκοῦ ἡ γένεσις. Such an article would certainly improve this hypothetical reading. However, the evidence that Alexander read this article is thin. In line 329.4 (Ross's evidence) the article preceding γένεσις (τὴν γένεσιν...) could be the result of Alexander's syntactical restructuring. For Bonitz's and Christ's view on the matter see below.

is (in a certain state)"²³⁶ (as Alexander's paraphrase in 329.36 indicates) does not at all improve the intelligibility of the β-reading. Quite to the contrary, Alexander's suggestion would make the β-reading worse. It would result either in the sentence εἰ δ' ἔστι μεταξύ, καὶ οὕτως ἡ ἀντίφασις ἔχει †... † εἰς λευκὸν οὐκ ἐκ μὴ λευκοῦ γένεσις ("But if there is an intermediate, in this way too the contradiction is (in a certain state) ... a coming to be white which was not from not-white");²³⁷ or, when the apodosis starts only after ἔχει, in the sentence εἰ δ' ἔστι μεταξὺ καὶ οὕτως ἡ ἀντίφασις ἔχει, †... † εἰς λευκὸν οὐκ ἐκ μὴ λευκοῦ γένεσις ("But if there is an intermediate and the contradiction is such, (then) ... a coming to be white which was not from not-white"). Could Alexander have seriously suggested something with such unhappy consequences?

The situation is similarly unsatisfactory when we take into consideration Alexander's second suggested solution (330.2–3). This time Alexander leaves the phrase καὶ οὕτως ἡ ἀντίφασις as it is. Alexander's alternative understanding is captured in the following paraphrase: καὶ οὕτως ἐστὶ τὸ εἶναι τῇ ἀντιφάσει (330.3). According to this understanding the sequence καὶ οὕτως ἡ ἀντίφασις is to be taken as "such is the being of the contradiction." This reformulation seems to suggest that Alexander wants to transport in thought the ἐστὶ from the protasis into the apodosis. However, it is more reasonable to assume that Alexander wants to relocate the comma such that the phrase καὶ οὕτως ἡ ἀντίφασις becomes part of the protasis. The purpose of the added ἐστὶ is then to signal that the phrase οὕτως ἐστὶ τὸ εἶναι τῇ ἀντιφάσει is equal in status to the εἰ δ' ἔστι μεταξύ-clause. Yet, if we suppose that Alexander had the β-reading, then not even this suggested understanding solves the problem of the β-reading: εἰ δ' ἔστι μεταξὺ καὶ οὕτως ἐστὶ τὸ εἶναι τῇ ἀντιφάσει, †... † εἰς λευκὸν οὐκ ἐκ μὴ λευκοῦ γένεσις ("But if there is an intermediate and the being of the contradiction is such, (then) ... a coming to be white which was not from not-white"). Can we seriously claim that this was Alexander's solution?

It is too unlikely that Alexander presents two solutions that each quite clearly fail to solve the problem of the β-reading, and so we must reconsider the question whether Alexander had at all the β-reading, and once more confront the question of what Alexander actually found in his own text. When we turn to the commentary lines that follow the passage we have addressed so far we learn that Alexander has, in fact, treated up to this point (330.3) only the first part of Aristotle's sentence (εἰ δ' ἔστι ... ἀντίφασις, 1011b35). Only beginning at 330.3 does Alexander comment on the second part of the sentence (1011b35–1012a1). In these lines of his

---

[236] The English translation of ἔχει as "it is" is slightly misleading since the verb ἔχειν (intransitive) means "to be" in the sense of a *full verb* ("to be in a certain state," often combined with an adverb denoting the state), but does *not* function like εἶναι as an *auxiliary verb* ("X is Y").

[237] Also the alternative translation (no comma after μεταξὺ, the apodosis beginnig after ἔχει) cannot solve the problem: "But if there is an intermediate and the contradiction is in this way too, (...) a coming to be white which was not from not-white."

commentary Alexander does not quote the *Metaphysics* text, but his paraphrase and his formulations—συμβήσεται ... τὴν γένεσιν ... εἰς λευκὸν γίγνεσθαι ἐξ οὐχὶ οὐ λευκοῦ (330.3) and καὶ γίνοιτο ἄν τι λευκὸν ἐξ οὐχὶ οὐ λευκοῦ (330.7–8)—clearly speak in favor of Bonitz's speculation[238] that Alexander *did* read the words εἴη ἄν τις, which we know as the α-reading. Alexander's comments also show that he read in his text the α-words *in addition* to the β-expression ἡ ἀντίφασις. Therefore we can reconstruct the following reading of lines 1011b35–1012a1 in ω^AL: εἰ δ' ἔστι μεταξὺ καὶ οὕτως **ἡ ἀντίφασις, εἴη ἄν τις** εἰς λευκὸν οὐκ ἐκ μὴ λευκοῦ γένεσις.[239] This reading, which combines the α- and the β-reading, clarifies the function of the words καὶ οὕτως in line 1011b35: the words καὶ οὕτως, now supplemented by ἡ ἀντίφασις, belong to the protasis ("But if there is an intermediate and such is the contradiction, then there would be..."). They do not, as suggested by the α-reading and believed by the editors, belong to the apodosis, which rather begins only at εἴη ἄν τις.

When we accept this reconstructed reading as the reading in ω^AL, Alexander's approach (330.2), which previously puzzled us, makes perfect sense. He wanted to add ἔχει (or supplement an ἐστί in thought) in a phrase which we now recognize as part of the protasis: εἰ δ' ἔστι μεταξὺ καὶ οὕτως ἡ ἀντίφασις <ἔχει>, εἴη ἄν τις .... ("But if there is an intermediate and the contradiction is such, then there would be..."). Alexander's proposal to add ἔχει in the second part (καὶ...) of the protasis is now quite unproblematic, but it can safely be regarded as a cosmetic and unnecessary correction. After all, it did not bother Aristotle in b32 to formulate the protasis without ἔχει: εἰ μὲν οὖν οὕτως ... .[240]

Bonitz considers the question whether or not Alexander's reading is preferable, and decides against it.[241] Bonitz follows the α-reading in his *Metaphysics* edition, and so do the editors after him.[242] This is perplexing given the fact that it is easier to explain the emergence of the α- and the β-readings from a misreading of the ω^AL-reading, than to explain the genesis of α by a misreading of β or to explain the genesis of β by a misreading of α.[243] When we take the letter sequence ἡ ἀντίφασις, εἴη ἄν τις εἰς (ΗΑΝΤΙΦΑϹΙϹΕΙΗΑΝΤΙϹΕΙϹ) as original to ω^αβ we can arrive at both readings α and β by way of two slightly different scribal errors, likely trig-

---

[238] Bonitz 1849: 213. Christ 1886a: *in app. crit.* follows Bonitz's assumption about what Alexander read in his text.

[239] It is surprising that Cassin/Narcy 1989: 262 do not even mention this reading of Alexander, especially since they intensively discuss Bonitz's commentary on the passage.

[240] The proposed correction shows how well Alexander knows Aristotle's diction, since phrases like εἰ (...) οὕτως ἔχει,... are rather common in Aristotle: see e.g. *EN* 1103b12; 1106a21; 1113a19.

[241] Bonitz 1849: 213 and *app. crit. ad loc.* The formulation in his apparatus does not make it clear whether Bonitz thinks we are dealing with a conjecture of Alexander or a testimony about his exemplar: *post* οὕτως *add.* ἡ ἀντίφασις *Alex., fort. recte.*

[242] Christ, Ross, Jaeger, Cassin/Narcy and Hecquet-Devienne.

[243] Bonitz 1849: 213.

gered by the repetition of a sequence of letters:

ΗΑΝΤΙΦΑCΙCΕΙΗΑΝΤΙCΕΙC > ΕΙΗΑΝΤΙCΕΙC = α-reading (either jump from ΗΑΝΤΙ to ΕΙΗΑΝΤΙ or jump from ΗΑΝΤΙ to ΗΑΝΤΙ and later adjustment of Η to ΕΙΗ, possibly invited by the following ἄν)

ΗΑΝΤΙΦΑCΙCΕΙΗΑΝΤΙCΕΙC > ΗΗΑΝΤΙΦΑCΙCΕΙC = β-reading (jump from ΙCΕΙ to ΙCΕΙ and later duplication of Η to ΗΗ).

The similarity among these letters makes it likely that the same type of error occurred twice at almost the same passage in the text.

Speaking now in terms of content, how does the reconstructed reading of ω^{AL} differ from the α-reading? Bonitz's first criticism of the α-version (εἰ δ' ἔστι μεταξύ, καὶ οὕτως εἴη ἄν τις...) is that the apodosis (beginning with καὶ οὕτως) states an unjustified inference.[244] Bonitz's other point of critique is that the μεταξὺ (b35) is not sufficiently specified. From Bonitz's point of view, this charge holds also for the text he suspects Alexander to have read.[245] Bonitz's remarks on the passage conclude with the claim that Aristotle's argument is in general unpersuasive.[246] Following Bonitz's criticism, but not accepting his conclusion, I want to look closely at the α-reading and the reading of ω^{AL}. Special attention should be given to the possibility that the ω^{AL}-reading, which I claim to be the original reading, actually makes good sense.

Line b35 in the α-version (εἰ δ' ἔστι μεταξύ, καὶ οὕτως ...) appears questionable because it does not sufficiently determine the type of intermediate that is at the center of attention at this point of the argument (the type grey between black and white). The protasis simply consists of the words εἰ δ' ἔστι μεταξύ—there is no specification given about what kind of μεταξύ we are dealing with.[247] It must however be conceded that the context and the train of thought suggest that Aristotle is now treating the *second* type of intermediate introduced at the beginning of the section (ἤτοι ... ἤ, 1011b30–31). By contrast, the first type of μεταξύ, which was treated in line b32 (εἰ μὲν οὖν οὕτως ...), is marked more clearly by the word οὕτως (b32). This οὕτως in b32 points back to the aforementioned type of inter-

---

[244]Bonitz 1849: 213: *Accedit quod in vulgata scriptura, ubi* καὶ οὕτως *ad apodosin trahendum est, parum apte dictum videtur «sic etiam fiat album non ex non albo»; neque enim, quod ex his verbis iure colligas, ex altero dilemmatis membro idem concluserat Aristoteles.* See also Ross 1924: 285. Bonitz probably argues from the usage of the formula εἰ ..., καὶ οὕτως in syllogisms, as e.g. in *APr.* 49b27–31: ἐν δὴ τοῖς τρισὶν ὅροις δῆλον ὅτι τὸ καθ' οὗ τὸ Β παντὸς τὸ Α λέγεσθαι τοῦτ' ἔστι, καθ' ὅσων τὸ Β λέγεται, κατὰ πάντων λέγεσθαι καὶ τὸ Α. καὶ εἰ μὲν κατὰ παντὸς τὸ Β, καὶ τὸ Α οὕτως ("If then we take three terms it is clear that the expression 'A is said of all of which B is said' means this, 'A is said of all the things of which B is said.' And if B is said of all of a third term, so also is A" [transl. by Jenkinson]).

[245]Bonitz 1849: 213: *etiamsi Alexandri lectionem receperimus, alterum* μεταξύ *genus ita describi, ut divinari magis quam cognosci possit.*

[246]Bonitz 1849: 213: *Universa autem argumentatio admodum est artificiosa.*

[247]That the bare μεταξὺ is unsatisfying was already seen by Schwegler (1847c: 183), who suggests adding ἐκείνως.

mediates and, more specifically, to the type of intermediate between a human being and a horse (τὸ μηδέτερον ἀνθρώπου καὶ ἵππου, b31). Once this type has been excluded (1011b32–35),[248] it seems a matter of basic logic that the next intermediate to be treated is the only one remaining. Still, the specification of the μεταξύ in b35 is weak. As I briefly mentioned above, one way of specifying the μεταξύ is to shift the comma to the position after καὶ οὕτως so that these words become part of the protasis (εἰ δ' ἔστι μεταξὺ καὶ οὕτως, .../ "But if there is an intermediate <u>also of such a kind</u>, ...").[249] However, this proposal is not persuasive, as the late position makes καὶ οὕτως appear syntactically unconnected.

This shortcoming of the α-reading shows up the superiority of the ω^AL-reading. Thanks to the addition of ἡ ἀντίφασις the phrase καὶ οὕτως ἡ ἀντίφασις as a whole is easily understood as part of the protasis. Moreover, this phrase offers a sufficient determination of the type of μεταξύ we are dealing with. As a result the protasis now reads: εἰ δ' ἔστι μεταξὺ <u>καὶ οὕτως ἡ ἀντίφασις</u>, ... ("But if there is an intermediate and <u>such is the contradiction</u>, then ..."). Only now does the protasis label the μεταξύ as the one mentioned in the preceding line (τὰ ἀντικείμενα καὶ μεταξύ, b34–35), that is, a μεταξύ that allows for change on the scale between two opposites (ἀντικείμενα). The phrase καὶ οὕτως ἡ ἀντίφασις not only makes this fact clear; it even specifies the μεταξύ in a way that is exactly parallel to the way in which Aristotle specified the other type of μεταξύ in line b32: ... ὡς τὸ μηδέτερον ἀνθρώπου καὶ ἵππου. εἰ μὲν οὖν <u>οὕτως</u>, ... (b31–32). These indications make the reconstructed reading in ω^AL preferable to the α-reading and, of course, the β-reading. The ω^AL-reading brings us to the reading in ω^αβ and is most likely the original reading.

When the reading of ω^αβ is restored, our passage reads thus:

Aristotle, *Metaphysics* Γ 7, 1011b35–1012a1

εἰ δ' ἔστι μεταξὺ καὶ οὕτως **ἡ ἀντίφασις, εἴη ἄν** [1012a1] τις εἰς λευκὸν οὐκ ἐκ μὴ λευκοῦ γένεσις, νῦν δ' οὐχ ὁρᾶται.

But if there is an intermediate and such **is the contradiction, then there would be** a [process of] coming to be white which was not from not-white; but as it is, this is never seen.

This case is then another example of the reconstruction of ω^αβ out of α and β with the help of ω^AL. It is special in that both versions α and β have been corrupted differently. Neither of the two readings in α and β is simply confirmed by the reconstructed reading in ω^AL. Since, however, each of the two versions preserves a *different* piece of the *complete* reading in ω^αβ, which is preserved uncorrupted in ω^AL, we are allowed to conclude that ω^αβ was identical to the reading in ω^AL. The

---

[248] This type of intermediate can hardly be called an intermediate: Alex. *In Metaph.* 329.12–14.
[249] Cf. Cassin/Narcy 1989: 262.

respective errors in α and β must then have occurred after ω^{αβ} has split into α and β.[250]

The seven passages analyzed in 4.3.1–4.3.3 have shown that the agreement of ω^{AL} with either α or β leads us to the reading in ω^{αβ}, and that this is the preferable reading. In none of the cases discussed is the reading that Alexander shares with α or β respectively inferior to the reading in the other version; there has been no conjunctive error in ω^{AL} and α or ω^{AL} and β.[251] Apart from these seven, there are of course many more cases in which the agreement of (what seems to be the reading in) ω^{AL} with either α or β is decisive for determining the reading in ω^{αβ} (at the end of this chapter is a highly selective two-part list of cases where the difference between α and β is especially apparent, and Alexander's testimony for ω^{AL} shows which of the two was most likely in ω^{αβ}). In many cases, the evidence in Alexander's commentary, confirming one or the other of the two versions, has not been taken seriously enough by the *Metaphysics* editors.

All in all, chapter 4's analysis of the relation between Alexander's text (ω^{AL}) and the directly transmitted versions of the *Metaphysics* shows that ω^{AL} and ω^{αβ} are two independent textual witnesses and that the evidence available in Alexander's commentary is therefore of utmost importance for the reconstruction of the

---

[250] Perhaps there have been certain material conditions in the manuscript of ω^{αβ} that facilitated the errors in α and β. Was the passage difficult to read? Even on the assumption that the readability was affected it remains an open question why the corruption in α and β occurred in the same passage yet affected different letters.

[251] The seven cases discussed here can be taken as representative of the entire evidence, in so far as conjunctive errors between ω^{AL} and α or ω^{AL} and β are extremely rare, if they exist at all. In all the cases where α and β differ in the first five books of the *Metaphysics*, I could only find four instances (i–iv), where it seems at least possible that ω^{AL} shares an erroneous reading with α or β. (i) In Δ 6, 1016b9–11, α includes the phrase ἢ ὧν ὁ λόγος μὴ εἷς, which is absent in β and in ω^{AL} (Al. 367.36–37). However, one can argue here that the seemingly shortened version in ω^{AL} and β is what Aristotle actually wrote and the amplification we find in α was prompted by Alexander's comment in 367.36–37. This scenario then would be parallel to the case analyzed in 5.3.3. (ii) The situation is similar in Γ 5, 1010b32. The amplified phrase μήτε τὰ αἰσθητὰ εἶναι μήτε τὰ αἰσθήματα given in α, which reads μηδὲ τὰ αἰσθητὰ εἶναι in ω^{AL} (Al.^p 315.35–316.2, 19–21) and β, seems to be preferable, yet it is not necessarily the original reading. Aristotle's own comment in the following line 1010b33 and Alexander's remarks in 315.35–316.1 seem to speak in favor of the α-reading, but they could also be the reason for the α-text having been amplified. (iii) In case of Α 5, 987a16, the agreement of ω^{AL} (Al.^p 47.11) and β (ἄπειρον καὶ τὸ ἕν) clearly attests to an erroneous interpolation preserved in ω^{AL} and ω^{αβ}, which would have been prompted by the reading in 987a18. In α this addition has been deleted, perhaps prompted by Alexander's comments. (iv) Finally, in Α 9, 991b29, ω^{AL} (Al.^c 113.8–9) seems to share an error with α (ἁπλῶς), whereas β (ἃ πῶς) seems to give the correct reading. Yet, Alexander's testimony in 113.7–15 is far from straightforward. In his edition of the commentary, Bonitz conjectured that the transmitted ἁπλῶς φησι, λέγων περὶ τῶν μαθηματικῶν (also in O) was in fact ἃ πῶς, φησί, ..., following the β-text of the *Metaphysics*. This conjecture, however, seems unjustified. What seems possible to me is that the original in ω^{αβ} and ω^{AL} (see πῶς in Al. 113.11 and 13!) read both ἁπλῶς and ἃ πῶς, and that ἁπλῶς was dropped in β and ἃ πῶς in α (cf. the case just discussed in 4.3.3).

ancient version of the *Metaphysics*. This importance can be summed up in the following rule: wherever the reading of ω^AL can be securely reconstructed (see 3.2–3.4) and it agrees with the reading in either α or β, this reading is most likely the reading of ω^αβ, which in turn is likely, but not necessarily, the correct reading.

Selective List of Agreements of ω^AL and β

981a4–5: ὀρθῶς λέγων α : om. β ω^AL (Al.^p 5.11–13)
981a11–12: οἷον ... καύσῳ α : om. β ω^AL (Al.^p 4.13–5.13)
981b2–5: τοὺς ... ἔθος α : om. β ω^AL (Al.^p 5.16–6.12)
983a17: τῶν οὐκ ἐλαχίστων α : τῷ ἐλαχίστῳ β ω^AL (Al.^p 18.22 Al.^c 18.20)
984a32–33: τοῦτο ... ὡμολόγησαν α : om. β ω^AL (Al.^p 30.9–10)
985a19–20: διὰ ... τότε α : om. β ω^AL (Al.^p 35.1–4)
985b7: τε καὶ μανὸν α : om. β ω^AL (Al.^p 36.1)
985b27: τοῖς ἀριθμοῖς α : τούτοις β ω^AL (Al.^p 37.22 38.5–6)
986a9: εἶναι ... φύσιν α : om. β ω^AL (Al.^p 40.28)
987b6: λόγον α : ὅρον β ω^AL (Al.^p 50.12)
987b10: πολλὰ τῶν συνωνύμων ὁμώνυμα α : πολλὰ τῶν συνωνύμων β ω^AL (Al.^l 50.17 Al.^c 50.22)
989a26–30: ὅλως ... φησιν α : om. β ω^AL (Al. 68.4)
992b3–4: τῆς ὕλης α : τῆς ὕλης ἢ ὕλην β ω^AL (Al.^c 122.16)
993b22: οὐ τὸ αἴτιον καθ' αὑτό α : οὐκ ἀίδιον β ω^AL (Al.^c 145.19)
997a23: τὸ ὅτι α : ὃ β ω^AL (Al.^c 192.6–7; 193.21; 194.12)
998b2: ἐστὶ α : συνέστηκε β ω^AL (Al.^p 202.28)
1000a29: ἦν α : ἦν ὅσα τ' ἐστὶν β ω^AL (Al.^c 220.5)
1002b31: ἀριθμῷ α : ἓν ἀριθμῷ β ω^AL (Al.^p 235.2)
1005a8: καὶ διὰ τοῦτο α : om. β ω^AL (Al.^p 263.9–17)
1008b15: βαδίζειν α : βαδίζειν δεῖν β ω^AL Al.^p 299.7–9)
1010b32: μήτε τὰ αἰσθητὰ εἶναι μήτε τὰ αἰσθήματα α : μηδὲ τὰ αἰσθητὰ εἶναι β ω^AL (Al.^p 315.35–316.2 et 316.19–21)
1011b19: ἀπόφασίς α : ἡ δὲ στέρησις ἀπόφασίς β ω^AL (Al.^c 327.10–11)
1013a23: κακόν α : καλόν β ω^AL (Al.^c 347.21)
1013b12: ἐνίοτε τῶν α : τῶν β ω^AL (Al.^p 350.31–32)
1015b16: ἕν, οὐδὲν γὰρ διαφέρει ἢ Κορίσκῳ τὸ μουσικὸν συμβεβηκέναι α : ἓν β ω^AL (Al.^p 362.33–363.3)
1016b24: ποσὸν καὶ ᾗ ποσὸν α : ποσὸν β ω^AL (Al.^p 368.34)
1019b16: ἀρχῆς ἄρσις τις α : ἀρχῆς β ω^AL (Al.^p 392.38)
1022a26–7: Καλλίας α : Καλλίας καθ' αὑτὸν καλλίας β ω^AL (Al.^p 416.3)
1022b35: καὶ τῷ φαύλως α : om. β ω^AL (Al. 419.32–420.1)
1023a29–31: τῆς πρώτης κινησάσης ἀρχῆς (οἷον ἐκ τίνος ἡ μάχη; ἐκ λοιδορίας α : τοῦ πρώτου κινήσαντος, οἷον ἐκ τῆς λοιδορίας ἡ μάχη β ω^AL (Al. 421.36–422.1).

Selective List of Agreements of ω^{AL} and α

986a20: καὶ ... περιττόν α ω^{AL} (Al.^p 41.30–31) : om. β
986b24: τὸν θεόν α ω^{AL} (Al.^p 44.9) : om. β
988a13–14: ὅτι ... μικρόν α ω^{AL} (Al.^p 59.20–23) : om. β
988b25–26: ὄντων καὶ ἀσωμάτων α ω^{AL} (Al.^p 64.23) : om. β
990b6: ἐπ' ἐκεῖνα α ω^{AL} (Al.^p 77.11) : ἐκεῖ β
990b9: δείκνυμεν α ω^{AL} (Al.^c 77.35 Al. 78.1–4) : δείκνυται β
992b7: καὶ ἔλλειψις α ω^{AL} (Al.^c 122.22 Al.^p 122.19) : om. β
994a29–30: καὶ ... ἐπιστημῶν α ω^{AL} (Al.^p 156.16–18) : om. β
995b33: καθ' αὑτὸ ἢ οὔ α ω^{AL} (Al.^l 178.4 Al.^p 178.14–16) : om. β
996a11: δυνάμει ἢ ἐνεργείᾳ α ω^{AL} (Al.^p 180.13–15) : om. β
996b1: κακῶν α ω^{AL} (Al.^p 182.38) : καλῶν β
996b4: τοῦ ζητουμένου α ω^{AL} (Al.^p 183.20 Al.^c 184.9) : om. β
1000b5: τὰ στοιχεῖα πάντα α ω^{AL} (Al.^p 220.23) : ἅπαντα β
1003a31: διὸ ... ὄν α ω^{AL} (Al.^p 240.28) : om. β
1004b15–16: οὕτω ... ἴδια α ω^{AL} (Al.^p 259.4; 20) : om. β
1005a5: ἕν α ω^{AL} (Al.^p 263.1–2) : ἕνα β
1007a21: εἶναι α ω^{AL} (Al.^l 285.2 Al.^c 285.11–12) : εἶναι μὴ εἶναι β
1007b33: καὶ α ω^{AL} (Al.^p 292.15–16) : καὶ ἡ τριήρης ἢ β
1009a26: ἄμφω ὄν α ω^{AL} (Al.^p 303.27–28) : ἄμφω ὄν, τούτεστιν ὂν καὶ μὴ ὄν β
1009b31: εἰ α ω^{AL} (Al.^p 307.12) : om. β
1012a12–13: τὰ ὄντα α ω^{AL} (Al.^l 332.16–17 Al.^p 332.19) : ταῦτα β
1012b31: αὐτό α ω^{AL} (Al.^p 343.8–10) : αὐτὸ ἀρχὴ λέγεται β
1014b21: συμπεφυκέναι ἢ α ω^{AL} (Al.^p 358.17 Al.^c 358.18; 27) : om. β
1015b16–17: τὸ δὲ ... μὲν α ω^{AL} (Al.^l 362.12–13) : om. β
1015b18–19: ταὐτὸ ... καὶ α ω^{AL} (Al.^p 362.15–16) : om. β
1015b22–23: τὸ ... συμβέβηκεν α ω^{AL} (Al.^p 362.22–23) : om. β
1017b17: ἐνυπάρχοντά ἐστιν ἐν τοῖς τοιούτοις α ω^{AL} (Al.^p 373.26 Al.^c 374.1–2) : ἔστιν β
1020b34: ὡρισμένος α ω^{AL} (Al.^c 403.18) : ὡρισμένος πρὸς ἕν β
1022b9–10: εἰς ... ἕξιν α ω^{AL} (Al.^p 417.34–35 Al.^c 417.37–418.1) : om. β
1022b21: συμφορῶν α ω^{AL} (Al.^p 418.31) : ἡδέων β
1024b10: ὧν τε α ω^{AL} (Al.^p 429.38) : ὧν β.

Cf. also the lists in the appendices B–D.

CHAPTER 5

Contamination of the Direct Transmission by Alexander's Commentary

The analysis of separative errors in $\omega^{\alpha\beta}$ against ω^{AL} led to the conclusion that ω^{AL} is independent of $\omega^{\alpha\beta}$ (see 4.1). This means that in the case of a textual divergence between α and β the agreement of one of the two with ω^{AL} leads most likely to the reading in $\omega^{\alpha\beta}$. Apart from this valuable use of reconstructed ω^{AL}-readings, Alexander's commentary stores other types of information. This information can be gathered through an investigation into the relationship between Alexander's *commentary* and the text of the direct transmission. Here, the first question is: how does Alexander's commentary relate to $\omega^{\alpha\beta}$? In section 5.1, I want to show that Alexander's commentary influenced, that is, contaminated, $\omega^{\alpha\beta}$ at a point before its split into the traditions α and β. Such contamination would not only rule out Jaeger's assumption that Alexander already had at his disposal both versions α and β,[1] but it would also allow for a more precise dating of $\omega^{\alpha\beta}$'s split into α and β and hence $\omega^{\alpha\beta}$ itself. In sections 5.2 and 5.3, I will ask how Alexander's commentary relates to the β-version and the α-version respectively. In 5.4, I will analyze two passages, where Alexander appears to know both readings in α and β.

5.1 CONTAMINATION OF $\omega^{\alpha\beta}$ BY ALEXANDER'S COMMENTS

5.1.1 Alex. *In Metaph*. 206.9–12 on Arist. *Metaph*. B 3, 998b22–28

The seventh aporia of book B is closely connected to the sixth. In the sixth aporia, Aristotle raises the question whether the genera or the primary constituents of a thing should be taken as its elements and principles (B 1, 995b27–29; B 3, 998a20–b14). The seventh aporia proceeds as though the sixth had been answered in favor

---

[1] Jaeger 1957: x. See also 1 above.

of genera and asks:[2] Is it the first and most remote genera or the lowest and most proximate genera that are the principles of things?[3]

Aristotle, *Metaphysics* B 3, 998b14–17

[998b14] πρὸς δὲ τούτοις εἰ καὶ ὅτι μάλιστα ἀρχαὶ τὰ γένη εἰσί, [15] πότερα δεῖ νομίζειν τὰ πρῶτα τῶν γενῶν ἀρχὰς ἢ τὰ [16] ἔσχατα κατηγορούμενα ἐπὶ τῶν ἀτόμων; καὶ γὰρ τοῦτο ἔχει [17] ἀμφισβήτησιν.

Besides this, even if the genera are in the highest degree principles, should one regard the first of the genera as principles, or those which are predicated directly of the individuals? This also admits of dispute.

998b15 πότερα α Al.¹ 204.24 : πότερον β Al.ᴾ 204.26 edd.

Aristotle will present five arguments against the first option that the first genera are in the highest degree principles. Aristotle's arguments target in particular the position that Being and One are among the first genera, a claim that seems to be Platonic in origin: Being and One (τὸ ὂν καὶ τὸ ἕν) are among the first genera because the first genera are most universally predicated and Being and One are most universally predicated (998b17–21).[4] The first genera are taken to be principles because what is most universal is in the highest degree a principle. In lines 998b22–28, Aristotle presents an argument against the view that Being and One are first genera and consequently principles.[5]

Aristotle, *Metaphysics* B 3, 998b22–28

οὐχ οἷόν τε δὲ τῶν ὄντων ἓν εἶναι γένος οὔτε τὸ ἓν οὔτε τὸ ὄν· [23] ἀνάγκη μὲν γὰρ τὰς διαφορὰς ἑκάστου γένους καὶ εἶναι καὶ [24] μίαν εἶναι ἑκάστην, ἀδύνατον δὲ κατηγορεῖσθαι ἢ τὰ εἴδη τοῦ [25] γένους ἐπὶ τῶν οἰκείων διαφορῶν ἢ τὸ γένος ἄνευ τῶν **αὐτοῦ** [26] εἰδῶν, ὥστ' εἴπερ τὸ ἓν γένος ἢ τὸ ὄν, οὐδεμία διαφορὰ οὔτε [27] ὂν οὔτε ἓν ἔσται. ἀλλὰ μὴν εἰ μὴ γένη, οὐδ' ἀρχαὶ ἔσονται, [28] εἴπερ ἀρχαὶ τὰ γένη.

But it is not possible that either One or Being should be a genus of things; for the differentiae of any genus must each of them both have being and be one, but it is not possible either for the species of the genus to be predicated of the proper differentiae or for the genus to be predicated [*sc.* of the differentiae] taken apart from the species; so that if One or Being is a genus, no differentia will either be one or have being. But

---

[2] Berti 2009: 119–20 understands aporia 7 as a particular case or sub-aporia of aporia 6. Already Schwegler 1847c: 131 treated the seventh aporia as part of the sixth.
[3] For aporia 7 and especially for its background in Academic discussions lead by Xenocrates see Berti 2009. See also Madigan 1999: 68–80.
[4] On the wording of lines 998b14–19 see 4.1.4.
[5] Aristotle, in accordance with his method in the third book of the *Metaphysics*, does not resolve the aporia. It is further questionable whether he even answers this aporia in *Metaph.* Z 12, 1038a19, as some modern commentators assume (cf. Jaeger 1912: 105). Madigan 1999: 80 offers a short discussion. On the present passage see Ross 1924: 235, Madigan 1999: 72–75, Barnes 2003: 329–36, and Berti 2009: 121–26.

if One and Being are not genera, neither will they be principles, if the genera are the principles.

998b17 ἀεὶ Al.ᵖ 204.29 Bonitz Ross Jaeger : δεῖ β : ὅτι α Bekker || ἀρχαί α Al.ᵖ 204.29 edd. : ἀρχάς β || 21 κατὰ πάντων μάλιστα λέγεται τῶν ὄντων] μάλιστα λέγεται κατὰ πάντων Al.ᶜ 204.34 || 22 τῶν ὄντων α Al.ᵖ 205.5 edd. : om. β || ἓν εἶναι γένος οὔτε τὸ ἓν οὔτε τὸ ὂν α Ross Jaeger : οὔτε τὸ ἓν οὔτε τὸ ὂν εἶναι γένος β Bekker Bonitz, cf. Al.ᵖ 205.5 || 24 τοῦ γ β Al.ᶜ 206.7 edd. : ἄνευ τοῦ EI^b || 25 ἐπὶ τῶν β Al.ᶜ 206.7 edd.⁶ : καὶ τῶν E : τῶν γ ζ Al.ᶜ 205.20 || γένος ἄνευ τῶν α ζ Al.ᶜ 206.9 edd. : γένος ἄνευ τούτων A^b : γένος τούτων ἄνευ τῶν M || αὐτοῦ Al. sua sponte proponit 206.10 ω^αβ edd. : om. ω^AL (Al.ᶜ 206.9) || 27 ὂν οὔτε ἓν β edd. (cf. Al.ᵖ 206.4 ἓν οὔτε ὄν) : τὸ ἓν οὔτε τὸ ὂν α : ἓν οὔτε ὂν ζ

Aristotle's argument that Being and One cannot be genera has three premises. First, the differentiae each have being and are one (998b23–24). Second, the species cannot be predicated of the differentiae (998b24–25). Third, the genus cannot be predicated of the differentiae when taken apart from its species (998b25–26). The conclusion follows directly: Being and One cannot be genera.

The two rules of predication that Aristotle brings forward as premises bear the weight of the argument. Aristotle justifies these rules in the *Topics* (Z 6, 144a31–b11) with five arguments in total. For the first rule that species cannot be predicated of the differentiae Aristotle gives three reasons: according to the first, the differentiae have a wider scope than the species.[7] According to the second, it is impossible to predicate the species of the differentiae because then the differentiae would belong to the species,[8] that is, the differentia "rational" would be a human being. As a third reason Aristotle points out that the species would be prior to the differentiae if they were predicated of the differentiae. But, differentiae are prior to the species (as in: the species is *defined by* the differentia).[9] So much can be said to the argumentative background of the first rule of predication (ἢ..., b24–25).[10]

---

[6] The connection of κατηγορεῖσθαι with the preposition ἐπί (as in the β-version and the citation in Alex.) is "less frequent" (LSJ s.v. κατηγορεῖσθαι III.). But, see, for example, the occurrences in lines 998b16 and 999a15 (both instance are in this aporia). More frequently used is κατηγορεῖσθαι + genitive or + κατά. I follow the β-reading as *lectio difficilior* (instead of the reading in γ ζ).

[7] *Top*. Z 6, 144b5–6: ἀδύνατον γάρ, ἐπειδὴ ἐπὶ πλέον ἡ διαφορὰ τῶν εἰδῶν λέγεται. / "for this is impossible, because the difference is a term with a wider range than the species." Cf. Alex. *In Metaph*. 205.17–19.

[8] *Top*. Z 6, 144b6–9; 6–8: ἔτι συμβήσεται τὴν διαφορὰν εἶδος εἶναι, εἴπερ κατηγορεῖταί τι αὐτῆς τῶν εἰδῶν. / "the result will be that the difference is a species, if any of the species is predicated of the difference." Cf. Alex. *In Metaph*. 205.21–27, where he makes the argument that the whole cannot be predicated of its parts.

[9] *Top*. Z 6, 144b9–11: πάλιν εἰ μὴ πρότερον ἡ διαφορὰ τοῦ εἴδους· / "Again, [it is wrong,] if the difference fails to be prior to the species."

[10] One might ask why Aristotle mentions at all the rule that the species cannot be predicated of the differentiae. It seems that all Aristotle needs to complete his argument, is to show that Being and One cannot be genera, and the rule that the genus cannot be predicated of the differentiae is sufficient for that. Ross (1924: 235) attributes the inclusion of the rule concerning species to argumentative thoroughness, and Berti (2009: 123 n. 48) hypothesizes that "there was someone who maintained that the

As for the second rule of predication (ἤ..., b25-26) Aristotle seems to say that it is impossible to predicate the genus (τὸ γένος) of the differentiae, if these are taken as separate from the species (ἄνευ τῶν αὐτοῦ εἰδῶν, b25-26). The reason why the genus cannot be predicated of its differentiae is stated in *Topics* Z 6, 144a31-b3. He gives two reasons. First, if the genus (e.g., animal) is predicated of its differentiae the absurdity follows that as many animals are predicated of each species as the differentiae are predicated of the species.[11] Second, and parallel to what was said against predicating the species of the differentiae, if the genus is predicated of the differentiae, it follows that the difference, *e.g.* rational, is an animal.[12]

Two issues arise in *Metaphysics* B 3, 998b23-28 as it has been transmitted down to us. First, Aristotle expresses the rule that forbids predicating the genus of the differentiae by simply saying it is impossible to predicate τὸ γένος ἄνευ τῶν αὐτοῦ εἰδῶν (998b25-26). Aristotle does not explicitly mention the (logical) subject of predication, i.e., a complement for the verb κατηγορεῖσθαι, as he does in the case of the other rule by stating τῶν οἰκείων διαφορῶν as the subject of predication.[13] According to the standard interpretation the subject of predication must be supplied from the preceding part of the sentence.[14] The only available candidate is τῶν οἰκείων διαφορῶν.[15] The translation of the *Metaphysics* passage quoted above

---

Being and the One are species." Thomas Johansen suggested at the annual meeting of the ESAP in 2012 that Aristotle might have mentioned the rule concerning the species because a genus can be a species. Concerning the interchangeability of the terms γένος and εἶδος see also Alexander 204.28: γένη γὰρ καὶ τὰ εἴδη νῦν λέγει. Cf. also Bonitz 1870: s. v. εἶδος 2., p. 218a7-8 and s. v. γένος 2., p. 151b34-35. I will briefly come back to this issue below: see p. 187 n. 37.

[11]*Top.* Z 6, 144a31-144b1; 36-37: εἰ γὰρ καθ᾽ ἑκάστης τῶν διαφορῶν τὸ ζῷον κατηγορηθήσεται, πολλὰ ζῷα τοῦ εἴδους ἂν κατηγοροῖτο· / "For if animal is to be predicated of each of its differentiae, then many animals will be predicated of the species." For a discussion about the correct understanding of the absurdity of "many animals" see Berti 2009: 124-25. He argues from an understanding of it that differs from Alexander's suggestion (Alex. *In Top.* 452.2-11 Wallies).

[12]*Top.* Z 6, 144b1-3: ἔτι αἱ διαφοραὶ πᾶσαι ἢ εἴδη ἢ ἄτομα ἔσονται, εἴπερ ζῷα· ἕκαστον γὰρ τῶν ζῴων ἢ εἶδός ἐστιν ἢ ἄτομον. / "Moreover, the differentiae will be all either species or individuals, if they are animals; for every animal is either a species or an individual."

[13]The β-reading (see apparatus on 998b23) probably goes back to the attempt to make explicit (τούτων) the implied subject of predication (τῶν οἰκείων διαφορῶν).

[14]This understanding of the passage is found in Bonitz 1849: 151-52 (*In verbis*: τὸ γένος ἄνευ τῶν αὐτοῦ εἰδῶν *b25 repetendum est ex antecedentibus*: ἐπὶ τῶν οἰκείων διαφορῶν, *genus non praedicatur de suis differentiis, si hae differentiae per se spectentur, seiunctae ab iis, quae inde efficiuntur, speciebus*), in Ross's translation in Barnes 1984 (see Barnes 1984: 1577: "it is not possible for the genus to be predicated of the differentiae taken apart from the species"; but cf. Ross 1924: 235 and Ross 1908 *ad loc.*), and in Berti 2009: 123 ("Rather Aristotle is saying that the genus cannot be predicated of its differentiae when they are considered as being separate from the species to which they belong, that is, considered as being other species of the same genus"). See also Barnes 2003: 330: "(P2) A genus is not predicated of its divisive differences taken apart from its species."

[15]This term included, lines 25 and 26 read: ἀδύνατον δὲ κατηγορεῖσθαι ... τὸ γένος (*sc.* τῶν οἰκείων διαφορῶν) ἄνευ τῶν αὐτοῦ εἰδῶν. / "it is impossible for the genus to be predicated of the proper differentiae taken apart from its own [i.e., the genus's] species."

follows this understanding. As for the phrase ἄνευ τῶν αὐτοῦ εἰδῶν (b25–26), it could be taken as follows: One cannot predicate a genus of a differentia when the differentia is separate from the species that it generates.[16] For example, "animal" cannot be predicated of "rational" if "rational" is considered to be something separate from the species "human being."[17]

The second issue is the word αὐτοῦ in the phrase ἄνευ τῶν αὐτοῦ εἰδῶν (b25–26). Its function is to mark the species as the genus's *own* species.[18] But it seems to state the obvious, namely, that Aristotle's rule of predication applies *only* to a genus's *own* species.[19] However, since Aristotle here does not speak of the relation of a genus to the species *of other genera*, the stress that comes with the αὐτοῦ might appear strange. One could add that the normal function of the article in Greek often has by itself the force of an English possessive adjective.[20] So when Aristotle says τὸ γένος κατηγορεῖται τῶν εἰδῶν, he means "the genus can be predicated of *its* species." When Aristotle speaks about the relation of genus and species, he typically does not underline the relation by means of a possessive pronoun.[21] Still, this evidence does not seem to be strong enough to cast serious doubt on the authenticity of the innocent αὐτοῦ. Alexander's comments on this passage store important information on this very word.

Alexander's paraphrase suggests that he read a slightly different text, but also that he understands it in similar fashion to the understanding given above:

Alexander, *In Metaph.* 205.28–33 Hayduck

ἀλλ' οὐδὲ τὰ γένη κατηγορεῖται τῶν [29] οἰκείων διαφορῶν, ὅταν αἱ διαφοραὶ χωρὶς τῶν εἰδῶν λαμβάνωνται καὶ μὴ [30] ἐν αὐταῖς τὰ εἴδη περιέχωνται. τὸ γὰρ ζῷον ὅταν λογικοῦ κατηγορῆται, [31] τοῦ ζῴου λογικοῦ (τοῦτο γὰρ ἐξακούεται τότε ἐκ τοῦ λογικοῦ) κατηγορεῖται, [32] ἐπεὶ αὐτῆς γε καθ' αὑτὴν τῆς διαφορᾶς ἄνευ τοῦ εἴδους λαμβανομένης οὐ [33] κατηγορεῖται, οἷον τῆς λογικότητος·

---

[16] The alternative interpretation of the line pairs the phrase ἄνευ τῶν αὐτοῦ εἰδῶν with τὸ γένος rather than with the supplemented τῶν οἰκείων διαφορῶν: one cannot predicate a genus of a differentia when it (i.e., the genus) is separate from the species. For this understanding see Ross 1924: 235; de Haas 1997: 239 with n. 232; Madigan 1999: 73–74.

[17] Cf. Alex. *In Metaph.* 205.28–33, Bonitz 1849: 151–52, and Berti 2009: 123.

[18] Ross's revised translation in Barnes 1984 silently slides over the αὐτοῦ: "but it is not possible for the genus to be predicated of the differentiae taken apart *from the species*" (my italics). In his original translation (Ross 1908), however, Ross wrote: "taken apart from *its* species" (my italics).

[19] The situation is different in the case of τῶν οἰκείων διαφορῶν / "the proper differentiae" in b25. Here it makes sense to underline the fact that we are talking about those differentiae that define the species.

[20] As e.g., in X. *An.* I.1,1: ὑπώπτευε τελευτὴν τοῦ βίου / "he was expecting the end of *his* life." For the possessive function of the definite article see Kühner/Gerth I: 461.2; p. 539.

[21] See esp. 999a10–11: σχολῇ τῶν γε ἄλλων ἔσται τὰ γένη παρὰ τὰ εἴδη. But also e.g. *Cat.* 5, 2b17–21: ὡς δέ γε αἱ πρῶται οὐσίαι πρὸς τὰ ἄλλα ἔχουσιν, οὕτω καὶ τὸ εἶδος πρὸς τὸ γένος ἔχει· ὑπόκειται γὰρ τὸ εἶδος τῷ γένει· τὰ μὲν γὰρ γένη κατὰ τῶν εἰδῶν κατηγορεῖται, τὰ δὲ εἴδη κατὰ τῶν γενῶν οὐκ ἀντιστρέφει· and 13, 15a4–5: τὰ δὲ γένη τῶν εἰδῶν ἀεὶ πρότερα. *Top.* Δ 3, 123a34–35: πᾶν γὰρ γένος κυρίως κατὰ τῶν εἰδῶν κατηγορεῖται.

But neither are genera predicated of their proper differentiae, when the differentiae are considered apart from the species and the species are not included in them. For example, when animal is predicated of rational, it is predicated of a rational animal (on this occasion this is what is meant by rational), whereas it is not predicated of the differentia taken by itself without the species, for example of rationality.

---

29 λαμβάνωνται A P^b : λαμβάνονται O ‖ 30 περιέχωνται P^b : περιέχονται A O ‖ κατηγορῆται P^b : κατηγορεῖται A O ‖ 31 τοῦτο A P^b : τοῦ O ‖ κατηγορεῖται addidit A$^{m.a.}$: om. A O : post ζῴου λογικοῦ habet P^b ‖ 32 γε P^b : τε A O ‖ καθ' αὐτὴν Ascl. Hayduck : καθ' αὐτῆς A O P^b LF ‖ τῆς διαφορᾶς P^b A$^{m.a.}$: τῶν διαφορῶν A O ‖ τοῦ P^b : τ' A O

Alexander twice paraphrases lines b25–26. The genus cannot be predicated of the differentiae when the differentiae are understood without the species they generate.[22] Notice that in lines 205.29 (χωρὶς τῶν εἰδῶν) and 205.32 (ἄνευ τοῦ εἴδους) Alexander formulates Aristotle's expression without either the αὐτοῦ or an equivalent term that would fit Alexander's syntax. This raises the question whether Alexander read at all the word αὐτοῦ.

After summarizing Aristotle's argument that concludes that being and One are neither genera nor principles, Alexander takes a closer look at Aristotle's expressions (206.6) and gives special attention to the phrase of our concern (998b25-26). Alexander quotes the phrase and entertains two possible interpretations of it. Here we find evidence that Alexander most likely has read a different text, but also that he now no longer agrees with the standard understanding.

Alexander, *In Metaph.* 206.9–12 Hayduck

τὸ δὲ ἢ τὸ γένος ἄνευ τῶν εἰδῶν ἤτοι [10] ἴσον ἐστὶ τῷ ἢ τὸ γένος ἄλλου τινὸς χωρὶς τῶν αὐτοῦ εἰδῶν κατηγορεῖ-[11]σθαι, ἢ τὸ γένος τῶν διαφορῶν κατηγορεῖσθαι, μὴ λαμβανομένων τῶν δια-[12]φορῶν ὡς εἰδῶν ἤδη καὶ συναμφοτέρων.

The statement "or [it is not possible] for the genus [to be predicated] apart from the species" is either equivalent to 'or [it is not possible] for the genus to be predicated of *anything else* except for *its own* species' or equivalent to 'or [it is not possible] for the genus to be predicated of the differentiae, unless the differentiae are taken as already being species and complex entities.'

In line 206.9, Alexander quotes the Aristotelian text, as the article τὸ indicates. This quotation confirms what his paraphrase (in 205.29 and 205.32) suggested: Alexander read ἄνευ τῶν εἰδῶν instead of ἄνευ τῶν αὐτοῦ εἰδῶν.[23] As it happens, Alexander was not alone in reading ἄνευ τῶν εἰδῶν;[24] it appears that the text of the Neoplatonic commentator Syrianus did not contain the αὐτοῦ either.[25]

---

[22] Cf. Berti 2009: 123.
[23] Cf. Madigan 1992: 144 n. 245 and Lai 2007: 543 n. 381.
[24] The Arabic version of the *Metaphysics*, however, seems to have included the αὐτοῦ. Scotus writes: *Neque genus predicatur nisi de formis que sunt ei*.
[25] Syrianus died in AD 437. Given that he wrote his *Metaphysics* commentary at the beginning of

Syrianus, *In Metaph.* 32.34–36 Kroll

... τοῦ [35] Ἀριστοτέλους ἐν τούτοις λέγοντος, ὅτι τὸ γένος οὐ κατηγορεῖται τῶν δια-[36]φορῶν <u>χωρὶς τοῦ εἴδους</u> ....

... Aristotle says here that the genus is not predicated of the differentiae without the species...

This raises the suspicion that the word αὐτοῦ is a later addition that occurred in ω^{αβ}. From where could this addition have come? Alexander's two interpretations (ἤτοι ... ἤ) of the rule stating that the genera cannot be predicated taken apart from the species (ἄνευ τῶν εἰδῶν) offer a clue. One of Alexander's interpretations diverges considerably from the standard interpretation, while the other is rather close to it.

According to the first interpretation (206.9–11), the genus cannot be predicated of *anything else* (ἄλλου τινὸς), other than its own species (τῶν <u>αὐτοῦ</u> εἰδῶν). Unlike modern scholars, Alexander does not draw the subject of predication, τῶν οἰκείων διαφορῶν, from the previous part of Aristotle's sentence. Instead, he supplies his own, which was evidently chosen in order to contrast what the genus cannot be predicated of with what it surely can be predicated of: As subject of predication, he supplies ἄλλου τινὸς, which he contrasts with τῶν αὐτοῦ εἰδῶν. Clearly, the genus can be predicated of its *own* species, whereas it cannot be predicated of anything else. Here in Alexander's *interpretation*, we find unexpectedly the αὐτοῦ.

In his second interpretation (206.11–12), Alexander comes back to the understanding that is implicit in his paraphrase at 205.30–34 and is also similar to the standard interpretation we already encountered. Here Alexander (206.11–12) brings in as the subject of the predication "the differentiae" (τῶν διαφορῶν) from the previous part of Aristotle's sentence. Under these circumstances the phrase ἄνευ τῶν εἰδῶν indicates that the differentiae are treated as separate from the species. We remember the example that Alexander suggested in 205.30–33: within the genus "animal" the difference "rational" should not be taken as meaning "rational animal" (i.e. human). In formulating this second interpretation, Alexander does not supply the word αὐτοῦ in order to specify the genus's *own* species.

We have two interpretations. In one (206.11–12), the missing subject of predication is clearly drawn from within Aristotle's text itself. In the other (206.9–11), the missing subject is supplied from without. It is in this *other* interpretation that we find an αὐτοῦ. We know that Alexander's text did not read the unidiomatic αὐτοῦ, and we suspect that it is a later addition. Thus the question arises whether there is some connection between the αὐτοῦ of the first of Alexander's interpreta-

---

the fifth century AD his *Metaphysics* copy can be dated to the end of the fourth or the beginning of the fifth century AD. Further study is needed to determine the relation of Syrianus's *Metaphysics* exemplar and his commentary to the text of our transmission. For an analysis of Syrianus's comments on our passage see de Haas 1997: 246–49.

tions and the αὐτοῦ of our manuscript tradition. If so, did someone want to foster Alexander's interpretation by bringing in the αὐτοῦ into the *Metaphysics* text or its margins? If so, why would someone want to follow Alexander on this issue?

To answer this last question, it might help to ask why Alexander would have at all proposed the interpretation that requires bringing into Aristotle's sentence the complementary parts ἄλλου τινὸς and αὐτοῦ. There does seem to be a reason why Alexander proposes this interpretation and why he highlights the fact that the genus cannot be predicated of anything else except its *very own* species. When we have a look into Alexander's treatise *De differentiis specificis*, preserved only in the Arabic,[26] we see that Alexander holds, contrary to Aristotle, that the differentiae are not mere determinants, but are in fact *species*, and what is more, species that are classed under the same genus under which the species, which the differentiae generate, are classed.[27] Alexander holds, contrary to Aristotle, that the genus *can* be predicated of the differentiae. When Alexander in his *Metaphysics* commentary suggests taking ἄνευ τῶν εἰδῶν to mean that the genus cannot be predicated of anything except its *very own species*, he implicitly undermines Aristotle's prohibition on predicating genera of differentiae, and he does this in order to accommodate his own understanding of differentiae.

In his subsequent comments on this section (206.12–33), Alexander attacks Aristotle's arguments calling them "dialectical," which is to say verbal or empty of value.[28] He then states in summary fashion the motivation for his disagreement. Before we look at the reasons stated in the commentary, let us take our bearings by the reasons Alexander provided in his treatise *De differentiis specificis*, for they are here stated more fully.[29] Here Alexander systematically examined not merely the *logical* status of the differentia, but also their *categorial* status.[30] The differentiae each have being and are one, and therefore, Alexander argues, must be classed under some genus or another. The question was: do the differentiae belong to *the*

---

[26] There is a German translation by Dietrich 1964 and a French translation by Rashed 2007: 56–65. (The English translation given in some of the subsequent notes is my own, based on the German and the English translations. However, it has been checked by Andreas Lammer against the Arabic original.) For an examination of Alexander's argument concerning the status of the differentiae see de Haas 1997: 211–19 and Rashed 2007: 66–81. See also Kupreeva 2010: 219–25.

[27] See de Haas 1997: 212–17.

[28] Alex. *In Metaph.* 206.12–13: δοκεῖ δέ μοι ἡ ἐπιχείρησις λογικωτέρα εἶναι, ὥσπερ οὖν καὶ αἱ πλεῖσται τῶν λεγομένων ὑπ' αὐτοῦ. / "The argument appears to me to be rather verbal, as indeed do most of the arguments that Aristotle mentions." See Madigan 1992: 96 n. 34. Madigan 1992: 144 n. 248 points to other commentary passages (210.20–21 and the closing words on book B in 236.26–29), in which Alexander criticizes Aristotle's argumentation in the book on the aporiae. See also Berti 2009: 122 and (concerning Aristotle) Bonitz 1870: 432b9–11 (e.g. *EN* A 1, 1217b21: λέγεται λογικῶς καὶ κενῶς).

[29] It is an open question whether Alexander in his *Metaphysics* commentary draws from the treatise or whether the commentary states an earlier version of the matter, which was later further developed in the treatise. Cf. de Haas 1997: 241.

[30] de Haas 1997: 188–89; 213–19 and 214–45.

same genus they divide, or do they belong to *another* genus?[31] Alexander answers: the differentiae belong to the *very same* genus they divide. Accordingly, the genus must be predicable of the differentiae.

In his commentary on our *Metaphysics* passage Alexander gives the following reasons:[32] Suppose that the differentiae are all qualities. It seems obvious that the differentiae of the genus "quality" are themselves under the genus "quality."[33] Alexander says that it would be absurd to say that all differentiae except those under the genus quality were qualities, and from this he infers that the genus *can* be predicated of the proper differentiae. (One might want to object to this reasoning that in the case of the genus "quality" the differentiae are not taken as differentiae but as species.[34])

Alexander, however, is aware of Aristotle's rules of predication given in the *Topics* (see above). He knows Aristotle's argument, stating that the genus "animal" cannot be predicated of the differentia "rational," because that would render "rational" an animal, which is absurd. Alexander answers that this problem first applies only to substances, and second, that it admits of a solution.[35] For Alexander, the reason that "animal" cannot be predicated of "rational" is because "animal" is a composite substance whereas "rational" is a simple substance, and it is absurd to class simples under composites. However there is a higher genus that embraces both composite substances like "animal" as well as simple substances like "rational": namely, the genus substance.[36] This genus can indeed be predicated of its proper differentiae.

---

[31] Alex. *De diff*. 136a = § 1 Dietrich; p. 123 (cf. Rashed 2007: 56): "We want to examine the differentiae and inquire into them exhaustively: So we ask: Under what genus should we subsume the differentia? Should we subsume it under the genus that it divides, or under another genus?" See also de Haas 1997: 214–17.

[32] Alex. *In Metaph*. 206.13–19 Hayduck: τὰς γὰρ διαφορὰς ὑφ' ὃ ἂν τάξωμεν γένος (δεῖ γὰρ ἔκ τινος γένους αὐτὰς εἶναι, εἴ γε τῶν ὄντων καὶ αὐταί), τῶν ἐκείνου τοῦ γένους διαφορῶν τὸ οἰκεῖον γένος κατηγορεῖσθαι ἀνάγκη. ἢ γὰρ οὐχ ἕξει γένος ὂν διαφοράς, ἢ ἔσονται ὑπὸ τὸ γένος ὃ διαιροῦσιν. εἰ γὰρ εἶεν αἱ διαφοραὶ ὑπὸ τὴν ποιότητα πᾶσαι, αἱ τῆς ποιότητος αὐτῆς διαφοραὶ δῆλον ὅτι ὑπὸ τὴν ποιότητα ἂν εἶεν. οὕτως τε ἂν κατηγοροῖτο τὸ γένος τῶν οἰκείων διαφορῶν·. / "For under whatever genus we range the differentiae—for they must be from some genus, given that they too are among beings—it is necessary for the proper genus of those differentiae to be predicated of the differentiae of that genus. For either it will be a genus but will not have differentiae, or the differentiae will be under the genus they divide. For if all differentiae were under quality, it would be clear that the differentiae of quality itself would be under quality. And thus the kind would be predicated of its proper differentiae." See also de Haas 1997: 241–45.

[33] Cf. Alex. *De diff*. 136a–b = § 4 Dietrich (cf. Rashed 2007: 57): "I say: they (*sc*. differentiae of quality) are simultaneously differentiae and species."

[34] Berti 2009: 123. Cf. also de Haas 1997: 192–94.

[35] Alex. *In Metaph*. 206.23–33 Hayduck.

[36] Cf. Alex. *De diff*. 137a = § 10 Dietrich (cf. Rashed 2007: 63): "The differentiae of the highest genera belong to their (i.e. the highest genera's) genera and species. Therefore we find that the differentiae of the highest genera are none other than the species, for their differentiae and species are one and the same."

This background information helps to explain in some measure the origin of Alexander's new interpretation of our *Metaphysics* passage. It appears that for Alexander the differentiae are species of the genus they divide; thus we are able to predicate the genus of them. This reasoning seems to be behind Alexander's seemingly harmless but in fact revolutionary alternative interpretation. The word αὐτοῦ, which Alexander adds in his reformulation (206.10),[37] encapsulates exactly this.

To sum up: the word αὐτοῦ given in the ω^{αβ} of the *Metaphysics* is unidiomatic in Aristotle. It does not appear in ω^{AL}. It does not seem to appear in Syrianus's text. It can be shown to be very idiomatic in Alexander. It appears in Alexander's idiosyncratic interpretation of the very phrase in which it appears in our manuscript tradition. As we saw, it is rooted in Alexander's own understanding of the categorical status of the differentiae and even encapsulates the very point of disagreement between Alexander and Aristotle. I therefore find it likely that the αὐτοῦ was a later addition to our text of the *Metaphysics*, ω^{αβ}. The most likely story is that some scholar or scribe, who wanted to endorse Alexander's own interpretation of the passage and perhaps also his understanding of the status of the differentiae, inserted αὐτοῦ either into the margin or the very body of Aristotle's text.[38]

5.1.2 Alex. *In Metaph.* 438.14–17 on Arist. *Metaph.* Δ 30, 1025a21–25

In the 30th chapter of book Δ, Aristotle examines the term accident (συμβεβηκός). An accident is what is said of something but neither of necessity nor for the most part. This is illustrated by two examples. First, if someone digs a hole in the ground to plant a tree and finds a treasure then this discovery is accidental to the digger since it happens neither necessarily nor for the most part (1025a15–19). Second, it

---

[37] Alexander's reformulation and syntactical understanding of our passage (apart from Alexander's idiosyncratic view on the differentia) has the potential to solve the problem of the unnecessary premise that the species cannot be predicated of the differentia. We could take up Alexander's formulation and take lines 998b24–26 (without changing anything in the Greek text) as "It is not possible to predicate the species of the differentiae *or to predicate the genus except of the species*" (cf. the translation, but not the commentary, *ad loc.*, in Madigan 1999). When we do not supply τῶν οἰκείων διαφορῶν as the subject of predication to the phrase τὸ γένος ἄνευ τῶν εἰδῶν, but take the verb κατηγορεῖσθαι absolutely, that is, without limiting its application through a stated subject of predication, then Aristotle rules out the predication of the genus of the differentiae in two steps: first, Aristotle prohibits the predication of the species of the differentiae and then states that the genus can be predicated *only of the species*. To say it again: the genus can be predicated only of the species and these cannot be *predicated of the differentiae*. Thus the rule regarding predication of species, which the accepted interpretation had rendered superfluous, can now serve a purpose.

[38] What we find in our text is, so to speak, a curious mixture of Aristotle and anti-Aristotle—a product of ancient scholarly work on the *Metaphysics* that had gone unnoticed by modern scholars.

is possible for a musically educated (μουσικός) human being to be pale, while it is by no means necessary or for the most part the case that musical education is accompanied by paleness (1025a19–21). As the following passage indicates, something is called an accident when it is present in something else without its presence being due to a determinate cause.[39]

Aristotle, *Metaphysics* Δ 30, 1025a21–25

ὥστ' ἐπεὶ [22] ἔστιν ὑπάρχον τι καὶ τινί, καὶ ἔνια τούτων καὶ ποῦ καὶ ποτέ, [23] ὅ τι ἂν ὑπάρχῃ μέν, ἀλλὰ μὴ διότι τοδὶ ἦν ἢ νῦν ἢ ἐν-[24]ταῦθα, συμβεβηκὸς ἔσται. οὐδὲ δὴ αἴτιον ὡρισμένον **οὐδὲν** [25] τοῦ συμβεβηκότος ἀλλὰ τὸ τυχόν· τοῦτο δ' ἀόριστον.

Therefore since there are attributes and they attach to a subject, and some of them attach in a particular place and at a particular time, whatever attaches to a subject, but not because it is this subject, at this time or in this place, will be an accident. Therefore there is no definite cause for an accident, but a chance cause, i.e. an indefinite one.

22 τι **α** Al.ᴾ 437.33 edd. : om. β ‖ 24 οὐδὲν ω^{αβ}, Al. οὐδὲ δὴ [a24] pro οὐδὲν interpretans 438.16, edd : om. ω^{AL} (Al.ᶜ 438.14) ‖ δ' **α** edd. : δὲ β Al.ᶜ 438.15

My focus will be on lines a24–25: οὐδὲ δὴ αἴτιον ὡρισμένον οὐδὲν τοῦ συμβεβηκότος. The negations in this sentence are remarkable in three respects. First, the negation οὐδὲ is followed by the particle δή. This particle[40] connects the thought with the preceding sentence and emphasizes the word οὐδέ. Generally, an emphatic δή seldom occurs after οὐδέ.[41] Second, strictly speaking, οὐδέ presupposes a negative preceding clause.[42] The word οὐδέ thus signals a connection to an *additional* negation. Yet in this case the preceding sentence (a21–24) *as a whole* is not negated ("Therefore ... whatever attaches to a subject will be an accident"). It does, however, contain a causal dependent clause (introduced by ἀλλὰ μὴ διότι) which contains a negation: "there is an accident, but <u>not</u> because it is this subject...." The οὐδὲ therefore seems to refer back to this negation.

The third feature of the sentence that calls for comment is the fact that after αἴτιον ὡρισμένον is negated by οὐδὲ δή another negation follows, οὐδέν, which itself also bears upon αἴτιον ὡρισμένον. In Greek, the accumulation of negatives is nothing unusual in itself. Multiple negatives can follow one upon the other and be

---

[39] Cf. the discussion of the accidental in *Metaph.* E 2, 1026b27–1027a20, *Top.* A 5, 102b4–7, and in the discussion of chance (τύχη) in *Ph.* B 4–6, 195b31–198a13, and especially 196b27–28. Cf. also Kirwan 1971: 180–82.

[40] See Denniston 1954: s.v. δή I, pp. 204–27 and IV, pp. 236–40.

[41] Denniston 1954: s.v. δή I.10, p. 222: "δή is not very often used to strengthen negatives."

[42] Kühner/Gerth II: § 535, 4b); p. 293: "In der attischen Prosa jedoch nur nach vorangegangenem negativem Gliede." The adversative meaning is confined to poetry: Kühner/Gerth II: § 535.4a); p. 293.

combined, especially when they are compound.[43] This leads to typical phrases like οὐδὲ μὴν οὐδέ, often found at the beginning of a sentence (Plato, *Alc. I* 107a7), and to combinations such as τἆλλα τῶν μὴ ὄντων οὐδενὶ οὐδαμῇ οὐδαμῶς οὐδεμίαν κοινωνίαν ἔχει (Plato, *Prm.* 166a1-2).[44] In the latter example every aspect in which there could be a κοινωνία is individually negated.

In the *Metaphysics* passage under discussion the situation is somewhat different. The negations οὐδὲ δή and οὐδέν both negate the existence of αἴτιον ὡρισμένον. This is made clear by the facts that αἴτιον does not have an article and, relatedly, that ὡρισμένον cannot function as a predicative nominal (in the sense of "the cause is not *something definite*..."). The additional οὐδέν can then be justified only as a negated indefinite pronoun (τι) whose function it is to supply strong emphasis: "Therefore there is *not any* definite cause for an accident...."

There is no parallel passage in the Aristotelian *corpus* in which οὐδὲ δή and a form of οὐδείς both negate one and the same sentential element in the same respect. In all cases in which οὐδὲ δή and a form of οὐδείς occur together in *one* sentence they negate either different phrases or different aspects of the same phrase.[45] We can therefore conclude that the word οὐδέν in line a25, although not impossible, is at least uncharacteristic of Aristotle's diction, and on the whole superfluous.[46]

Having considered the negation in terms of its three suspicious features we should now turn to Alexander's commentary, which contains important information on this passage.

Alexander, *In Metaph.* 438.14-17 Hayduck

οὐδὲ δή, φησίν, αἴτιον ὡρισμένον ἐστὶ [15] τοῦ συμβεβηκότος, ἀλλὰ
τὸ τυχόν· τοῦτο δὲ ἀόριστον. δύναται [16] τὸ οὐδὲ δὴ καὶ ἀντὶ τοῦ οὐδὲν

---

[43] Cf. Kühner/Gerth II: §514, 1; pp. 203-205.

[44] "The other things have no communion in any way whatsoever with anything which is non-existent."

[45] In *Metaph.* Z 1040a8, οὐδὲ δὴ ἰδέαν οὐδεμίαν ἔστιν ὁρίσασθαι the negation οὐδεμίαν bears upon ἰδέαν in the infinitive clause, whereas οὐδὲ δή negates ἔστιν. In *EN* 1142b6-7 (οὐδὲ δὴ δόξα ἡ εὐβουλία οὐδεμία), the negation οὐδεμία belongs to the predicative nominal. In *HA* Δ 8, 534b9-10, Aristotle writes about fishes, Οὐδὲ δὴ τῆς ὀσφρήσεως αἰσθητήριον οὐδὲν ἔχει φανερόν, ὀσφραίνεται δ' ὀξέως ("and although it neither has a visible organ for smell, its sense of smell is remarkably keen"). The word οὐδὲ here negates τῆς ὀσφρήσεως so that οὐδὲν alone negates αἰσθητήριον φανερόν. That οὐδὲ negates the sense of smell (τῆς ὀσφρήσεως) is made clear by the preceding sentence, in which a visible sense of hearing was negated (οὗτοι γὰρ τῆς ἀκοῆς αἰσθητήριον μὲν οὐδὲν ἔχουσι φανερόν ..., 534b7-8). A further parallel passage in *HA* Δ 8, 535b11-12 reveals itself to be obsolete from a text-critical point of view (see Balme/Gotthelf 2002).

[46] Perhaps *Metaph.* E 2, 1027a5-7 offers a helpful parallel. In discussing the accidental, Aristotle states that there is no *techne* and no determinate capacity in accidental results. Aristotle says τῶν μὲν γὰρ ἄλλων ἐνίοτε δυνάμεις εἰσὶν αἱ ποιητικαί, τῶν δ' οὐδεμία τέχνη οὐδὲ δύναμις ὡρισμένη (and not ...οὐδὲ δύναμις ὡρισμένη <οὐδεμία>).

εἰρηκέναι· ἤδη γὰρ εἶπεν ὅτι μή ἐστιν [17] αἴτιον τοῦ συμβεβηκότος διὰ τοῦ ἀλλὰ μὴ διότι τοδὶ ἦν.

"Therefore," he says, "there is no definite cause of an accident, but a chance cause, i.e. an indefinite one." The words "therefore there is no" could also be taken to mean "there is no," for Aristotle has already stated that an accident does not have a cause by saying "but not because it is this subject" [a23].

17 ἦν **A O S** : om. **P**ᵇ

Alexander's explanation of the Aristotelian sentence is motivated by his astonishment over the negation οὐδὲ δή. Three parts of Alexander's comments are important for my purposes: (i) Alexander's quotation of Aristotle's text; (ii) his alternative formulation of οὐδὲ δή; (iii) his justification for the use of οὐδὲ δή.

First let us look at the citation in lines 438.14–15: This quotation of Aristotle's sentence from ωᴬᴸ departs from the directly transmitted text in a way that seems insignificant but is nevertheless highly important for my investigation. For, in line 1025a21 ωᴬᴸ does not contain the somewhat superfluous negation οὐδέν, which we find in ωαβ (αἴτιον ὡρισμένον <u>οὐδὲν</u> τοῦ συμβεβηκότος). Instead of the οὐδέν ωᴬᴸ reads the word ἐστί (αἴτιον ὡρισμένον <u>ἐστὶ</u> τοῦ συμβεβηκότος).⁴⁷ I will return to this significant fact shortly.

Let us now turn to the second part of the comment, lines 438.15–16, where Alexander offers an *alternative* formulation of Aristotle's expression. He presents it using the typical formula "δύναται X (καὶ) ἀντὶ τοῦ Y εἰρηκέναι." With this formula Alexander usually replaces (at least in thought) an unclear expression in the *Metaphysics* with a more or less equivalent alternative whose meaning is clearer. In doing this Alexander's intention is not to emend the transmitted text but to bring out what Aristotle meant to say.⁴⁸ The expression ἀντὶ τοῦ in this sense always introduces Alexander's alternative formulation and should therefore be understood as "instead of" or "in the sense of."⁴⁹ In the present passage, Alexander interprets the phrase οὐδὲ δή, which entails a connection to the negation of the preceding sentence, as equivalent to simply negating αἴτιον ὡρισμένον with οὐδέν. Accordingly, the strongly inferential *"therefore there is no"* is to be taken

---

⁴⁷ See Dooley 1993: 184 n. 603.

⁴⁸ Alexander uses the word δύναται to introduce viable interpretations of a given sentence or phrase. His explication of an Aristotelian phrase then often comes in the form of a reformulation. See for example *In Metaph.* 141.29 (δύναται ἴσον εἶναι τὸ εἰρημένον τῷ X); 157.35–36 (δύναται καὶ τὸ X εἰρηκέναι ὡς ἴσον τῷ Y); See also *In APr.* 237.37; 313.18; 342.25. His interpretation of an Aristotelian phrase is often introduced by δύναταί τις ἀκοῦσαι (or a similar formula): *In Metaph.* 171.14; 180.23; 217.18; (for the usage of the verb ἀκούειν in Simplicius's commentaries see Baltussen 2008: 45–46); Alexander also combines these formulas: e.g. *In Metaph.* 255.19; 368.11.

⁴⁹ See e.g., *In Metaph.* 192.16–17: δύναται δὲ τὸ καὶ ἐξ ὧν μιᾶς εἰρηκέναι ἀντὶ τοῦ καὶ ἐξ ὧν αἱ περὶ ἐκείνου τοῦ ὑποκειμένου ἀποδείξεις γιγνόμεναι. 256.27: δύναται τὸ ἔχειν καὶ ἀντὶ τοῦ ἔχεσθαι εἰρῆσθαι. 380.30–31: δύναται τὸ ἢ αὐτὰ ἢ ἐξ ὧν ἐστιν εἰρηκέναι ἀντὶ τοῦ καὶ αὐτὰ καὶ ἐξ ὧν ἐστιν. See also 286.2–6, analyzed in 5.2.2. See further examples in Alex. *In APr* 144.20–21; 323.32–33; 325.25–26; 362.11–14.

as the simple phrase "there is no." For Alexander, all Aristotle meant to say is that "there is no (*sc.* definite cause)."

We turn now to the third part of this commentary passage, lines 438.16–17. There we find Alexander's explanation (γὰρ) for Aristotle's use of οὐδὲ δή. In Alexander's view, Aristotle wrote "therefore there is no" (οὐδὲ δή) instead of simply "there is no" (οὐδέν) because the preceding sentence, although positive, contains a negative dependent clause. In this dependent clause the formulation "but not because it is this subject" already indicates that an accident has no (definite) cause. This is why, according to Alexander's understanding, Aristotle chose a negative particle that signals a connection to a preceding negation.

To sum up the evidence available in Alexander's commentary: Alexander did not read the suspicious οὐδέν in his copy of the *Metaphysics* (ω^{AL}). The reading in ω^{AL} is very likely the original reading of the *Metaphysics*. Alexander is surprised, just as we are, by the expression οὐδὲ δή, which he takes from a lexical point of view to mean οὐδέν, but from a grammatical point of view to function as a connection to the preceding dependent clause. The facts about Alexander's text and his interpretation support each other: for, if Alexander, contrary to the evidence of his citation in 438.14–15, had actually read the word οὐδέν rather than ἐστί, then he could not have suggested taking οὐδὲ δή in the sense of οὐδέν.

This evidence forces me to conclude that the doubling of the negations οὐδὲ δή and οὐδέν in ω^{αβ} is a later corruption, which probably emerged from a marginal note in ω^{αβ} (or an ancestor of it). I suspect this note to have reported that οὐδὲ δή should be taken as οὐδέν, as Alexander recommended in his commentary. At some point in the transmission this note was incorporated into the *Metaphysics* text, where it replaced the syntactically superfluous ἐστί.[50] If this conclusion is correct, then our *Metaphysics* text contains a distorted picture of an explanatory interpretation that Alexander offered in order to clarify the inferential character of one of Aristotle's opening phrases.

5.1.3 Alex. *In Metaph.* 372.10–17 on Arist. *Metaph.* Δ 7, 1017a35–b6

In Δ 7, Aristotle discusses the different meanings of "that which is" (τὸ ὄν), and thereby also the meanings of "to be" (τὸ εἶναι).[51] The chapter can be divided into

---

[50] The Latin translation by Scotus does not allow for certainty regarding whether the *Vorlage* of the Arabic version read the ω^{αβ}-reading or the ω^{AL}-reading: *Et accidens non habet causam terminatam omnino nisi casu.* Nor does Asclepius's commentary offer conclusive evidence about his own text. In his paraphrase 357.20–21 (erroneously marked as a citation by Hayduck) he writes: ἄλλως τε δὴ οὐδέν ἐστιν αἴτιον ὡρισμένον τοῦ συμβεβηκότος, ἀλλὰ τὸ τυχόν,... . It is tempting to jump to the conclusion that οὐδέν ἐστιν shows that Asclepius's exemplar, just as ω^{αβ}, contained the οὐδέν, but that the ἐστί had not dropped out. However, it is also possible that Asclepius followed Alexander's suggestion and wrote δὴ οὐδέν instead of οὐδὲ δή in his paraphrase.

[51] See Kirwan 1971: 140–47.

two parts: the discussion of things that are in an accidental sense (τὸ ὂν λέγεται τὸ μὲν κατὰ συμβεβηκὸς..., 1017a7–19) and the discussion of things that are in their own right (... τὸ δὲ καθ' αὑτό, 1017a19–b9). Aristotle discusses things that are in their own right with regard to the categories (1017a19–30), to truth and falsity (1017a31–35), and to the distinction between being potentially and being actually (1017a35–b9). The discussion of being potentially and being actually begins thus:

Aristotle, *Metaphysics* Δ 7, 1017a35–b6

ἔτι τὸ εἶναι ση-[1017b1]μαίνει καὶ τὸ ὂν τὸ μὲν **δυνάμει** [**ῥητὸν**] τὸ δ' ἐντελεχείᾳ [2] τῶν εἰρημένων τούτων· ὁρῶν τε γὰρ εἶναί φαμεν καὶ τὸ δυ-[3]νάμει [**ῥητῶς**] ὁρῶν καὶ τὸ ἐντελεχείᾳ, καὶ [τὸ] ἐπίστασθαι [4] ὡσαύτως καὶ τὸ δυνάμενον χρῆσθαι τῇ ἐπιστήμῃ καὶ τὸ [5] χρώμενον, καὶ ἠρεμοῦν καὶ ᾧ ἤδη ὑπάρχει ἠρεμία καὶ [6] τὸ δυνάμενον ἠρεμεῖν.

Again, 'being' and 'that which is,' in these cases we have mentioned, sometimes mean being potentially, and sometimes being actually. For we say[52] both of that which sees potentially and of that which sees actually, that it is seeing, and both of that which can use knowledge and of that which is using it, that it knows, and both of that to which rest is already present and of that which can rest, that it rests.

δυνάμει ῥητὸν **α** Bekker Ross (ῥητὸν secl. Bonitz Christ Jaeger) : ῥητὸν δυνάμει **β** : δυνάμει ω^AL (Al.^p 372.10–12) Ar^u (*Scotus*), sed Al. proponit ῥητὸν explicandi causa 372.12–13 ‖ 3 ῥητῶς **α** secl. Bekker Bonitz Christ Jaeger : om. **β** Ross ‖ τὸ secl. Bonitz Christ Ross Jaeger

Lines 1017a35–b2 cause some difficulties. First, it is peculiar that the two-part subject τὸ εἶναι ... καὶ τὸ ὂν is separated by the verb σημαίνει.[53] Nevertheless this construction does prove helpful in that it places τὸ ὂν next to the object, which consists of a τὸ μὲν ... τὸ δ'-construction, both parts of which require τὸ ὂν as a complement. Just a few lines earlier, Aristotle positioned the subject of a sentence, which similarly consisted of two coordinated parts, in a similar way (1017a31): ἔτι τὸ εἶναι σημαίνει καὶ τὸ ἔστιν ὅτι ἀληθές... .

Second, the expression τῶν εἰρημένων τούτων, which stands at the end of the sentence, might cause some concern, since it seems difficult to integrate the phrase into the rest of the sentence. Alexander, as we will see presently, seems to have been especially concerned with the correct understanding of these three words, going so far as even to adjust his comments on this passage to his understanding of τῶν εἰρημένων τούτων. What does this genitive depend on? The

---

[52] In book Δ, Aristotle looks at the different senses that the terms he investigates have in common parlance. Cf. the phrases used in the context of our passages in Δ 7: 1017a7 τὸ ὂν λέγεται, a7 εἶναί φαμεν, a10 λέγοντες, a14 λέγωμεν, a20 εἶναι λεγόμενα οὕτω λέγεται. As Kirwan 1971: 122 puts it, "asking e.g. 'how many senses has the word "falsehood"?' rather than e.g. 'how is falsehood possible?'" has raised doubt whether book Δ should be taken as genuine part of the *Metaphysics* at all. See also Ross 1924: 308. Cf. Rosemann 1989: 96–98, who cites Averroes' view on Aristotle's method in book Δ and his reference to the πρὸς ἕν-structure ("focal meaning") of the terms under discussion.

[53] Cf. Schwegler 1847c: 213–14.

answer seems to be that it depends on τὸ μὲν δυνάμει τὸ δ' ἐντελεχείᾳ, signifying that the sense of "to be" introduced by ἔτι (1017a35) is not a new sense—that is, not a sense different from the ones already discussed in Δ 7: the differentiation of being potentially and being actually refers to "those cases we have already mentioned" in this chapter.[54]

But the main textual problem of the present passage is that in line b1 of both families α and β there appears the (verbal-)adjective ῥητὸν next to the term δυνάμει. What is more, in line b3 of the α-version, we find the suspicious adverb ῥητῶς placed behind the word δυνάμει.[55] Aristotle nowhere else uses the adverb ῥητῶς, and so editors unanimously athetize it. In the following, I will concentrate on the word ῥητὸν in line b1, since the additional ῥητῶς in line b3 occurs only in the α-version and is explicable as a reaction to the ῥητὸν in b1. The two families differ slightly in regard to the position of ῥητὸν (b1): the α-text reads δυνάμει ῥητόν, and the β-text ῥητὸν δυνάμει.

The word ῥητόν characterizes "that which is potentially" (δυνάμει) as something that can be said (ῥητὸν) to be, but *is* not actually.[56] However, the expression ῥητόν itself is quite suspicious.[57] The expression does not occur elsewhere in Aristotle in the sense that is required here, and its deletion would not cause any problems for the passage.[58] Furthermore, it oddly disrupts the unity of the phrase τὸ μὲν δυνάμει τὸ δ' ἐντελεχείᾳ. Both Bonitz and Ross see in Alexander's commentary proof of the old age of ῥητόν.[59] I will argue, on the other hand, that Alex-

---

[54]Schwegler 1847c: 214. Bonitz 1849: 242: *Quarta significatio* τοῦ ὄντος, *quod vel* τὸ δυνάμει ὄν *vel* τὸ ἐντελεχείᾳ *significat, superiores omnes complectitur*. Ross 1924: 309: "While the first three senses seem to answer to three types of judgment, [...] the fourth answers not to a type of statement co-ordinate with these, but to two senses in which each of the them may be taken." See also Kirwan 1971: 146. One might here refer to *De anima* B 2, 413b11–13 as a parallel passage of this usage of the genitive, wherein τῶν εἰρημένων τούτων refers back to what was said before: νῦν δ' ἐπὶ τοσοῦτον εἰρήσθω μόνον, ὅτι ἐστὶν ἡ ψυχὴ τῶν εἰρημένων τούτων ἀρχὴ καὶ τούτοις ὥρισται, θρεπτικῷ, αἰσθητικῷ, διανοητικῷ, κινήσει ("At present we must confine ourselves to saying that soul is the source of these phenomena and is characterized by them, viz. by the powers of self-nutrition, sensation, thinking and movement").

[55]Cf. LSJ s.v. ῥητός adv. "expressly, distinctly," occurs first in Polybios. Ross 1924: 309: "ῥητῶς before ὁρῶν in l. 3 seems to be spurious; it is not found, as ῥητόν l.1 is, in A^b, Al., Asc."

[56]The adjective ῥητὸν (b1) can be integrated more easily into the given context than the adverb ῥητῶς (in b3 of the α-text). The expression δυνάμει ῥητὸν could simply be taken as "said to be potentially." Cf. Ross 1924: 309: "*can be said* by virtue of a potentiality...."

[57]Bonitz 1849: 242: *Sententia Aristotelis nihil habet obscuri, sed de ipsis verbis plus una oritur dubitatio. Primum* ῥητόν *et* ῥητῶς *antiquitus iam in textum irrepsisse testis est Alexander p. 332,22 [= 372,12 Hayduck], non genuinum illud essefacile sibi persuadebit, qui omnes Aristotelis de potentia locos contulerit; sed qui potuerit inferri in textum non video.* See also Schwegler 1847c: 213–14. Ross 1924: 309: "ῥητόν has caused much difficulty to the editors. Elsewhere in Aristotle the word occurs only in its ordinary meaning of 'stated, fixed,' which cannot be the meaning here."

[58]*Pace* Ross 1924: 309: "It seems quite possible to retain it [ῥητόν], and it even makes the construction more natural."

[59]Bonitz 1849: 242; Ross 1924: 309: "...it occurs in all the manuscripts and as a variant in Alexander...."

ander's words in fact do not indicate that he knew the reading ῥητὸν as a variant. Rather, there is good reason to assume that Alexander only introduced the term ῥητόν as part of a thought experiment. So we should have a look at Alexander's comments on the passage.

Alexander, *In Metaph.* 372.10–17 Hayduck

λέγει δὲ τὸ [11] εἶναί τε καὶ τὸ ὂν σημαίνειν πρὸς τοῖς εἰρημένοις καὶ τὸ δυνάμει τε καὶ [12] ἐντελεχείᾳ. ἂν δὲ ᾖ γεγραμμένον ἀντὶ τοῦ δυνάμει <u>τὸ ῥητόν</u>,[60] λέγοι ἂν [13] ῥητὸν τὸ δυνάμει, ὅτι ῥηθῆναι μὲν ἀληθές ἐστιν, οὐ μὴν ἔστιν ἤδη. [14] τὸ δὲ ἐντελεχείᾳ τῶν εἰρημένων τούτων, τουτέστι τῶν ῥητῶν τε [15] καὶ δυνάμει. ὁρῶν γὰρ εἶναι λέγεται καὶ τὸ ἤδη ὁρῶν καὶ τὸ δυνάμενον, [16] ὡς τὸ κοιμώμενον, τοῦ εἶναι καὶ ἐπ' ἐκείνου καὶ ἐπὶ τούτου κατηγορου-[17]μένου·

Aristotle says, furthermore, that 'to be' and 'being' also signify that which is potentially and that which is in actuality. If instead of "potentially" the expression 'can be said' is written [in Aristotle's text], he should mean that what is potentially is *sayable*, because it is correct that it be said although it does not already exist. [The following words] "and sometimes being actually of those cases that are said [to be],"[61] that is, of those that are *sayable* and potentially. For he who is actually seeing and he who is capable of seeing (in the way what is asleep [is capable of seeing]) are alike said to be 'seeing,' 'to be' being predicated both of the former and of the latter.

13 ἔστιν **A P**[b] **S** : καὶ ἔστιν **O LF** ‖ 14–15 ῥητῶν τε καὶ δυνάμει **A O S** : καθ' αὐτὸ καὶ κατὰ συμβεβηκὸς ὄντων **P**[b] ‖ 16 ἐπ' ἐκείνου καὶ ἐπὶ τούτου **A O** : ἐπὶ τούτου καὶ ἐπ' ἐκείνου **P**[b]

Alexander's comments on *Metaphysics* Δ 7, 1017a35–b6 begin with a paraphrase of lines a35–b2. Since this paraphrase (372.11–12) renders Aristotle's words rather faithfully, we can infer from it that Alexander was confronted with lines a35–b2 in ω^αβ *without* the word ῥητόν, which in the direct transmission stands either in front of δυνάμει (β) or behind it (α).[62] It can further be noted that Alexander does not render the words τῶν εἰρημένων τούτων in his paraphrase. His quotation at 372.14–15, however, makes it clear that he did read these words in his text. Alexander's expression πρὸς τοῖς εἰρημένοις ("furthermore") in line 372.11 should not be taken as paraphrase of the words τῶν εἰρημένων τούτων (b2), but rather as paraphrase of the word ἔτι (1017a35).[63]

---

[60]My presentation of Alexander's text differs from Hayduck's typographical marking of these words.

[61]I translate the Aristotelian words that Alexander quotes according to *Alexander's* understanding of them, which differs from the understanding expressed in the translation of Aristotle given above. According to Alexander's interpretation, the genitive τῶν εἰρημένων τούτων means that it holds for *both* ways of being (potential and actual) that they are said to be and that they are potential.

[62]The Arabic translation is also based on an exemplar that did not contain ῥητόν. Scotus writes: *Et etiam quedam entia sunt potentia et quedam actu.*

[63]Sepúlveda also understands Alexander's πρὸς τοῖς εἰρημένοις in this way and translates *praeterea* (f. z.i.v.).

In the lines following his paraphrase (372.12–15), Alexander takes a closer look at the text he just paraphrased. He focuses on the words τὸ μὲν δυνάμει τὸ δ' ἐντελεχείᾳ τῶν εἰρημένων τούτων (b1–2), which he quotes and intersperses with his own remarks: δυνάμει (372.12) ... τὸ δ' ἐντελεχείᾳ τῶν εἰρημένων τούτων (372.14). In order to understand Alexander's comments we have to identify what it is in Aristotle's text that disturbs him: it is the late-positioned genitive τῶν εἰρημένων τούτων (in b2). Apparently, Alexander does not understand the genitive to function as a back-reference to the previously discussed meanings of "to be," but rather as an additional determinant of the terms δυνάμει and ἐντελεχείᾳ. As he explains (τουτέστι) in lines 372.14–15, he takes the phrase τῶν εἰρημένων τούτων to indicate that *both* ways of being (δυνάμει and ἐντελεχείᾳ) are *said* (... εἰρημένων ...), and that they are potential. This understanding is manifested in his paraphrase of the genitive: as τῶν <u>ῥητῶν</u> τε καὶ δυνάμει (14–15).[64]

What Alexander intends to express with this explication is already clear from lines 372.12–13. There, Alexander considered whether the word δυνάμει could be replaced with the word ῥητόν (ἂν δὲ ᾖ γεγραμμένον ἀντὶ ... λέγοι ἄν). According to Alexander's hypothesis that both ways of being, the actual and the potential, can be subsumed under the inclusive notion "that which is *sayable* and potential" (τῶν ῥητῶν τε καὶ δυνάμει), something that is potential is also ῥητόν (372.12), that is, it *can be said* to be (ῥηθῆναι μὲν ἀληθές ἐστιν, 372.13) without needing to be actually (οὐ μὴν ἔστιν ἤδη, 372.13).[65] According to this interpretation then, the two terms δυνάμει and ἐντελεχείᾳ, which *both* refer to a being that is sayable and potential, differ from each other in that what is δυνάμει is *sayable* (ῥητόν) and *potential* only, whereas what is ἐντελεχείᾳ is actual in addition. Since Alexander's suggestion to take δυνάμει to mean ῥητόν matches exactly with his own idiosyncratic interpretation of the phrase τῶν εἰρημένων τούτων, and since nothing in the text indicates that Alexander knew ῥητόν as a variant reading from another manuscript, it seems reasonable to conclude that the proposal to substitute δυνάμει with ῥητόν goes back to Alexander himself.[66]

---

[64] The reading in **P**[b] (καθ' αὑτὸ καὶ κατὰ συμβεβηκὸς ὄντων in place of ῥητῶν τε καὶ δυνάμει) can be understood as an attempt to do away with the admittedly daring interpretation suggested here by Alexander. The reading in **P**[b] makes Alexander explain Aristotle's words τῶν εἰρημένων τούτων in a way that is more in tune with our understanding. The weakness of the **P**[b]-reading is that it squares considerably less well with the preceding explanatory argument than the reading attested to by **A** and **O**.

[65] The words εἰρημένων and ῥητόν have the same root: (ϝ)ρη-. The relation between these two terms might have been self-evident to Alexander. In other contexts, Alexander uses the term ῥητόν in the sense of "what was said," e.g. in Aristotle's text: 311.27.

[66] How could Alexander have arrived at the term ῥητόν? I did not find any other passage in Alexander where he uses the term ῥητόν to denote a potential being. One is perhaps reminded of the Stoic usage of the term λεκτόν (on the Stoic influence on Alexander see Sharples 1987: 1178 with further literature). Are the two terms connected? On the basis of the Stoic fragments, I cannot see a direct connection of λεκτόν to the *potential* being in the sense in which Alexander uses ῥητόν here. On the meaning of the Stoic term see Frede 1994.

By contrast, Ross holds that Alexander, like Asclepius, refers to ῥητόν as a variant reading;[67] it is "a variant in Alexander and Asclepius." Asclepius, in contrast to Alexander, does speak explicitly of two different versions of the text in his commentary. He reports that instead of δυνάμει some other manuscripts read ῥητόν: ἐπὶ τούτοις σημαίνει τὸ ὂν τὸ μὲν δυνάμει τὸ δὲ ἐντελεχείᾳ· ἔνια γὰρ τῶν ἀντιγράφων τὸ δυνάμει ἔχουσιν, ἄλλα δὲ τὸ ῥητόν, ὥσπερ καὶ τὰ ἐνταῦθα. τὸ αὐτὸ δὲ δηλοῖ· ῥητὸν γάρ ἐστι τὸ δυνάμενον λέγεσθαι (*In Metaph.* 318.32–34 Hayduck).[68] The words ὥσπερ καὶ τὰ ἐνταῦθα suggest that Asclepius's own manuscript(s) read the word ῥητόν, but this seems to contradict the words Asclepius had just quoted (τὸ μὲν δυνάμει τὸ δὲ ἐντελεχείᾳ). On the assumption, therefore, that the (transmitted) wording of the quotation is authentic, we are forced to assume that not *all* of the manuscripts to which Asclepius refers by τὰ ἐνταῦθα contained the ῥητόν.[69]

In any case, it is clear that Asclepius speaks about two different traditions, *the one* reading δυνάμει, *the other* reading ῥητόν. The same testimony is given in a *scholium* (dependent on Asclepius or Alexander?) in the *Metaphysics* manuscript E (*Parisinus gr.* 1853).[70] Yet Alexander's remarks are quite different from Asclepius's and the scholium's testimonies. Alexander does not speak about another copy of the text that read the ῥητόν. He does not use signal words like ἔν τισιν ἀντιγράφοις. Nor does Asclepius's mention of a variant reading in the sixth century AD by any means imply that Alexander, writing around AD 200, also knew of this variant. Consequently, I do not follow Ross's view on the matter, but rather think it likely that Alexander himself, in commenting on Aristotle's phrase τῶν εἰρημένων τούτων, coined the term ῥητόν as an explanatory paraphrase of δυνάμει.[71]

---

[67] Ross 1924: 309. Bonitz 1849: 242 does not state clearly whether he thinks Alexander found ῥητὸν in his own *Metaphysics* copy or as a variant reading in another manuscript: *Primum ῥητόν et ῥητῶς antiquitus iam in textum irrepsisse testis est Alexander p. 332,22* [= 372.12 Hayduck]...." Schwegler 1847c: 213–14 treats it as self-evident that Alexander found ῥητὸν in his text. Hayduck, by putting the words δυνάμει and ῥητόν into spaced letters, creates the impression that these words are quoted from Alexander's *Metaphysics* text.

[68] Ascl. *In Metaph.* 318.32–34: "In these cases the 'that which is' means 'sometimes that which is potentially and sometimes that which is actually.' For, some manuscripts read 'potential,' but others 'sayable,' as do those here. However, both mean the same: for, 'sayable' means to say that something is potential."

[69] The idea that Asclepius here builds on Alexander's commentary conflicts with the fact that, as Luna 2001: 108 shows, Asclepius does not draw on Alexander's commentary in book Δ. On the possibility of Asclepius using a different *Metaphysics* text than his teacher Ammonius, on whose lectures his commentary is based, see Kotwick 2015.

[70] Brandis 1836 prints the scholium next to the excerpt from Asclepius's commentary: 701b6–7: γρ. "τὸ μὲν ῥητόν, τὸ δ' ἐντελεχείᾳ," καὶ οὕτω, "τὸ μὲν δυνάμει, τὸ δὲ ἐντελεχείᾳ." On the scholia in E see Golitsis 2014a with further literature.

[71] That said, we cannot exclude the possibility that Alexander here marks a *varia lectio* more laxly, or that he adopts the idea of substituting δυνάμει by ῥητόν from another commentary. The termi-

The question that matters most for my purposes is the following: is there a causal connection between Alexander's comments on the passage and the *Metaphysics* text transmitted in our manuscripts? The directly transmitted text does not square well with the story Asclepius tells about the alternative reading ῥητόν, because in that story ῥητόν *replaces* δυνάμει. In our text, on the other hand, ῥητόν comes in addition to δυνάμει: in the α-text, ῥητόν stands after δυνάμει, in the β-text, we find it in front of δυνάμει.

That both families share the same error, but in slightly different form, suggests that the errors was present already in their ancestor. The following explanation seems plausible: the word ῥητόν was written either in the margin or between the lines of ω^{αβ}, but in such a way that it did not precisely signal to the scribe(s) copying the text whether it is meant as addition or correction and where exactly it is supposed to be put into the text. The gloss in ω^{αβ} was very likely intended to offer a *varia lectio* for δυνάμει, but was then mistakenly understood as a correction for an omission. Since the gloss furthermore lacked sufficient clarity as to where the word should be inserted, the placement of ῥητόν differs slightly in α and β. In α, ῥητόν was inserted after δυνάμει, in β in front of it. Given that we find ῥητόν *added* in both versions (instead of being used as substitution for δυνάμει[72]), it is unlikely that this happened in both versions independently of each other and without having been suggested by ω^{αβ}.

In this scenario, Alexander's commentary does not play any role. The fact, however, that in Alexander's commentary on this passage the word ῥητόν appears *next to* the word δυνάμει rather than *instead of* it could be seen as causally related to the way in which α and β are corrupted. In other words, the presence of the *addition* of ῥητόν in ω^{αβ} leads us back to Alexander's comments on the passage. And so we might answer the question of why someone would *add* the word ῥητόν to δυνάμει rather than replace δυνάμει by saying that this is how it appears in Alexander's commentary. We twice (372.12–13) find the words δυνάμει and ῥητόν standing so close to each other (δυνάμει τὸ ῥητόν … ῥητὸν τὸ δυνάμει) that the erroneous addition of ῥητόν to δυνάμει in ω^{αβ} is easily explained by a

---

nology used, however, speaks in favor of the view that Alexander brings forward his *own* suggestion. Alexander often uses the preposition ἀντί to introduce his substitution of an Aristotelian term by an equivalent that, from Alexander's perspective, expresses the thought just more clearly. So, e.g., in 286.2–6: Alexander proposes to interpret Aristotle's wording in the sense of (ἀντὶ τοῦ) his own slightly modified version of it. Cf. also the passages enumerated in section 5.1.2, on p. 190 n. 49 above. It is certainly impossible to understand ἀντί in the sense of πρό (in a local sense) "before" as Dooley 1993: 45 takes it in our passage (Δ 7, 1017b1). According to this untenable view, Alexander was just concerned with changing the position of the word ῥητόν to standing *before* the word δυνάμει. Dooley (1993: 45) translates: "If 'can be spoken of' were written *before* 'potentially' in Aristotle's text…" (my emphasis). Borgia (2007: 915) follows Dooley in this ("Se *prima* di «in potenza» ci fosse scritto «che può dirsi»…." my emphasis).

[72] As it is the case with the reading reported by Asclepius (see above).

rushed adoption from the commentary.[73] And so the odd mistake that occurred in α and β and that was provoked by an ambiguous gloss in ω^{αβ} appears to have had its origin after all in Alexander's commentary.

If the suspicious combination of the terms ῥητόν and δυνάμει in ω^{αβ} actually emerged from an adoption from Alexander's commentary, then the question whether Alexander himself developed the term ῥητόν or borrowed it from an ongoing discussion on the *Metaphysics* passage loses its relevance. The key factor in Alexander's influence on ω^{αβ} is simply the way in which the alternative expression ῥητόν visually appears in the text of the commentary. It is just this peculiarity that connects the error in α and β with Alexander's wording.

5.1.4 Alex. In Metaph. 164.15–165.5 on Arist. Metaph. α 2, 994b21–27

In α 2, 994b21–27, Aristotle states that it is impossible for an infinite series of causes either to exist or to be thought of or grasped mentally. He illustrates the impossibility of an infinite series of causes with the infinite divisibility of a line: it is impossible to count the line's infinite sections, and so it is impossible to think or mentally grasp the line as one is counting the sections. Yet there is a crucial difference between an infinite series of causes and the line's infinite divisions: one can think the line when one ceases to count the sections (νοῆσαι δ᾽ οὐκ ἔστι μὴ στήσαντα, 994b24), "but further, the matter in a changeable thing must be cognized"—ἀλλὰ καὶ τὴν ὕλην **ἐν κινουμένῳ** νοεῖν ἀνάγκη.[74] Aristotle's specification beginning with ἀλλὰ καὶ… ("but further…") as we find it in ω^{αβ} is puzzling and has troubled readers and commentators since antiquity.[75] Corrections and conjectures have been proposed: ancient commentators known to Alexander (*In Metaph.* 164.24–165.5) wanted to read ἀλλὰ καὶ τὴν ὕλην κινουμένην νοεῖν ἀνάγκη ("but further it is necessary to recognize the matter in motion"). More recently, Ross changed the text to ἀλλὰ καὶ τὴν ὕλην οὐ κινουμένῳ νοεῖν ἀνάγκη ("but the

---

[73]That this adoption is likely to happen can be seen in Dooley's mistranslation: 1993: 45: "If 'can be spoken of' were written *before* 'potentially' in Aristotle's text…." Dooley notes (1993: 145 n. 169): "Alexander evidently read *to men dunamei rhēton*."

[74]Arist. *Metaph.* α 2, 994b21–27: καὶ τὸ γιγνώσκειν οὐκ ἔστιν, τὰ γὰρ οὕτως ἄπειρα πῶς ἐνδέχεται νοεῖν; οὐ γὰρ ὅμοιον ἐπὶ τῆς γραμμῆς, ᾗ κατὰ τὰς διαιρέσεις μὲν οὐχ ἵσταται, νοῆσαι δ᾽ οὐκ ἔστι μὴ στήσαντα (διόπερ οὐκ ἀριθμήσει τὰς τομὰς ὁ τὴν ἄπειρον διεξιών), ἀλλὰ καὶ τὴν ὕλην [ω^{αβ}: ὅλην ci. Ross] **ἐν κινουμένῳ** [ω^{αβ} ci. Al. 164.23 : κινουμένῳ ω^{AL} : κινουμένην Al.¹ 164.15 Al. 164.24] νοεῖν ἀνάγκη. καὶ ἀπείρῳ οὐδενὶ ἔστιν εἶναι· εἰ δὲ μή, οὐκ ἄπειρόν γ᾽ ἐστὶ τὸ ἀπείρῳ εἶναι. / "And knowledge becomes impossible; for how can one think things that are infinite in this way? For this is not like the case of the line, to whose divisibility there is no stop, but which we cannot think of if we do not make a stop; so that one who is tracing the infinitely divisible line cannot be counting the sections. But further, the matter must be recognized **in something that changes**. Again, nothing infinite can exist; and if it could, at least being infinite is not infinite."

[75]Alex. *In Metaph.* 164.15–165.5; Bonitz 1849: 134; Ross 1924: 219–20.

whole line also must be apprehended by something in us which does not move (in thought) from part to part").[76]

Alexander's commentary offers the solution to this textual puzzle. Alexander's paraphrase (see 164.18–20 and an earlier reference in his commentary in 148.12–13) testifies that the original reading is κινουμένῳ. This reading had been preserved in ω^AL but had been corrupted into ἐν κινουμένῳ in ω^αβ. Furthermore, Alexander's comments reveal themselves to be the source for the corruption in ω^αβ. The reading ἐν κινουμένῳ in ω^αβ stems from *Alexander's* proposal for expressing more clearly what, according to Alexander's understanding, Aristotle's means to say. According to Alexander's understanding, Aristotle compares the unknowability of the infinite (τὸ ἄπειρον) to the unknowability of matter (παρατίθεται σημεῖον τὴν ὕλην, 164.16–18), since matter, as it is without shape and so is in a way infinite, is also unknowable—at least in the scientific sense of knowledge (164.18–19). Accordingly matter can only be recognized through something that changes (κινουμένῳ), or as Alexander expresses it, *in* something that changes (ἐν κινουμένῳ, 164.18–20; 22–23). This alternative formulation of Alexander was adopted into ω^αβ at some point before its split into α and β.

Marwan Rashed (2007: 315–16 n. 861) first suspected that what we find in our manuscripts of this *Metaphysics* passage might actually be Alexander's interpretation of it. Christian Pfeiffer and I (in an article in progress) analyze extensively the *Metaphysics* passage in respect to both the evidence in Alexander's commentary and Aristotle's account of the infinite as given in the *Physics* and show that Rashed's suspicion is indeed correct. We demonstrate not only that the reading in ω^αβ has been corrupted by Alexander's commentary; we also demonstrate that the reading in ω^AL, which is to be reconstructed from Alexander's comments, is certainly the *original* reading authored by Aristotle. Reading κινουμένῳ (ω^AL) instead of Alexander's conjecture ἐν κινουμένῳ (ω^αβ) decisively changes the meaning of the sentence. What Aristotle in fact says is: ἀλλὰ καὶ τὴν ὕλην **κινουμένῳ** νοεῖν ἀνάγκη / "but it is also necessary that the one who moves thinks the matter." This statement coheres well not only with the preceding part of the sentence in α 2, 994b24–25, but also and especially with what Aristotle says about both the infinite and the infinitely devisible line in *Physics* Γ 6 (207a21–26) and Θ 8 (263a23–b9).

From *Physics* Γ 6 (207a21–26) we learn that Aristotle calls the property of the line by which it is infinitely divisible its *matter* (ὕλη). Thus it makes sense for him to say (in α 2, 994b25–26) that in order to think the infinite divisibility of the line, one has to think the matter of the line (τὴν ὕλην … νοεῖν ἀνάγκη). Furthermore, as we learn from *Physics* Θ 8 (263a23–b9) and Aristotle's answer to Zeno's paradox as to how it is possible to move along a continuous line, a line is not actually but rather potentially (οὐκ ἐντελεχείᾳ ἀλλὰ δυνάμει, 263a28–29) infinitely divisible. And so in *moving* along a continuous line (ὁ γὰρ συνεχῶς κινούμενος, 263b7)

---

[76] Ross 1924: 219–20.

one only potentially stops at infinitely many points. Against this background, it makes good sense for Aristotle to compare thinking the infinite to moving in thought along a continuous line, which on account of its *matter*, could potentially be divided at infinitely many points. This is what Aristotle means, according to the reading in ω^AL and freed from Alexander's influential misinterpretation, when he says in α 2, 994b25-26 τὴν ὕλην **κινουμένῳ** νοεῖν ἀνάγκη / "the one who moves has to think the matter."

5.1.5 Alex. Fr. 12 Freudenthal (Averroes *Lām* 1481-82) on Arist. *Metaph.* Λ 3, 1070a13-19

The third chapter of book Λ seems to combine several thematically disparate parts. Judson suggests seeing the chapter's unifying thought in the priority of form (εἶδος) over composite substances and other principles.[77] In the *Metaphysics* passage quoted below (1070a13-19), Aristotle discusses the question of whether a form (τόδε τι)[78] exists separately from composite substances (παρὰ τὴν συνθετὴν οὐσίαν).[79] In addressing this question, Aristotle distinguishes between natural and artificial substances. First (1070a13-14), he denies the existence of a separate form of artificial substances (for example, a house), although he adds parenthetically that one could speak of a form (τὸ εἶδος) that exists separately in the craftsman's mind (a15-17).[80] Then, he states that a separate *this* (τόδε τι), i.e. a separate form, if it can exist at all, can exist only for natural substances (a17-18). He continues with a reference to Plato, or, according to what is most likely the correct reading (namely, that of ω^AL—see 3.5.2.2), to "those who postulate the Forms."[81] This latter group is correct in that Forms—if they exist at all—exist only of natural things.

Aristotle, *Metaphysics* Λ 3, 1070a13-19[82]

ἐπὶ μὲν οὖν τινῶν τὸ τόδε τι [14] οὐκ ἔστι παρὰ τὴν συνθετὴν οὐσίαν, οἷον οἰκίας τὸ εἶδος, εἰ [15] μὴ ἡ τέχνη (οὐδ' ἔστι γένεσις καὶ φθορὰ τούτων, ἀλλ' ἄλ-[16]λον

---

[77] Judson 2000: 125.

[78] The "this" (τόδε τι) here means "form." For this meaning of τόδε τι see *Metaph.* Z 12, 1037b26-27. On the more general meaning of the term τόδε τι see Weidemann 1996: 91-93.

[79] Judson 2000: 131-33.

[80] Judson 2000: 133: "The artefact-form can exist in a way which makes it both causally prior to and independent of the composite; but, although it cannot undergo a process of coming to be, it only exists from time to time (i.e. either whenever there is someone who is master of the art of building, or, more probably, whenever the form of the house is thought of); and this sort of transient and dependent existence disqualifies the form from being separate."

[81] On the text in line 1070a18 and its reconstruction see 3.5.2.2.

[82] The section that I treat here ends before lines 1070a19-20, the transmission of which is also problematic (discussions of the text can be found in Ross 1924, II: 356-57, Judson 2000: 133 n. 61 and Fazzo 2012b: 251-53), but irrelevant to my argument. Alexander, followed by Averroes, also addresses lines 1070a19-20 separately from the lines in question here.

τρόπον εἰσὶ καὶ οὐκ εἰσὶν οἰκία τε ἡ ἄνευ ὕλης καὶ [17] ὑγίεια καὶ πᾶν τὸ κατὰ τέχνην), ἀλλ᾽ εἴπερ, ἐπὶ τῶν φύ-[18]σει· διὸ δὴ οὐ κακῶς Πλάτων ἔφη **ὅτι εἴδη ἔστιν ὁπόσα** [19] **φύσει, εἴπερ ἔστιν εἴδη** ....

Now in some cases the 'this' does not exist apart from the composite substance, e.g. the form of house does not so exist, unless the art of building exists apart (nor is there generation and destruction of these forms, but it is in another way that the house apart from its matter, and health, and all things of art, exist and do not exist); but if it does it is only in the case of natural objects. And so Plato was not far wrong when he said that there are as many Forms as there are kinds of natural things, if there are Forms at all ...

---

13 τι A^b Christ Ross Jaeger : om. α ε || 14 συνθετὴν Ross : συνθέτην α A^b C : σύνθετον M ρ || 16 τε A^b Christ Ross Jaeger : om. α ε || 18 δὴ α ε : om. A^b || Πλάτων ἔφη (α Michael^p 677.12–13) Ross Jaeger Fazzo vel ὁ Πλάτων ἔφη (A^b ε) Bekker Bonitz Christ, Ar^u Al.^γρ Fr. 12 F : οἱ τὰ εἴδη τιθέντες ἔφασαν ω^AL (Fr. 12 F), Ar^m || 18–19 ὅτι εἴδη ἔστιν ὁπόσα φύσει, εἴπερ ἔστιν εἴδη ci. Al. (Fr. 12 F) ω^αβ : ὅτι εἴπερ ἔστιν εἴδη, ἔστιν ὁπόσα φύσει ω^AL Ar^m

In the following analysis I will focus on lines 1070a18–19 and in particular on the phrase ὅτι εἴδη ἔστιν ὁπόσα φύσει, εἴπερ ἔστιν εἴδη / "that there are as many Forms as there are kinds of natural things, if there are Forms at all." Averroes—our only source of Alexander's commentary on book Λ (see 2.5)—reports some important information about the wording of these lines. This information suggests that Alexander had discussed them extensively in his commentary.

In the first place, we want to know what Alexander read in ω^AL. To find that out, we need to look into the *textus* that introduces Averroes' commentary and his report of Alexander's comments.[83] We read (*Lām* 1481): "Therefore, those who postulated the Forms were not wrong in saying that these, if they exist at all, are all things existing by nature."[84] As pointed out earlier (see 3.5.2.2), Averroes' text differs from ours in that instead of "Plato" it reads "those who postulate the Forms." More importantly for our present concerns, though, is the fact that the text also differs from ours in regard to the position of the conditional clause "if the Forms exist at all."[85] In the Arabic text, the conditional clause *precedes* the mention of Plato's statement about the ideas: "if there are Forms at all, then they are all things existing by nature." Freudenthal reconstructs the following Greek text as the model for the Arabic version:

---

[83]For this part of his commentary Averroes uses the Arabic translation by Abū Bišr Mattā, the edition of which included Alexander's commentary. We do not know how this edition combined Aristotle's *Metaphysics* and Alexander's commentary. See Bertolacci 2005: 253–57.

[84]Genequand 1986: 100. Freudenthal 1885: 86.18–21: "Und aus diesem Grunde haben nicht übel gethan die, welche die Ideen annehmen; denn wenn diese auf irgendeine Weise vorhanden sind, so sind sie Alles, was von Natur ist." Scotus: *Et ideo non fecerunt male illi qui posuerunt formas. Quia ista si fuerunt aliquo modo, sunt omnia que existunt secundum naturam.*

[85]Andreas Lammer has personally assured me that the position of the clause in the English translation corresponds to the position of the clause in the Arabic original.

Freudenthal 1885: 86 n. 3: "Alexander las also wahrscheinlich":
διὸ δὴ οὐ κακῶς ἔφασαν οἱ τιθέμενοι⁸⁶ τὰ εἴδη ὅτι <u>εἴπερ ἔστιν εἴδη</u> ἐστὶν ὁπόσα φύσει....

Freudenthal seems to assume that the text in Averroes' lemma corresponds to the text Alexander read. As it happens, this assumption proves true when one looks at Averroes' excerpts from Alexander's commentary. These reveal that Alexander's text agrees with the reading in Averroes' lemma.

As Averroes reports, Alexander made two suggestions for emending the sentence in question. His first suggestion is to rephrase the sentence slightly, in order to flesh out what Aristotle expresses rather tersely.

> Genequand 1986: 101
> It would be easier to understand if it was put in this way: 'therefore, those who postulated the Forms were right, if they exist at all, in assuming all that comes from them to be by nature.'

> Fr. 12 Freudenthal (87.3–6)
> Er sagt: Dieser Satz würde folgendermaßen deutlicher sein: ‚Und aus [4] diesem Grunde haben nicht übel gethan die, welche die Ideen annehmen [5] (wenn diese auf irgend eine Weise eine Existenz haben), indem sie Alles, [6] was aus ihnen entsteht, der Natur zuerkannten'.

> Scotus
> *Et iste sermo erit manifestior si legatur ita: Et ideo non male fecerunt ponentes formas si habent esse aliquo modo affirmando omne quod ex eis fit per naturam.*

Alexander does not change the order of the clauses (see lines 87.3–6 Freudenthal). What he does is merely rephrase the end of the sentence and in place of "then they are all things existing by nature" propose "in assuming all that comes from them to be by nature."⁸⁷ According to this formulation, Aristotle praises those who postulate the Forms for holding that all things that come from Forms are natural. Alexander's concern is to change slightly the ending of the sentence, but not to alter the sequence of the sentence's parts. Since the sentence confirms the order of the clauses as they are found in Averroes' lemma, we may safely assume that Alexander encountered them in this order in ω^{AL}.

Freudenthal suggests the following Greek formulation as an execution of Alexander's first suggestion:

Freudenthal 1885: 87 n. 1: "Alexander will lesen":
διὸ δὴ οὐ κακῶς ἔφασαν οἱ τιθέμενοι τὰ εἴδη, εἴπερ ἔστιν <u>ταῦτα</u>, ὅτι φύσει ἐστὶν ὁπόσα <u>ἐξ αὐτῶν γίγνεται</u> ....

---

⁸⁶Freudenthal 1885 reconstructs οἱ τιθέμενοι τὰ εἴδη. Since Aristotle nowhere uses exactly this formula, I propose to reconstruct οἱ τὰ εἴδη τιθέντες instead (see section 3.5.2.2; p. 77 n. 229).

⁸⁷Cf. Freudenthal 1885: 87.5–6: "indem sie Alles, was aus ihnen entsteht, der Natur zuerkannten" and Michael Scotus: *affirmando omne quod ex eis fit per naturam.*

Here we find Alexander making a slight change in the focus of the sentence. According to our text, the words εἴδη ἔστιν ὁπόσα φύσει (1070a18–19) express the idea that Forms can only be postulated of natural things. According to Alexander's slight reformulation, the postulation of Forms offers a criterion for deciding which things count as natural and which do not. Although the domain of things remains the same, Alexander's rephrasing emphasizes that natural things originate from forms.[88]

Alexander seems unsatisfied with his first suggestion. As a way of preparing the reader for his second suggestion, he offers a paraphrase of the sentence that points in the direction he wants to go.

> Genequand 1986: 101
> He (Alexander) says: it is also possible to understand his statement in this way: therefore, he who postulated a Form for all these things which exist by nature, if this form exists at all, was not wrong.

> Fr. 12 Freudenthal (87.6–9)
> Er sagt: Es ist mög-[7]lich, diese seine Worte so zu verstehen: Aus diesem Grunde haben nicht [8] übel gethan die, welche Ideen für alle Dinge annahmen, die von Natur [9] sind – wenn anders die Idee in irgend einer Weise Existenz hat.

> Scotus
> Dixit: Et potest intelligi sic: Et ideo non male fecerunt ponentes formas istorum omnia que sunt secundum naturam si forma habeat esse aliquo modo.

As far as the position of the phrase ἐστὶν ὁπόσα φύσει is concerned, Alexander stays close to what he (most likely) found in ω[AL]. He paraphrases: "… a Form for all these things which exist by nature…." The innovation of this reformulation is that Alexander postpones the clause that contains the condition that Forms exist at all (εἴπερ ἔστιν εἴδη, 1070a19) until the end of the sentence, which closes with "if this form exists at all."[89]

Consonant with the changes implemented in this paraphrase, Alexander proposes his second suggestion for emending the text. This emendation involves transposing the conditional clause "if the Forms exists at all."

> Genequand 1986: 101
> He says: it would be easier to understand what he means <u>if the word 'existing' was</u>

---

[88]That this is the point Alexander wants to make here becomes clear in the paraphrase that precedes his reformulation: "he (Aristotle) does not say that they are right in an absolute way, but merely that it was right to suppose them to be the natural things" (Genequand 1986: 101).

[89]There is a change from the plural "those who postulate the Forms" to the singular "he who postulates a Form" in Genequand's English translation. Freudenthal's German translation still gives the plural: "die, welche Ideen … annahmen"; as does Scotus's Latin version: *ponentes formas*. The Arabic text in Bougyes edition reads the singular form (as was confirmed to me by Andreas Lammer). The context makes it clear that this change to the singular is irrelevant to Alexander's actual point.

transposed from its place near 'the forms' and taken together with 'they' (*fa-hiya*), so that the sentence would be 'therefore, he who postulated the Forms to be all things that exist by nature was not wrong, if the Forms exist at all.'

Fr. 12 Freudenthal (87.10–14)
Er sagt, es ist möglich, den Sinn dieser Stelle einfacher zu ge-[11]winnen, <u>wenn wir die Worte umkehren</u>, so dass sie lauten würden:[90] ‚Und [12] darum haben nicht unrecht gethan die, welche die Ideen annehmen; denn [13] sie sind Alles, was von Natur ist, wenn es überhaupt eine Existenz für [14] die Ideen giebt'.

Scotus
Dixit: Et erit manifestior iste sermo <u>si mutaverint hanc particulam 'ens,'</u> ab hac particula 'forme,' et fuerit posita cum hac particula 'sunt,' ita quod sic legatur: Et ideo non male fecerunt ponentes formas quod ista sunt omnia que sunt secundum naturam si forme habent esse.

Freudenthal's Greek version of Alexander's transposition of the conditional clause reads thus:[91]

Freudenthal 1885: 87 n. 2: "Es sollte gelesen werden":
διὸ δὴ οὐ κακῶς ἔφασαν οἱ τιθέμενοι τὰ εἴδη, ὅτι εἴδη ἐστὶν ὁπόσα φύσει, <u>εἴπερ ἔστιν εἴδη</u> ....

The conditional clause, stating that Aristotle's approval for the theory of Forms is conditional on the existence of the Forms, is moved to the end of the sentence ("if the word 'existing' was transposed from its place near 'the forms'"). This disentangles the somewhat tortuous sentence in ω^{AL}, in which the conditional clause (εἴπερ ἔστιν εἴδη) follows directly upon the first mention of the Forms.

As Freudenthal suggests in a footnote,[92] Alexander's proposal to transpose the conditional clause results in precisely the construction we find in our text, ω^{αβ}. This strongly suggests that Alexander's proposed reading found its way into ω^{αβ} or one of its ancestors. It is possible that this happened accidentally; Alexander's conjecture could have been placed in the margins of an earlier version of ω^{αβ}, from where it was later inserted into the text of ω^{αβ}. Or it is possible that someone consciously followed Alexander's lead in this regard and incorporated the reading into the text. The second possibility is endorsed by the fact that the interpolation was carried out cleanly—there is no collateral damage in the adjacent lines of text. Yet, what speaks against this and in favor of the first possibility is that, if the interpolation was a conscious editorial decision, one might wonder why the corrector did not also accept Alexander's reading in the previous part of the sen-

---

[90] Andreas Lammer confirmed to me that Freudenthal's translation here is less accurate than the translation by Genequand.

[91] See also Martin 1984: 117 n. 2.

[92] Freudenthal 1885: 87 n. 2: "Die Conjectur Alexanders ist also in der letzten Hälfte des Satzes zur Vulgata geworden." See also Freudenthal 1885: 46.

tence (1070a18), where he has οἱ τὰ εἴδη τιθέντες ἔφασαν instead of ω^{αβ}'s Πλάτων ἔφη (see 3.5.2.3).

In any case, Alexander's emendation makes perfect sense and improves the structure of the sentence,[93] yet by no means suggests itself as the only correct reading of the passage. We can therefore confidently rule out the possibility that Alexander's own suggested version of the text and the reading of the direct transmission conform with each other simply because ω^{αβ} preserved the original reading and Alexander happened to hit upon it.[94] Alexander's conjectured reading is idiosyncratic enough to be identified for what it is when we encounter it in the text of our manuscripts. This leads me to conclude that we are dealing here with yet another example of Alexander's commentary influencing the *Metaphysics* text: his emendation, recommending that the conditional clause εἴπερ ἔστιν εἴδη be transposed, was adopted into ω^{αβ} or one of its ancestors.[95]

The five case studies discussed in 5.1 show that Alexander's commentary, written around AD 200, influenced the ω^{αβ} text either directly or indirectly through one of ω^{αβ}'s ancestors. In all five cases, Alexander's *Metaphysics* text (ω^{AL}) offers a satisfactory reading that differs from the reading of our manuscript tradition in ω^{αβ}, while his own suggested emendations and/or interpretative proposals coincide with the wording in ω^{αβ}.[96]

---

[93] Since Alexander's commentary is not preserved in the original Greek we can only speculate as to whether the additional mention of εἴδη in our *Metaphysics* text (1070a19) was prompted by Alexander's transposition alone. It appears as if the second εἴδη might have been left out in the (presumably) original version found in ω^{AL}. The position of the εἴπερ-sentence would then make it possible for εἴδη to function as the subject for both the verb in the εἴπερ-sentence (εἴπερ ἔστιν) and the verb in the ὅτι-sentence (ὅτι ... ἐστίν): ὅτι εἴπερ ἔστιν εἴδη ἐστὶν ὁπόσα φύσει. Perhaps Aristotle even moved the εἴπερ-sentence forward in order to avoid repeating εἴδη. Alexander would then have slightly expanded this dense formulation in order to make it clearer; this change would have later become popular and then even been adopted into our *Metaphysics* text.

[94] Freudenthal does not even mention this as a possibility.

[95] Fr. 13b F (88.17–22) contains Alexander's report of a variant reading of lines 1070a18–20. The text Alexander quotes here as a variant seems to contradict Freudenthal's and my interpretation of Fr. 12 F, because Alexander's quotation contains the transposition of the conditional clause that Alexander, we hold, himself conjectured. Freudenthal (1885: 88 n. 2) points to a corruption of the Arabic and the Hebrew version of Averroes' commentary in just this passage (cf. Genequand 1986: 102). But, according to Bouyges' apparatus, the corruption by no means affects the whole sentence, and so we are not permitted to call into question the authenticity of the whole sentence. Still, Fr. 13b should not undermine our interpretation of Fr. 12. After all, it is quite conceivable that the *Metaphysics* quotation in the later passage in Averroes was corrupted such that it adopted Alexander's suggestion (cf. Martin 1984: 119 n. 12). What clearly speaks in favor of questioning the reliability of the quote of our line (1070a18–19) in Fr. 13b is that the line is irrelevant in this part of the commentary, which is instead concerned with lines 1070a19–20. In Fr. 12, by contrast, Alexander clearly marks his emendation as *his own suggestion* for improving the text. Furthermore, Alexander usually says quite directly when he knows of a preferable reading from another manuscript (see Fr. 4b and 12; and section 3.6, p. 91).

[96] Apart from the five passages analyzed here, the following three seem possible or likely candidates

These traces of contamination of ω^αβ by Alexander's commentary make it possible to date ω^αβ more precisely than ever before. Thus far we have been able to determine AD 400 as the *terminus ante quem* of ω^αβ's spilt into the two branches α and β and hence the *terminus ante quem* for ω^αβ.[97] Now we can even provide a *terminus post quem* for ω^αβ. This terminus is set by Alexander's commentary on the *Metaphysics* or, more precisely, its rise to fame.[98] Taking AD 225 as the starting point of the circulation of Alexander's commentary and dating the emergence of the β-version to no later than the second half of the fourth century AD, we can now date ω^αβ to the period between AD 225 and 400.

How can we explain the contamination of ω^αβ by Alexander's comments? We can assume that Alexander's commentary, because of its comprehensive (and from the perspective of later generations rather orthodox)[99] account of Aristotle's text, quickly became the standard commentary for teaching and studying the *Metaphysics*. On this basis it seems reasonable to assume that ω^αβ goes back to an exemplar in which a teacher or student marked down some of Alexander's remarks and suggestions. From there they found their way into the text itself.

Most of the aforementioned changes to ω^αβ on the basis of Alexander's commentary (5.1.1–4) can be attributed to mechanical or accidental incorporations of marginal notes into the body of the *Metaphysics* text. In the case of the last case study (5.1.5), however, one might suppose that someone had deliberately adopted Alexander's emendation into the text because he thought it would make the sentence clearer.

The conclusion to be drawn from the analysis in 5.1.1–5 is that, contrary to what the current state of research suggests, the influence of Alexander's commentary on the text of Aristotle's *Metaphysics* is not confined to the β-version of book A,[100] but rather extends further back in time to ω^αβ and indeed covers all parts of the *Metaphysics* text to which Alexander's extant commentary refers.

---

for further examples of contamination of ω^αβ by Alexander's comments: A 5, 986a3: εἶναι ἀριθμόν ω^AL (Al.ˢ 39.23–24), Al. 39.24 διό τι ἐξ ἀριθμῶν καὶ κατ' ἀριθμὸν καὶ ἁρμονίαν. ἁρμονίαν εἶναι καὶ ἀριθμόν (cf. also Al.ᴾ 40.23–24) ω^αβ. – A 8, 988b26: περὶ γενέσεως ω^AL (Al. 64.26–27: τὸ δὲ καὶ περὶ γενέσεως ἀντὶ τοῦ καὶ γενέσεως καὶ φθορᾶς, …), cf. *Ph*. B 7, 198a31–35 : περὶ γενέσεως καὶ φθορᾶς ω^αβ. – Γ 2, 1004a12: ἢ <γὰρ> ἁπλῶς λέγομεν ω^AL (Al.ᶜ 253.1–2), Al.ᴾ 253.3–7 ἢ … γὰρ ἁπλῶς λέγουσα … ἁπλῶς λεγομένη … : ἢ ἁπλῶς λεγομένη β : ἢ ἡ ἁπλῶς λεγομένη α.

[97] Cf. 1. Primavesi 2012b: 457–58 confirms this date by drawing from observations made by v. Christ (1886a: VII), Jaeger (1912: 181) and Alexandru (2000) on the catchwords preserved by the β-manuscripts.

[98] Primavesi 2012b: 457–58 also speaks of Alexander's commentary as *terminus post quem* for the split into α and β. This rests on the assumption that the revision process undergone by the β-version *coincides* with ω^αβ's spilt into α and β. This assumption is not warranted; it is by no means necessary for the β-revision to have occurred historically at the same time as the (perhaps completely mechanical) copying of ω^αβ into further manuscripts, two of which became what we reconstruct as α and β. (cf. sections 1 and 5.2 above).

[99] Cf. Fazzo 2004: 6–7.

[100] Primavesi 2012b: 457–58.

5.2 CONTAMINATION OF β BY ALEXANDER'S COMMENTS

Frede/Patzig 1988 suggest that the majority of the divergences between the α- and the β-version in book Z can be attributed to a revision of the β-text; Primavesi 2012 argues that the same goes for book A.[101] In the course of this revision, the wording in the β-text was regularly modified in order to make it clearer, for instance, by filling out dense phrases.[102] While Frede/Patzig arrived at this view by comparing the α- and the β-readings, Primavesi corroborates this claim by evaluating two witnesses that go beyond α and β. These witnesses are the doublet in M 4–5 for the section of A 9,[103] and Alexander's commentary for the whole of book A.[104] Concerning Alexander's commentary, Primavesi demonstrates that the β-text of book A contains traces of Alexander's comments on the text.[105] Primavesi connects this observation with the revision thesis and concludes that the β-reviser used Alexander's commentary as a source of inspiration for his changes in the *Metaphysics* text.[106]

---

[101] Frede/Patzig 1988: 14–17 and Primavesi 2012b: 409; 457–58.

[102] Frede/Patzig 1988: 14: "Wir haben bei der Überprüfung vieler Stellen den Eindruck gewonnen, daß A^b in vielen Fällen einen glatteren Text als EJ bietet. Dieser Befund scheint jedoch in charakteristischen Fällen auf regulierende Eingriffe in den aristotelischen Text zurückzugehen. Diese Eingriffe sind nur verständlich, wenn man voraussetzt, daß die Urheber der Tradition β, möglicherweise die Editoren jener vermuteten antiken Textausgabe, in manchen Fällen eine für Aristoteles charakteristische, aber etwas ungewöhnliche Ausdrucksweise nicht verstanden haben und daher meinten, der Text müsse entsprechend verändert werden." Primavesi 2012b: 439: "So far, the hypothesis by Frede & Patzig, our starting point, has been corroborated to a remarkable degree. Book A has supplied ample evidence for the following rules of thumb: whenever the wording of a passage transmitted by both α and β diverges between α and β in a way which is obviously due to conscious intervention, the change is most likely to have been produced by the β-reviser...."

[103] As Primavesi 2012b: 412–20 shows, the first person plural forms in the context of the critique of Forms in A 9 have been exchanged in β with third person (plural or singular) forms, with the result that book A was brought into conformity with book M.

[104] Frede/Patzig could not have used these witnesses, since Alexander's commentary on book Z is not extant, nor is there a doublet of book Z in any other part of Aristotle's works.

[105] Primavesi 2012b: 424–39. Text 7 (pp. 424–28) is a clear example of a β-reading that incorporates Alexander's comments into the *Metaphysics* text. Texts 10–12 (pp. 434–36) are possible, but less secure adoptions of β from Alexander's commentary, since the reading of ω^{AL} cannot be reconstructed apart from the paraphrase on which β is supposed to have based its revision. In other words, it cannot be ruled out that the paraphrase in Alexander simply represents what he read in ω^{AL} and its agreement with β points to the older reading, which was given in ω^{αβ}. Regarding text 12 (990a33–990b2), Alexander's lemma (76.6) suggests that he found the β-reading in his text. In the case of text 14 (pp. 437–39), it seems more reasonable to understand the α-reading as a later interpolation, both β and ω^{AL} then preserving the correct reading (cf. 4.3.1), than to assume that the β-reviser deleted the (quite fitting and innocuous) addition on the basis of the evidence in Alexander's commentary.

[106] Primavesi 2012b: 457: "The β-reviser's main source of inspiration for his dealings with the *common text* was Alexander's commentary."

In the following, I will analyze six new cases that indicate that Alexander's commentary influenced the β-text. In five of the six cases the passages in question come from a book other than book A.[107] This points to the conclusion that the contamination of the β-text by Alexander, which Primavesi discovered for book A,[108] is in fact not restricted to this book. Whether these cases of contamination should all be attributed to one specific revision process that the β-version underwent, or whether they should rather be seen as traces of an influence that Alexander's commentary exerted over a longer period of time, are questions that I will only touch upon.[109]

5.2.1 Alex. *In Metaph.* 421.7–15 on Arist. *Metaph.* Δ 23, 1023a17–21

In the 23rd chapter of book Δ, a book that can be described as an encyclopedia of philosophically relevant terms, Aristotle examines the term ἔχειν, meaning "to have," "to hold." Having discussed ἔχειν in respect to the meanings "to treat a thing according to one's own nature or impulse" (1023a8–11), "to be a recipient of something" (1023a11–13), and "to contain" (1023a13–17), Aristotle considers as a fourth option "to prevent something from moving or acting according to its own impulse."

Aristotle, *Metaphysics* Δ 23, 1023a17–21

> ἔτι τὸ κωλῦον κατὰ τὴν αὑτοῦ [18] ὁρμήν τι κινεῖσθαι ἢ πράττειν ἔχειν λέγεται τοῦτο αὐτό, [19] οἷον καὶ οἱ κίονες τὰ ἐπικείμενα βάρη, καὶ ὡς οἱ ποιηταὶ [20] τὸν Ἄτλαντα ποιοῦσι τὸν οὐρανὸν ἔχειν ὡς συμπεσόντ᾽ ἂν [21] ἐπὶ τὴν γῆν, ὥσπερ καὶ τῶν φυσιολόγων τινές **φασιν**·
>
> That which hinders a thing from moving or acting according to its own impulse is said to *hold* it, as pillars *hold* the incumbent weights, and as the poets make Atlas

---

[107]The case from book A that I discuss is not mentioned by Primavesi 2012b as a case that displays Alexander's influence on β. Apart from my six cases and the cases in Primavesi 2012b: 424–35, there are more passages in β where contamination by Alexander is possible or likely (see, for instance, α 2, 994b5: Al.^{c} 159.10 and Al.^{P} 157.33; Γ 2, 1003b21: Al.^{l} 245.20–21, Al.^{P} 245.24–25 and Al.^{c} 251.5). I chose the six cases because they offer secure evidence for the readings in α, β, and ω^{AL} and Alexander's interpretation. That is to say, (i) the difference between α and β is substantive enough, (ii) the reading of ω^{AL} can be reconstructed according to my rule of thumb stated on p. 57, and (iii) Alexander's own contribution to the passage is idiosyncratic enough (cf. 1). Apart from the cases discussed in Primavesi 2012b or mentioned in the present study, one may think that there are more passages in the β-version of the *Metaphysics* where the agreement between β and Alexander's comments may be due to the contamination of β by Alexander, but which I have either overlooked or which cannot be identified simply because we cannot reconstruct the reading in ω^{AL} as a touchstone and thus cannot determine the causal relation of the agreement between the β-reading and the evidence in Alexander's commentary.

[108]Cf. Primavesi 2012b: 457 n. 165.

[109]Whether or not the β-text of the rest of the *Metaphysics*, that is apart from book A and Z, shows clear traces of a revision process can only be determined through a study of the complete β-text itself, something that clearly lies beyond the scope my the present study.

hold the heavens, implying that otherwise they would collapse on the earth, as some of the natural philosophers also say.

18 αὐτό **α** edd. : ταῦτα **β** ‖ 20 τὸν ἄτλαντα ποιοῦσι **α** edd. : ποιοῦσιν ἄτλαντα **β** ‖ 21 καὶ **β** edd. : om. **α** ‖ φασίν **α** Ascl.ᵖ 348.32-34 Arᵘ (*Scotus*) edd. : φασίν. ἄτλας δ᾽ οὐρανὸν εὐρὺν ἔχει κρατερῆς ὑπ᾽ ἀνάγκης **β** ex Al. 421.11-12

Here Aristotle discusses the fourth meaning of ἔχειν, which is "to hold back and keep something from moving according to its own impulse." He illustrates this sense of ἔχειν with two examples: firstly, columns *hold* the weight of the part of the building that is resting upon them and thereby hinder its downward motion. Secondly, poets use the verb ἔχειν to describe the task of the mythical figure Atlas, who holds up the heavens with his hands and thereby prevents them from falling down onto the earth.[110] Aristotle adds that the idea of an active power, which hinders the heavens from collapsing, is not restricted to the realm of poetry. Some natural philosophers, whom he does not name, share this idea. Such, in any case, is the text according to the **α**-version.[111]

In the **β**-version, the word φασίν (a21) is directly followed by a verse from Hesiod's *Theogony* about Atlas: Ἄτλας δ᾽ οὐρανὸν εὐρὺν ἔχει κρατερῆς ὑπ᾽ ἀνάγκης (*Theogony* 517). This verse fits remarkably well with Aristotle's description of Atlas, so well in fact that we may even assume that Aristotle had exactly this verse in mind when he described Atlas in our passage: οἱ ποιηταὶ τὸν Ἄτλαντα ποιοῦσι τὸν οὐρανὸν ἔχειν (a19-20).[112] Thus the crucial question is: did Aristotle actually *quote* here the verse he had in mind? That he wrote down the verse verbatim is an unlikely scenario given that he had already paraphrased the verse's content in lines a19-21. This makes a quotation of the verse superfluous.

Additionally, the verse following the verb φασίν in the **β**-text (a21) appears ill-fitting, for the subject of the form φασίν ("they say") is no longer the poets (οἱ ποιηταί) but the natural philosophers (τῶν φυσιολόγων τινές), a group that does

---

[110] Aristotle refers in two other works to Atlas and the cosmological idea connected with this figure. In these passages, too, Aristotle does not support his reference to Atlas with a verse quotation from a mythical story. He seems to take for granted that his readers are familiar with the verses he is alluding to. In *MA* 3, 699a27-b11, he criticizes the idea of an Atlas who moves the universe without an external unmoved point. See Nussbaum 1978: 300-304. In *Cael.* B 1, 284a18-26, Aristotle criticizes, as he does in our passage, the assumption implied in the figure of Atlas, namely, that there is a force which acts upon the universe and on which the universe ultimately depends (*Cael.* B 1, 284a18-20: οὔτε κατὰ τὸν τῶν παλαιῶν μῦθον ὑποληπτέον ἔχειν, οἵ φασιν Ἄτλαντός τινος αὐτῷ προσδεῖσθαι τὴν σωτηρίαν.../ "we must not believe the old tale which says that the world needs some Atlas to keep it safe... "). In the *Cael.* passage, as here in Δ 23, the critique of the idea of Atlas is combined with criticism of a presocratic thinker, whom Aristotle this time even calls by name: Empedocles (see also below).

[111] Asclepius's paraphrase (348.30-34) agrees with the **α**-text. As far as the sentence in 1023a20-21 is concerned, the Arabic tradition, too, agrees with the **α**-reading.

[112] Cf. the description of Atlas in the first book of the *Odyssey*: α 53b-54: ἔχει δέ τε κίονας αὐτὸς / μακράς, αἳ γαῖάν τε καὶ οὐρανὸν ἀμφὶς ἔχουσι. Other differences aside, Homer agrees with Hesiod in using the verb ἔχειν to describe what Atlas is doing.

not include Hesiod.[113] Even if we consider the possibility that Aristotle had inserted the verse into the text at a position other than the one it occupies in the β-text, we still run into problems. It is equally impossible from a syntactic standpoint to connect the verse with the verb ποιοῦσι (sc. τὸν Ἄτλαντα ... ἔχειν) in line a20, the subject of which are the poets. Therefore, the verse addition, which is transmitted by the β-version only, should be deleted as a later interpolation into the text.[114]

If the verse is an interpolation, how might it have come into the β-text? Alexander's commentary gives us a clue. He writes:

Alexander, *In Metaph.* 421.7–15 Hayduck

ἔχειν λέγεται καὶ τὰ κω-[8]λύοντά τινα κατὰ τὴν αὐτῶν ὁρμὴν ἢ φύσιν πράττειν τι ἢ κινεῖσθαι· οὕτως [9] οἱ κίονες ἔχειν λέγονται τὰ ἐπικείμενα αὐτοῖς, ἀνέχοντες αὐτὰ καὶ κω-[10]λύοντες κατὰ τὴν αὐτῶν φύσιν φέρεσθαι κάτω. οὕτως καὶ τὸν Ἄτλαντα [11] οἱ ποιηταὶ τὸν οὐρανὸν ἔχειν λέγουσιν· "<u>Ἄτλας δ' οὐρανὸν εὐρὺν ἔχει</u> [12] <u>κρατερῆς ὑπ' ἀνάγκης</u>"· ὡς γὰρ συμπεσουμένου ἐπὶ τὴν γῆν κατὰ τὴν [13] αὐτοῦ φύσιν, εἰ μὴ ἔχοιτο καὶ ἀνέχοιτο ὑπ' αὐτοῦ. οὕτω καὶ τῶν φυσι-[14]κῶν ὅσοι διὰ τὴν δίνην μένειν τὸν κόσμον λέγουσι καὶ μὴ συμπίπτειν, [15] λέγοιεν ἂν αὐτὸν ὑπὸ τῆς δίνης ἔχεσθαι.

Things are also said 'to hold' if they prevent a thing from doing something or from moving according to its own impulse or nature; thus pillars are said to hold the parts resting on top of them, since they hold these parts up and prevent them from tumbling down according to their own nature. So too the poets say that Atlas holds the heavens: <u>But Atlas, under strong compulsion, holds up the wide heavens</u>, as if the [whole] heaven would, according to its own nature, collapse onto the earth if it were not held, i.e. held up, by him. And those natural philosophers who assert that because of its whirling motion the earth remains [in position] and does not collapse would also say that it is 'held,' in this sense, by the vortex.

8 ἢ φύσιν] ἡ φύσις **A**[a.c.] || 11 εὐρὺν **A P**[b] : om. **O** || 12–13 κρατερῆς ... μὴ **A O** : ὡσανεὶ **P**[b] || 12 ἐπὶ **A**[p.c.]**S** : ὑπὸ **O** || 15 λέγοιεν **P**[b] **LF S** : λέγοι **A O**

Alexander paraphrases Aristotle's words, following them closely. However, his paraphrase includes the verse from Hesiod's *Theogony*, which Aristotle, we are led to believe, must have had in mind and which, we find, is a later interpolation into the β-text. In Alexander's commentary the verse quotation squares well with the rest of Alexander's paraphrase. Alexander does not use Aristotle's expression (οἱ ποιηταὶ ... ποιοῦσι, a19–20) to describe what the poets are doing. He rather uses a saying verb to introduce the verse as the contents of what is said (οἱ ποιηταὶ ...

---

[113] In *Metaph.* B 4, 1000a9–10 Aristotle explicitly includes Hesiod among the group of *theologoi*. Further, Aristotle's treatment of Hesiod in A 4, 984b23 and A 8, 989a10 does not at all imply that he classes him with the natural philosophers. Finally, Alexander (see below) indicates clearly that according to the ancient understanding Hesiod belongs on the side of the poets.

[114] The words are already classified as a later addition in Christ 1853: 22. Since Bekker, editors have been treating the verse as an interpolation.

λέγουσιν· [verse quotation]). The fact that the verse blends in naturally with Alexander's comments, while it fits clumsily at best in the β-text of the *Metaphysics*, strongly suggests that the verse was not taken from Aristotle's text by Alexander, but rather found its way from Alexander's commentary into the β-text.

Notice that in this commentary passage Alexander supplements Aristotle's sparse information with additional data. This holds for both the Hesiod quotation and the extra information Alexander provides about the presocratic teaching. When Aristotle refers to some φυσιολόγοι who also believe in the existence of a force that prevents the earth from collapsing, Alexander infers, probably on the basis of *De caelo* B 1 (284a18–26),[115] that Aristotle here has in mind Empedocles, whose whirl (δίνη) holds the earth in the middle position by centripetal force.[116] In *De caelo* B 1, Aristotle in the same breath speaks about the mythical conception of an Atlas holding the heavens and explicitly mentions Empedocles' theory of the whirl (δίνη).[117] We see that Alexander here supplements Aristotle's *Metaphysics* text with explications, examples, and further material from other sources. This type of commentatory initiative on Alexander's part fits with my suspicion that he did not find the Hesiodic verse in ω^{AL}, but added it himself to illustrate Aristotle's argument.[118]

Jaeger writes in his *apparatus criticus* regarding the verse in question: *affert Al^p unde sumpsit A^b*. He, too, believes that the β-reading goes back to an adaptation based on Alexander's commentary. But for Jaeger the influence of Alexander's commentary on the *Metaphysics* text is restricted to manuscript **A^b**. New collations of the *Metaphysics* manuscripts[119] now show that the verse appears in the β-manuscript **M** (Ambros. F 113 sup.) as well, which is independent of **A^b**, thus making the verse a feature of the *hyparchetype* β. Although Jaeger recognized **A^b**

---

[115] Cael. B 1, 284a18–26: Διόπερ οὔτε κατὰ τὸν τῶν παλαιῶν μῦθον ὑποληπτέον ἔχειν, οἵ φασιν Ἄτλαντός τινος αὐτῷ προσδεῖσθαι τὴν σωτηρίαν· ... οὔτε δὴ τοῦτον τὸν τρόπον ὑποληπτέον, οὔτε διὰ τὴν δίνησιν θάττονος τυγχάνοντα φορᾶς τῆς οἰκείας ῥοπῆς ἔτι σώζεσθαι τοσοῦτον χρόνον, καθάπερ Ἐμπεδοκλῆς φησιν. / "Hence we must not believe the old tale which says that the world needs some Atlas to keep it safe .... We must no more believe that than follow Empedocles when he says that the world, by being whirled round, received a movement quick enough to overpower its own downward tendency, and thus has been kept from destruction all this time" (transl. by Stocks). See also Perilli 1996: 56–58.

[116] See Empedocles, DK 31 A 67; Arist. *Cael.* B 13, 295a16–21: "Others [say], with Empedocles, that the motion of the heavens, moving about it at a higher speed, prevents movement of the earth, as the water in a cup, when the cup is given a circular motion, though it is often underneath the bronze, is for the same reason prevented *from moving with the downward movement which is natural to it* (my emphasis)." See also DK 31 B 35.4.

[117] Cf. Nussbaum 1978: 300–301. Also Simplicius (374.25–31 Heiberg), drawing from Alexander's now lost commentary on *De caelo*, quotes Hesiod's verse about Atlas together with a verse on Atlas from the *Odyssey* (both verses are ascribed to Homer).

[118] Additionally one might expect that Alexander, had he found the verse in the Aristotelian text as we find it in the β-version, would have commented on the suspicious position of the verse.

[119] Conducted by Pantelis Golitsis and Ingo Steinl (Aristoteles-Archiv, Berlin).

as a representative of an independent family,[120] he did not realize that this version of the *Metaphysics* has been influenced by Alexander's commentary and that therefore Alexander could not have used this family, as he wants to claim.[121]

The fact that the verse in the β-text does not fit well suggests that it was first written for illustrative purposes in the margin of the text and only later and perhaps accidentally incorporated into the body of the text. If one wants to ascribe the presence of the verse to a β-reviser, then one has to assume that something went wrong during the revision process,[122] resulting in the clumsy positioning of the verse in the text.

5.2.2 Alex. *In Metaph.* 285.32–36; 286.2–6 on Arist. *Metaph.* Γ 4, 1007a20–23

In Γ 3, 1005b17–23, Aristotle introduces the principle of non-contradiction, which prohibits a thing from being both F and not F at the same time, as the most secure of all principles. Aristotle engages with the deniers of this principle (Γ 4,1005b35–1006a5) by pointing out that the principle of non-contradiction neither requires proof nor can, in fact, be proved (1006a5–11). However, its validity can be demonstrated negatively by refuting its denial (ἀποδεῖξαι ἐλεγκτικῶς) (1006a11–15).[123] Later in chapter 4, Aristotle shows that several absurdities follow from the denial of the principle. In Γ 4, 1007a20–23, he shows that those who say that something is simultaneously both a man and not a man do away with substance (οὐσία) and essence (τὸ τί ἦν εἶναι)[124] (Γ 4, 1007a20–23). The οὐσία determines the essence of a thing, i.e. what it *is*. To be essentially a man means to be what it is for a man to be; to be essentially a man precludes the possibility of not being a man or of being a non-man.[125] This entails that it is impossible to say of a thing that it both is and is not a man.[126] When the opponents of the principle of non-contradiction claim that this is nevertheless possible, they turn all things into accidents. To say that

---

[120] Jaeger builds on the discovery made by Christ 1886 that A^b contain *reclamantes* at the end of certain books and therefore go back to an ancient papyrus-edition (Jaeger 1912: 181). See 1.

[121] Jaeger 1957: x–xii.

[122] As we learn from Primavesi's first example (2012b: 424–28), a passage that underwent revision is likely to show signs of unintended collateral damage.

[123] See Kirwan 1971: 90–92 and Rapp 1993.

[124] I translate the Aristotelian formula τὸ τί ἦν εἶναι as "essence." The more literal rendering, "what it is to be" (as translated by Kirwan), points to the original gist of the expression, but is impractical for my purpose. On the equation of the τί ἦν εἶναι with the essence and definition of a thing see *Metaph.* Δ 18, 1022a24–27: ὥστε καὶ τὸ καθ' αὑτὸ πολλαχῶς ἀνάγκη λέγεσθαι. ἓν μὲν γὰρ καθ' αὑτὸ τὸ τί ἦν εἶναι ἑκάστῳ, οἷον ὁ Καλλίας καθ' αὑτὸν Καλλίας καὶ τὸ τί ἦν εἶναι Καλλίᾳ; *Metaph.* Ζ 4, 1029b14: ὅτι ἐστὶ τὸ τί ἦν εἶναι ἑκάστου ὃ λέγεται καθ' αὑτό; 1030a3: ὅπερ γάρ τί ἐστι τὸ τί ἦν εἶναι.

[125] See Weidemann 1980 and Kirwan 1971: 100–101, who refer to Aristotle's theory of predication in *APo* A 22, 83a24–32.

[126] Cf. the following lines Γ 4, 1007a23–31.

something is essentially a man and not a man is to argue that being a man is a mere accident and so does not constitute the essential being of the thing of which it is predicated.[127]

In the following I focus on the opening lines of the argument, 1007a20–23, and compare the three (directly and indirectly transmitted) versions ω^AL, α, and β with respect to those lines. The readings in α and β are represented in the manuscripts, but the reading of ω^AL has to be reconstructed from Alexander's comments on the passage.[128]

Aristotle, *Metaphysics* Γ 4, 1007a20–23

[ω^AL]

ὅλως δ' ἀναιροῦσιν οἱ τοῦτο λέ-[21]γοντες οὐσίαν καὶ **τὸ τί ἦν εἶναι**. πάντα γὰρ ἀνάγκη συμ-[22]βεβηκέναι φάσκειν αὐτοῖς, καὶ τὸ ὅπερ ἀνθρώπῳ εἶναι ἢ [23] ζῴῳ εἶναι **τί ἦν εἶναι μὴ εἶναι**.

And in general those who say this do away with substance and **essence**. For they must say that all things are accidents, and that to-be-precisely-what-it-is-to-be-a-man or an animal is **not an essence**.

[α-text]

ὅλως δ' ἀναιροῦσιν οἱ τοῦτο λέ-[21]γοντες οὐσίαν καὶ **τὸ τί ἦν εἶναι**. πάντα γὰρ ἀνάγκη συμ-[22]βεβηκέναι φάσκειν αὐτοῖς, καὶ τὸ ὅπερ ἀνθρώπῳ εἶναι ἢ [23] ζῴῳ εἶναι **μὴ εἶναι**.

And in general those who say this do away with substance and **essence**. For they must say that all things are accidents, and that **there is no** such thing as to-be-precisely-what-it-is-to-be-a-man or an animal.

[β-text]

ὅλως δ' ἀναιροῦσιν οἱ τοῦτο λέ-[21]γοντες οὐσίαν καὶ **τὸ τί ἦν εἶναι μὴ εἶναι**. πάντα γὰρ ἀνάγκη συμ-[22]βεβηκέναι φάσκειν αὐτοῖς, καὶ τὸ ὅπερ ἀνθρώπῳ εἶναι ἢ [23] ζῴῳ εἶναι **μὴ εἶναι τί ἦν εἶναί τινος**.

And in general those who say this do away with substance and **the not-to-be-an-essence**. For they must say that all things are accidents, and that to-be-precisely-what-it-is-to-be-a-man or an animal is **not the essence of anything**.

---

21 εἶναι ω^AL (Al.¹ 285.2 Al.ᶜ 285.11–12) α Ar^u (*Scotus*) edd. : εἶναι μὴ εἶναι β || 22–23 ἢ … εἶναι (tert.) om. E || 23 τί ἦν εἶναι μὴ εἶναι ω^AL (Al.ᶜ 286.3) Bonitz : μὴ εἶναι α Ar^u (*Scotus*) Bekker Schwegler Christ Ross Jaeger Cassin/Narcy Hecquet-Devienne : μὴ εἶναι τί ἦν εἶναί τινος β (ex Al.ᵖ 286.33–34) || μὴ … εἶναι) om. V^d

---

[127] Weidemann 1980: 78–79. Cf. also Kirwan 1971: 100–101 with a different view on the argument.

[128] This was already done by Bonitz, who even follows the reading of ω^AL in his edition of the *Metaphysics*. Since the reconstruction is fairly complex, and Bonitz does not mention it in his commentary on the passage (1849: 193–94), and since I will draw important consequences from it, extensive comments on the reconstruction of ω^AL will be necessary.

Before comparing the three versions with regard to their origin and value, I will justify the reconstruction of the reading in ω^AL which is given above.[129] The first sentence (ὅλως ... τὸ τί ἦν εἶναι. a20–21) is quoted in the lemma (285.1–2) and agrees with the α-reading (see above). In the course of his comments Alexander returns to this sentence and cites its concluding phrase (285.11–12): εἰπὼν γὰρ ἀναιροῦσιν οὐσίαν, ποίαν οὐσίαν ἐδήλωσε προσθεὶς καὶ τὸ τί ἦν εἶναι ("Having said 'they do away with substance,' he makes clear what kind of substance he means by adding 'that is, essence'"). Lemma and citation thus testify that the reading of the first sentence in ω^AL is identical with the reading in α.

The reconstruction of the reading in the second sentence (πάντα ... μὴ εἶναι. [α] / πάντα ... μὴ εἶναι τί ἦν εἶναί τινος [β], a21–23) in ω^AL is more complicated.[130] In 285.21–31 Alexander recapitulates Aristotle's argument that the deniers of the principle of non-contradiction do away with substance and essence:

Alexander, *In Metaph.* 285.32–36 Hayduck

εἰπὼν δὲ καὶ τὸ ὅπερ ἀνθρώπῳ εἶναι ἢ ζῴῳ οὐκέτι προσέθηκε τὸ [33] ἀναιροῦσιν ἦν γὰρ προειρημένον καὶ κείμενον· προεῖπε γὰρ ὅλως δὲ [34] ἀναιροῦσιν οἱ τοῦτο λέγοντες. τὸ δὲ τί ἦν εἶναι μὴ εἶναι ἐνδε-[35]έστερον ἔχειν δόξει· λείπει γὰρ αὐτῷ ὁ 'καί' σύνδεσμος, ἵνα ᾖ 'καὶ τὸ τί [36] ἦν εἶναι μὴ εἶναι,' τουτέστι καὶ τὸν ὁρισμὸν μὴ εἶναι.

Having said "and to-be-precisely-what-it-is-to-be-a-man or an animal" he does not add 'they do away with [it],' for this has already been said and posited; for he has already said "in general those who say this do away with." The phrase [a23] "that ... essence is not" will seem to be incomplete; for it is missing the conjunction "and," so as to read "*and* that the essence does not exist," that is, and that the definition does not exist.

32 δὲ P^b S : om. A O || 33 καὶ κείμενον A P^b S : om. O || 35 αὐτῷ A P^b S : αὐτὸ O

Alexander is dissatisfied with the second sentence he finds in his text (a21–23). As his quotation of the phrase τί ἦν εἶναι μὴ εἶναι (285.34) shows,[131] ω^AL had: πάντα γὰρ ἀνάγκη συμβεβηκέναι φάσκειν αὐτοῖς, καὶ τὸ ὅπερ ἀνθρώπῳ εἶναι ἢ ζῴῳ εἶναι τί ἦν εἶναι μὴ εἶναι ("For they must say that all things are accidents, and that to-be-precisely-what-it-is-to-be-a-man or an animal is not an essence"). The cause

---

[129] Madigan's reconstruction (Madigan 1993: 162 n. 478) of Alexander's reading as well as Alexander's remarks on how to understand it are confused. First, Alexander (286.3) *does* read εἶναι after ζῴῳ (a23). Second, Alexander's first proposal for how to tackle this passage *does not* aim at making the second sentence grammatically dependent on the first (see below). Third, Alexander *does* propose his second interpretation as a reformulation of the Aristotelian sentence; he certainly had not already encountered this interpretation as a variant reading (see below; for Alexander's labeling of variant readings see 3.6).

[130] Bonitz 1848 app. crit. *ad loc.*

[131] This divergence from the text of the direct transmission is confirmed by another quotation in 286.3.

of Alexander's dissatisfaction seems to be the phrase "that to-be-precisely-what-it-is-to-be-a-man or an animal is not an essence." Perhaps this appeared odd to him because in the previous sentence Aristotle had just said that the opponents *do away* with essence, a concept which would seem to encompass what-it-is-to-be-a-man. In other words, as far as Alexander was concerned, the second sentence failed to observe the rule stated in the first sentence.[132] This is probably why Alexander's first attempt to fix the problem includes a reminder of the first sentence and a repetition of the fact that Aristotle had already said that they do away with essence (προεῖπε γὰρ ὅλως δὲ ἀναιροῦσιν, 285.33–34). Alexander holds that this result is also implied in the second sentence, even though Aristotle does not state it explicitly (οὐκέτι προσέθηκε τὸ ἀναιροῦσιν, 285.32–33).

Alexander's second attempt to cope with the seemingly unsatisfactory predicative τί ἦν εἶναι entails adding the conjunction καί. Alexander suggests that by adding καί the phrase τί ἦν εἶναί becomes a subject in its own right, standing beside the other subjects τὸ ὅπερ ἀνθρώπῳ εἶναι and [τὸ] ζῴῳ εἶναι (285.35–36). The negation μὴ εἶναι then applies to all three subjects:[133] "and that to-be-precisely-what-it-is-to-be-a-man or to-be-precisely-what-it-is-to-be-an-animal <u>and</u> essence do not exist." This solution would remedy Alexander's dissatisfaction regarding the missing negation of being a man and being an animal. Yet it does not completely satisfy him. He therefore makes a third proposal (δύναται ... 286.2), this time reformulating Aristotle's sentence in order to express the intended meaning more clearly.

Alexander, *In Metaph*. 286.2–6 Hayduck

δύναται [3] καὶ τὸ ὅπερ ἀνθρώπῳ ἢ ζῴῳ εἶναι τί ἦν εἶναι μὴ εἶναι εἰρῆσθαι [4] ἀντὶ τοῦ τὸ ὅπερ ἀνθρώπῳ εἶναι ἢ ζῴῳ εἶναι μὴ εἶναι τί ἦν εἶναί τινος, [5] τουτέστι μὴ ἐν τῇ οὐσίᾳ μηδὲ ἐν τῷ τί ἐστι κατηγορεῖσθαι, ὡς εἰρῆσθαι [6] τὸ τί ἦν εἶναι ἀντὶ τοῦ τί ἐστιν.

The phrase "and that to-be-precisely-what-it-is-to-be-a-man or an animal is not an essence" may be said in place of 'and that to-be-precisely-what-it-is-to-be-a-man or an animal is not the essence *of anything*,' that is, is not predicated in the category of substance or what it is essentially, 'essence' being used in place of what a thing is.

4 τοῦ τὸ **A O** : τοῦ **P**b

Alexander's third solution for dealing with the sentence is not so much a conjecture imposed on the Aristotelian text as it is a new formulation, suggested by a commentator and designed to facilitate the understanding of what Aristotle

---

[132] If this is what Alexander's dissatisfaction amounts to, then one can object to Alexander's view that the abolition of essence consists simply in turning all essences into accidents. Aristotle's two thoughts need not contradict each other.

[133] This yields the following text in lines a22–23: "...καὶ τὸ ὅπερ ἀνθρώπῳ εἶναι ἢ ζῴῳ εἶναι <u>καὶ</u> τί ἦν εἶναι μὴ εἶναι."

means to say. Alexander quotes from ω^AL (confirming our reconstruction) and then replaces the sentence (ἀντὶ, 286.4) with an explicit reformulation.[134] He inverts the order of the phrases τί ἦν εἶναι and μὴ εἶναι and adds the word τινός. By placing the predicate (μὴ εἶναι) between the subject and predicative (τί ἦν εἶναι) Alexander presents the latter in a new light. And by adding the indefinite τινός, Alexander highlights his new understanding of the τί ἦν εἶναί: the opponents do not do away with any features that the animals have, for example, being a man or being an animal; rather, all they do is deny that these features constitute the essence of these beings. The opponents cannot put the to-be-precisely-what-it-is-to-be-a-man under the category of essence or definition (286.5–6), but, as Aristotle said in the first part of the second sentence, only in the category of accidents.[135]

Having reconstructed the reading of ω^AL and reproduced the way in which Alexander understands and tries to clarify it, I now will turn to a comparison of the three versions. Here, I deal separately with the section's first sentence (1007a20–21: ὅλως ... εἶναι,) and with the section's second sentence (1007a21–23: πάντα γὰρ ...). With respect to the first sentence, the ω^AL-text and the α-text agree. In these two versions, the sentence ends in 1007a21 with τὸ τί ἦν εἶναι ("essence"). The β-version, by contrast, has the sentence end with τὸ τί ἦν εἶναι μὴ εἶναι, that is, augmented by the words μὴ εἶναι. This augmentation does not fit with the rest of the sentence. In order to make sense of the words μὴ εἶναι in the given syntax, one has to understand them as part of an articular infinitive; but then the phrase τὸ τί ἦν εἶναι μὴ εἶναι has to be taken as "the not-to-be-an-essence," which results in the implausible assertion that the deniers do away with substance as well as "to be not-an-essence." The ending of the first sentence as transmitted by the α-version and confirmed by ω^AL is clearly preferable.

The second sentence concludes differently in all three versions. The ω^AL-text and the α-text disagree in the following way: according to the α-version, the deniers of the principle of non-contradiction must say that all things are accidents (a21–22) and that essences (μὴ εἶναι, a23), such as to-be-precisely-what-it-is-to-be-a-man (τὸ ὅπερ ἀνθρώπῳ εἶναι)[136] and to-be-precisely-what-it-is-to-be-an-animal ([τὸ] ζῴῳ εἶναι), do not exist. By contrast, in the version of the ω^AL-text, Aristotle asserts that the deniers must say that all things are accidents, and that things like to-be-precisely-what-it-is-to-be-a-man or to-be-precisely-what-it-is-to-be-an-animal are *not essences* (τί ἦν εἶναι μὴ εἶναι, a23).

Two arguments speak in favor of the authenticity of the ω^AL-version against

---

[134] The formula "δύναται Χ εἰρῆσθαι ἀντὶ τοῦ Υ" is used by Alexander in many other places in his commentary. He applies it in order to introduce his own explanatory reformulation of Aristotle's thought. See also 5.1.2 and esp. p. 190.

[135] For Alexander's explication of Aristotle's argument see also *In Metaph.* 285.2–31.

[136] For the expression infinitive + *dativus aristotelicus* (cf. Bonitz 1870 s. v. "*Dativus*," p. 166b38–39; εἶναι, p. 221a34ff;) and its relation to τὸ τί ἦν εἶναι (cf. Bonitz 1870: τὸ τί ἦν εἶναι, p. 763b49ff) see Bassenge 1960 and also Weidemann 1980: 78–80.

the α-version.¹³⁷ From the perspective of the sentence as a whole, the α-reading is unsatisfying: in the first part of the sentence (a21–22), Aristotle says that all things (πάντα, a21) are turned into accidents, while in the second part, he excludes essences from the group of "all things." That is to say, essences are not even accidents but are not (anything) at all (μὴ εἶναι). By contrast, the ω^AL-reading is much to be preferred. For Aristotle would then be saying that the opponents turn all things into accidents, thus making to-be-precisely-what-it-is-to-be-a-man, or to-be-precisely-what-it-is-to-be-an-animal, no longer essences (τί ἦν εἶναι μὴ εἶναι) but, as stated in the sentence's first part, accidents. The second argument is paleographical. The genesis of the α-reading can be easily explained as originating from the ω^AL-reading. Given the accumulation of the εἶναι infinitives in the present passage, a *saut du même au même* could have easily happened in line a23 (εἶναι τί ἦν εἶναι).¹³⁸

What can be said about the second sentence as transmitted by the β-version? Since the β-text shares the phrase τί ἦν εἶναι with the ω^AL-version (although they are positioned differently in their respective texts,) one might be inclined to adduce the β-reading as evidence for the authenticity of the ω^AL-reading against the α-reading. However, this would leave β's additional τινός unexplained. Having explored Alexander's commentary on the passage, we can explain the ending of the sentence in β as an adoption of Alexander's proposal for making the sentence clearer. As we saw, Alexander proposed rephrasing the second sentence to τὸ ὅπερ ἀνθρώπῳ εἶναι ἢ ζῴῳ εἶναι μὴ εἶναι τί ἦν εἶναί τινος (286.4), thereby transposing the words μὴ εἶναι and τί ἦν εἶναι and adding τινός (see above). This is precisely what we find in the β-version. Since Alexander's *own* suggested reformulation matches the β-version exactly, we can safely assume that someone revised the phrase in the β-text according to Alexander's words. Alexander's clarification of Aristotle's expression became an emendation of the *Metaphysics* text.¹³⁹

---

¹³⁷ If the preference for the ω^AL-version rather than the α-version is justified, then this is another example of ω^AL preserving the correct reading while our tradition had been corrupted (cf. 4.1). Among the editors of the *Metaphysics*, Bonitz 1848 alone prefers the reading of ω^AL. In Bonitz 1842: 166, following Bekker, he assumed the α-version to be correct. Yet, having edited Alexander's commentary in 1847, Bonitz then changed his mind about this passage. Unfortunately, in his commentary (1849: 193–94) he is silent about Alexander's testimony and his treatment of it.

¹³⁸ The risk of an error due to *saut du même au même* is obviously increased in this passage: in V^d this error occurred in line a23: Instead of (the β-reading) ζῴῳ εἶναι μὴ εἶναι τί ἦν εἶναί τινος, V^d has ζῴῳ εἶναι τινος (see apparatus). E also seems to have suffered an error due to *saut du même au même*: a22–23 εἶναι ἢ ζῴῳ εἶναι τί ἦν εἶναι μὴ εἶναι (see apparatus).

¹³⁹ We can only speculate about the text in β *prior* to this Alexander-based revision. Was the prior version identical to the wording of ω^AL or of α? Put differently, what represents the older version, ω^AL or α? For the first sentence, both read the same text. For the second sentence, the wording in ω^AL appears to be the older one; we can explain the reading in α as the result of a *saut du même au meme*, and its content does not fit within the context as well as the reading in ω^AL. But since we do not know if the β-reading prior to the revision coincided with ω^AL or α, we cannot determine the reading of ω^αβ.

What about the first sentence? According to the β-version, the first sentence reads (a20–21): ὅλως δ' ἀναιροῦσιν οἱ τοῦτο λέγοντες οὐσίαν καὶ <u>τὸ τί ἦν εἶναι μὴ εἶναι</u>. As argued above, the meaning of this reading is unsatisfactory. But, once again, an examination of Alexander's commentary may allow us to explain how this version came about. In 285.34 (see above), Alexander quotes in isolation the phrase τί ἦν εἶναι μὴ εἶναι, taking it from line a23 (!) in *his* copy of the *Metaphysics*.[140] It is exactly this phrase that we find as the ending of the first sentence, line a21 (!), in the β-text. Thus one might wonder whether Alexander's citation of line a23 was misunderstood as a reference to line a21 of Aristotle's text. This misunderstanding could have occasioned the wrong reading in the β-version. Someone could have expanded the original reading in β, line a21, τὸ τί ἦν εἶναι (preserved in α and ω^{AL}), with the words μὴ εἶναι, mechanically following Alexander's allegedly alternative reading. This could explain why at the end of the first sentence (a21) in the β-version, we find exactly the ending of the second sentence as given in ω^{AL} (a23). Such confusion was made possible because ω^{AL} (in contrast to the α-version) contains the words τί ἦν εἶναι in line a23. Alexander's isolated quotation of these words + μὴ εἶναι almost invites this error of association. What lends plausibility to this explanation is the fact that we have already found a *clear* indication that Alexander's commentary influenced the wording in the β-text in respect to the *second* sentence (a23). It happened once; it easily could have happened twice.

Still, one could raise an objection to the claim that the first sentence in β derives from a misunderstood quotation of Alexander: it seems possible that the wording of the second sentence (a22–23) in the β-version *prior to* its corruption (possibly through Alexander's commentary) was identical to ω^{AL} (... καὶ τὸ ὅπερ ἀνθρώπῳ εἶναι ἢ ζῴῳ εἶναι τί ἦν εἶναι μὴ εἶναι). The ω^{αβ}-reading would then have been just this, and the error that occurred in the α-text (the dropping of τί ἦν εἶναι due to *saut du même au même*) would have happened *after* the split. If that was the case, then β's corruption in the *first* sentence might have been caused not by a (misguided) assimilation of Alexander's commentary, but by a mechanical error in which the ending of the second sentence (... τί ἦν εἶναι μὴ εἶναι, a23, hypothetically written in ω^{αβ}) was written incorrectly as the ending of the first sentence (... τὸ τί ἦν εἶναι μὴ εἶναι, a21, according to β). In this case, the corruption of the first sentence in β would not be due to Alexander's commentary but to the context of the *Metaphysics* itself. But although the explanation of the corruption of the first sentence in β that this objection presupposes is attractive in so far as it does not depend on an external source (such as Alexander's commentary), it is unattractive in that it entails speculative assumptions about the date of the corruption of the α-reading (the dropping of τί ἦν εἶναι in a23) and the original reading in ω^{αβ} (line

---

[140]That this is his point of reference is made clear in the second instance in which Alexander quotes the phrase τί ἦν εἶναι μὴ εἶναι (285.34–35). There, Alexander's remarks indicate the original context of the phrase (285.35–36).

a23). In the end, it seems reasonable to stick with the first explanation of the corruption in the first sentence and ascribe it, too, to Alexander's influence on this passage.

We can now state the following conclusion about the genesis of the three transmitted versions of this *Metaphysics* passage: in ω^AL we find the original text. The α-version agrees in the first sentence with the correct reading in ω^AL; in the second sentence the words τί ἦν εἶναι have dropped out. This loss might have already happened in the ω^αβ-version. The β-version shows traces of an adoption of words from Alexander's commentary. To begin with, we find at the end of the second sentence the exact words of Alexander's own suggested reformulation of the *Metaphysics* text. Further, at the end of the first sentence, we can spot an error that is explicable either as a misplaced quotation of the *Metaphysics* text stemming from Alexander's commentary or as a misplaced phrase stemming from what could be supposed to have been the wording in the adjacent lines of the *Metaphysics*. The first instance of contamination in β by Alexander does not tamper with the meaning of the *Metaphysics* text, but merely follows Alexander's clarification of it. This alteration could be attributed to a reviser who consciously adopted Alexander's correction. In the second instance of contamination in β, the mechanical adjustment causes an error in the *Metaphysics* text. Although this instance allows for a less certain reconstruction of its origin, we can say that it derives from an unintended adoption either from the commentary or the *Metaphysics* text. However, since we can assume that someone reworked the β-version in this very passage[141] based on Alexander's comments, it is quite reasonable to further assume that, in the process of copying a phrase out of the commentary into the *Metaphysics* text, a mistake occurred.[142]

5.2.3 Alex. *In Metaph.* 262.37–263.5 on Arist. *Metaph.* Γ 2, 1005a2–8

There is another passage in book Γ (Γ 2, 1005a2–13) where the difference between the β-reading and the α-reading suggests that the β-text has been influenced by Alexander's comments. In Γ 2, Aristotle defines "that which is" as something that can be said in many ways (τὸ δὲ ὂν λέγεται μὲν πολλαχῶς..., 1003a33), but with reference to one nature (... ἀλλὰ πρὸς ἓν καὶ μίαν τινὰ φύσιν, 1003a33–34) or one principle (πρὸς μίαν ἀρχήν, 1003b6). Accordingly, "that which is" is homonymous not in a general, but in a specific sense: all things that are bear some πρὸς ἕν-relation to οὐσία (καὶ οὐχ ὁμωνύμως ἀλλ' ὥσπερ καὶ τὸ ὑγιεινὸν ἅπαν πρὸς ὑγίειαν, 1003a34–35).

Aristotle's standard example of such a πρὸς ἕν-relation is health: we use the term "healthy" to refer to many different things that are related to health.

---

[141] For further traces of the β-revision making use of Alexander's commentary in Γ 4 see 5.2.4.
[142] Cf. the case in Primavesi 2012b: 424–28.

"Healthy" can be said of a certain way of living, a certain human being or a certain complexion. The same is true of being and the things that are said to exist (οὕτω δὲ καὶ τὸ ὂν λέγεται πολλαχῶς μὲν ἀλλ' ἅπαν πρὸς μίαν ἀρχήν, 1003b5–6). Substances, as well as affections of substances and even things that are not, can all be said to stand in relation to οὐσία (1003b5–10). In Γ 2, Aristotle uses this understanding of "that which is" to solve the first aporia of book B.

Before embarking on an analysis of the passage at the end of Γ 2, the passage on which I will focus in the remainder of this chapter, it would be helpful to have a closer look at the Aristotelian πρὸς ἕν schema and its implications, specifically with respect to the two concepts of homonymy and synonymy. At the beginning of the *Categories* (*Cat.* 1, 1a1–4), Aristotle seems to understand "homonymous" as applying to things that have the same name (ὄνομα κοινόν), but different definitions (λόγος τῆς οὐσίας ἕτερος).[143] This is to be distinguished from "synonymous," he says, which applies to things whose name and definition are the same. For instance, a human being and an ox are synonyms in so far as they are both animals (*Cat.* 1, 1a6–10).[144] Since synonymous things constitute a genus or species, they are predicated with reference to *one* kind of thing: in Aristotle's diction, they are said καθ' ἕν (according to *one*). The ἕν according to which synonymous things are said is, then, a γένος or an εἶδος.[145] Accordingly, Alexander writes, regarding the Aristotelian term καθ' ἕν in his commentary on Γ 2, 1003b12–16 (243.31–32): καθ' ἕν μὲν λεγόμενα λέγει τὰ συνώνυμα καὶ ὑφ' ἕν τι κοινὸν τεταγμένα γένος ("By 'things said in accordance with one thing' he means the synonyms, things ranged under some one common genus").

By contrast, the definitions of those things that are related πρὸς ἕν are not the same. A healthy diet and a healthy human being do not belong to one genus. Things that are said πρὸς ἕν (*with reference* to one) are not said καθ' ἕν (*according to* one genus).[146] Rather, the πρὸς ἕν-relation determines a kind of homonymy.[147]

---

[143] According to Shields (1999: 11), homonyms can generally be grouped in two classes: in the case of "discrete homonymy" the definitions do not correspond at all, while in the case of "comprehensive homonymy" the definitions do not overlap completely.

[144] The special group of paronymous things (*Cat.* 1, 1a12–15) can be left aside here. See, however, note 147.

[145] This does not exclude the possibility that Aristotle may speak of something "in reference to one" or "according to one" in the abstract, that is without presupposing a specific γένος or εἶδος.

[146] Cf. *Metaph.* Z 4, 1030b2–3: οὐδὲ γὰρ ἰατρικὸν σῶμα καὶ ἔργον καὶ σκεῦος λέγεται οὔτε ὁμωνύμως οὔτε καθ' ἕν ἀλλὰ πρὸς ἕν.

[147] Ross 1924: 256 equates the πρὸς ἕν-relation with the third class mentioned in the *Categories*: παρώνυμα (*Cat.* 1, 1a12–15). Bonitz 1870: 514b is more cautious (*incerta*) about equating πρὸς ἕν and παρώνυμα. Shields 1999: 103–27 offers a detailed analysis of this type of homonymy. See also Rapp 1992: 534–38 and Lewis 2004. Shields 1999 calls this type of homonymy "core-dependent homonymy." He says (106): "CDH2: x and y are homonymously in a core-dependent way F iff: (i) they have their name in common, (ii) their definitions do not completely overlap, and (iii) there is a single source to which they are related." Cf. also 124–25.

The definitions of the things are not the same, but the things receive their names in relation to one common reference point.

Aristotle uses the concept of πρὸς ἕν to answer the question whether "that which is" belongs to *one* science. As all things that are called "healthy" belong to one science, all things related πρὸς ἕν (1003b11–12) belong to one science. This is by reason of the fact that "even these [things related πρὸς ἕν] in a sense are said καθ' ἕν" (καὶ γὰρ ταῦτα τρόπον τινὰ λέγεται καθ' ἕν, 1003b14–15). Of course, although we have seen that things that are καθ' ἕν differ from things that are πρὸς ἕν, both concepts equally bring those things related to one another through those concepts under *one* science (1004a24–25).[148]

In the final part of Γ 2, Aristotle adduces a further argument[149] to show that "that which is" belongs, insofar as it *is*, to one science. This argument is based in part on arguments that Aristotle developed in the course of chapter Γ 2. There is general consensus that "that which is" and substance consist of contraries.[150] The principles of contraries are unity and plurality. Since contraries belong to *one* science, their principles belong to *one* science as well.[151] Immediately after this argument, Aristotle employs the καθ' ἕν- / πρὸς ἕν-distinction in an additional argument: even if what is said to be one is not said καθ' ἕν, it is said πρὸς ἕν, and therefore all that is said to be one belongs to one science anyway.

Aristotle, *Metaphysics* Γ 2, 1005a2–8

φανερὸν οὖν καὶ ἐκ [3] τούτων ὅτι μιᾶς ἐπιστήμης τὸ ὂν ᾗ ὂν θεωρῆσαι. πάντα γὰρ [4] ἢ ἐναντία ἢ ἐξ ἐναντίων, ἀρχαὶ δὲ τῶν ἐναντίων τὸ ἓν [5] καὶ πλῆθος. ταῦτα δὲ μιᾶς ἐπιστήμης, εἴτε **καθ'** ἓν λέγε-[6]ται εἴτε μή, ὥσπερ ἴσως ἔχει τἀληθές. ὅμως εἰ [7] καὶ πολλαχῶς λέγεται τὸ ἕν, πρὸς τὸ πρῶτον τἆλλα [8] λεχθήσεται καὶ τὰ ἐναντία ὁμοίως.

It is obvious then from these considerations too that it belongs to one science to examine being *qua* being. For all things are either contraries or composed of contraries, and unity and plurality are the starting-points of all contraries. And these belong to one science, regardless of whether they are or are not said *according to one common notion*, as is probably true. Yet even if 'one' has several meanings, the other meanings will be related to the primary[152] meaning—and similarly in the case of the contraries.

5 δὲ **α** edd. : δὲ καὶ **β** || καθ' ἓν **α** ω^AL (Al.^P 263.1-2) Ascl.^P 248,2 edd. : καθ' ἕνα **β** ex Al. 263.2

---

[148]*Metaph*. 1004a24–25: οὐ γὰρ εἰ πολλαχῶς, ἑτέρας, ἀλλ' εἰ μήτε καθ' ἓν μήτε πρὸς ἓν οἱ λόγοι ἀναφέρονται.

[149]Cf. Kirwan 1971: 85: "apparently *ad hominem*."

[150]*Metaph*. 1004b29–31: τὰ δ' ὄντα καὶ τὴν οὐσίαν ὁμολογοῦσιν ἐξ ἐναντίων σχεδὸν ἅπαντες συγκεῖσθαι· πάντες γοῦν τὰς ἀρχὰς ἐναντίας λέγουσιν. / "And nearly all thinkers agree that being and substance are composed of contraries; at least, they all name contraries as their first principles."

[151]Cf. *Metaph*. 1004a9–26; 1004a31–1004b4.

[152]That Aristotle does not use the usual term πρὸς ἕν here is most likely due to the fact that in this

τούτεστιν ἕνα λόγον ἔχοντα ‖ 6 ἔχει **α** Bonitz Cassin/Narcy Hecquet-Devienne : καὶ ἔχει **β** : ἔχει καὶ Bekker Christ Ross Jaeger ‖ 7 λέγεται τὸ ἕν **α** Al.ᶜ 263.9 edd. : τὸ ἓν λέγεται **β**

My efforts here focus on the assessment of the phrase καθ᾿ ἕν (1005a5) in the **α**-text in comparison to καθ᾿ ἕνα in the **β**-text. As I said above, the ἕν according to which (καθ᾿) things are synonymous is the neuter γένος or εἶδος (cf. Alex. 243.31–32).[153] By contrast, the term πρὸς ἕν seems to be used more flexibly. We should recall that at the beginning of Γ 2 Aristotle spoke of the πρὸς ἕν-relation as referring to one nature (πρὸς μίαν φύσιν) or one principle (πρὸς μίαν ἀρχήν). It is true that in the present chapter Aristotle is speaking of καθ᾿ ἕν, that is, of a ἕν that refers to a neuter only. Yet there is a passage in the *Eudemian Ethics* in which Aristotle labels synonymous things with the term καθ᾿ ἕνα λόγον. As the context in *EE* shows, he uses the expression καθ᾿ ἕνα λόγον synonymously with the expression καθ᾿ ἕν εἶδος.[154] Therefore, it would be wrong to decide against καθ᾿ ἕνα here in the *Metaphysics* solely because ἕνα is not neuter.[155] The actual problem with the καθ᾿ ἕνα reading is that the expressions καθ᾿ ἕν and πρὸς ἕν are *abbreviations* of καθ᾿ ἕν γένος λέγεται and πρὸς ἕν γένος λέγεται, and that whenever the concept of καθ᾿ ἕν and πρὸς ἕν is expressed by the abbreviated formula they appear exclusively in the neuter form. Thus, the phrase καθ᾿ ἕνα *without* λόγος or[156] τρόπος does not occur in any other passage in the whole Aristotelian corpus in the sense of "(said) according to one thing."[157] For this reason, the **β**-text, which in our passage reads not καθ᾿ ἕν but καθ᾿ ἕνα without a λόγον, must be corrupt.

Is what we find in the **β**-text the result of a slip of the pen or does it derive from a scholarly decision to emphazise that the ἕν, according to which unity and plurality are said, is the λόγος?[158] With this question in mind, let us turn to Alexander's commentary:

---

case we are dealing with the πρὸς ἕν-relation of the One (ἕν). The ἕν, in reference to which all things that are one are said is just the first ἕν, which Aristotle here calls τὸ πρῶτον.

[153] Aristotle only rarely states these nouns explicitly: e.g., *Top*. Z 10, 148a33 ὡς συνωνύμου καὶ καθ᾿ ἕν εἶδος; cf. also *Top*. A 7, 103a17 τοῖς καθ᾿ ἕν εἶδος ὁπωσοῦν (*sc.* ταὐτὸν) λεγομένοις.

[154] *EE* 1236b21–26: τὸ μὲν οὖν ἐκείνως μόνον λέγειν τὸν φίλον βιάζεσθαι τὰ φαινόμενα ἐστί, καὶ παράδοξα λέγειν ἀναγκαῖον· καθ᾽ ἕνα δὲ λόγον πάσας ἀδύνατον. λείπεται τοίνυν οὕτως, ὅτι ἔστι μὲν ὡς μόνη <ἡ> πρώτη φιλία, ἔστι δὲ ὡς πᾶσαι, οὔτε ὡς ὁμώνυμοι καὶ ὡς ἔτυχον ἔχουσαι πρὸς ἑαυτάς, οὔτε καθ᾽ ἓν εἶδος, ἀλλὰ μᾶλλον πρὸς ἕν. / "To speak, then, of friendship in the primary sense only is to do violence to the phenomena, and makes one assert paradoxes; but it is impossible for all friendships to come under one definition. The only alternative left is that in a sense there is only one friendship, the primary; but in a sense all kinds are friendship, not as possessing a common name accidentally without being specifically related to one another, nor yet as falling under one species, but rather as in relation to one and the same thing" (transl. by Solomon).

[155] In fact, it is the λόγος in the sense of definition, by way of which things are synonymous (*Cat*. 1, 1a7).

[156] Cf. *Metaph*. K 3, 1060b33.

[157] In those passages where καθ᾿ ἕνα occurs without specification it means "one at a time, individually" (e.g. *Mete*. A 8, 346a6–7) or "firstly" (e.g. *Top*. E 2, 130a35–36).

[158] That καθ᾿ ἕνα in the **β**-text means "individually" (LSJ s.v. κατά B.II.3 *of Numbers*) can safely be ruled out since the context of the whole chapter makes it clear that the expression καθ᾿ ἕν(α) refers to synonyms.

Alexander, *In Metaph.* 262.37–263.5 Hayduck

μιᾶς δέ φησιν [1] ἐπιστήμης εἶναι τὴν περὶ ἑνὸς καὶ πλήθους θεωρίαν, ἄντε ᾖ ταῦτα καθ᾽ [2] ἓν λεγόμενα ἑκάτερον αὐτῶν, τουτέστιν ἕνα λόγον ἔχοντα καὶ μίαν φύσιν [3] καὶ ὡς γένη τῶν ἄλλων τῶν ὑπ᾽ αὐτὰ κατηγορούμενα, ἄντε καὶ μὴ οὕτως [4] ἔχῃ ἀλλ᾽ ᾖ τῶν πολλαχῶς λεγομένων, ὥσπερ προείρηταί καὶ ἀληθές [5] ἐστιν.

He says that the consideration of unity and plurality belongs to one science, regardless of whether each of these is said "according to one common notion," i.e., they have *one* definition and *one* nature, and they are predicated, as genera, of the other things, the things that fall under them, or whether, on the contrary, they are among things said in many ways, as has been said earlier and is true.

262.37–263.1 φησιν ἐπιστήμης **A O** : ἐπιστήμης φησὶν **P**ᵇ ‖ 1 ταῦτα **P**ᵇ **F S** : ταὐτά **A O L** ‖ 2 λεγόμενα **P**ᵇ : λεγόμενον **A O S** ‖ 4 καὶ **A O P**ᵇ **S** : τε καὶ **LF** Ascl.

Alexander's paraphrase follows Aristotle closely: the consideration of unity and plurality falls under *one* science. This holds regardless of whether or not (in Aristotle εἴτε ... εἴτε μή, in Alexander ἄντε ᾖ ... ἄντε καὶ μή) everything that is one is synonymous and said according to one (καθ᾽ ἕν). Because Alexander sticks closely to Aristotle's formulation (263.1–2), we can assume that Alexander's text contained the formula καθ᾽ ἕν, which is the **α**-reading, and not the formula καθ᾽ ἕνα, which is the **β**-reading.[159] This assumption based on the paraphrase is confirmed indirectly by an explanatory addition that Alexander makes to the Aristotelian text in his commentary. Alexander explicates the meaning of Aristotle's formula καθ᾽ ἕν as follows (263.2–3): τουτέστιν ἕνα λόγον ἔχοντα καὶ μίαν φύσιν καὶ ὡς γένη τῶν ἄλλων τῶν ὑπ᾽ αὐτὰ κατηγορούμενα. With this, Alexander specifies what is meant by ἕν: a common definition (ἕνα λόγον), one nature (μίαν φύσιν) or a common genus (γένη).

In light of Alexander's explanatory reformulation it seems plausible to hypothesize that Alexander's explanation of the word ἕν and the corrupt reading in β are related.[160] It looks as though the neuter ἕν in the β-text was changed to ἕνα in order to adjust it to the masculine λόγος. But why would someone change ἕν to ἕνα without adding λόγον? Was the editorial intervention to be kept as slight as possible? Or was it taken as self-evident that ἕνα referred to λόγος? Assuming that someone intended to note Alexander's explanation down in the margins of the

---

[159] The same can be said about the text that Asclepius used when writing his commentary. See his paraphrase in *In Metaph.* 248.2. Concerning the *Vorlage* of the Arabic tradition, it seems more difficult to make a secure judgment. Scotus writes: *Et unius scientie est consideratio de istis, si dicuntur de <u>uno</u> et si non dicuntur.*

[160] As an explanation of the β-reading which does not suppose influence by Alexander's commentary one could entertain the following hypothesis: in the original sequence ΕΝΛΕΓΕΤΑΙ the Λ was corrupted into Α due to the similarity between the two letters in majuscule script. In order to restore the reading of ΛΕΓΕΤΑΙ an additional Λ was added later on (resulting in the sequence ΕΝΑΛΕΓΕΤΑΙ). (This hypothesis was brought to my attention by one of the anonymous referees.)

β-text and added (τουτέστιν) ἕνα λόγον as a gloss, it is possible that ἕνα was taken as a variant to ἕν later on in the course of the transmission and so adopted into the text. Another possibility is that someone indeed added ἕνα λόγον to the β-text, but that λόγον was omitted later on because of its similarity to λέγεται. Indeed, since the specifics of the adoption prove to remain obscure, the conclusion that ἕνα in the β-text goes back to contamination by Alexander's commentary should be drawn less confidently than in the two previously discussed cases.

5.2.4 Alex. In Metaph. 144.15–145.8 on Arist. Metaph. α 1, 993b19–23

The first chapter of book α ἔλαττον begins with general considerations regarding the "investigation of the truth" (ἡ περὶ τῆς ἀληθείας θεωρία, 993a2930).[161] Aristotle also touches upon the question regarding the proper name for this philosophical undertaking.

Aristotle, *Metaphysics* α 1, 993b19–23[162]

ὀρθῶς δὲ καὶ τὸ κα-[20]λεῖσθαι **τὴν φιλοσοφίαν ἐπιστήμην τῆς ἀληθείας**. θεωρητικῆς [21] μὲν γὰρ τέλος ἀλήθεια πρακτικῆς δ᾽ ἔργον· καὶ γὰρ ἂν [22] τὸ πῶς ἔχει σκοπῶσιν, οὐκ ἀίδιον ἀλλὰ πρός τι καὶ νῦν [23] θεωροῦσιν οἱ πρακτικοί.

It is right also that **philosophy** should be called **knowledge of the truth**. For the end of theoretical knowledge is truth, while that of practical knowledge is action (for even if they consider how things are, practical men do not study what is eternal but what is relative and in the present).

---

19 δὲ β Al.¹ 144.15 Ascl.¹ 118.17 edd. : δὴ α | καὶ β Al.¹ 144.15 : ἔχει καὶ α Ascl.¹ 118.17 edd. || 19–20 καλεῖσθαι β Al.¹ 144.15 edd. : καλέσαι α Ascl.¹ 118.17 || 20 τὴν φιλοσοφίαν ἐπιστήμην τῆς ἀληθείας α ω^AL (Al.¹ 144.15–16 Al.ᵖ 144.17–19) edd. : τὴν κατὰ φιλοσοφίαν ἐπιστήμην τῆς ἀληθείας θεωρητικήν β Ar^u fort. ex Al.ᵖ 144.17–18 : τὴν φιλοσοφίαν ἐπιστήμην τῆς ἀληθείας θεωρητικήν <V^d> J^c

My analysis will focus on the formulation τὴν φιλοσοφίαν ἐπιστήμην τῆς ἀληθείας in line b20. The β-version of this line differs in the following way from the α-version printed above: τὴν κατὰ φιλοσοφίαν ἐπιστήμην τῆς ἀληθείας θεωρητικήν.[163] Before looking at how the β-reading is to be explained, some other textual differences between α and β in lines b19–20 call for attention. The first difference concerns the beginning of the sentence. The α-text reads ὀρθῶς δὴ ἔχει καὶ τὸ καλέσαι... ("It is then right to call..."), while the β-text reads ὀρθῶς δὲ καὶ τὸ καλεῖσθαι... ("It is right that ... should be called"). Which of the two versions preserves the original wording? The editors Bekker, Bonitz, Christ, Ross, and Jaeger

---

[161]Cf. Szlezák 1983: 233–36.

[162]The information in the apparatus covers only lines 19–20, i.e. the lines that concern me at present. On the reconstruction of lines 993b20–22 see 5.4.2.

[163]The *Vorlage* of the Arabic tradition apparently also read the β-text. Scotus translates: *Et rectum est vocare scientiam veritatis philosophie philosophiam speculativam.*

read a blending of α and β, taking δὲ ... καλεῖσθαι from β and ἔχει from α: ὀρθῶς δὲ ἔχει καὶ τὸ καλεῖσθαι. There are parallel passages in Aristotle's writings for the α-formulation ὀρθῶς ... ἔχει καὶ + infinitive[164] as well as for the β-formulation ὀρθῶς ... καὶ + infinitive.[165] I follow the β-reading in lines b19–20 because, as we will see further below, Alexander confirms the middle-passive infinitive present of the β-text[166] (καλεῖσθαι) in his lemma[167] as well as in his paraphrase (καλεῖται, 144.18; καλεῖσθαι, 145.7).[168]

In order to decide which reading in line b20 should be preferred, it is useful first to get acquainted with Aristotle's classification of the sciences. This will help to determine the meaning of the terms ἐπιστήμη, θεωρητική (sc. ἐπιστήμη), πρακτική (sc. ἐπιστήμη) and φιλοσοφία, and how they relate to each other. In doing this, I orient myself by the classification Aristotle offers in the first chapter of book E (cf. 1025b18–28 and 1026a10–23).[169]

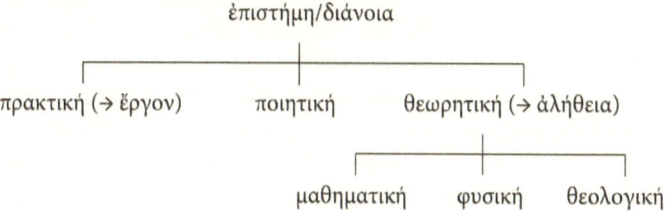

Placing the term philosophy (φιλοσοφία) in this scheme is not unproblematic. On the one hand, the term φιλοσοφία can have a broad meaning in Aristotle, being equivalent to the term ἐπιστήμη.[170] On the other, it can specifically denote the *theoretical* branch of the sciences.[171]

Such an analysis of the term φιλοσοφία is compatible with the passage in *Metaphysics* E 1 (1026a18–19): ὥστε τρεῖς ἂν εἶεν φιλοσοφίαι θεωρητικαί, μαθηματική, φυσική, θεολογική. Here, the noun φιλοσοφία is combined with the adjective θεωρητική, and so stands in place of the term ἐπιστήμη. Yet this combination of φιλοσοφία + θεωρητική is rare and appears in only one other passage in the Aris-

---

[164]Cf. *GA* E 4, 784b32–33: ὀρθῶς δ' ἔχει καὶ λέγειν and *Ph.* Θ 1, 251b33–34.
[165]Cf. *PA* Γ 2, 663a34: ὀρθῶς δὲ καὶ τὸ ἐπὶ τῆς κεφαλῆς ποιῆσαι.
[166]The α-text reads the active infinitive aorist καλέσαι.
[167]144.15–16: ὀρθῶς δὲ καὶ τὸ καλεῖσθαι....
[168]Or should we suspect that β took over the infinitive form from Alexander?
[169]The chart is based on Ross 1924: 353, but is slightly expanded for the present purpose. See also *Metaph.* K 7, 1064b1–3: δῆλον τοίνυν ὅτι τρία γένη τῶν θεωρητικῶν ἐπιστημῶν ἔστι, φυσική, μαθηματική, θεολογική and *Top.* Z 6, 145a15–16: καθάπερ καὶ τῆς ἐπιστήμης. θεωρητική γὰρ καὶ πρακτική καὶ ποιητική λέγεται.
[170]Cf. *Metaph.* Γ 2, 1004a2–3: καὶ τοσαῦτα μέρη φιλοσοφίας ἐστὶν ὅσαι περ αἱ οὐσίαι. Λ 8, 1074b11: ἑκάστης καὶ τέχνης καὶ φιλοσοφίας.
[171]Bonitz 1870 s.v. φιλοσοφία.

totelian corpus (*EE* A 1, 1214a13).¹⁷² The scarcity of the expression is due, perhaps, to the fact that the combination φιλοσοφία + θεωρητική comes close to being tautological, as the term φιλοσοφία mainly functions as an equivalent of the expression ἐπιστήμη θεωρητική¹⁷³ (or θεωρία).¹⁷⁴ Philosophy is the *theoretical* science that branches out into the three areas of mathematics, physics and theology, also called first philosophy or metaphysics.

Let us return to our passage in α ἔλαττον 1. In line b20, the α-reading τὴν φιλοσοφίαν ἐπιστήμην τῆς ἀληθείας ("[it is right to call] philosophy knowledge of the truth") stands in contrast to the β-reading τὴν κατὰ φιλοσοφίαν ἐπιστήμην τῆς ἀληθείας θεωρητικήν ("[it is right that] the philosophical knowledge of the truth [is called] theoretical"). The α-reading is favorable, because this reading raises no suspicions regarding grammar and content. By calling philosophy "knowledge of the truth" Aristotle links the present passage to the very beginning of book α, where he speaks about "the investigation of the truth" (ἡ περὶ τῆς ἀληθείας θεωρία, 993a29–30). Additionally, the α-reading offers an ideal transition to the subsequent sentence, in which theoretical knowledge is introduced as geared towards the truth and is placed in contrast to practical knowledge, which aims at action: θεωρητικῆς μὲν γὰρ τέλος ἀλήθεια πρακτικῆς δ' ἔργον (993b20–21).

By contrast, the β-reading reveals the following two problems: first, it contains an awkward heap of nouns for expressing the subject (τὴν κατὰ φιλοσοφίαν ἐπιστήμην τῆς ἀληθείας / "the philosophical knowledge of the truth"). Primarily responsible for the awkwardness is the addition of θεωρητικήν as a new predicative adjective.¹⁷⁵ The second and more severe problem is that the statement made by the sentence is unsatisfactory. This holds regardless of how one understands the word philosophy here. For, if the word philosophy is taken broadly to mean "knowledge" or "science" (ἐπιστήμη) the β-sentence does not make sense at all. For the expression τὴν κατὰ φιλοσοφίαν ἐπιστήμην would then mean something like "the scientific science." If, on the other hand, φιλοσοφία is taken more narrowly to mean "theoretical knowledge" (as in the α-version) then the following problem arises: when mention is made of "the philosophical knowledge of the truth" (τὴν κατὰ φιλοσοφίαν ἐπιστήμην) that happens to be theoretical, one is confronted with the perplexing questions as to what other *non*-philosophical types of knowledge exist that concentrate on the truth or what philosophical knowledge is *not* concentrated on the truth. These questions only become more perplexing when we find the subsequent sentence state that theoretical knowledge¹⁷⁶ is intrinsically

---

¹⁷²See Mueller-Goldingen 1991: 2 n. 4.

¹⁷³There is no passage in the corpus where Aristotle speaks, for instance, of a φιλοσοφία πρακτική.

¹⁷⁴See Bonitz 1870: s.v. φιλοσοφία; p. 821a8–32, "philosophia."

¹⁷⁵Placing a κατά + noun combination instead of an adverb between the article and the noun (see the β-reading τὴν κατὰ φιλοσοφίαν ἐπιστήμην) is not alarming in itself (cf. LSJ s.v. κατά B.VIII). There is one parallel expression in *Top.* A 2, 101a34: πρὸς δὲ τὰς κατὰ φιλοσοφίαν ἐπιστήμας.

¹⁷⁶Grammatically speaking, the adjectives θεωρητικῆς and πρακτικῆς in 993b20–21 could refer to

connected to the truth, whereas practical knowledge is linked to action. Since, then, theoretical knowledge, which here means philosophy, is equivalent to the investigation of the truth, it is not possible that there be philosophical knowledge that is not concentrated on the truth: the β-formulation in b20 ("the philosophical knowledge of the truth") reveals itself to be redundant.[177]

What is responsible for the changes in the β-version? They have very likely been occasioned by the addition of the adjective θεωρητική. Did this adjective emerge from a dittography? This seems possible indeed, given that the next sentence starts with the words θεωρητικῆς μὲν ... (993b20–21). Or are we dealing with a corruption occasioned by the intrusion of a marginal gloss, which contained θεωρητικήν as an explication of the words ἐπιστήμην τῆς ἀληθείας?

Alexander's commentary may help us to find an answer.

Alexander, *In Metaph.* 144.15–145.8 Hayduck

ὀρθῶς δὲ καὶ τὸ καλεῖσθαι τὴν φιλοσοφίαν ἐπιστήμην [16] τῆς ἀληθείας. [17] χρησάμενος τῷ τῆς ἀληθείας ὀνόματι ἐπὶ τῆς θεωρητικῆς φιλοσο-[18]φίας, νῦν ὅτι εὐλόγως οὕτως καλεῖται συνίστησι· φιλοσοφίαν γὰρ ἰδίως τὴν [19] θεωρητικὴν λέγει, ὡς δι' ὧν ἐπιφέρει δηλοῖ, λέγων θεωρητικῆς μὲν [145.1] γὰρ τέλος ἀλήθεια, καὶ ταύτης ἔτι μᾶλλον τὴν περὶ τῶν πρώτων ἀρχῶν [2] τε καὶ αἰτίων τῶν παντάπασιν αἰσθήσεως κεχωρισμένων καὶ τῇ αὐτῶν [3] φύσει ὄντων, ἣν καὶ σοφίαν καλεῖ. συνίστησι δὲ τὸ προειρημένον οὕτως. [4] ἐπεὶ τέλος τῆς θεωρητικῆς φιλοσοφίας ἐστὶν ἡ ἀλήθεια, ἀπὸ δὲ τοῦ τέ-[5]λους καὶ τοῦ σκοποῦ ἑκάστη μέθοδος χαρακτηρίζεταί τε καὶ τὸ εἶναι ἔχει, [6] εἰκότως παρὰ τῆς ἀληθείας αὕτη καλεῖται· τοῦτο γὰρ αὐτῆς τέλος. καὶ [7] τῇ τῶν πρακτικῶν ἐπιστημῶν δὲ παραθέσει ἔδειξε τὸ εὐλόγως καλεῖσθαι [8] παρὰ τῆς ἀληθείας τὴν θεωρητικήν·

It is right also that philosophy should be called knowledge of the truth. Aristotle has previously applied the term 'truth' to theoretical philosophy,[178] and now confirms that this designation is correct; for he says that theoretical knowledge is properly speaking 'philosophy,' as he makes clear by his next statement, saying: "for the end of theoretical knowledge is truth." And this is especially the case with that knowledge that has for its object the first principles and causes, which [exist] in complete separation from sense perception and in virtue of their own nature, the knowledge that Aristotle also calls 'wisdom.' He confirms his statement in this

---

either ἐπιστήμη or φιλοσοφία, both of which occur in the preceding sentence. Yet taking into consideration the cohesion of the two sentences, it becomes clear that Aristotle is not speaking about theoretical and practical *philosophy* (Bonitz's translation), but about theoretical and practical *knowledge* (Ross's translation). It does not suffice to say that philosophy (in general) is called "knowledge of the truth" *because* theoretical philosophy (which then would be a sort of subgroup of general philosophy) is geared towards the truth. The two terms philosophy and theoretical knowledge are equivalent from Aristotle's point of view (see above). Furthermore, as pointed out already, Aristotle never speaks about a "practical philosophy" (cf. Mueller-Goldingen 1991: 2 n. 4).

[177] Understandably, no editor has opted for the β-reading.
[178] *Metaph.* α 1, 993a29–30.

way. Since the end of theoretical philosophy is truth, and every scientific discipline derives its specific character and existence from its end and goal, it is fitting that theoretical philosophy should receive its name from truth, for truth is its end. He also shows that it is fitting that theoretical philosophy should receive its name from truth by his reference to the practical sciences.

---

15 δὲ **A O** *Metaph.*^{β} : δὲ ἔχει **P^b S** *Metaph.*^{a} ‖ 15–16 ἐπιστήμην τῆς ἀληθείας **A O** *Metaph.*^{a} : ἐπιστήμην τῆς ἀληθείας θεωρητικήν **P^b S**(*scientiam inquam philosophie, veritatis appellare contemplationem*) *Metaph.*^{β} | 145.1 τὴν Hayduck ex Ascl. : ἡ **A O** : ἤ **P^b** | τῶν **A O** : om. **P^b** ‖ 6 παρὰ **A P^b S** : περὶ **O** ‖ 8 παρὰ **A P^b S** : περὶ **O**

The text quoted in the lemma (144.15–16) reveals that Alexander read the correct reading in line b20, that is, the reading preserved by the **α**-tradition: τὴν φιλοσοφίαν ἐπιστήμην τῆς ἀληθείας.[179] Alexander's subsequent paraphrase and his comments on Aristotle's text confirm this as the reading of his text.

In his comments Alexander points out that Aristotle has previously used the term "truth" for "theoretical philosophy." Alexander thereby links this passage to the beginning of α 1 (993a29–30), where Aristotle describes his investigation into metaphysics as ἡ περὶ τῆς ἀληθείας θεωρία. It is worth noting that Alexander renders Aristotle's θεωρία (993a29) with the expression θεωρητικὴ φιλοσοφία (144.17–18). Already in his comments on the beginning of α 1 (*In Metaph.* 138.28–29) Alexander used the term θεωρητικὴ φιλοσοφία.[180] On several other occasions in his commentary he uses θεωρητικὴ φιλοσοφία to refer to what Aristotle himself calls θεωρία.[181] Thus, the expression θεωρητικὴ φιλοσοφία is an established element of Alexander's diction. By contrast, Aristotle himself uses it only twice and does not employ it in the larger context of the present passage.[182]

---

[179] The new collations by Pantelis Golitsis indicate that the lemma in manuscript **P^b** reads the sentence in question (993b15–16) as τὸ καλεῖσθαι τὴν φιλοσοφίαν ἐπιστήμην τῆς ἀληθείας θεωρητικήν and hence shares with the β-version the addition of the adjective θεωρητικήν. (Cf. also the reading in the lemma in Sepúlveda's translation; on divergent readings in Sepúlveda's lemmata see also 2.3). The reading in **P^b** thus does not completely coincide with β, where we find the preposition κατὰ preceding φιλοσοφίαν, but seems to be contaminated by it. I will argue in the following that the commentary by Alexander shows that he himself found the α-reading in his *Metaphysics* text (ω^{AL}), as preserved in the lemma by the commentary manuscripts **A** and **O**.

[180] Also in his proem to the commentary on book α (138.8–9) Alexander refers to the Aristotelian θεωρία as θεωρητικὴ φιλοσοφία.

[181] Alex. *In Metaph.* 139.5, 139.22–140.2, 141.37, 142.8–9, 143.5. Cf. also the later passages in 145.4, 146.6, 147.5, 149.16–18. See further 169.21–26.

[182] In his use of this expression, Alexander might be drawing on the terminology of *Metaph.* E 1, 1026a18–19: ὥστε τρεῖς ἂν εἶεν φιλοσοφίαι θεωρητικαί, μαθηματική, φυσική, θεολογική. Aristotle uses the expression θεωρητικὴ φιλοσοφία only here and in *EE* A 1, 1214a13. Rather than connecting the adjective θεωρητική to the noun φιλοσοφία, Aristotle connects it usually with ἐπιστήμη. See *Metaph.* E 1, 1026b22–23: αἱ μὲν οὖν θεωρητικαὶ τῶν ἄλλων ἐπιστημῶν αἱρετώταται, αὕτη (*sc.* ἡ πρώτη φιλοσοφία) δὲ τῶν θεωρητικῶν. *EE* A 5, 1216b10–11; B 3, 1221b5–6; 1227a9–10; *Top.* Z 6, 145a15–16 and Z 11, 149a9–10. See also *Metaph.* K 7, 1064b1–3 and 1064a16–18 for a comparison to the already quoted passage in *Metaph.* E 1, 1026a18–19. For Aristotle, theoretical knowledge or θεωρία is equivalent to

After his reference to the beginning of α 1 and the expression ἡ περὶ τῆς ἀληθείας θεωρία, Alexander comes back to the passage in question (νῦν, 144.18), more specifically, to the sentence in lines 993b19–20, where Aristotle connects the term "truth" with the term "theoretical philosophy" (144.17–18). In his summary of the sentence, Alexander uses the expression θεωρητικὴ φιλοσοφία. At first sight, one might take this as an indication that Alexander had found the latter term in his *Metaphysics* text (ω^AL) and hence that his *Metaphysics* text is closer to the β-version[183] than to the α-version. Upon closer inspection, however, it becomes quite clear that Alexander's comments square much better with the α-reading than with the β-reading.[184] For what Alexander in fact says is that it is legitimate (εὐλόγως οὕτως καλεῖται, 144.18) to give to philosophy the term "truth" (cf. φιλοσοφίαν γὰρ, 144.18), and this is what is expressed in the α-version.[185] Mere word choice distinguishes Aristotle and Alexander: φιλοσοφία in Aristotle, θεωρητικὴ φιλοσοφία in Alexander.

That the presence of the adjective θεωρητικὴ next to φιλοσοφία in his commentary is due to his own idiom and not to the reading of his *Metaphysics* text is made clear by the subsequent sentence of his commentary (144.18–19). In this sentence Alexander gives the reason (γὰρ) why he spoke of θεωρητικὴ φιλοσοφία. He justifies his words by claiming that Aristotle understands *theoretical knowledge* to be equivalent to philosophy (144.18–19), and he justifies this claim by quoting the *subsequent* sentence of the *Metaphysics* (993b20–1), in which truth is described as the τέλος of θεωρητική (sc. knowledge). This argumentation shows that Alexander inferred the identification of θεωρητική solely from *this* sentence of the *Metaphysics* (993a20–1) and not from the preceding sentence in 993a19–20. Therefore, it is reasonable to conclude that ω^AL's reading of 993a19–20, which is the sentence that concerns us, did not include the adjective θεωρητική.

Alexander's commentary reveals yet more. Based on the evidence in the commentary, it seems quite possible that Alexander's explication of the passage occasioned the addition of θεωρητικήν and the features associated with it to the β-version. When paraphrasing the Aristotelian text in his own words, Alexander uses somewhat idiosyncratic terminology: according to his exposition, Aristotle

---

philosophy, and so there is no need to combine the two terms. Alexander also knows that Aristotle uses the expression θεωρητικὴ φιλοσοφία only rarely and prefers instead σοφία (Alex. *In Metaph.* 146.5–6: ἡ προκειμένη πραγματεία περὶ τῆς ἀληθείας θεωρία… ἡ σοφία καὶ ἡ θεωρητικὴ φιλοσοφία; see also 145.3 and the quotation above).

[183] Yet, Alexander's comments (including the expression θεωρητικὴ φιλοσοφία) do not exactly match with the β-version, where the adjective θεωρητικὴ refers to ἐπιστήμη.

[184] The author of the *recensio altera*, however, most likely found the β-reading in his *Metaphysics* copy. His comments on the passage make it clear that he regarded the addition of θεωρητική to the term φιλοσοφία as superfluous (see *app.* in 144 Hayduck).

[185] Asclepius adopts lines 144.17–145.6 from Alexander almost verbatim (Ascl. *In Metaph.* 118.19–28). He adds a quotation of the relevant sentence from the *Metaphysics* text as it reads in the (correct) α-version (118.20–21).

claims that "*theoretical* philosophy" can rightly be called knowledge of truth. A few lines later, Alexander paraphrases the relevant sentence (ὀρθῶς δὲ καὶ τὸ καλεῖσθαι τὴν φιλοσοφίαν ἐπιστήμην τῆς ἀληθείας, 993b19–20) in the following way: τὸ εὐλόγως καλεῖσθαι παρὰ τῆς ἀληθείας τὴν θεωρητικήν (145.7-8). These words of Alexander may well have motivated the addition of the word θεωρητική into the margin of lines 993b19–20 of the β-text. Such an addition would have indicated how Alexander had explained the term philosophy (φιλοσοφίαν, 993b20). From there, the marginal gloss θεωρητική could have found its way into the body of the β-text, but then in order to render the addition grammatically acceptable, someone might then have added the preposition κατά.

To conclude, Alexander's commentary does not in the present case constitute a model that was directly copied into the β-text. And yet, despite the fact that we cannot rule out the possibility of a dittography that happened independently of any influence by Alexander, it seems quite plausible to assume that his explication of the passage occasioned a gloss in the margins of an ancestor of our β-text. From here it was incorporated into the β-version, and further adjustments to the sentence followed in turn.

5.2.5 Alex. *In Metaph*. 31.27–32.9 on Arist. *Metaph*. A 3, 984b8–13

In the third and fourth chapters[186] of *Metaphysics* A, Aristotle determines which of the four causes (given in his *Physics*) his Presocratic predecessors had already recognized and how they interpreted them. He begins with the material cause, which was the first to be recognized. Some natural philosophers proposed one, others more than one material cause (983b6–984a16).[187] Since a material cause alone cannot account for generation and destruction or change in material things, the search for a further cause was inevitable (984a18–22). Matter itself cannot be the cause of its own change: οὐ γὰρ δὴ τό γε ὑποκείμενον αὐτὸ ποιεῖ μεταβάλλειν ἑαυτό (984a21–22). Wood does not transform itself into a bed nor does bronze into a statue (984a22–25). The cause to seek after the material cause is, therefore, the efficient cause: ἡ ἀρχὴ τῆς κινήσεως (984a27).

In 984b5–8 Aristotle points to those thinkers who proposed several material causes and assigned to one of them an efficient role. He speaks of fire causing the other elements to move, showing that he has Empedocles in mind.[188] He then continues thus:

---

[186] Chapters 3 and 4 of book A are closely related: Ross 1924: 124–41; Betegh 2012: 105–106 with n. 2.
[187] Cf. Barney 2012: 76–95.
[188] See DK 31 B 62. On the position of fire as compared with the other elements see also *GC* B 3, 330b19–21 (= DK 31 A 36): ἔνιοι δ' εὐθὺς τέτταρα λέγουσιν οἷον Ἐμπεδοκλῆς. συνάγει δὲ καὶ οὗτος εἰς τὰ δύο· τῷ γὰρ πυρὶ τἆλλα πάντα ἀντιτίθησιν.

Aristotle, *Metaphysics* A 3, 984b8–13

μετὰ δὲ τούτους καὶ τὰς τοιαύτας ἀρχάς, [9] ὡς οὐχ ἱκανῶν οὐσῶν γεννῆσαι τὴν τῶν ὄντων φύσιν, πάλιν ὑπ' [10] αὐτῆς τῆς ἀληθείας, ὥσπερ εἴπομεν, ἀναγκαζόμενοι τὴν [11] ἐχομένην ἐζήτησαν **ἀρχήν**. τοῦ γὰρ εὖ καὶ καλῶς τὰ μὲν [12] ἔχειν τὰ δὲ γίγνεσθαι τῶν ὄντων ἴσως οὔτε πῦρ οὔτε γῆν οὔτ' [13] ἄλλο τῶν τοιούτων οὐθὲν οὔτ' εἰκὸς αἴτιον εἶναι ...

When these men and the principles of this kind had had their day, as the latter were found inadequate to generate the nature of things, men were again forced by the truth itself, as we said, to inquire into the next kind of cause. For surely it is not likely either that fire or earth or any such element should be the reason why things manifest goodness and beauty both in their being and in their coming to be ...

---

11 ἀρχήν **α** edd. : ἀρχὴν τουτέστι τὴν ποιητικὴν τούτων εὖ ἔχειν καὶ καλῶς **β** cf. Al. 32.8–9 ‖ 13 ἄλλο **α** : ἄλλο τι **β**

At first glance the reader might be inclined to think that since Aristotle just treated the material cause and mentioned the need for an efficient cause (μετὰ δὲ τούτους καὶ τὰς τοιαύτας ἀρχάς, 984b8), he is now proceeding with the treatment of "the next principle" (ἐχομένην ... ἀρχήν, 984b11). The search for this "next principle," Aristotle tells us, is motivated by the evident order and beauty in the world (εὖ καὶ καλῶς τὰ μὲν ἔχειν τὰ δὲ γίγνεσθαι 984b11–12). It seems, therefore, that Aristotle is hinting at the final cause.

Ross has presented two arguments that show this understanding of the passage to be problematic.[189] First, it is simply not true that the thinkers Aristotle will mention in the following lines and the principles ascribed to them—Anaxagoras's *Nous* (984b15–22), Hesiod's *Eros* (984b23–32), and Empedocles' Love and Strife (984b32–985a10)—can be taken as advocates of a final cause. Second, in his interim résumé in 985a10–18, Aristotle goes on to state explicitly that the aforementioned predecessors had only a rudimentary grasp of *two* causes: in Aristotle's terminology, the material and the efficient cause. In this summary, Aristotle does not even mention the (third) final cause. Therefore, it is clear that in the relevant passage in 984b11 "the *next* principle" (τὴν ἐχομένην ... ἀρχήν) refers not to the final but rather to the efficient cause.[190] The earlier philosophers were led to the notion of an *efficient* cause because truth itself drove them to ask why nature exhibits order—the question that led Aristotle (and only Aristotle) to discover the final cause.[191]

---

[189] Ross: 1924: 135–36. Cf. also Barney 2012: 96.

[190] This conclusion is further corroborated by the fact that the earlier thoughts on the efficient cause were not actually discussed in the preceding sentences. What we find there is just a passing mention of those thinkers who also saw a principle of movement in their material causes (984b3–8).

[191] Ross 1924: 136: "Thus, while the inquiry 'what set things changing?' did not lead to the notion of a distinct efficient cause, which is the proper answer to that inquiry, the question 'why are things well ordered?' did lead to that notion." See also Barney 2012: 96.

Looking at the passage as transmitted by the β-version, we read in line 984b11 after the words τὴν ἐχομένην ... ἀρχήν the following addition: τουτέστι τὴν ποιητικὴν τούτων εὖ ἔχειν καὶ καλῶς. These words express exactly the interpretation we have worked out so far: ἀρχήν does not, as might initially be suspected, refer to the final cause, but instead to the efficient cause. Approaching the β-reading with the above interpretation in mind, it appears reasonable and even justified. Yet, upon closer inspection the β-reading reveals peculiarities that make it doubtful in itself and that clash with the words following in Aristotle's text. Therefore, we do well to follow Bonitz, who understands this β-reading as a later addition to the text.[192]

The β-addition τουτέστι τὴν ποιητικὴν τούτων εὖ ἔχειν καὶ καλῶς consists of two parts: first, the words τουτέστι τὴν ποιητικὴν τούτων, which relate directly to the previously mentioned ἀρχή and specify it as the efficient cause of existing things (τούτων relates to τῶν ὄντων in line b9); second, the infinitive εὖ ἔχειν καὶ καλῶς, which is grammatically to be taken as the effected object or genitive attribute of the efficient cause (τὴν ποιητικὴν). The infinitive, however, is suspicious because it lacks the article that would normally precede an infinitive that functions as a genitive attribute, and that would be grammatically preferable when taken as an effected object. Consequently, Hayduck and Jaeger have suggested reading τοῦ instead of τούτων.[193] Even if one understands the infinitive εὖ ἔχειν καὶ καλῶς as a result that depends on the verbal adjective ποιητικός ("capable of making X to do..."), which then functions like the verb ποιεῖν, we still face the difficulty that the subject within the infinitive clause (τούτων) is in the wrong case (genitive instead of accusative). In addition to the difficulty of connecting the infinitive εὖ ἔχειν καὶ καλῶς to the preceding part of the sentence εὖ ἔχειν καὶ καλῶς occasions an odd repetition of the phrase εὖ καὶ καλῶς ... ἔχειν, which occurs once more in the subsequent sentence. The β-text reads in 984b11–12 τουτέστι τὴν ποιητικὴν τούτων εὖ ἔχειν καὶ καλῶς. τοῦ γὰρ εὖ καὶ καλῶς τὰ μὲν ἔχειν... .

Apart from these syntactical oddities, the usage of the word ποιητικός as a *terminus technicus* for the efficient cause arouses suspicion. The form ποιητικός is not too unusual, one might argue, given that Aristotle introduces our passage by describing the efficient cause as *producing* change (984a21–22): ποιεῖ μεταβάλλειν (*sc.* τὸ ὑποκείμενον). In this way, he connects the sought-for second principle with the image of a craftsman-like productive power conveyed by the verb ποιεῖν. The fact that in 984a27 Aristotle calls the efficient cause ἡ ἀρχὴ τῆς κινήσεως, which unlike ποιητικός is a typical expression of his, contravenes this argument. Although Aristotle does use the adjective ποιητικός to describe the efficient cause

---

[192] Bonitz 1848: XVI followed by Primavesi 2012c: 478 (*app. crit. ad loc.*). See also Christ 1853: 22.

[193] Hayduck proposed this reading for the text of the commentary by Asclepius, who quotes a *Metaphysics* text that includes the β-addition (Ascl. *In Metaph.* 27.31–32). Jaeger 1957 (*app. crit.*): vetus interpretamentum marginale fuisse vid. sed τοῦ pro τούτων legi debebat.

in several passages in his writings,[194] the word in these instances denotes the efficient cause in a general rather than in a classificatory sense, that is to say, not in respect to the four-cause theory, which is at issue in our passage.

Once more, Alexander's commentary provides helpful information about the origin of the β-addition.

Alexander, *In Metaph.* 31.27–32.9 Hayduck

Μετὰ δὲ τούτους καὶ τὰς τοιαύτας ἀρχάς.
[28] Τούτους τοὺς πάνυ παλαιούς, τὰς δὲ τοιαύτας ἀρχὰς τὰς ὑλικάς. [32.1] λέγοι δ' ἂν καὶ μετὰ τοὺς ἐν ταῖς ὑλικαῖς ἀρχαῖς καὶ <u>τὴν ποιητικὴν</u> [2] <u>αἰτίαν</u> θεμένους, ὡς οὐκ οὐσῶν τούτων ἱκανῶν τῶν ἀρχῶν πρὸς τὸ τὴν [3] τῶν ὄντων γεννῆσαι φύσιν· τὰ μὲν γὰρ τάξεως μετέχει καὶ κατά τινα [4] ἀκολουθίαν ὁρᾶται γινόμενα, ἐκείνων δὲ οὐδὲν τοιαύτης τάξεως οἷόν τε [5] αἰτίαν παρέχειν. ἀλλ' οὐδὲ τὸ αὐτόματον εὔλογον τούτων αἰτιάσασθαι· [6] διὰ τοῦτο ἐζήτησαν τὴν τῆς τοιαύτης γενέσεως αἰτίαν, ὥσπερ ὑπ' αὐτῶν [7] τῶν πραγμάτων καὶ τῆς ἐν τούτοις ἀληθείας ὁδηγηθέντες καὶ ἐπαχθέντες [8] καὶ ἀναγκασθέντες. <u>ἐχομένη δὲ ἀρχὴ γενέσεως μετὰ τὴν ὑλικὴν ἡ ποιη-</u>[9]<u>τική</u>.

When these men and the principles of this kind had had their day. "These men" are the very ancient philosophers, and "the principles of this kind" are the material ones. Aristotle might also mean, 'when those who counted the efficient cause too among the material principles had had their day,' because they realized that these latter principles are inadequate to generate the nature of the things that are; for these things participate in order and are seen to come into being according to a certain sequence, but none of those [material principles] could provide an explanation of such order. Nor was it reasonable to make spontaneity responsible for this order,[195] and therefore they were seeking the cause of this sort of generation, as if things themselves and the truth in them were showing them the way and forcibly leading them on. Now the principle of generation that follows the material cause is the efficient cause;

---

27 ἀρχάς **A O S** : om. **P**ᵇ ‖ 28 τὰς δὲ τοιαύτας Hayduck S?[et] : τὰς τοιαύτας **A O** : τοὺς τιθέντας **P**ᵇ

Alexander, too, understands Aristotle's τούτους (984b8) as referring both to those thinkers who postulated a material cause only (31.28) as well as to those who also integrated an efficient cause into the material cause (32.1-3). Therefore, the "next kind of cause" refers to the efficient cause (32.5-9). Alexander's term for the efficient cause differs from Aristotle's. Alexander calls it ἡ ποιητικὴ αἰτία. The adjective ποιητικός in the sense of "efficient (cause)" is quite common among later

---

[194] See *Metaph.* Λ 10, 1075b31; *GC* A 7, 324b13-14: Ἔστι δὲ τὸ ποιητικὸν αἴτιον ὡς ὅθεν ἡ ἀρχὴ τῆς κινήσεως, …; *de An.* Γ 5, 430a10-13.
[195] Here Alexander refers to Aristotle's words in 984b14, which are not part of the passage quoted above.

authors, including those writing commentaries on Aristotle.¹⁹⁶ Among them is Alexander, who seems to use the term in order to underline the productive function of this type of cause, a function that Aristotle captures in his picture of a craftsman.¹⁹⁷ In stark contrast to Aristotle's diction, in Alexander's commentary the phrase ἡ ποιητικὴ αἰτία is a standard term for the efficient cause, a fact that becomes particularly obvious in the context of our passage.¹⁹⁸

After summarizing Aristotle's train of thought (32.3-8) Alexander explains what Aristotle means by τὴν ἐχομένην … ἀρχήν ('the next kind of cause') in 984b10-11. Alexander says: ἐχομένη δὲ ἀρχὴ γενέσεως μετὰ τὴν ὑλικὴν ἡ ποιητική (32.8-9). According to Alexander, the phrase τὴν ἐχομένην … ἀρχήν can only refer to ἡ ποιητική, the efficient cause. This explanation coincides in function, content and not least in certain peculiar features of its expression with the addition we find in the β-version, which supplements τὴν ἐχομένην … ἀρχήν with the explication τουτέστι τὴν ποιητικὴν τούτων εὖ ἔχειν καὶ καλῶς (984b10-11).

Given that the expression ἀρχὴ ποιητική is firmly rooted in Alexander's commentary but appears in the β-text within an odd, syntactically challenging, explicatory addition, we are compelled to believe that the β-version contains a later interpolation that draws from Alexander's comments on the passage. It must be conceded, however, that Alexander's words alone are not sufficient to explain every feature of the β-reading. The β-addition contains the infinitive εὖ ἔχειν καὶ καλῶς, which specifies both the effect the cause produces and the motive for the discovery of the efficient cause. This infinitive does not have an equivalent in Alexander's commentary. It is true that a few lines later Alexander speaks once more about the ποιητικὴ αἰτία as a principle of τοῦ καλῶς καὶ τεταγμένως γίνεσθαι (32.16-17), but it remains more plausible to regard the β-words εὖ ἔχειν καὶ καλῶς as drawn from or inspired by the subsequent lines of the *Metaphysics*: τοῦ γὰρ εὖ καὶ καλῶς τὰ μὲν ἔχειν τὰ δὲ γίγνεσθαι τῶν ὄντων… (984b11-12). Thus, the following scenario is a viable reconstruction of what happened: first, a reader or scribe working on the β-text added (perhaps only in the margins) the explanatory gloss τουτέστι τὴν ποιητικὴν τούτων next to the word ἀρχήν, an addition which very likely draws from Alexander's commentary. Later, someone tried to integrate this gloss into the *Metaphysics* text by somehow connecting it with Aristotle's train of thought. The result of these attempts at clarification is the syntactically peculiar infinitive: εὖ ἔχειν καὶ καλῶς.

---

¹⁹⁶ LSJ s.v. ποιητικός, cf. Plot. VI, 7, 20,8. See also Simp. *In Phys.* 317.8-9.

¹⁹⁷ In our passage the term ἀρχὴ γενέσεως is used as a name for the efficient cause (32.6 and 8; cf. πρὸς τὸ τὴν τῶν ὄντων γεννῆσαι φύσιν in 32.2-3 where Alexander paraphrases Aristotle's wording in 984b9).

¹⁹⁸ See, for example: 29.1, 3, 6-7, 13; 30.14; 31.18, 19, 23; 32.16; 33.9, 13-16; 34.2. See also 181.33: κινήσεως δὲ ἀρχὴν λέγει (sc. Ἀριστοτέλης) τὸ ποιητικὸν αἴτιον and 220.1 (on this passage see 4.1.3).

5.2.6 Alex. *In Metaph.* 295.29–32 on Arist. *Metaph.* Γ 4, 1008a18–27

In Γ 4, Aristotle endeavors to establish the validity of the principle of non-contradiction by listing the absurd consequences that follow from its denial. In the course of the chapter Aristotle develops several arguments, one of which tries to derive the consequence that everything is one out of the denial of the principle of non-contradiction (1007b19–20).[199] The passage that concerns us at present is part of the fourth argument,[200] which demonstrates that the principle's deniers cannot assert anything (1008a7–34). In the text preceding our passage, Aristotle made the following steps: whenever an opponent holds that something is and at the same time is not, this *either* implies that everything asserted may also be denied and, similarly, that everything that is denied may also be asserted (1008a11–13), *or* it implies that everything that is asserted may also be denied but *not* that everything that is denied is also asserted (1008a14–15). If the latter option is the case then there is something that certainly is not the case; and if something is known with certainty not to be the case, then the opposite affirmation is knowable all the more (1008a15–18).

The following passage, from line 1008a18 onwards, continues this train of thought in the following way: if everything can be equally denied and asserted, then it is either true or untrue to state at one time that a thing is white and then at a later time that it is not white.

Aristotle, *Metaphysics* Γ 4, 1008a18–27

εἰ δὲ ὁμοίως καὶ ὅσα ἀποφῆσαι φά-[19]ναι,[201] ἀνάγκη ἤτοι ἀληθὲς διαιροῦντα λέγειν, οἷον ὅτι [20] λευκὸν καὶ πάλιν ὅτι οὐ λευκόν, ἢ οὔ. καὶ εἰ μὲν [21] μὴ ἀληθὲς διαιροῦντα λέγειν, οὐ λέγει τε ταῦτα καὶ [22] οὐκ ἔστιν οὐθέν (τὰ δὲ μὴ ὄντα πῶς ἂν φθέγξαιτο ἢ [23] βαδίσειεν;), καὶ πάντα δ' ἂν εἴη ἕν, ὥσπερ καὶ πρότερον [24] εἴρηται, καὶ ταὐτὸν ἔσται καὶ ἄνθρωπος καὶ θεὸς καὶ τριή-[25]ρης καὶ αἱ ἀντιφάσεις αὐτῶν (εἰ δ' ὁμοίως καθ' ἑκάστου, [26] οὐδὲν διοίσει ἕτερον ἑτέρου· εἰ γὰρ διοίσει, τοῦτ' ἔσται ἀληθὲς [27] καὶ ἴδιον)· ὁμοίως <u>δὲ καὶ εἰ</u> διαιροῦντα ἐνδέχεται ἀληθεύειν…

But if what is denied is equally asserted, necessarily it is either correct to state separately, for instance, that a thing is white, and again that it is not-white, or not. [i] And if it is not correct to state separately, our opponent is not really stating them, and nothing at all exists (but how could non-existent things speak or walk?). Also

---

[199] Ross 1924 counts this as the second argument, Kirwan 1971 as the third. See also 4.3.2.1.
[200] Cf. Ross 1924: 267; Kirwan 1971: 103–104; Cassin/Narcy 1989: 218–21.
[201] On the placement of the comma before ἀνάγκη see Ross 1924: 271 and his original translation in Ross 1908. In the revised Oxford Translation, known as the *Complete Works of Aristotle* (Barnes 1984; see 1591) the translation has been changed according to the position of the comma after ἀνάγκη. The translation above follows Ross's original punctuation (without adopting the exact wording of his original translation).

all things will on this view be one, as has been already said,[202] and man and God and trireme and their contradictories will be the same. (For[203] if it can be predicated alike of each subject, one thing will in no wise differ from another; for if it differs, this difference will be something true and peculiar to it.) [ii] And if one may with truth state it separately ...

18–19 φάναι **α** Bonitz Ross Jaeger Cassin/Narcy Hecquet-Devienne : κατὰ τούτων ἔστι φάναι **β** Bekker Christ || 21 λέγειν **α** edd. : λέγει **β** || 23 βαδίσειε **α** Al.P 295.17 Ascl.P 269.32 Bonitz Christ Ross Jaeger Cassin/Narcy Hecquet-Devienne : νοήσειε **β** Bekker || 25 εἰ δ' **α** ω^AL (Al.¹ 295.29 Al.P 295.30) Ascl.P 270.2-3 Bonitz Cassin/Narcy Hecquet-Devienne : εἰ γὰρ **β** Al. interpretans 295.30, Bekker Christ Ross Jaeger

In line 1008a18–20 Aristotle introduces the following alternative: if one states the two sentences "X is white" and "X is not-white" separately at different times, then one speaks either truly or falsely. The first alternative that Aristotle pursues is the latter, which holds that one does not speak the truth if one says at one time "X is white" and then later "X is not-white." He introduces his examination of this option with the words καὶ εἰ μὲν (a20). He then states that the opponent would then be saying nothing at all, a consequence that he connects to the result of his previous argument, in which he concluded that for the opponents everything must be one (a23–25). In a25–27, he drives the absurdity further home, pointing out that then all differences are annihilated. At a27 Aristotle finally turns to the second arm of the original disjunction, which holds that one does speak the truth if one says at one time "X is white" and then later "X is not-white." The words introducing the second option, ὁμοίως δὲ καὶ εἰ (a27) take up καὶ εἰ μὲν (a20).

The textual divergence found in the **α**- and the **β**-version in line a25 is intrinsically connected to the structure of this passage. According to the **α**-reading the sentence in line a25 begins with the words εἰ δ', whereas the **β**-text reads εἰ γὰρ. The **α**-reading is, as we will see presently, confirmed by Alexander's commentary as the reading of ω^αβ. Compared with the **β**-reading, the δέ in **α** is a *lectio difficilior*: its meaning here is "for."[204] The abolition of all differences, that everything can be affirmed and denied (a25–27), offers an explanation of the preceding statement that human and trireme and God would all be the same (a24–25). And so the γὰρ in the **β**-text would appear to be the result of an attempt to make this meaning clearer. The challenge posed by the **α**-reading is to not yield to the temptation to interpret the εἰ δ' in line a25 as a complement to εἰ μὲν in line a20; the actual complement to εἰ μὲν (a20) comes only in a27 in the form of δὲ καὶ εἰ.[205] In order

---

[202] Reference to Γ 4, 1007b20.

[203] The particle δ' here has the force of a γάρ, meaning "for" (Denniston 1954: s.v. δέ I.C.1(i), 169–70). This does not mean, however, that one should read with the β-tradition γὰρ instead of δ' attested by ω^AL and **α**. The issue will be discussed in more detail below.

[204] This function of the particle δέ is common. See previous note.

[205] The danger that τὰ δὲ in line a22 might be taken as a complement to εἰ μὲν (a20) is lower, but it nevertheless seems that Ross and Jaeger wanted to prevent such a misunderstanding by putting the

to avoid the danger of misconstruing the thought in the α-text one could put lines a25–27 into parentheses (see text above). The β-version seems to have opted for another strategy of avoiding the error by reading instead of the particle δέ the particle γάρ.

Turning to the passage in Alexander's commentary, we see that he quotes in his lemma the relevant protasis of line a25, and immediately afterwards comments on the particle at the beginning of the sentence.

Alexander, *In Metaph.* 295.29–32 Hayduck

1008a25 Εἰ δὲ ὁμοίως καθ' ἑκάστου.
[30] Ὁ δὲ σύνδεσμος ἀντὶ τοῦ γάρ κεῖται. ἔστι γὰρ τὸ ἀκόλουθον· εἰ [31] γὰρ ὁμοίως καθ' ἑκάστου πᾶσα ἀντίφασις ἀληθής, οὐδὲν ἄλλο ἄλλου [32] διοίσει, οὐδὲν ἔχον ἴδιον.

For if it can be predicated alike of each subject …
The conjunction δέ ['but,' 'for'] is used in place of γάρ ['for']; for the run of the argument is [as follows]:[206] For if in like manner every pair of contradictories is true of each thing, then nothing will differ from anything else, as it will have nothing distinctive.

---

29 καθ' ἑκάστου **A O** : καθ' ἕκαστον **P**[b]

This passage makes it clear that Alexander found the particle δέ in line a25 of his copy of the *Metaphysics*:[207] The reading in the lemma is immediately confirmed by what follows (Ὁ δὲ σύνδεσμος... 295.30). Alexander comments on the particle δέ, in order to make its meaning clear: he explains that it should be taken to mean γάρ, that is, to signal the consecutive and even explanatory character of the sentence. Alexander justifies this interpretation through his understanding of the thought expressed in εἰ δ' ὁμοίως ... ἑτέρου (a25–26) as a follow-up (τὸ ἀκόλουθον) to the preceding thought. With this comment Alexander also makes it clear that the words εἰ δ' ὁμοίως ... do not introduce the second of the alternatives Aristotle gives his opponent.

So Alexander suggests understanding the present particle δέ in the sense of

---

sentence in lines 1008a22–23 into brackets.

[206] In the sentence ἔστι γὰρ τὸ ἀκόλουθον in 295.30 the particle γάρ must not, as Madigan 1993: 82 takes it ("'for' is what follows"), be taken as a quotation from the Aristotelian text or as a word about which Alexander says something, but simply as a particle from Alexander's *own* sentence. Alexander simply states his explication: "for it (i.e. the sentence in question) is a sequel." Casu 2007: 697–98 translates correctly ("...poiché ciò che segue è").

[207] Apparently Asclepius, too, found in his *Metaphysics* copy the δέ that is preserved by the α-text (270.2–3). To decide on the basis of the Latin version what was in the *Vorlage* of the Arabic tradition is difficult. Scotus writes: *Et si sermo de unoquoque istorum fuerit idem*... . In order to make a clear decision we would either expect a *nam* for γάρ or an *autem* for δέ. Since the Greek particle δέ is generally additive in character (see Denniston 1954: s.v. δέ I.A.; pp. 162–65), it seems more accurate to translate a δέ with the Latin word *et* than a γάρ.

(ἀντὶ τοῦ …)²⁰⁸ γάρ. This does not imply, however, that he wants to *replace* the word δέ in the Aristotelian text with the word γάρ. Moreover, one might assume that this suggestion goes back to Alexander himself, but we cannot rule out that Alexander just reports what he has found as a gloss in his manuscript or in Aspasius's or another commentary. For, to explain or even substitute δέ by γάρ is a fairly common exegetical remark found also in the scholiastic tradition.²⁰⁹ Still, Alexander shows repeated interest in the question of whether δέ or γάρ is the more appropriate particle in several passages of his commentary, and this clearly indicates that Alexander is sensitive to this issue,²¹⁰ a fact that remains even if he shares this sensitivity with other commentators. Given that Alexander first confirms the α-reading to be the reading in ω^AL and then reveals that he wants this reading to be understood in just the way we find it in the β-text, we might be encountering here the intervention of a reader or reviser in the β-text, who changed the β-text according to Alexander's suggestion.

That we are dealing with a passage in the β-text where an intentional revision indeed occurred is corroborated by evidence from the surrounding sentences. There are several instances in which the β-reading shows traces of a revision process in which someone rewrote some of Aristotle's sparse expressions into more detailed formulations. Instances of this can be seen in 1008a17,²¹¹ a18,²¹² a18–

---

²⁰⁸ On the formula ἀντὶ τοῦ… as a way of introducing alternative formulations of what Aristotle says, see also 5.1.1, 5.1.2, and 5.2.2.

²⁰⁹ See the remark ὁ δέ ἀντὶ τοῦ γάρ in the scholia to Euripides, *Hec.* 94, 644; *Or.* 196, 702; *Pho.* 250, 817 Schwartz.

²¹⁰ Alexander often asks whether a transmitted δέ should be taken as (54.11–12) or even substituted by a γάρ (37.20–21; 172.13–15). See also 270.12–17, where Alexander suggests deleting δέ without substitution. The result is that the half sentence εἰ [δὲ] μὴ … τἀναντία (1005b26–27) belongs to the preceding and not to the subsequent sentence (see also 3.6).

²¹¹ 1008a17 ἂν α : γὰρ ἂν β. We can infer on the basis of Alexander's close paraphrase that his text agrees with the α-reading. The reading in β (γὰρ ἂν) suggests that someone wanted a new sentence to start here. Thus, the preceding sentence in the β-version (a16–17) already ends with γνώριμον. As Bonitz 1847: 86–87 states, the sentence καὶ εἰ … γνώριμον is unsatisfactory when shortened in this way, because it only repeats what was already said in lines a15–16.

²¹² 1008a18 ἢ ἡ ἀντικειμένη α Al.ᵖ 294.24 : ἢ ἡ ἀντικειμένη ἀπόφασις β. While the α-version reads γνωριμωτέρα ἂν εἴη ἡ φάσις ἢ ἡ ἀντικειμένη ("the opposite assertion will be more knowable"), β reads γνωριμωτέρα γὰρ ἂν εἴη ἡ φάσις ἢ ἡ ἀντικειμένη ἀπόφασις ("For the assertion will be more knowable than the opposite negation"). Cf. *APo* A 25, 86b34; *Int.* 5, 17a8; *Metaph.* Γ 4, 1007b34–1008a2, on this see 4.3.2.1). The particle γάρ in the β-formulation expresses an unjustified corollary. The statement of the preceding sentence, that non-being is knowable when it is determined, does not *follow* from the fact that the affirmation is more knowable than the negation.

a19,²¹³ and a23²¹⁴.²¹⁵ Alexander's commentary confirms the α-reading as the older one in all these cases. In these cases Alexander's commentary may or may not be the source or at least a model for the reading in the β-version. Yet what about the γάρ in 1008a25? Here the changed wording in β exactly matches with Alexander's interpretation. Therefore it seems natural to combine the two observations about the revised character of the passage as a whole on the one hand and the match of β's γάρ with Alexander's comments on the other and assume that, in the case of γάρ, the supposed reviser had recourse to Alexander's commentary.²¹⁶ However, there remains the *caveat* that the interpretation given by Alexander, which agrees with the reading found in β, is not idiosyncratic enough to supply this supposition with secure evidence.

On the basis of the six cases analyzed here (5.2.1–6), we can conclude, as Primavesi 2012 did with respect to the first book of the *Metaphysics*, that the β-version is contaminated by Alexander's commentary in the later books of the *Metaphysics*. (Naturally, this conclusion can only be drawn for those books for which Alexander's commentary has been preserved.) How is the contamination of the β-version to be explained? What kind of contamination is it?

Primavesi 2012b argues that the traces of Alexander's commentary in the β-version stem from a revision process that created the β-text as a version distinct from the α-version.²¹⁷ In other words, there was a moment in the textual history of the *Metaphysics* when a revision of text ω^{αβ} (what Primavesi calls the "*common text*") resulted in a version that is (in its main character, i.e. apart from minor textual changes that occurred during the later transmission) to be identified with what we call β. According to the conclusions drawn by Primavesi 2012b, the revision pro-

---

²¹³1008a18–19 ὅσα α Al.ᵖ 295.1: ὧν ἔστιν β. The α-text preserves the reading: εἰ δὲ ὁμοίως καὶ ὅσα ἀποφῆσαι φάναι... / "But if what is denied is equally asserted...." Alexander's paraphrase confirms the α-reading. The β-version contains the same statement as the α-version, but in a more elaborate rendering. Here, the suspicion that the β-version (ὧν ἔστιν ἀποφῆσαι κατὰ τούτων ἔστι φάναι) has been expanded according to a model provided by Alexander's commentary (ὅσα ἀποφῆσαι ταῦτα καὶ καταφῆσαι, 295.1) seems justified. Still, the β-reviser might also have oriented himself towards Aristotle's own alternative formulations in the context of the passage (1008a12–14).

²¹⁴1008a23 βαδίσειε α Al.ᵖ 295.17 : νοήσειε β. Aristotle means to show the absurdity of the opponent's position by asking τὰ δὲ μὴ ὄντα πῶς ἂν φθέγξαιτο ἢ βαδίσειεν; The examples, speaking and walking, are examples of ordinary human behavior. If nothing definite exists there cannot be anyone who speaks (φθέγξαιτο) or walks (βαδίσειε). The β-text reads instead of the verb form βαδίσειε / "could walk" (α) the verb νοήσειε / "could think." On the topic of this reading, Alexander's text also agrees with α (295.12–14 and 16–17).

²¹⁵That the β-text in this section exhibits features similar to those described by Frede/Patzig 1988: 13–17 (for book Z) and Primavesi 2012b: 457–58 (for book A) may indicate that parts of the *Metaphysics* other than just book A and Z also underwent revision.

²¹⁶This indeed fits the description that Primavesi 2012b: 457 gives of the β-revision that is based on Alexander.

²¹⁷Primavesi 2012b: 457.

cess resulted in some of Aristotle's sparse phrases being replaced by more readable ones, as already suggested by Frede/Patzig 1988, and relied on the guidance of authoritative models, most prominent of which being Alexander's commentary.[218] This revision most likely took place before the end of the fourth century AD.[219]

Are the six β-passages that I discussed above and that exhibit contamination by Alexander's commentary to be explained as a result of this revision process? One could argue that since some of the passages in β exhibiting signs of contamination are "rough" rather than "smoothed,"[220] not all such readings peculiar to the β-text can be attributed to the intention of a careful reviser. One might argue in return, however, that such oddities—especially those in a contaminated passage—indicate that intervention indeed did occur,[221] because errors are an unavoidable, if unintentional, byproduct of any revision process. Nevertheless, the reading in β for some of the discussed passages seems to have resulted from a scenario that very likely consisted of two steps: first, the addition of a marginal gloss containing Alexander's interpretation and, second, the more or less accidental incorporation of this marginal gloss into the text.[222]

Since we know that some of Alexander's formulations and corrections had already found their way into the *Metaphysics* text at the $\omega^{\alpha\beta}$-stage (see 5.1), we can legitimately assume that Alexander's commentary contaminated the *Metaphysics* text at more than one stage during the transmission process. Why then should we restrict the influence that Alexander's comments exerted on the β-text to one revision process? In the case of the β-version, there is one further piece of evidence to be taken into account. Our two most important witnesses to the β-text, A^b and M, contain in their margins Alexander's commentary in the *recensio altera* version.[223] One is free to speculate that the β-version already included the marginal commentary some time before the transliteration process in the ninth century AD.[224] Such close transmission of text and commentary makes contamination

---

[218] According to Primavesi 2012b, the β-reviser drew inspiration or particular phrases from other sources and incorporated them into the *Metaphysics* text.

[219] Cf. 1. Since the β-version contains *reclamantes* that go back to an ancient edition on papyrus scrolls, the revision is most likely to have happened before AD 400, when papyrus editions were no longer produced and there was no need for *reclamantes*.

[220] See the examples in 5.2.1, 5.2.2 (first of the two cases), 5.2.4, 5.2.5.

[221] See the first example in Primavesi 2012b: 424-28 ("Text 7"). The very fact that the additional words taken over from Alexander's commentary do not exactly fit to the syntactical context shows that this is not what Aristotle originally wrote.

[222] See the case studies in 5.2.3-4.

[223] On the *recensio altera* see 2.4.

[224] One might object that the independent β-witness fragment Y (*Paris. Suppl.* 687) contains the text of books I and K and none of Alexander's commentary in the margins. In this case, one might then reply, the marginal commentary that might have been present in the parent of this manuscript had been left out of Y.

more likely.²²⁵ In five of the six commentary passages analyzed above, the text of the *recensio altera* is identical to the authentic version of the commentary.²²⁶ Thus it is at least theoretically possible that these instances of contamination in the β-version stem from the *recensio altera* in the margins. At this point, of course, this remains speculation.

Given that it is beyond the scope of this study to discuss or even evaluate the impact of the so-called β-revision itself, it seems best to conclude, on the basis of the evidence that scholars have analyzed so far, that some of the traces that Alexander's comments left behind in the β-version are to be attributed to a revision process that this version underwent some time before the end of the fourth century AD. This is especially likely in cases such as 5.2.2 and 5.2.6, where the text in β appears to have been consciously revised on the basis of a suggestion made by Alexander. However, given the evidence of Alexander's widespread impact on the text (see also 5.3) it seems unjustified to attribute the contamination of β exclusively to one revision process. Thus, some of the words or phrases incorporated from Alexander's commentary into the β-text could very well stem from glosses that had been added to the text's margins over time and from where they found their way into the text more or less accidentally (5.2.3–5).

5.3 CONTAMINATION OF α BY ALEXANDER'S COMMENTS

Having analyzed the contamination of ω^{αβ} and β with Alexander's commentary, I now turn to the α-version and the question of how it relates to Alexander's commentary. Is there evidence that Alexander's commentary influenced the text of the α-version, as there is in the case of the ω^{αβ}-version or the β-version (see 5.1 and 5.2)? Does the α-text contain later "corrections" that were based on Alexander's paraphrase or occasioned by his critical remarks on the text?²²⁷ I will answer these questions with analyses of five different *Metaphysics* passages and Alexander's comments on them (5.3.1–5). In the first two case studies I will analyze Alexander's paraphrase as a possible source of the α-reading, and in the subsequent three case studies I will investigate whether Alexander's critical remarks on Aristotle's text and argument occasioned a textual change in the α-version.

---

²²⁵ This is what Bonitz 1848: XVI suspects to have happened. He states cautiously: *Necessitudinem quandam intercedere codici A^b cum commentario Alexandri in eius margine scripto, saepius quum utrumque inter se conferrem suspicabar, nec tamen certi quidquam de ea re statuerim.* And also Primavesi 2012b: 457 briefly draws attention to this peculiarity (cf. also Primavesi 2012b: 389 n.12).

²²⁶ The exception is 5.2.4. In the case of 5.2.2, the relevant commentary section (285.32–286.6) of the text of the *recensio altera* is extant only in L; in F a larger section of the commentary is missing (cf. *app. crit.* in Hayduck's edition).

²²⁷ In his analysis of book A of the *Metaphysics* Primavesi 2012b limits his treatment of Alexander's influence to the β-version of the text.

5.3.1 Alex. *In Metaph.* 26.14–18 on Arist. *Metaph.* A 3, 983b33–984a3

In A 3 Aristotle examines his predecessors' accounts of the causes and principles, specifically with the purpose of finding out whether and in what way earlier thinkers had treated one or more of the four causes.[228] First to be discovered, he finds, was what he calls the material cause (983b6–18). Thales claimed that the material cause of all things is water (983b18–27). Some mythical accounts also speak of the world as emerging from water (εἰσὶ δέ τινες οἵ ... πρώτους θεολογήσαντας οὕτως οἴονται ... ὑπολαβεῖν). Figures like Oceanus and Tethys[229] and the idea that the gods customarily swore oaths to Styx[230] indicate that for the oldest poets, as for Thales, water held a position of fundamental importance (983b27–33). As can be seen in the passage below, Aristotle makes no commitments as to the antiquity of this view or the validity of the history he gives of it.[231]

Aristotle, *Metaphysics* A 3, 983b33–984a3

εἰ μὲν οὖν [1] ἀρχαία τις αὕτη καὶ παλαιὰ τετύχηκεν οὖσα περὶ τῆς φύ-[2]σεως ἡ δόξα, τάχ' ἂν ἄδηλον εἴη, Θαλῆς μέντοι λέγεται [3] **οὕτως** ἀποφήνασθαι περὶ τῆς πρώτης αἰτίας.

It may perhaps be uncertain whether this opinion about nature is primitive and ancient, but Thales at any rate is said to have declared himself **thus** about the first cause.

3 οὕτως β Al.ˡ 26.14 et Al.ᶜ 26.16 Ascl.ᵖ 225.14 Bekker Bonitz Christ Ross Jaeger : τοῦτον τὸν τρόπον α ex Al.ᵖ 26.17–18, Primavesi

Aristotle speaks cautiously about the possible origin and age of the view that water is the first principle. Where myth is concerned, this may be because the mythical way of speaking makes a clear assessment impossible, as he points out in B 4, 1000a5–19. Furthermore, myth does not make its claims in the form of arguments that can be accepted or refuted.[232] Where Thales is concerned, Aristotle's lack of commitment in attributing such a theory to his Milesian predecessor ("Thales ... is said to have declared himself thus," 984a2–3) is more striking. However, a cautious attitude towards Thales is also visible in other passages in the Aristotelian

---

[228] For an analysis of *Metaph.* A 3, 983a24–984b8 see Barney 2012.

[229] In the *Iliad* Oceanus and Tethys are called the origins of the world: 14.201 (=14.302): Ὠκεανόν τε θεῶν γένεσιν καὶ μητέρα Τηθύν and 14.246: Ὠκεανοῦ, ὅς περ γένεσις πάντεσσι τέτυκται. Plato also mentions a Homeric Theogony which begins with Oceanus and Tethys: *Tht.* 152e; *Cra.* 402b.

[230] In Hesiod's *Theogony* 361 Styx is the daughter of Tethys and Oceanus. Zeus decrees that the gods make their oaths to Styx, because of her commitment to him in the fight against the Titans: *Theogony* 383–403. See West 1966: 275–76.

[231] Cf. Barney 2012: 88–90.

[232] Cf. Barney 2012: 88–90. On other occasions, Aristotle speaks respectfully of myth and the ancient knowledge mythical stories may contain (*Metaph.* Λ 8, 1074a38–b14). Concerning Aristotle's attitude towards myth see also Palmer 2000: 184–91.

corpus.²³³ Alexander himself recognizes Aristotle's caution (26.16–18) and attributes it to the fact that no written evidence of Thales' view existed in Aristotle's day.²³⁴

The divergence between the α- and β-version in line 984a3 seems slight as far as content is concerned. The α- and the β-versions point back to the view Thales is supposed to have held in slightly different yet still quite similar ways: β points back with οὕτως ("thus") and α points back with τοῦτον τὸν τρόπον ("in this way"). A clue to how this change in expression came about is provided in Alexander's commentary. The evidence in the commentary (found both in the lemma and in a citation) tells us that ω^AL also read οὕτως. In Alexander's paraphrase, we find that in place of the words οὕτως (which is given in his own text and preserved by our β-version) is the phrase τοῦτον τὸν τρόπον, which we find in the α-version.²³⁵

Alexander, *In Metaph.* 26.14–18 Hayduck

984a2 Θαλῆς μέντοι λέγεται <u>οὕτως</u> ἀποφήνασθαι περὶ [15] τῆς πρώτης αἰτίας.

[16] Εἰκότως τὸ λέγεται <u>οὕτως</u> ἀποφήνασθαι· οὐδὲν γὰρ προφέρεται [17] αὐτοῦ σύγγραμμα, ἐξ οὗ τις τὸ βέβαιον ἕξει τοῦ ταῦτα λέγεσθαι <u>τοῦτον</u> [18] <u>τὸν τρόπον</u> ὑπ' αὐτοῦ.

Thales at any rate is said to have declared himself thus about the first cause.

The statement 'is said to have declared himself thus' is reasonable, for no writing of his is preserved from which one can be certain that these things were said by him in this way.

14–16 περὶ τῆς πρώτης ... ἀποφήνασθαι **O P^b S** : om. **A**

Alexander compliments Aristotle for speaking cautiously in light of the fact that there were no extant writings of Thales that could substantiate the claim. The expression λέγεται οὕτως ("he is said to have declared himself") conveys that there is much uncertainty whether Thales spoke in exactly this way (τοῦτον τὸν τρόπον).

Alexander's words allow us to infer, first, that he read οὕτως in ω^AL and, second, that he himself chose the expression τοῦτον τὸν τρόπον to reformulate Aristotle's wording.²³⁶ But why does Alexander render οὕτως as τοῦτον τὸν τρόπον?

---

²³³ Ross 1924: 129 points to *Cael.* B 13, 294a29–30, *de An.* A 2, 405a19–21, A 5, 411a8 and *Pol.* A 11, 1259a18–19. See also Barney 2012: 86, who describes Aristotle's attitude in our passage as one of "scrupulous modesty about the evidence."

²³⁴ This reason certainly does not apply to the mythical accounts at least as far as Homer's *Iliad* and Hesiod's *Theogony* are concerned. On the loss of all written works of Thales see also Simp. *In Ph.* 23.29–33.

²³⁵ In his Latin translation of the commentary, Sepúlveda renders οὕτως by *hoc pacto* and τοῦτον τὸν τρόπον by *ad hunc modum*.

²³⁶ We have no reason for suspecting that someone adjusted Alexander's lemma *and* citation to the β-text. What is more, we see clearly from the other parts of his reformulation that Alexander avoided

Simply for the sake of variation? Or is there a difference, however slight, between these two expressions? The first palpable difference between the two expressions is that Alexander and Aristotle both make far more frequent use of the word οὕτως than the expression τοῦτον τὸν τρόπον:

> Aristotle, *Metaph.*: οὕτω(σ)(ι) 284 times / τοῦτον [...] τὸν τρόπον 22 times
>
> Alexander, *In Metaph.* Α–Δ: οὕτω(ς) 916 times / τοῦτον [...] τὸν τρόπον 13 times.

Next, it seems that the expression τοῦτον τὸν τρόπον is more explicit than the simple word οὕτως in conveying that something is happening in exactly *this way*.[237] The two expressions seem to differ in respect to the insistence with which something is said to occur in a certain way. In our case the formula τοῦτον τὸν τρόπον squares well with the pointed tone that Alexander adopts in his explication. Alexander reformulates and thereby accentuates Aristotle's remark "Thales is said to have expressed himself thus." Alexander writes: it is simply impossible to determine with any certainty (βέβαιον) that Thales has spoken *in exactly this way*. Alexander reformulates Aristotle such that he turns the positive statement about the uncertainty of the sources into a negative statement ("no writing ... is preserved") about the unavailability of any reliable statement. This enables Alexander to emphasize that there is no certainty about the *precise content* of Thales' view ("...that these things were said by him in this way"). By comparison, οὕτως fits quite well into the positive formulation expressing uncertainty over Thales' view ("he is said to have declared himself thus") that we encounter in the *Metaphysics* passage (according to the β-version and ω^AL).

If we were to replace οὕτως in the *Metaphysics* text with τοῦτον τὸν τρόπον then the *uncertainty* concerning Thales' view would be strangely coupled with the determinacy of the "*in exactly this way.*" Since we do not know what Thales actually said, it would come as a surprise to hear that he is said to have spoken in *just this way*. Whereas the words τοῦτον τὸν τρόπον fit well into Alexander's own explication, the reading οὕτως is clearly preferable in the *Metaphysics* passage.

The agreement of ω^AL and β indicates that the reading in ω^αβ was οὕτως. The fact that the α-version contains the expression that Alexander uses in his *own* reformulation (τοῦτον τὸν τρόπον) suggests that these words found their way

---

repetition of Aristotle's terms. Thus τοῦτον τὸν τρόπον confirms the οὕτως of the lemma and citation. Therefore we can surmise that he would not have said τοῦτον τὸν τρόπον, if this phrase had already been present in his ω^AL-text.

[237] This can be seen in the parallel passages in which Alexander uses τοῦτον τὸν τρόπον: 26.18; 57.13; 135.26; 152.13; 156.21; 157.27; 159.15; 386.16; 391.14–15; 422.9. Oftentimes Alexander will then specify the way or manner (τρόπος) in the following lines. Apart from this meaning of the expression one can see in Aristotle's *Metaphysics* that τοῦτον τὸν τρόπον is used to stress the fact that something is meant in exactly this sense: e.g. 987a20, 1015b35, 1018b30, 1023a22, 1039a8, 1061a1. But no such confidence or certainty is conveyed in Aristotle's remark about Thales: λέγεται οὕτως ἀποφήνασθαι ("he is said to have declared himself thus").

from Alexander's commentary into the α-version of the *Metaphysics*. A copyist or scholar may have regarded them as a correction or clarification of what Aristotle says, without recognizing that the remainder of Alexander's sentence is spoken from a slightly different perspective, which does not completely coincide with Aristotle's wording.

It is far less likely that β took over οὕτως from Alexander's citation and (or) lemma and (supposedly) used it in place of τοῦτον τὸν τρόπον. Since Alexander himself writes τοῦτον τὸν τρόπον two lines later, the β-reviser would be in the awkward position of having both followed Alexander's authority and disregarded it. But even if one were to consider this explanation viable, one would still run into the difficulty that Alexander's reformulation (τοῦτον τὸν τρόπον) accords by sheer coincidence with the α-reading, which, after all, fits the *Metaphysics* passage less than the β-reading.[238]

In sum, it is quite reasonable to assume that the coincidence of the α-version and Alexander's own reformulation came about because someone adjusted the wording in α in accord with Alexander's paraphrase. By contrast the reading attested to by the β-version and ω^AL leads us back to the original reading, which was also found in ω^αβ.

5.3.2 Alex. *In Metaph.* 38.5–7 on Arist. *Metaph.* A 5, 985b23–29

In the fifth chapter of book A, not far from the *Metaphysics* passage just analyzed, Aristotle discusses the Pythagorean theory of principles.[239] Aristotle's inquiry is part of his attempt to confirm or correct his own four cause theory. According to Aristotle, the Pythagoreans were the first to establish and develop mathematical disciplines (μαθήματα).[240] On account of their intimate familiarity with mathematics, they extended the application of mathematical principles to all other things. Aristotle attempts to reconstruct the development of the Pythagorean theory of principles.[241] It starts with the following two premises: numbers are by nature primary among the μαθήματα (985b24); they bear more resemblances to things than do fire, earth and water.

Aristotle, *Metaphysics* A 5, 985b23–29

Ἐν δὲ τούτοις καὶ πρὸ τούτων οἱ καλούμενοι Πυθαγόρειοι [24] τῶν μαθημάτων ἁψάμενοι πρῶτοι ταῦτα προήγαγον καὶ [25] ἐντραφέντες ἐν αὐτοῖς τὰς τούτων ἀρχὰς τῶν ὄντων ἀρχὰς [26] ᾠήθησαν εἶναι πάντων. ἐπεὶ δὲ τούτων οἱ ἀριθμοὶ φύσει

---

[238]There is no evidence to suggest that Alexander borrowed from another tradition when writing his explanation.

[239]See Schofield 2012: 141–55 and Primavesi 2014.

[240]The scope of these disciplines was much wider than what we understand as mathematics; it embraced astronomy and music theory. See Primavesi 2014: 229.

[241]See Primavesi 2014: 230–36.

[27] πρῶτοι, ἐν δὲ **τούτοις** ἐδόκουν θεωρεῖν ὁμοιώματα [28] πολλὰ τοῖς οὖσι καὶ γιγνομένοις, μᾶλλον ἢ ἐν πυρὶ καὶ γῇ [29] καὶ ὕδατι ...

Contemporaneously with these philosophers[242] and before them, the so-called Pythagoreans devoted themselves to mathematics; they were the first to advance this study, and having been brought up in it they thought its principles were the principles of all things. Since of these [i.e. mathematical sciences] numbers are by nature the first, and in **these** [i.e. numbers] they supposed they could see many resemblances to the things that exist and come into being—more than in fire and earth and water ...

---

24 ταῦτα α Ascl.ᴾ 35.31 Bekker Bonitz Christ Primavesi : ταῦτά τε β Ross Jaeger ‖ προῆγον α Ascl.ᴾ 35.31 Primavesi : προήγαγον β Bekker Bonitz Christ Ross Jaeger ‖ 25 τῶν ὄντων ἀρχὰς α Al.ᴾ 37.13; 19 Ascl.ᴾ 35.32 edd. : om. β ‖ 27 τούτοις β ωᴬᴸ (Al.ᴾ 37.22-23; 38.5-6) Ross Jaeger : τοῖς ἀριθμοῖς α fort. ex Al.ᴾ 38.5-6 (Ascl.ᶜ 35.33-34) Bekker Bonitz Christ Primavesi

The following analysis will focus on line b27. According to the β-version, this line contains the demonstrative pronoun τούτοις ("[in] these"). But the α-version reads τοῖς ἀριθμοῖς ("numbers") instead, thus spelling out what the antecedent of τούτοις would be. In order to assess the status of τούτοις in line b27 we first have to have a look at the demonstrative pronoun τούτων ("of these") in line b26. The antecedent of τούτων is found in the preceding sentence (985b23-26), although it is not easily found, as there is more than just one possible candidate:[243] the principles of all things (τῶν ὄντων ἀρχὰς ... πάντων, b25-26),[244] all things (τῶν ὄντων ... πάντων, b25-26),[245] the principles of mathematics (τὰς τούτων ἀρχὰς, b25), and mathematics itself (τῶν μαθημάτων, ... ταῦτα ... τούτων b24-25). I agree with Primavesi 2014 that the pronoun τούτων (b26) refers to μαθήματα ("mathematics," b24), and that Aristotle thus starts his reconstruction of the theory from the universally accepted position that mathematics starts with numbers.

In any case, the frequent use of demonstrative pronouns[246] in this passage corresponds to the condensed exposition Aristotle gives here of the Pythagorean theory of principles. These pronouns allow Aristotle to refer briefly to the aforementioned terms without extending and burdening the exposition through

---

[242]These are the Atomists, Leucippus and Democritus, whom Aristotle treats in A 4, 985b4-22. On the question whether Parmenides and Empedocles should be included, too, see Primavesi 2014: 228. See also Alexander (37.6-12), who takes further options into consideration.

[243]See the discussion in Primavesi 2014: 234-35.

[244]This possibility can be excluded. The assumption that numbers are first among the principles of all things cannot be the *starting point* from which the Pythagoreans, according to Aristotle, or Aristotle himself could begin. Ross's translation, however, inclines the reader to this interpretation (Barnes 1984: 1559): "...they thought its principles were the principles of all things. Since of these principles..." (my emphasis).

[245]Schofield 2012: 144 n. 8 follows Alexander (37.21-22) in taking "all things" to be the antecedent.

[246]Ἐν δὲ τούτοις (b23), πρὸ τούτων (b23), ταῦτα (b24), τὰς τούτων ἀρχὰς (b25), τούτων (b26), ἐν δὲ τούτοις (b27).

repetition or detailed description. Aristotle speaks in summary fashion, he hints to us, because he has given a comprehensive account of the Pythagorean theory "elsewhere."[247]

By contrast, the antecedent of the demonstrative pronoun τούτοις in line b27 is instantly clear. The pronoun τούτοις refers back to οἱ ἀριθμοί in line b26. Yet, in light of the high number of pronouns in this passage and the ambiguous τούτων in line b26, it is easy to imagine that someone had been confused by the flurry of demonstratives and had taken τούτοις (b27) to refer to τούτων (b26) and its antecedent. It is just as easy to imagine that someone, hoping to prevent such a misunderstanding had changed the τούτοις (β) to the explicative τοῖς ἀριθμοῖς, which we find in the α-version. By just such an intervention we can account for the difference between the two versions α and β. By contrast, it is quite unimaginable that someone would have changed τοῖς ἀριθμοῖς to τούτοις (to avoid repeating οἱ ἀριθμοί?),[248] given the abundance of demonstrative pronouns in the passage. Following the rule *utrum in alterum*, it seems more likely that the β-reading τούτοις was changed into the α-reading τοῖς ἀριθμοῖς.

With these considerations in mind, we look at the evidence in Alexander's commentary. In 37.21–38.1, Alexander summarizes Aristotle's statement about the two premises of the Pythagorean theory of principles (985b26–29),[249] and then turns to the second premise in particular (38.5–7), which asserts that numbers bear more resemblances to things than the elements do. At the beginning of the commentary passage in 37.21–38.1 we read the following:

Alexander, *In Metaph.* 37.21–38.1 Hayduck

ἐπεὶ γὰρ τῶν ὄντων οἱ ἀριθμοὶ φύσει πρῶτοι [22] (ἐξ ἀφαιρέσεως γὰρ οὗτοι) καὶ ὅτι ἐδόκουν <u>ἐν αὐτοῖς</u> ὁμοιώματα πολλὰ [23] πρὸς τὰ ὄντα ὁρᾶν καὶ πρὸς τὰ γιγνόμενα, καὶ μᾶλλον <u>ἐν τούτοις</u> ἢ ἐν [38.1] τοῖς ἁπλοῖς σώμασιν...

For since numbers are by nature first among the things that are[250] (for they are from abstraction), and they (i.e. the Pythagoreans) supposed they could see <u>in them</u> (i.e.

---

[247] *Metaph.* 986a12–13: διώρισται δὲ περὶ τούτων ἐν ἑτέροις ἡμῖν ἀκριβέστερον. / "we have discussed these matters more exactly elsewhere." In his commentary on A 5, Alexander excerpts from Aristotle's lost monograph on the Pythagoreans (see Primavesi 2011c: 170–71). On fragments of this monograph in Alexander see Wilpert 1940. On the application of these fragments to the explication of Aristotle's account in A 5 see Primavesi 2014: 236–46.

[248] One could argue that the β-reviser shuns repetition (cf. Patzig/Frede 1988: 14; Primavesi 2012b: 457–58) and accordingly suspect that the more repetitive α-version preserves the authentic text. This is, however, unlikely as in this passage the β-reviser's intention would clash with the condensed style of the presentation and the already existing repetition of pronouns.

[249] Here, Alexander uses the particle δέ, which he proposes as a correction for Aristotle's γάρ (37.20–21). Cf. 3.6.

[250] As pointed out above (p. 246 n. 245), Alexander understands τούτων in 985b26 to refer back to the things that are (τῶν ὄντων).

> numbers) many resemblances to the things that are and that come into being, and [this] in them (i.e. numbers) rather than in the simple bodies …

---

23 πρὸς alt. **A** : om. **O P**ᵇ

In his paraphrase, Alexander stays close to the Aristotelian original, referring back to the numbers (οἱ ἀριθμοί) by means of demonstrative pronouns (37.22–23) in agreement with the β-reading.

A second reference to Aristotle's words, occurring a few lines later in the commentary, once more indicate that ω^AL, like the β-text, read ἐν τούτοις (b27). Alexander again paraphrases Aristotle's formulation by, again, referring to the numbers with ἐν τούτοις. The agreement of ω^AL with the β-reading confirms the above assumption that β is the older reading.[251] Alexander's second paraphrase reads:

> Alexander, *In Metaph.* 38.5–7 Hayduck
>
> ὁμοιώματα δὲ μᾶλλον πρὸς τὰ ὄντα καὶ γινόμενα ἡγοῦντο [6] ἐν τούτοις εἶναι, τουτέστι τοῖς ἀριθμοῖς, ἢ ἐν τούτοις τοῖς σώμασιν ἃ στοι-[7]χεῖά φαμεν…
>
> They thought that resemblances to the things that are and that come into being are in them, i.e. in numbers, rather than in those bodies we call 'elements.'

This passage reveals more than just the aforementioned agreement between ω^AL and β. For Alexander not only repeats the demonstrative pronoun he finds in Aristotle's text; he also adds his own explication of τούτοις, so as to make clear that it refers back to the numbers: τουτέστι τοῖς ἀριθμοῖς (38.6). This clarifying addition shows quite definitely that Alexander did not have the α-reading τοῖς ἀριθμοῖς (985b27) in his *Metaphysics* text, and furthermore points to a possible origin for the α-reading. A reader or scribe of the α-version could have followed Alexander in his insistence on stating the antecedent of τούτοις unambiguously, and replaced τούτοις in the α-text with τοῖς ἀριθμοῖς.[252]

As pointed out above, Aristotle's concise presentation and repeated usage of demonstrative pronouns make it likely that someone would have changed τούτοις to τοῖς ἀριθμοῖς. This holds irrespective of Alexander's comments on the passage—in fact, the replacement of τούτοις with τοῖς ἀριθμοῖς appears so natural that it is quite possible for the substitution to have occurred here without the influence of Alexander. This passage therefore stands in contrast to the previous one

---

[251] It would be unreasonable to suggest that β adopted τούτοις from Alexander's paraphrase and used it to replace τοῖς ἀριθμοῖς in the *Metaphysics* text. First, the τούτοις makes the sentence more difficult to understand, as we saw above. Second, this adaptation would be at odds with Alexander's own explications of the passage, as we will see below. Primavesi 2012c nevertheless follows the α-reading.

[252] Jaeger 1917: 490 assumes that τοῖς ἀριθμοῖς is a gloss that intruded into the α-family (which Jaeger calls "recension Π"). Jaeger cites Alexander as evidence for the authenticity of the β-reading, but he does not link the evidence in Alexander's commentary to the contamination that occurred in α. Both Jaeger und Ross follow the β-text in their editions.

(5.3.1), where it was not obvious that οὕτως should be replaced with τοῦτον τὸν τρόπον.²⁵³ Nevertheless, it would here be a remarkable coincidence if Alexander had reformulated Aristotle's words in exactly the same way in which a reader or scribe of the α-text, independently of Alexander's commentary ended up revising them. It might further be added that the type of correction we encounter here in α (i.e. change from pronoun to noun) is by no means a common feature of α,²⁵⁴ a fact that makes Alexander's influence in this case perhaps more likely. All in all, then, the conclusion that we are dealing here with a further trace of Alexander's influence on the α-text might seem justified.

5.3.3 Alex. In Metaph. 33.17–19; 23–26 on Arist. Metaph. A 4, 985a4–10

In A 4, Aristotle gives the following explanation of how his predecessors, the material cause having been discovered, went on to formulate a second principle: they recognized that the material cause alone could not account for the order and beauty in the world, and so they searched for a further principle, an efficient cause (A 3, 984b8–15; cf. also 5.2.5). Parmenides and Hesiod made love and desire (ἔρως) to be such causes (A 4, 984b23–31), while Empedocles, recognizing that the world also contains disordered and bad things, introduced two principles, love (φιλία) and strife (νεῖκος) (984b32–985a4).

Aristotle, *Metaphysics* A 4, 985a4–10

εἰ γάρ τις ἀκολουθοίη καὶ λαμβάνοι πρὸς τὴν διά-[5]νοιαν καὶ μὴ πρὸς ἃ ψελλίζεται λέγων Ἐμπεδοκλῆς, εὑρή-[6]σει τὴν μὲν φιλίαν αἰτίαν οὖσαν τῶν ἀγαθῶν τὸ δὲ νεῖκος [7] τῶν κακῶν· ὥστ' εἴ τις φαίη τρόπον τινὰ καὶ λέγειν καὶ [8] πρῶτον λέγειν τὸ κακὸν καὶ τὸ ἀγαθὸν ἀρχὰς Ἐμπεδοκλέα, [9] τάχ' ἂν λέγοι καλῶς, εἴπερ τὸ τῶν ἀγαθῶν ἁπάντων αἴτιον [10] αὐτὸ τὸ ἀγαθόν ἐστι [**καὶ τῶν κακῶν τὸ κακόν**].

For if we were to follow out the view of Empedocles, and interpret it according to its meaning and not to its lisping expression, we should find that Love [*Philia*] is the cause of good things, and Strife [*Neikos*] of bad. Therefore, if someone said that Empedocles in a sense both mentions, and is the first to mention, the Bad and the Good as principles, he should perhaps be right, given that the cause of all good things is the Good itself [**and of the bad things the Bad**].

10 καὶ τῶν κακῶν τὸ κακόν α ex Al. 33.25–26 (Ascl.ᶜ 31.9) Bekker Bonitz : om. β ω^AL Ascl.ᵖ 31.9-11 del. Ross Jaeger Primavesi

My focus will be on the words καὶ τῶν κακῶν τὸ κακόν in line 985a10, which

---

²⁵³ Are these two passages (the one discussed in 5.3.1 and the one discussed presently) related?
²⁵⁴ There is only one comparable case in A 1, 980a28–29: τῆς αἰσθήσεως α : ταύτης β. Here, Alexander's commentary (2.22–4.11) unfortunately does not offer any evidence as to whether ω^AL goes with α or β.

Primavesi 2012b identified as an "α-supplement," that is, a secondary addition to the α-text.[255] In the passage quoted above, Aristotle argues that Empedocles' principles Love and Strife are the causes of good and bad things (a4–7). He further entertains the argument that Empedocles was the first to introduce[256] the Good (τὸ ἀγαθόν) and the Bad (τὸ κακόν) as principles (a7–9),[257] remarking that such holds especially "if indeed" or "given that" (εἴπερ)[258] the Good itself is the cause of all good things. According to the β-version, Aristotle's sentence stops at the mention of the Good itself, but the α-text goes further, adding to the conditional clause (which here has causal force: "given that…," "since…") that the Bad is the cause of bad things. Since Aristotle denies the existence of the Bad as a principle,[259] the β-version is clearly preferable to the α-reading. Moreover, the α-version exhibits a symmetry, which although implied in the duality of Empedocles' principles,[260] does not at all square with "the Good itself" as principle.

In his commentary Alexander paraphrases the passage and criticizes the argument it contains. From the commentary we are able to gather that Alexander did not find the α-supplement καὶ τῶν κακῶν τὸ κακόν in his *Metaphysics* text. But in addition to this, Alexander's comments also provide clues as to the origin of the α-supplement.

Alexander, *In Metaph.* 33.17–19; 23–26 Hayduck

ἐπειδὴ γὰρ ἐν τοῖς οὖσιν ἔστι καὶ τὰ κακά, Ἐμπεδοκλῆς ἐν τοῖς αἰτίοις [18] ἔθετο οὐ τὴν τῶν ἀγαθῶν ἀρχὴν μόνον, ἥτις ἐστὶ φιλία, ἀλλὰ καὶ τὴν [19] τῶν κακῶν, ὅ ἐστι τὸ νεῖκος. [20–23] … εἰ δὲ τὸ τῶν ἀγαθῶν αἴτιον ἀγαθὸν [24] καὶ τὸ τῶν κακῶν κακόν, ἀρχὰς ἂν εἴη Ἐμπεδοκλῆς τὸ ἀγαθὸν καὶ τὸ [25] κακὸν τιθέμενος, ἀγαθὸν μὲν τὴν φιλίαν, τὸ νεῖκος δὲ κακόν. εἰπὼν δὲ [26] οὕτως περὶ τοῦ ἀγαθοῦ, περὶ τοῦ κακοῦ ἡμῖν προσθεῖναι κατέλιπε.

For since bad things, too, exist among the things that are, Empedocles included among the causes not only the principle of good things, which is Love, but also the

---

[255] Primavesi 2012b: 440–43. *Metaphysics* editors preceding Primavesi have, since Christ 1886, followed the β-reading.

[256] Primavesi 2012b: 443 points to the passage in A 8, 989b4–21 where Aristotle records Anaxagoras's theory of primeval mixture as an earlier equivalent of Plato's second principle of the ἀόριστος δυάς.

[257] Cf. also Aristotle's comments on Empedocles' principles Love and Strife in Λ 10, 1075b1–7. See also Beere 2009: 326.

[258] LSJ s.v. εἴπερ II.

[259] In *Metaph.* Θ 9, 1051a17–18, Aristotle explicitly says that there is no Bad over and above things: δῆλον ἄρα ὅτι οὐκ ἔστι τὸ κακὸν παρὰ τὰ πράγματα. See Beere 2009: 325–28 and 344–47. Primavesi 2012b: 443 calls "the Bad itself" an "inexcusable blunder" by "Platonic standards." Consider passages like *Rep.* III 402c, V 476a and *Tht.* 176e, however, where Plato has Socrates speak about the Form of the Bad. For the Bad itself in the Platonic tradition see Plot. I 8,3,1–4,5 (see also Dörrie/Baltes 1996: 123.8, pp. 190–94 and 516).

[260] See the discussion in Primavesi 2012b: 440–42.

principle of bad things, which is Strife. ... But if the cause of good things is good and that of bad things bad, Empedocles would seem to make the Good and the Bad his principles, Love the good, Strife the bad. Yet having spoken in this way about the Good, he has left it to us to supply the point about the Bad.

18 ἐστὶ **A O** : ἐστὶν ἡ **P**^{b} || 19 ὅ **LF** Ascl. : ἥ **A O P**^{b} **S** || 23 δὲ **A O** : γὰρ **P**^{b} **S** || καὶ τὸ **A O** : τε καὶ **P**^{b} || 25 νεῖκος δὲ **A O** : δὲ νεῖκος **P**^{b}

Alexander puts forward a version of Aristotle's argument that has been subject to three relevant alterations: Firstly, he changes the order in which the argument is presented. Secondly, he does not speak of the Platonic idea of "the Good itself," but rather of the principle that the cause of good things is itself good. This allows him, thirdly, to also mention a bad principle as the cause of bad things. I will now look more closely at these differences.

Alexander follows Aristotle (985a4-7) in identifying Empedocles' Strife with the principle of bad things and Love with the principle of good things (33.17-19). Then, in 33.23-24, he puts forth a slightly modified version of the thought that Aristotle expressed conditionally (εἴπερ / "given that") at the end of the passage (985a9-10). In Aristotle's version, Empedocles is credited with the discovery of the Good as principle on the condition of the assumption (εἴπερ) that the Good itself is the principle of good things. By contrast, Alexander's condition (εἰ), which he puts at the beginning of the sentence, does not mention "the Good itself" but rather applies the so-called "Causal Resemblance Principle"[261] to the case of good and bad things. According to this principle the cause of good things is something that itself is good and the cause of bad things something itself bad. Consequently, in speaking of a principle of good things that itself is good, Alexander makes no mention whatsoever of the Platonic principle of "the Good itself." This allows him, without further ado, to include a principle of bad things that itself is bad. In 33.24-25, Alexander returns to a close proximity to Aristotle's thought and adopts his conclusion that Empedocles made the Good and the Bad his principles (985a8-9) and that he identifies Love with the Good and Strife with the Bad (33.25).

In 33.25-26, Alexander reflects on the way in which he modified Aristotle's argument. He highlights his most obvious alteration and points out that Aristotle speaks about the Good only, while leaving it up to the reader to supplement the argument with the point about the Bad (33.25-26).[262] As we have just seen, in

---

[261] This principle seems to have been widely accepted by ancient philosophers. See Makin 1990: 138, who calls it "Causal Resemblance Principle" and also "Degree of Reality Principle." See also Sedley 1998. Betegh 2012: 126 speaks of a "principle of causational synonymy" in connection with Aristotle's interpretation of Empedocles' theory that a good principle causes good things and a bad principle causes bad things. See also *Metaph.* α 1, 993b24-31 and for the principle *propter quod alia, id maximum tale* see Lloyd 1976 and Rashed 2007: 312.

[262] There are other passages in Alexander's commentary where he notes that Aristotle did not explicitly express a point that would naturally follow from what had been said. Cf. for example: 192.5-6: οὐκέτι τὸ ἑξῆς προστιθεὶς αὐτῷ (on this see Madigan 1992: 120 n. 142); 193.14: ἐπαύσατο μηδὲν

Alexander's version of the passage the point about the bad is included, although in a different way. The purpose of this reflection seems to be to justify his adjustment made to Aristotle's argument through the addition of the principle of bad things that is itself bad (καὶ τὸ τῶν κακῶν κακόν, 33.24).

We can infer two things from Alexander's concluding reflection on the Aristotelian text. First, as recently demonstrated by Primavesi 2012b: 442, Alexander's words make it abundantly clear that ω^AL did not have the α-supplement καὶ τῶν κακῶν τὸ κακόν.[263] The presence of the bad as principle of bad things in Alexander's paraphrase is due entirely to his own modification of the argument and is not based on his *Metaphysics* text (ω^AL).[264] Since ω^AL and β agree with each other, we can assume that ω^αβ did not have the α-supplement either.[265]

Second, we see that what Alexander describes as a (perhaps even intentional) omission by Aristotle is filled in precisely by the α-supplement. But there is more. Alexander's own filling in as presented in his paraphrase of the Aristotelian passage is—apart from the position of the article—identical with the α-supplement. Alexander says καὶ τὸ τῶν κακῶν κακόν, while the α-supplement reads καὶ τῶν κακῶν τὸ κακόν. When we view the two formulations and their respective syntactical context side by side the close parallel is plain to see:[266]

Arist. A 4, 985a9–10	εἴπερ τὸ τῶν ἀγαθῶν ἁπάντων αἴτιον αὐτὸ τὸ ἀγαθόν ἐστι [καὶ τῶν κακῶν τὸ κακόν].
Alex. 33.23–24	εἰ δὲ τὸ τῶν ἀγαθῶν αἴτιον ἀγαθὸν καὶ τὸ τῶν κακῶν κακόν

ἐπενεγκών. Whenever Alexander recognizes an omission by Aristotle and suggests a supplement I take this to be a conjecture (see 3.6): 193.32–33; 264.17–18; 321.1. There is no exact parallel to Alexander's present diagnosis that Aristotle left it to the reader to fill in the gap.

[263] One could object that Alexander's remark does not tell exactly where in Aristotle's argument the bad is left out and that we therefore cannot know for sure that ω^AL did not read the α-supplement. However, as Primavesi 2012b: 442 demonstrated, the argumentative step that is made explicit in the α-supplement is the only one which Aristotle could have left out without threatening the parallelism between Empedocles' Love and Strife and the Good and the Bad, which is Aristotle's main point.

[264] Asclepius provides contradictory information in his commentary. He quotes (31.8–9) the α-version of the *Metaphysics* text (viz. including the words καὶ τῶν κακῶν τὸ κακόν), yet adopts Alexander's comment about the omission and supplementation of the bad. That Asclepius clearly understood the words he copied from Alexander can be inferred from his own remark in 31.10–11. Has the quote in Asclepius's commentary been subsequently adjusted to the α-version? For a discussion of the textual evidence in Asclepius's commentary see Kotwick 2015.

[265] From the content alone of the two readings it is clear that ω^AL and β have the correct text and the α-version contains a later addition. There is therefore no reason to speculate that the β-reading is the result of a deletion based on the model given in Alexander's commentary. It is more reasonable to assume that β preserved the correct reading, which was corrupted in α by a later supplement. Cf. the analysis by Primavesi 2012b: 442.

[266] The similarity is so striking that we would be forced to assume that Alexander here gives an exact paraphrase of the *Metaphysics* text, had he not made it crystal clear that in his *Metaphysics* text there is no mention of the bad in the conditional clause and were we unable to rule out the possibility that Aristotle mentions the "bad itself" as a principle.

Despite the obvious parallelism between these two phrases, still evident is where Alexander's exposition of the argument diverges from the Aristotelian counterpart. This divergence is encapsulated in the positioning of the article τό. In the α-supplement the article stands next to the noun to which it belongs: τῶν κακῶν τὸ κακόν / "The bad (*scil.* is the cause) of bad things." But in Alexander's commentary the article τό stands far away from the adjective κακόν. The reason is this, that in Alexander's commentary the article τό does not belong to κακόν, but to αἴτιον, which has to be supplied in thought from the preceding part of the sentence: τὸ (*scil.* αἴτιον) τῶν κακῶν κακόν / "the cause of bad things (is) bad." This is visible in the above analysis: Alexander slightly changed Aristotle's argument and speaks about the principle according to which the cause of something is itself what it causes.

We further see that the only difference between Alexander's paraphrase and Aristotle's text is that Alexander does not speak about "the Good itself" but about the Good simply. As I argued above, this shift is what allows Alexander to position the Bad alongside the Good as a principle. While the words of the α-supplement are a disturbing appendage in the context of the *Metaphysics* passage they are entirely appropriate to the context of Alexander's commentary. Leaving aside all questions of philosophical meaning, the counterpart to the Good itself (αὐτὸ τὸ ἀγαθόν) we would expect to find in the *Metaphysics* passage is "the Bad itself" (αὐτὸ τὸ κακόν). But this is not what the α-supplement offers. Therefore, it is reasonable to conclude that the formulation was taken over directly from Alexander's commentary and inserted into the α-text. Since Alexander speaks only about the bad (*scil.* principle) (τὸ ... κακόν) and not about the Bad itself, the α-supplement reads τὸ κακόν and not αὐτὸ τὸ κακόν. However, since a direct copy of Alexander's words (καὶ τὸ τῶν κακῶν κακόν) made little sense in the Aristotelian context, the article τό was placed next to κακόν, thus changing the meaning to "the bad ... of the bad things."

We see that the striking points of similarity and the interesting incongruities between the two phrases reveal Alexander's commentary to be the origin of the α-supplement. It is likely that a reader of the commentary interpreted Alexander's reflection about Aristotle's leaving it to us to supply the bad differently than I did. I take Alexander's reflection to be a justification of the slight alteration Alexander made to Aristotle's argument, but the hypothetical reader seems to have taken it as an invitation to supplement[267] and correct the *Metaphysics* text.[268] It could very

---

[267] Sepúlveda rendered Alexander's words περὶ τοῦ κακοῦ ἡμῖν προσθεῖναι κατέλιπε into the simple words *omisit mentionem de malo*. Here, the inviting character of Alexander's phrase is lost. For Sepúlveda's inaccuracies in translating Alexander's comments on textual issues see 2.3 and 5.3.4.

[268] This is how I understand (*pace* Primavesi 2012b: 442 n. 136) the comment in Ross 1924: 137: "καὶ ... κακόν, omitted by A^b, Alexander, and Asclepius, was probably suggested to some copyist by Alexander's remark that something of the sort *must be supplied* to complete the sense [my emphasis]." Also Jaeger 1957 *ad loc.* supposes such an influence on the manuscript E: "Al aliquid huiusmodi desiderabat,

254 ALEXANDER AND THE TEXT OF ARISTOTLE'S *METAPHYSICS*

likely have been that a reader who interpreted Alexander's words in this way also drew inspiration from his paraphrase and copied directly from his commentary.

5.3.4 Alex. *In Metaph.* 67.20–68.4 on Arist. *Metaph.* A 8, 989a22–26

In chapter A 8, Aristotle critically engages the theories of causation put forward by the Presocratic philosophers Empedocles, Anaxagoras, and the so-called Pythagoreans.[269] The discussion of Empedocles begins in line 989a18.[270] Aristotle first classes the mistakes Empedocles makes: some of the mistakes he makes he shares with the monists, even though he posited four material causes, and some of the mistakes he makes are peculiar to him (989a21–22). He focuses his critique with the following two points. First, Empedocles' theory of the four elements denies that the elements change into and out of each other (989a22–24).[271] Second, it is unreasonable to assume two efficient causes rather than just one (989a25–26).[272] The following section contains the two points of critique in Aristotle's words:

Aristotle, *Metaphysics* A 8, 989a22–26

γιγνόμενά τε γὰρ ἐξ [23] ἀλλήλων ὁρῶμεν ὡς οὐκ ἀεὶ διαμένοντος πυρὸς καὶ γῆς τοῦ [24] αὐτοῦ σώματος (εἴρηται δὲ ἐν τοῖς περὶ φύσεως περὶ αὐτῶν), [25] καὶ περὶ τῆς τῶν κινουμένων αἰτίας, πότερον ἓν ἢ δύο θετέον, [26] οὔτ' ὀρθῶς οὔτε **εὐλόγως** οἰητέον εἰρῆσθαι παντελῶς.

For we see these bodies produced from one another, which implies that the same body does not always remain fire or earth (we have spoken about this in our works on nature);[273] and regarding the moving cause and the question whether we must suppose one or two, he must be thought to have spoken neither **correctly** nor altogether reasonably.

---

26 εὐλόγως ω^AL β <E>γρ Ar^n Bekker Bonitz Christ Ross Jaeger Primavesi : ἀλόγως ci. Al. 68.3–4 α Ascl.^c 60.25

Lines a25–26 summarize Aristotle's critique of Empedocles' thoughts on the efficient cause. The two branches α and β offer divergent readings. According to the β-version, Aristotle says that Empedocles spoke neither correctly (ὀρθῶς, a26) nor altogether reasonably (εὐλόγως … παντελῶς, a26). In the α-version, we read

---

unde supplevit E." See also Jaeger 1917: 486. The new collations by Pantelis Golitsis and Ingo Steinel show that the supplement is not just present in the ms. E, but in the whole α-family. Cf. also Betegh 2012: 125 with n. 47.

[269] See Primavesi 2012a: 225–63. On the differences among Aristotle's treatments of these thinkers in *Metaph.* A 3–5 and A 8 see Primavesi 2012a: 226–27.

[270] Primavesi 2012a: 229–32.

[271] Primavesi 2012a: 232–35.

[272] Primavesi 2012a: 235–39.

[273] According to Alexander (*In Metaph.* 67.13–15), this refers to *Cael.* Γ, while according to Asclepius (*In Metaph.* 60.11–12) it also refers to *GC* B. See Primavesi 2012a: 233–34.

in place of οὔτε εὐλόγως ("nor reasonably") the words οὔτε ἀλόγως ("nor unreasonably"). The small difference in letters results in a great difference in meaning. Whereas in the β-reading Aristotle describes Empedocles' theory with two negated terms, saying that he spoke "neither correctly nor altogether reasonably" (οὔτ' ὀρθῶς οὔτε εὐλόγως), Aristotle's assessment in the α-version is more positive: he says that Empedocles spoke "neither correctly nor altogether unreasonably" (οὔτ' ὀρθῶς οὔτε ἀλόγως).

The doubly negative condemnation of Empedocles in the β-version (οὔτ' ὀρθῶς οὔτε εὐλόγως) is not tautological, as it might at first appear. A close look reveals that the two negated terms mean two different things. The οὔτ' ὀρθῶς expresses that Empedocles' assumption of two moving causes is simply mistaken. The subsequent words οὔτε εὐλόγως ... παντελῶς point to another flaw in Empedocles' theory: on the whole and in respect to his entire cosmic system (παντελῶς) he does not make proper use (οὔτε εὐλόγως) of his moving causes, Love and Strife. The two negated terms in the β-reading thus make good sense.

The authenticity of the β-reading is supported by two other passages in the *Metaphysics*. In A 4, 985a21–31, we find Aristotle use a similar expression (οὔτε ... οὔτε) to criticize Empedocles' employment of the moving causes within his cosmic system. There, Aristotle's says that Empedocles uses his causes "neither sufficiently nor does he attain consistency in their use" (οὔτε ἱκανῶς, οὔτ' ἐν τούτοις εὑρίσκει τὸ ὁμολογούμενον, 985a23). In the second parallel passage, not far from the passage that presently concerns us, Aristotle uses, again, two negated terms, this time in his critique of Anaxagoras. He says of Anaxagoras in A 8, 989b19 that he expresses himself "neither correctly nor clearly" (οὔτ' ὀρθῶς οὔτε σαφῶς). As with the β-reading under examination, this paired negative appears at first redundant, yet is not: the critique hits the content of the theory and the form of presentation.

In light of the above, the β-reading in our passage appears preferable to the α-reading. This conclusion is reinforced by the fact that all *Metaphysics* editors unanimously opt for the β-reading, and cemented by the fact that Alexander's commentary (67.18–68.4) confirms the β-reading. We can infer from his words that he read εὐλόγως (β) in ω^AL. Alexander's commentary gives us yet more: it provides important information concerning the origin of the α-reading.

Alexander, *In Metaph.* 67.20–68.4 Hayduck

φαίνεται γὰρ καὶ ἡ φιλία τὰ τοῦ νεί-[21]κους ποιοῦσα· διαιρεῖ γὰρ καὶ διακρίνει τὰ κατ' ἰδίαν ὅλα ὄντα, ἵνα [22] συγκρίνῃ καὶ ἓν σῶμα ποιήσῃ. ἀλλὰ καὶ τὸ νεῖκος αὐτῷ οὐ μόνον δια-[68.1]κρίνει, ἀλλὰ καὶ συγκρίνει καὶ συνάγει τὰ ὅμοια πρὸς ἄλληλα, χωρίζον ἐκ [2] τῆς ἑνώσεως αὐτά· ὥστε οὐδὲν θάτερον θατέρου μᾶλλον τῶν ἀντικειμένων [3] ἐστὶν αὐτῷ ποιητικόν. ἢ ἄμεινον γεγράφθαι τὸ οὔτε ἀλόγως, ἵνα ᾖ τὸ [4] λεγόμενον οὔτε πάντῃ ὀρθῶς οὔτε ἀλόγως πάντῃ.

For Love obviously produces the effects of Strife, too, since it divides and separates things that by themselves are wholes, in order to combine them and make one body. But Strife too, in his view, not only separates, but also combines and brings together things that are like, separating them from their unified state [with things unlike them], so that for him neither one of these two opposed principles is in any way a more efficient cause than the other. Or, the text would be better written thus: "[neither correctly] nor [altogether] unreasonably," so that the sense would be that Empedocles spoke neither altogether correctly nor altogether unreasonably.

20 τοῦ **A O** : om. **P**ᵇ ‖ 3 αὐτῷ **A O S** : αὐτὸ **P**ᵇ ‖ ἢ ἄμεινον γεγράφθαι **A O** : ἢ ἄμεινον γεγράφθω **P**ᵇ : *quanquam nescio an rectius sit, quod in quibusdam exemplaribus legitur ad hunc modum* **S**

Alexander accomplishes two things in this commentary passage. First, he spells out what Aristotle's assessment οὔτ' ὀρθῶς οὔτε εὐλόγως means. His focus in doing so falls on the expression οὔτε εὐλόγως. Second, he offers an alternative reading for this expression.

In explicating Aristotle's critique of Empedocles, Alexander repeats both what Aristotle said in A 4 (985a21–31) and also his own comments on that passage (35.6–23). Love, in order to unify all the elements, must break down combinations of elements that already exist. Strife, by dissolving all mixtures and combinations of the elements, also causes the elements to group together according to their kinds. Thus, the effect of unification is not restricted to Love's action, nor is the effect of separation restricted to Strife's action. As a result, neither action is clearly defined.[274]

In the end, in his commentary on the passage in A 8 Alexander seems unsatisfied with Aristotle's negative résumé of Empedocles' principles of movement and thinks Aristotle's judgment should be milder. Perhaps this is because Alexander expects Aristotle to be gentler on a thinker he tends to view favorably, or perhaps it is because Alexander himself thinks that Empedocles deserves a more positive assessment. In any case, Alexander expresses doubts on whether the reading he finds in his *Metaphysics* text constitutes the best possible summary of the critique on Empedocles (68.3). He does not tell us whether he thinks the reading is a corruption of Aristotle's original or whether he believes Aristotle's (original) expression could be improved upon. He simply states that in place of the phrase οὔτ' ὀρθῶς οὔτε <u>εὐ</u>λόγως it would be better to read the phrase οὔτ' ὀρθῶς οὔτε <u>ἀ</u>λόγως. Thus, Alexander advocates toning down Aristotle's judgment to the statement that although Empedocles did not speak correctly, he did not on the whole speak unreasonably either (68.4).

What kind of textual change does Alexander's suggestion constitute? The editors Christ (1886a), Bonitz (1848)[275] and Primavesi (2012)[276] take Alexander's sug-

---

[274] For the question whether this critique of Empedocles is appropriate see Primavesi 2012a: 235–39.
[275] Christ and Bonitz in their apparatus: γρ *Alex*.
[276] Primavesi 2012c *ad loc.*: *Al*. 68, 3–4 *ex alio libro citans*. See also Primavesi 2012b: 408 n. 84.

gestion to be a report of a *varia lectio*.²⁷⁷ Ross (1924), however, notes it in his apparatus as Alexander's own conjecture.²⁷⁸ Indeed, the phrase ἄμεινον γεγράφθαι (or, when following **Pᵇ**, the imperative form γεγράφθω)²⁷⁹ does not at all suggest that Alexander reports a *varia lectio* found in another version of the text. Most likely Alexander would have introduced such a variant reading by using one of his standard expressions (φέρεται δὲ ἔν τισι... / γράφεται ...).²⁸⁰

The phrase ἄμεινον γεγράφθαι first of all suggests (ἤ, 68.3) that in Alexander's opinion the text could be improved by a slight adjustment.²⁸¹ Whether the suggested correction is his own idea or whether it was in fact borrowed from another commentator cannot be determined for certain (despite the absence of any reference to τινές, cf. 3.5). Yet, given the evidence of two parallel passages in which Alexander introduces his own emendations with ἄμεινον γεγράφθαι (186.31; 233.26)²⁸² it can

---

²⁷⁷ We also find this interpretation expressed in the Latin translation or paraphrase by Sepúlveda (*f.* e.i.r*): <u>quanquam nescio an rectius sit, quod in quibusdam exemplaribus legitur ad hunc modum</u>, neque penitus absque ratione, ut sit sensus. Neque recte prorsus, nec penitus absque ratione*. Sepúlveda expands Alexander's often vague hints at an alternative reading into a more detailed suggestion. So Sepúlveda's words do not necessarily represent what he found in his Greek manuscripts. It seems more likely that Sepúlveda himself added these details to Alexander's short and rather vague ἄμεινον γεγράφθαι (or ἄμεινον γεγράφθω as in **Pᵇ**): ἄμεινον became *nescio an rectius sit* and γεγράφθαι became *quod in quibusdam exemplaribus legitur*. There are several instances where Sepúlveda does something similar. In 186.31-32, Alexander writes: ἤ ἄμεινον γεγράφθαι... . Sepúlveda "translates" (*f.* k.iv.v*): Aut certe <u>melius in quibusdam exemplaribus scriptum est</u> ad hunc modum... *. Here, the commentary context clearly shows that we are dealing with Alexander's *own suggestion* for an alternative reading. In yet another parallel passage, Alexander, according to the evidence in the Greek manuscripts, introduces his own conjecture with the expression ἄμεινον γεγράφθαι. This time Sepúlveda seems to provide a more literal translation (*f.* n.v.v*): quanquam melius ad hunc modum scriptum est.*

The *recensio altera*, which Golitsis 2014b dates to the sixth or seventh century AD, could (once again) have had access to both versions **α** and **β**. It reads (*app.* 67 Hayduck): οὕτως γὰρ οὔτ' ὀρθῶς οὔτ' εὐλόγως οἰητέον εἰρῆσθαι αὐτῷ παντελῶς. <u>γράφεται δὲ ἐν ἄλλοις</u> οὔτε ὀρθῶς οὔτε <u>ἀλόγως</u>.

²⁷⁸ Ross 1924: *ci. Al.* Jaeger (1957) leaves the matter undecided and says merely: *Al.*

²⁷⁹ The reading in **A** and **O** (ἄμεινον γεγράφθαι) is preferable to the reading in **Pᵇ** (ἄμεινον γεγράφθω) simply because we have several parallel passages in Alexander (and other authors, e.g. Galen), where the formula ἄμεινον γεγράφθαι is used, but no parallel passage for the formula ἄμεινον γεγράφθω.

²⁸⁰ Cf. 3.6.

²⁸¹ On the expression ἄμεινον (ἐστίν) + infinitive see Kühner/Gerth II: § 482, 9; p. 60 and § 484, 31; p. 76.

²⁸² In 186.11-187.6 (on *Metaph.* B 2, 996b22-26) Alexander argues that the reading transmitted in his *Metaphysics* text is unsatisfactory. He then proposes his own solution, introducing it with ἄμεινον γεγράφθαι. The direct transmission brought down to us the following reading for lines B 2, 996b24-26: ὥστ' ἄλλης ἂν δόξειεν ἐπιστήμης εἶναι τὸ θεωρῆσαι τῶν αἰτίων τούτων ἕκαστον. / "therefore it would seem to belong to different sciences to investigate these cases severally." Since Aristotle had already described the efficient and the final cause as opposed to each other (ἀντικείμενον, 996b24), he now, Alexander reasons (οὐδαμῶς κατάλληλον, 186.15), contradicts his own position that contraries belong to *one* science (996a20-21). Therefore, after a detailed discussion (186.11-31), Alexander suggests as *ultima ratio* that an οὐκ should be added to the sentence in 996b24-26: ὥστ' οὐκ ἄλλης ἂν ... . The

be safely assumed that we are dealing here with Alexander's own conjecture.

Is there any evidence either in the text of the *Metaphysics* or in Alexander's commentary that could back up the view on Empedocles implied in Alexander's conjecture? In the first place, we find that in several passages in book A and α Aristotle praises aspects of Empedocles' theory[283]—Aristotle does not take Empedocles to be an unreasonable man. Next is the telling passage B 4, 1000a22–b21 and Alexander's corresponding comments. In B 4, 1000a5–1001a3, Aristotle is concerned with the question whether the principles of perishable and imperishable things are the same (tenth aporia). In section 1000a22–b21, he extensively discusses the two Empedoclean principles Love and Strife. Although Aristotle here repeats his criticism from A 4, namely, that the functions of Love and Strife are not clearly distinguished (1000a26–b17), his introduction credits Empedocles' theory with some degree of internal coherence (1000a24–25): καὶ γὰρ ὅνπερ οἰηθείη λέγειν ἄν τις μάλιστα ὁμολογουμένως αὑτῷ, Ἐμπεδοκλῆς... / "even the man whom one might suppose to speak most consistently—Empedocles—...." This respectful attitude towards Empedocles' theory matches that of Aristotle's concluding remarks on the question of the principles' perishability. He says (1000b17–18): ἀλλ' ὅμως τοσοῦτόν γε μόνον λέγει (sc. Ἐμπεδοκλῆς) ὁμολογουμένως. / "But yet in this regard alone at least he speaks consistently."[284]

In his commentary on the tenth aporia, Alexander offers a summary of Aristotle's critical engagement with Empedocles. In this overview (219.15–37) he focuses on examining what Aristotle means when he speaks about the internal coherence of Empedocles' theory (ὁμολογουμένως αὑτῷ, 1000a25). To this end, he provides a review of Empedocles' consistency, and thereby anticipates three points Aristotle makes in the discussion later on: the indistinct assignment of Love and Strife's functions (and the failure to state the cause of their effects) is criticized as incon-

---

result of his grappling with the problem is a solution introduced with the words ἄμεινον γεγράφθαι. The *Metaphysics* manuscript E presents Alexander's suggestion as a "variant reading," added *in margine* by the second hand (γρ. καὶ οὐκ ἄλλης); manuscript E^s, also by the second hand, attributes this "variant" explicitly to Alexander: γρ(άφει) ὁ Ἀλέξανδρος οὕτως· ὥστ' οὐκ ἄλλης ... . For the issue in the *Metaphysics* text that Alexander addresses, see Ross 1924: 229 and Crubellier 2009: 60–61.

In 233.21–28 Alexander suggests changing the text from καὶ εἴδει / "and in kind" (1002b24) to ἀλλ' εἴδει / "but in kind." He introduces what is most likely his own suggestion with the words ἄμεινον γεγράφθαι (233.26). See Ross 1924: 250 and Mueller 2009: 207 n. 29.

[283] Cf. for example 984a11–13 and Alexander's commentary in 27.28–28.7 (cf. Dooley 1989, 51 n. 104). According to Alexander, Aristotle prefers Empedocles' theory to Anaxagoras's because it is better to assume a definite than to assume an indefinite number of principles. In A 10 (993a11–27) Aristotle credits Empedocles with a principle equivalent to the formal cause, but goes on to criticize Empedocles' inconsequent application of the principle. See Alexander 134.15–136.17.

[284] The β-version reads μόνος in place of the μόνον of the α-version (1000b18). In the β-version (preferred by Ross and Jaeger), Aristotle's praise of Empedocles is even stronger, since here he does not say that Empedocles is consistent in *one single* aspect, but that *he alone* is consistent while everyone else is not.

sistent (219.18-19; 29-37).²⁸⁵ Despite that, Empedocles' theory is consistent in so far as, first, all things that come to be are accounted for from the same principles, and, second, the things that come to be are perishable whereas the principles (the four elements and Love and Strife) are imperishable (219.21-29).²⁸⁶ Thus, Empedocles' theory proves itself to be consistent in one respect, yet inconsistent in another.

Returning to our *Metaphysics* passage in A 8 (989a26) we find in the reading proposed and favored by Alexander (οὔτ' ὀρθῶς οὔτε ἀλόγως) the same pairing of consistent and inconsistent aspects that Alexander highlights in his overview of Aristotle's critique of Empedocles. According to Alexander's alternative reading, the text would say that Empedocles spoke "neither correctly nor altogether unreasonably" (68.3-4), thus signaling that the assessment of Empedocles includes both a positive and a negative aspect. And so Alexander's proposed reading for line 989a26 matches exactly with how he describes Aristotle's view on Empedocles' theory in his commentary on the tenth aporia in B 4.

To conclude: Alexander's suggested and preferred reading, οὔτε ἀλόγως, is what we find in the **α**-text. Yet, Alexander did not read this, as ω^AL and **β** confirm our preferred reading, οὔτε εὐλόγως, and this confirmation proves the reading to be the older one. This all leads to the conclusion that someone incorporated Alexander's recommended reading into the **α**-text.²⁸⁷

5.3.5 Alex. *In Metaph*. 380.25-30; 381.1-4 on Arist. *Metaph*. Δ 10, 1018a20-25

In Δ 10 Aristotle discusses the meaning of "opposites" (ἀντικείμενα). He starts off with the following observations:

Aristotle, *Metaphysics* Δ 10, 1018a20-25

Ἀντικείμενα λέγεται ἀντίφασις καὶ τἀναντία καὶ τὰ [21] πρός τι καὶ στέρησις καὶ ἕξις καὶ ἐξ ὧν καὶ εἰς ἃ ἔσχατα [22] αἱ γενέσεις καὶ φθοραί· καὶ ὅσα μὴ ἐνδέχεται ἅμα [23] παρεῖναι τῷ ἀμφοῖν δεκτικῷ, ταῦτα ἀντικεῖσθαι λέγεται [24] ἢ αὐτὰ ἢ ἐξ ὧν ἐστίν.

---

²⁸⁵Cf. *Metaph*. B 4, 1000a26-b12.

²⁸⁶Alex. *In Metaph*. 219.23-27: τοῦτο γὰρ ὁμολογούμενον δοκεῖ λέγεσθαι ὑπ' αὐτοῦ τὸ πάντα τὰ γινόμενα ἐκ τῶν αὐτῶν ἀρχῶν ὄντα ὁμοίως καὶ φθείρεσθαι πάντα, ἀλλ' οὐ τὰ μὲν φθείρεσθαι τὰ δὲ μένειν ἀίδια. μόνα δὲ ἄφθαρτα τὰ στοιχεῖα ὑποτίθεται, ταῦτα δέ ἐστι τὰ δ' σώματα καὶ τὸ νεῖκος καὶ ἡ φιλία. / "This statement of his seems to be consistent, that all things that come to be, being from the same principles, likewise all perish, not that some perish while others remain and are eternal (he supposes that only the elements are imperishable, that is, the four bodies and Strife and Love)." Cf. *Metaph*. B 4, 1000b17-20.

²⁸⁷That someone revised this section of text (989a26) in the **α**-version becomes even more likely when taking into consideration that in the subsequent lines 989a26-30 another intervention occurred in the **α**-text, as Primavesi 2012a: 454-56 demonstrated. Are these textual interventions related?

φαιὸν γὰρ καὶ λευκὸν ἅμα τῷ [25] αὐτῷ οὐχ ὑπάρχει· διὸ[288] ἐξ ὧν ἐστὶν ἀντίκειται [**τούτοις**].

We call opposites contradiction, and contraries, and relative terms, and privation and possession, and the extremes from which and into which generation and dissolution take place; and the attributes that cannot be present at the same time in that which is receptive of both, are said to be opposed—either themselves or their constituents. For grey and white do not belong at the same time to the same thing: therefore their constituents are opposed [**to these**].

22 αἱ β Al.ᵖ 380.14 edd. : οἷον αἱ **α** ǁ 25 ἀντίκειται β ω^AL (Al.ᵖ 381.3) edd. : ἀντίκειται τούτοις **α**, cf. Al.ᵖ 381.2-4

Aristotle begins his investigation of opposites by enumerating five types of opposites (1018a20-22), the first four of which (i.e., contradiction, contraries, relative terms, and privation and possession) are identical to those mentioned in Categories 10 (11b17-19).[289] In addition to these four, Aristotle introduces in our Metaphysics passage as a fifth type of opposites the extremes from which and into which generation and destruction take place (1018a21-22). This list of five terms exhausts the possible kinds of opposites. Aristotle then, in lines a22-23, gives a criterion that all five types meet. This criterion is not itself a further type of opposite.[290] The criterion states: opposed to one another (ἀντικεῖσθαι) are those attributes that cannot be present at the same time in the same thing that is capable of receiving both (μὴ ἐνδέχεται ἅμα παρεῖναι).

This criterion, however, ranges over more than just the five types of ἀντικείμενα stated so far. In fact, the criterion extends so far that it holds not just for ἀντικείμενα but for other things as well. As Aristotle points out in line a24 the criterion holds for the five types of ἀντικείμενα presented above (a24 ἢ αὐτὰ) as well as for those attributes whose constituents (a24 ἢ ἐξ ὧν ἐστὶν) are opposed. What are these constituents? Aristotle answers with an example.

As we know from the Categories, grey and white are not opposites and so they do not correspond with any of the five types of ἀντικείμενα. Yet, the criterion that they cannot be present in the recipient at the same time indeed holds for them, too (a24-25), and so there is some sense in which they are said to be opposed to one another. How, then, do grey and white relate to one another? In chapter 10

---

[288]Jaeger changes the transmitted διὸ ("therefore") into διότι ("because") (app. crit.: διότι correxi). Kirwan, following Jaeger's text, writes in his translation "because" (Kirwan 1971: 43). I retain the transmitted text and I ascribe the awkward "therefore" (διὸ never means "because" in Aristotle: Bonitz 1870: s.v. διό, p. 198b16) to Aristotle's terse writing style. The διὸ makes sense when in thought it is preceded by "but grey and white are not called opposites" (cf. Alex. In Metaph. 380.27).

[289]Cat. 10, 11b16-18: Λέγεται δὲ ἕτερον ἑτέρῳ ἀντικεῖσθαι τετραχῶς, ἢ ὡς τὰ πρός τι, ἢ ὡς τὰ ἐναντία, ἢ ὡς στέρησις καὶ ἕξις, ἢ ὡς κατάφασις καὶ ἀπόφασις. / "Things are said to be opposed to one another in four ways: as relatives or as contraries or as privation and possession or as affirmation and negation" (transl. by Ackrill).

[290]Kirwan 1970: 152; cf. Ross 1924: 314.

of the *Categories* (11b32–12a25) we read that grey is an intermediate (ἀνὰ μέσον, see 12a2–11 and 12a17–20)²⁹¹ between the contraries (τἀναντία) black and white.²⁹² Since grey and white, however, are not opposed to each other, but it is true that they cannot be predicated of the same substratum at the same time, Aristotle infers (διό, a25) that it is the constituents of grey and white (ἐξ ὧν ἐστὶν), i.e., black and white, that are opposed to each other.²⁹³

Let us now have a look at the transmission of the passage. In line 1018a25, the α-text contains an additional word, τούτοις, which is absent from the β-text. (We will see that ω^{AL}, agreeing with β, does not read the word τούτοις either.) By this addition, the α-reading states that the constituents of grey and white, i.e., black and white, are opposed to "these," which grammatically must refer to grey (!) and white. The β-version is clearly preferable. It states that the *constituents* of grey and white, i.e., black and white, are opposed (to each other). The α-addition τούτοις determines the constituents (ἐξ ὧν ἐστὶν), black and white, to be opposite to the attributes φαιὸν and λευκὸν, named in the previous sentence. Consequently the constituents are opposed to the pair they constitute. This is confusing and does not at all square with the scheme outlined in Aristotle's discussion, which is about the relation of two opposite poles. The word τούτοις in the α-text certainly is a later addition.²⁹⁴

Alexander addresses the *Metaphysics* passage in detail. At first he quotes the relevant sentence and then paraphrases and explicates it bit by bit:

Alexander, *In Metaph.* 380.25–30 Hayduck

ἔτι φησὶν ἀντικεῖσθαι καὶ ταῦτα ἃ μὴ ἐνδέχεται ἅμα παρεῖναι [26] τῷ ἀμφοῖν δεκτικῷ, ἢ αὐτὰ ἢ ἐξ ὧν ἐστιν· ἐπεὶ γὰρ οὐδὲ τὰ μεταξὺ [27] τῶν ἐναντίων ἅμα παρεῖναι τῷ αὐτῷ οἷόν τε, καὶ οὐ λέγεται ἀντικείμενα, [28] διὰ τοῦτο εἶπεν ἢ αὐτά, εἰ ὡς ἐναντία εἴη, ἢ τὰ ἐξ ὧν ἐστιν· τὰ γὰρ [29] μεταξὺ τῷ τὰ ἐξ ὧν ἐστι καὶ αὐτὰ ἀντικείμενα εἶναι, τούτῳ οὐ δύναται [30] ἅμα τῷ αὐτῷ ὑπάρχειν, ᾧ τὸ ἕτερον τῶν ἀμίκτων.

---

²⁹¹Alexander in his commentary to our *Metaphysics* passage calls the intermediate τὰ μεταξύ (380.26), staying close to Aristotle's word usage in *Metaph.* Γ 7, 1011b23–1012a1 (μεταξύ).

²⁹²According to Aristotle, only on the condition that contraries are such that it is not necessary that one or the other belong to the thing that they are predicated of is there an intermediate between them. There is no intermediate between sickness and health, but there is between black and white, namely, grey (*Cat.* 10, 12a8–21). Aristotle discusses the difference between contradictories, of which there cannot be an intermediate, and contraries, of which there always is an intermediate, also in Γ 7, 1011b23–1012a1. The context here is his discussion of the principle of excluded middle (cf. also I 5, 1056a15–b2).

²⁹³Cf. Ascl. *In Metaph.* 322.3–6.

²⁹⁴Instead of τούτοις we would rather expect ἀλλήλοις as addition to the text. See Asclepius's paraphrase (322.4–6): φαιὸν γὰρ καὶ λευκὸν ἅμα οὐχ ὑπάρχουσι τῷ αὐτῷ, ἐπειδὴ ἐξ ὧν ὑπάρχουσι, ταῦτα ἀντίκεινται ἀλλήλοις, [ὡς] τουτέστι τὸ λευκὸν καὶ τὸ μέλαν. / "For grey and white are not simultaneously present in the same thing, because their constituents are opposite *to each other*, that is to say, the white and the black."

He says further that also those things are opposed, "that cannot be present at the same time in that which is receptive of both, either themselves or their constituents." For since not even the intermediates between contraries can be present at the same time in the same subject, and are not called opposites, he says "either the attributes themselves," if these were to be [regarded] as contraries, "or their constituents"; for because the constituents of the intermediates are themselves opposites, the intermediates cannot belong at the same time to the same subject to which one or other of the unmixed [attributes] belongs.

26–29 ἐπεὶ ... τούτῳ **A O S** : καὶ γὰρ καὶ αὐτὰ ἀντικείμενα· ταῦτα γὰρ **P**[b]

Alexander's quote (380.26) as well as his paraphrase (380.28) show that his text (ω[AL]) agrees with the β-reading, that is, does not contain the additional τούτοις. What is more, Alexander's exposition of Aristotle's words agrees with our understanding of it.[295]

In the subsequent passage (380.30–40), however, Alexander examines the possible ways of understanding the phrase ἢ αὐτὰ ἢ ἐξ ὧν (1018a24) and its implications for the passage. In doing so, he develops an interpretation that diverges distinctly from ours. According to the understanding Alexander develops here, the phrase ἢ αὐτὰ ἢ ἐξ ὧν indicates that attributes that cannot be present at the same time in the same subject are opposite either to each other or *to their intermediates*.

> Alexander, *In Metaph.* 380.33–37
>
> δύναται τὸ ἢ αὐτὰ ἢ ἐξ ὧν ἐστι δηλωτικὸν εἶναι τοῦ ἢ αὐτὰ [34] ἀλλήλοις ἀντικεῖσθαι τὰ μὴ δυνάμενα ἅμα παρεῖναι τῷ ἀμφοτέρων δεκ-[35]τικῷ, <u>ἢ τούτοις ἐξ ὧν ἐστιν</u>. λευκὸν μὲν γὰρ καὶ μέλαν αὐτὰ ἀλλήλοις [36] ἀντίκειται (ἐναντία γάρ), ἀντίκειται δὲ καὶ τὸ ἐξ ἀμφοῖν ἑκατέρῳ αὐτῶν, [37] τῷ μὴ δύνασθαι μηδὲ αὐτὰ ἅμα τινὶ ἐκείνων ὑπάρχειν.
>
> "Either the attributes themselves or their constituents" could also indicate that [attributes] that cannot be present at the same time in a subject capable of receiving both of them are opposed either to each other or <u>to their constituents</u>. For white and black are themselves opposed to each other (since they are contraries), and what is constituted out of both of them is also opposed to each of them because they cannot belong at the same time to any of them.

According to Alexander's understanding, Aristotle expresses with the words ἢ ἐξ ὧν ἐστὶν ("or their constituents") not, as understood above, that white and grey are opposed because their constituents, black and white, are opposed, but that

---

[295] On the whole and in the most important respects. The καί in line 380.25 might point to a divergence between Alexander's understanding and ours: the καί indicates that Alexander takes the criterion of being unable to be present in the same subject at the same time to introduce a new type of ἀντικείμενα. (The καί cannot be understood to refer to Aristotle's καί in 1018a22, for this καί Alexander renders as ἔτι.)

white and grey are opposed because the intermediate grey is itself opposed to its constituents black and white.²⁹⁶

It is striking that Alexander, in spelling out his interpretation of Aristotle's phrase in line 1018a24, comes astonishingly close to the reading that we find in line 1018a25 of the α-text. In his comments Alexander reformulates Aristotle's expression ἢ ἐξ ὧν ἐστὶν (a24) into ἢ (sc. ἀντικεῖσθαι) τούτοις ἐξ ὧν ἐστιν (380.35). In line 1018a25 of the α-text, we read ἐξ ὧν ἐστὶν ἀντίκειται τούτοις. What are we to make of this similarity? Is it just that Alexander's comments and the α-reading only appear to be almost identical, or do they in fact express identical thoughts? As explained above, the corrupt α-reading says that the constituents of grey and white (black and white) are opposed to grey and white. As we just saw, this is the thought Alexander voices in his comment: intermediate attributes (such as grey) are opposed to their constituents (black and white). Thus, the α-reading in line 1018a25 is almost identical with Alexander's reformulation of line 1018a24 not only in regard to the addition of τούτοις, but also regarding the idea that there is an opposition between the constituents (black and white) and the thing that they constitute (the intermediate grey). The near identity in thought and the similarity in expression suggest that there is a causal connection between the two readings such that someone adopted Alexander's reformulation of 1018a24 by adding τούτοις to line 1018a25 of the α-text.

Alexander's interpretation, which we have been examining and seems to appear in the α-text, finds reflection in lines 381.1–4 of his commentary. Alexander cites and then expands Aristotle's expressions in his own words as follows:

Alexander, *In Metaph.* 381.1–4 Hayduck

καὶ εἴη ἂν τὸ καὶ ὅσα μὴ [2] ἐνδέχεται ὡς ἴσον τῷ καὶ καθόλου ὅσα μὴ ἐνδέχεται εἰρημένον. τὸ δὲ [3] διὸ ἐξ ὧν ἐστιν ἀντίκειται <u>ἐλλιπῶς εἰρημένον</u> ἴσον ἂν εἴη τῷ διὸ ἐξ [4] ὧν ἐστι τὰ μεταξὺ <u>ἐκείνοις</u> ἀντίκειται.

And the words, "and the attributes that cannot be present" may be taken as equivalent to, 'and in general whatever attributes cannot be present.' "Therefore their constituents are opposed" is an <u>elliptical statement</u> equivalent to, 'therefore the constituents of the intermediate are opposed <u>to these latter</u>.'

---

3 ἐξ **A O S** : καὶ ἐξ **Pᵇ**

Once more, the quotation in line 381.3 confirms that ω^AL agrees with β in not reading τούτοις (1018a25). Following his quotation, Alexander adds his own expanded reformulation, in which we find an understanding of the sentence that is in perfect accord with the interpretation he gave above (cf. 380.33–25). Alexander

---

²⁹⁶ Although Alexander is not firmly committed to his own interpretation (380.37–38), the aspect of his interpretation that interests us here, namely that according to Alexander the intermediates are opposed to their constituents, is dear to Alexander throughout his considerations in 380.38–381.1.

describes the quoted sentence as elliptical (ἐλλιπῶς εἰρημένον, 381.2), and offers an alternative version that displays two changes: διὸ ἐξ ὧν ἐστι τὰ μεταξὺ ἐκείνοις ἀντίκειται. In the first change the intermediates are named explicitly, τὰ μεταξὺ, with whose constituents we are dealing. In the second, Alexander adds the demonstrative pronoun ἐκείνοις, which expresses his view that the intermediates are opposed to their constituents. This second addition is identical in content (though not in verbal expression) to Alexander's earlier explanation (380.35) of the sentence, in which he introduced τούτοις into Aristotle's argument.

Let us consolidate the evidence in Alexander's commentary: Alexander rephrases Aristotle's ἢ ἐξ ὧν ἐστὶν (1018a24) as ἢ τούτοις ἐξ ὧν ἐστιν (380.35). He then declares Aristotle's sentence διὸ ἐξ ὧν ἐστὶν ἀντίκειται in 1018a25 (β) to be elliptical, and reformulates it as διὸ ἐξ ὧν ἐστι τὰ μεταξὺ ἐκείνοις ἀντίκειται (381.3–4). In each case Alexander's clarifications amount to adding a demonstrative pronoun to Aristotle's terms for the constituents, ἐξ ὧν ἐστὶν (in a24) and ἐξ ὧν ἐστὶν ἀντίκειται (in a25), thereby injecting into the text his interpretation, stating that intermediates are opposed to their constituents. To the first description he adds the pronoun τούτοις (writing ἀντικεῖσθαι … τούτοις ἐξ ὧν ἐστι in 380.35), and to the second he adds ἐκείνοις (writing διὸ ἐξ ὧν ἐστι τὰ μεταξὺ ἐκείνοις ἀντίκειται in 381.3–4). This invites me to draw the following conclusion concerning the *Metaphysics* text: in the α-text, which differs from the certainly correct β-text, we find Alexander's reformulation of the phrase ἢ ἐξ ὧν ἐστὶν (a24) incorporated into the phrase ἐξ ὧν ἐστὶν ἀντίκειται (a25).

This state of evidence suggests that the additional τούτοις in the α-version stems from Alexander's commentary. It appears that someone was pleased with Alexander's interpretation of Aristotle's terse argument, and modified the α-text accordingly. One might ask why this someone added τούτοις in line a25, thereby following the reformulation Alexander gave of line a24, instead of adding ἐκείνοις, which appears in Alexander's rendition of the sentence in a25. There might be a number of reasons for this. One is that the addition of τούτοις obviates the need to add τὰ μεταξὺ. The pronoun τούτοις in the α-text (a25) refers back to the terms φαιὸν and λευκὸν (a24), that is, to the attributes that are said to be opposed to their constituents. Once τούτοις is added, the mentioning of τὰ μεταξὺ becomes unnecessary, yet were τὰ μεταξὺ nevertheless added, it would be more natural to refer to them with the pronoun ἐκείνοις[297] rather than τούτοις. To put it differently, the person who wanted to reproduce Alexander's interpretation in the α-text had the choice of adding, on the one hand the longer τὰ μεταξὺ … ἐκείνοις or, on the other, the shorter τούτοις. He understandably opted for the shorter option.

Nevertheless the addition of τούτοις does yield a slight difference between Alexander's understanding and the sentence as it appears in the α-version. Accord-

---

[297] LSJ s.v. ἐκεῖνος I.1. "generally with reference to what has gone immediately before."

ing to Alexander, the *intermediates* (of which grey is only one)[298] are opposed to their constituents (black and white). According to the α-version, the *intermediate* grey and the attribute white (being contrary to black) are opposed to their constituents (black and white). It seems that the person who incorporated Alexander's comments into the α-text either did not recognize this difference or was indifferent to it.

Let the motivation and understanding of the scholar or scribe be what they may, what matters for the present investigation is this: in this case study we find an instance of a corruption in the α-text most likely occasioned by Alexander's comments that occurred in a passage of the *Metaphysics* text that is beyond book A.

In view of the case studies presented in 5.3.1–5, it is possible to state the following about the contamination of the α-version by Alexander's commentary. As far as the evidence I have discussed here is concerned, some of Alexander's comments on the *Metaphysics* do seem to have found expression in the α-version. The cases of α's contamination by Alexander can be characterized as reactions to Alexander's discussions of a *Metaphysics* passage, such that the α-text contains either direct incorporations of Alexander's reformulations or responses to his criticisms. Four of the five cases concern book A of the *Metaphysics* (5.3.1–4), and only one concerns a book other than A, namely book Δ (5.3.5).

A comparison of the influence of Alexander's commentary on the α-version to the influence it had on the β-version suggests that Alexander's influence on the α-text was less widespread. I have analyzed the traces of contamination in the β-version throughout books A–Δ (excepting B). These traces may be attributed, but do not need to be restricted, to the revision process that the β-version likely underwent. In the α-text, however, contamination through Alexander's comments appears to be concentrated on the first book, with, so far, one exception occurring in the fifth book.[299]

Alexander of Aphrodisias was referred to by later generations as *the* commentator. Alexander's commentary was of major importance for the reception of Aristotle's *Metaphysics* throughout late antiquity.[300] As we have already seen Alexander's commentary influenced $ω^{αβ}$ between AD 225 and 400,[301] and then influenced

---

[298] Alexander alone speaks of them in the plural. Cf. Dooley 1994: 149 n. 220.

[299] Again, I chose these five cases to show α's contamination because the evidence for $ω^{αβ}$, α, and β as well as for Alexander's own interpretation of the passage allow for secure conclusions. Besides these secure cases, the contamination of α by Alexander's commentary seems possible in A 3, 983a28–29: Alex. *In Metaph.* 21.11–15; 28–33; A 6, 987b9–10: Alex. *In Metaph.* 50.17–1.25; B 3, 998b21: Alex. *In Metaph.* 204.34–205.5; Δ 6, 1016b7–11: Alex. *In Metaph.* 367.29–37; A 5, 986a4: Alex. *In Metaph.* 40.21; Al.^{p} 40.23; Γ 5, 1010b32: Alex. *In Metaph.* 315.35–316.2. Again, if my conclusions about the α-version are correct, we are allowed to expect some invisible cases of contamination.

[300] On Alexander's influence on later authors see Sharples 1987: 1220–24.

[301] See section 5.1.

the β-version from the point of its separation from α onwards (likely including the point of departure) until possibly as late as AD 850.[302] The above five case studies enable us to conclude now that Alexander's comments influenced also the α-version during the time period between its separation from the β-version and AD 850.

5.4 CONTAMINATION OF β BY ω^AL OR OF α BY ALEXANDER'S REPORT OF A *VARIA LECTIO*?

There are two cases that cannot be easily allocated to one of the types of contamination discussed so far. In both cases the β-version agrees with the reading in ω^AL, while the α-version agrees with a *vario lectio* that Alexander cites in his commentary. In the first case, the reading shared by β and ω^AL is certainly correct, and likely to be the original reading. In the second case, it is not possible to reach a definitive decision about which of the two readings is the original one. In both cases, more than one plausible answer can be given to the question regarding the identity of the source and the target of contamination.

5.4.1 Alex. *In Metaph.* 347.19–25; 348.5–8 on Arist. *Metaph.* Δ 1, 1013a17–23

Aristotle begins his encyclopedia of philosophically significant terms, book Δ, with an entry on ἀρχή (Δ 1, 1012b34–1013a23).[303] While the primary meaning of ἀρχή[304] in the first books of the *Metaphysics* is a "principle" or "cause" (αἰτία), here in Δ 1 Aristotle covers ἀρχή's whole spectrum of meaning, a spectrum that encompasses such meanings as a "starting point on a road" or a "rule." A final meaning of ἀρχή that Aristotle treats is "the point from which one first gets acquainted" with a given thing[305] (γνωστὸν τὸ πρᾶγμα πρῶτον, 1013a14–15). Aristotle concludes the chapter in the following way:

Aristotle, *Metaphysics* Δ 1, 1013a17–23

πασῶν μὲν οὖν κοι-[18]νὸν τῶν ἀρχῶν τὸ πρῶτον εἶναι ὅθεν ἢ ἔστιν ἢ γίγνεται ἢ [19] γιγνώσκεται· τούτων δὲ αἱ μὲν ἐνυπάρχουσαί εἰσιν αἱ δὲ [20] ἐκτός. διὸ ἥ τε φύσις ἀρχὴ καὶ τὸ στοιχεῖον καὶ ἡ διάνοια [21] καὶ ἡ προαίρεσις καὶ οὐσία καὶ τὸ οὗ ἕνεκα· πολλῶν γὰρ [22] καὶ τοῦ γνῶναι καὶ τῆς κινήσεως ἀρχὴ τὸ ἀγαθὸν καὶ τὸ [23] **καλόν**.

It is common, then, to every origin to be the first point from which a thing either is or comes to be or from which one gets acquainted with it; but of these some are

---
[302] See section 5.2.
[303] See Kirwan 1971: 123–24.
[304] Lumpe 1955 offers a discussion of the meaning of ἀρχή from the Presocratics to Aristotle.
[305] Kirwan 1971: 123.

immanent in the things and others are outside. Therefore the nature [of a thing] is an origin, and so are the elements [of a thing], and thought and decision, and substance, and that for the sake of which—for the good and the **beautiful** are the origin both of the knowledge and of the movement of many things.

17–18 κοινὸν τῶν ἀρχῶν α edd. : τῶν ἀρχῶν κοινὸν β ‖ 21 ἡ β Al.ᶜ 347.7 Bekker Bonitz Ross Jaeger : om. α Christ ‖ 22 τὸ ἀγαθὸν β Al.ᶜ 347.21 : τἀγαθὸν α edd. ‖ 23 καλόν β ωᴬᴸ (Al.ᶜ 347.21 Al.ᵖ 347.23–24) Arᵘ (Scotus) edd. : κακόν α Al.ʸᵖ 348.7–8 Ascl.ᶜ 305.15–16

I will focus especially on the last sentence of this passage (1013a22–23). In this sentence Aristotle clarifies (γάρ) what it means for "that for the sake of which" (τὸ οὗ ἕνεκα) to be a principle. The β-text characterizes "that for the sake of which" as the good and the beautiful, which is the aim and hence the starting point (ἀρχή) of every action. Understanding the aim of every action as "the good and the beautiful" (τὸ ἀγαθὸν καὶ τὸ καλόν) has Platonic origins. Aristotle uses it here as a way of characterizing the final cause.³⁰⁶ The α-text reads, however, τὸ ἀγαθὸν καὶ τὸ <u>κακόν</u>. In our passage the final cause (οὗ ἕνεκα) is being described, and so the β-reading in lines 1013a22–23 is preferable to the α-reading.³⁰⁷ There are a few passages in the *Politics* and the *Rhetoric* in which the good and the bad are mentioned in the context of human action and its origin,³⁰⁸ yet here the good retains its status

---

³⁰⁶ Annas 1976: 212 gives a short overview of Aristotle's stance regarding the Platonic Form of the Good. Aristotle dissociated himself clearly from the *Form* of the Good in *EE* A 8 (1217b1–1218a1), while avowing that the good is the τέλος and cause of human action. *EE* A 8, 1218b4–6; 9–12: ἀλλὰ πολλαχῶς <u>τὸ ἀγαθόν, καὶ ἔστι τι αὐτοῦ καλόν</u>, καὶ τὸ μὲν πρακτὸν τὸ δ' οὐ πρακτόν. πρακτὸν δὲ τὸ τοιοῦτον ἀγαθόν, τὸ οὗ ἕνεκα. οὐκ ἔστι δὲ τὸ ἐν τοῖς ἀκινήτοις. ... <u>τὸ δ' οὗ ἕνεκα ὡς τέλος ἄριστον</u> καὶ αἴτιον τῶν ὑφ' αὐτὸ καὶ πρῶτον πάντων. ὥστε τοῦτ' ἂν εἴη αὐτὸ τὸ ἀγαθὸν τὸ τέλος τῶν ἀνθρώπῳ πρακτῶν ("But 'good' is ambiguous, and there is in it a noble part, and part is practicable but the rest not so. The sort of good that is practicable is an object aimed at, but not the good in things unchanging. ... But the object aimed at as end is best, and the cause of all that comes under it, and first of all goods. This then would be the good *per se*, the end of all human action" [transl. by Solomon]). See also *Ph.* B 3, 195a23–25: τὰ δ' ὡς τὸ τέλος καὶ τἀγαθὸν τῶν ἄλλων· τὸ γὰρ οὗ ἕνεκα βέλτιστον καὶ τέλος τῶν ἄλλων ἐθέλει εἶναι. *Rh.* A 6, 1362b5–9: καὶ τὴν ἡδονὴν ἀγαθὸν (sc. ἀνάγκη) εἶναι· πάντα γὰρ ἐφίεται τὰ ζῷα αὐτῆς τῇ φύσει. ὥστε καὶ τὰ ἡδέα καὶ τὰ καλὰ ἀνάγκη ἀγαθὰ εἶναι· τὰ μὲν γὰρ ἡδονῆς ποιητικά, τῶν δὲ καλῶν τὰ μὲν ἡδέα τὰ δὲ αὐτὰ καθ' ἑαυτὰ αἱρετά ἐστιν. Cf. also *MA* 6, 700b25–35 and *Metaph.* N 4–5, 1091a29–1092a21. For the good as final cause see *EN* A 1, 1094a1–3, A 7, 1097a18–24 and *Metaph.* A 3, 983a31–32. Ross 1924: 291 references M 3, 1078a31, where Aristotle differentiates between the ἀγαθόν and the καλόν (1078a31–32): τὸ ἀγαθὸν καὶ τὸ καλὸν ἕτερον (τὸ μὲν γὰρ ἀεὶ ἐν πράξει, τὸ δὲ καλὸν καὶ ἐν τοῖς ἀκινήτοις).

³⁰⁷ Bonitz 1849: 220 declares the α-reading impossible: *malum per se nunquam nec potest dici nec dicitur ab Aristotele causa finalis*. The Arabic translation also seems to have read the preferable β-reading. We can infer this from 's translation, although the syntax at the end of the sentence seems to have suffered on the long journey between languages: *Et similiter natura est principium et elementum etiam. Et cogitatio et voluntas et substantia et illud propter quid <u>et bonum et largum est</u> principium plurium. Et cognitio et motus etiam.*

³⁰⁸ Schwegler 1847c: 190 refers to *Pol.* H 13, 1332a16–17 and *Rh.* B 2, 1378b11–13. The bad as a principle is explicitly ruled out in *Metaph.* Θ 9, 1051a19–21. Cf. Beere 2009: 344–47. See also *Metaph.* N 4, 1091b30–32: ταῦτά τε δὴ συμβαίνει ἄτοπα, καὶ τὸ ἐναντίον στοιχεῖον, εἴτε πλῆθος ὂν εἴτε τὸ ἄνισον καὶ

as principle even for actions that aim simply to avoid the bad.[309] The editors of the *Metaphysics* unanimously follow the β-reading.

In Alexander's commentary on this passage we encounter the very rare case of Alexander being familiar with both the α- and the β-reading. Below we see Alexander quoting (347.20–21), paraphrasing (347.23–24), and commenting (347.21–348.7) on the text as he finds it in his exemplar ω^AL. The reading in ω^AL is identical with the correct reading preserved by the β-version.

Alexander, *In Metaph*. 347.19–21; 23–24 Hayduck

προσέθηκε δὲ καὶ ὑπὸ [20] τίνα τῶν εἰρημένων ταῦτα ὑπάγεται, εἰπὼν πολλῶν γὰρ καὶ τοῦ γνῶναι [21] καὶ τῆς κινήσεως ἀρχὴ τὸ ἀγαθὸν καὶ τὸ <u>καλόν</u>. ... τῆς δὲ κινήσεως ἀρχὴν τὸ τέλος εἶπεν, ὅπερ ἐστὶ [24] τὸ ἀγαθὸν καὶ τὸ <u>καλόν</u>.

He states to which of the [types of beginning] that have been mentioned these are reduced, saying: 'for the good and the beautiful are the origin both of the knowledge and of the movement of many things.' ... but he calls the beginning of movement the end, that which is the good and the beautiful.

A few lines later, Alexander concludes his remarks on the τέλος as the principle of human action:

Alexander, *In Metaph*. 348.5–8 Hayduck

ἐν γὰρ τοῖς πρακτοῖς καὶ τὸ τέλος ἀρχή. ἔστι δέ, ὥσπερ τὸ [6] ἀγαθὸν ἀρχὴ πράξεως, οὕτω πολλάκις καὶ τὸ κακόν· φεύγοντες γὰρ αὐτὸ [7] πράσσομέν τινα. διὸ ἔν τισι γράφεται πολλῶν γὰρ [8] καὶ τοῦ γνῶναι καὶ τῆς κινήσεως ἀρχὴ τὸ ἀγαθὸν καὶ τὸ κακόν.

for in the case of things that are to be done the end too is a starting point. But as the good is a beginning of action, so too, in many instances, is evil, for in attempting to avoid it we perform certain actions. Hence some manuscripts have this reading: 'For good and evil are the origin both of the knowledge and of the movement of many things.'

5 πρακτοῖς **A P**^b **S** : πρακτικοῖς **O**

Here Alexander quotes a *varia lectio* (ἔν τισι γράφεται) that reads the text we find in the α-version. He introduces the variant reading with the following consider-

---

μέγα καὶ μικρόν, τὸ κακὸν αὐτό. / "These absurdities follow, and it also follows that the contrary element, whether it is plurality or the unequal, i.e. the great and small, is the bad-itself." See Annas 1976: 216. Plotinus, too, mentions the "bad itself": I 8, 3,1–4,5; see also Dörrie/Baltes 1996: 123.8, pp. 190–94 and 516. Cf. Dörrie/Baltes 1996: 123.4–9, pp. 186–97 and 506–20.

[309] *Rh*. A 6, 1362a34–37: τούτων δὲ κειμένων ἀνάγκη τάς τε λήψεις τῶν ἀγαθῶν ἀγαθὰς εἶναι καὶ τὰς τῶν κακῶν ἀποβολάς· ἀκολουθεῖ γὰρ τῷ μὲν τὸ μὴ ἔχειν τὸ κακὸν ἅμα, τῷ δὲ τὸ ἔχειν τὸ ἀγαθὸν ὕστερον. / "All this being settled, we now see that both the acquisition of good things and the removal of bad things must be good; the latter entails freedom from the evil things simultaneously, while the former entails possession of the good things subsequently" (transl. by Roberts).

ation: just as the good can serve as the starting point (ἀρχή) of our actions, so too can the bad motivate our behavior; while we sometimes act to attain the good, we also sometimes act to avoid the bad. Therefore (διό), there is a *varia lectio* that expresses this thought, and reads τὸ ἀγαθὸν καὶ τὸ κακόν. Alexander's rationale certainly makes the variant reading look like a plausible alternative. Whether this mirrors Alexander's own understanding of the passage or whether he found the justification combined with the variant reading in his source, we do not know. We also do not know if the reading can be traced back to a conjecture. It might also be that the variant reading derives from a scribal confusion,[310] given that the two readings differ from each other in only one letter.

The pressing question for the present purpose is this: how are we to understand the fact that Alexander here seems to have knowledge of the two divergent versions α and β, knowledge that he quite simply could not have had? After all, Alexander's commentary is the *terminus post quem* of ω^{αβ}'s split into α and β (see 5.1). Is it plausible to suppose that the agreement between Alexander's *varia lectio* and our α-reading is due purely to a scribal error that just happened to have occurred twice in the same passage?[311] This is of course theoretically possible, though ascribing the phenomenon to coincidence is not a satisfactory explanation. There are three other possible, more satisfactory explanations for the facts:

(i) The agreement of the (correct) β-reading with the reading in ω^{AL} testifies to the reading in ω^{αβ} (for parallel cases see 4.3.1). The coincidence of the (incorrect) α-reading with the variant reading known by Alexander goes back to an adjustment of the α-text according to the variant reading attested to in Alexander's commentary. Such a "correction" could have been motivated and facilitated by Alexander's somewhat positive discussion of the variant.

(ii) The agreement of the (correct) β-reading with the reading in ω^{AL} testifies to the reading in ω^{αβ} (for parallel cases see 4.3.1). The agreement of the (incorrect) α-reading with the variant reading known by Alexander testifies to the reading in a version φ, of which Alexander had sporadic knowledge through *variae lectiones* or other commentaries such as the one by Aspasius (see 3.5.1),[312] and which at a later point in the transmission (that is after ω^{αβ} split into α and β) influenced the α-text.

(iii) The agreement of the (correct) β-reading with the reading in ω^{AL} goes back to the adjustment of the β-text to the ω^{AL}-reading Alexander attests to (for examples of an adjustment of the β-text to Alexander's own comments see 5.2). The agreement of the (incorrect) α-reading with the variant reading known by Alex-

---

[310] The two terms καλὸν–κακόν just like ἀγαθὸν–κακόν occur together often and in different contexts. Cf. the case B 2, 996b1 (κακῶν α vs. καλῶν β) in 4.3.2.2.

[311] Cf. Primavesi's (2012b: 408 n. 84) explanation of the (seemingly parallel) case εὐλόγως vs. ἀλόγως in 989a26 (see 5.3.4).

[312] Such a version φ could be among the versions that I call ω^{ASP2-n}, that is all possible versions that were known (either completely or sporadically) by Aspasius.

ander testifies to the reading in ω^αβ (for parallel cases see 3.5.2).

In deciding between these three possible explanations, we should be guided by the following question: which of them is most likely in light of the other evidence we have acquired so far about the relation between Alexander's commentary and the direct transmission of the *Metaphysics*?

A defense of the first option might explain the agreement of Alexander's *varia lectio* with the α-version as another instance of the contamination of α by Alexander (cf. 5.3). However, in this case it would not have been Alexander's own interpretation that prompted the contamination but the variant reading he reports and discusses (favorably). The fact that according to this scenario, the agreement of ω^AL with β testifies to the reading in ω^αβ (which is likely to be the original one, cf. 4.3.1), speaks for the first option: this conclusion is in tune not only with the general fact that ω^AL is independent of ω^αβ, but also with the specific fact that καλόν is the correct reading.

The advantage of the first option holds also for the second one. The difference in the second scenario is that the contamination of α was not exerted by Alexander's commentary and the *varia lectio* he reports, but rather was triggered directly by the version φ (or ω^ASP2-n), from which Alexander's variant ultimately stems. This seems to be a viable explanation, since nobody would expect Alexander to be the source of *every* contamination in α (or β). In fact, as Primavesi 2012b shows, the α-version did later on incorporate additions ("α-supplements") independent of Alexander's commentary.[313]

According to the third scenario, however, the agreement of ω^AL and β does not testify to the reading in ω^αβ, but rather to the contamination of β by the reading preserved in Alexander's commentary. This scenario is similar to the cases of β-contamination by Alexander's commentary discussed in 5.2. Yet it differs from these cases in that it is not Alexander's interpretation that was here incorporated into the β-text, but rather the reading he attested to as the reading in ω^AL. Furthermore, according to this option, α would have preserved the reading of ω^αβ, which must have been corrupted after it split from its common ancestor with ω^AL (which preserved the correct reading), and before Alexander wrote his commentary, because he shows signs of already knowing this corrupted reading as a *varia lectio* (3.5.2).

Given that in this case we can determine the reading in β and ω^AL (καλόν) as the correct and therefore most likely original reading, the first option appears to be the most plausible explanation. Yet options two and three certainly remain possible. In the next case to be discussed, which exhibits features parallel to the present one, it is not clear which of the two possible readings is the original one. This makes matters more complicated.

---

[313] On the possible influence that Asclepius's commentary had on the α-version see Kotwick 2015.

5.4.2 Alex. *In Metaph.* 145.8-12; 19-146.4 on Arist. *Metaph.* α 1, 993b19-23

The first chapter of book α ἔλαττον begins with introductory remarks on the "investigation of the truth" (ἡ περὶ τῆς ἀληθείας θεωρία: 993a29-b19). Aristotle determines that truth is the subject and the goal of theoretical, in contrast to practical, science. According to Aristotle's classification of the sciences in *Metaph.* E 1, the terms θεωρία and θεωρητικὴ ἐπιστήμη refer to mathematics, physics and theology. Among these three, theology (θεολογική), i.e., the field that includes metaphysics, is ranked the highest.[314]

Aristotle, *Metaphysics* α 1, 993b19-23[315]

ὀρθῶς δὲ καὶ τὸ κα-[20]λεῖσθαι τὴν φιλοσοφίαν ἐπιστήμην τῆς ἀληθείας. θεωρητικῆς [21] μὲν γὰρ τέλος ἀλήθεια πρακτικῆς δ' ἔργον· καὶ γὰρ ἂν [22] τὸ πῶς ἔχει σκοπῶσιν, οὐκ **ἀΐδιον** ἀλλὰ πρός τι καὶ νῦν [23] θεωροῦσιν οἱ πρακτικοί.

It is right also that philosophy should be called knowledge of the truth. For the end of theoretical knowledge is truth, while that of practical knowledge is action (for even if they consider how things are, practical men do not study **what is eternal** but what is relative and in the present).

22 οὐκ ἀΐδιον β ω^AL (Al.^p 145.10 Al.^c 145.19) Ar^u vel οὐ τὸ ἀΐδιον Brandis Bekker Bonitz Christ Ross : οὐ τὸ αἴτιον καθ' αὑτό α Al.^yp 145.21-22, Ar^j Jaeger : οὐ τὸ αἴτιον οὐ καθ' αὑτό ζ

In line b22, α and β differ in the following way: according to the β-reading, "what is eternal" (οὐκ ἀΐδιον) is ruled out as the subject of practical (in contrast to theoretical) ἐπιστήμη.[316] According to the α-reading (οὐ τὸ αἴτιον καθ' αὑτό)[317] Aristotle rules out "the cause in itself" as the subject of practical science.[318] Alexander's commentary offers two pieces of information regarding the divergent readings: firstly, Alexander's paraphrase and his comments on the Aristotelian text reveal that ω^AL read οὐκ ἀΐδιον in accordance with the β-version. Secondly, Alexander

---

[314] *Metaph.* E 1, 1026a18-23.

[315] The information given in the apparatus refers to lines 993b21-23 only, i.e., to the lines that I will examine in this section. On the reading in lines 993b19-21 see 5.2.4.

[316] Following the edition by Brandis (1823), the *Metaphysics* editors Bekker, Bonitz, Christ, and Ross add an article to ἀΐδιον. However, the other abstract terms πρός τι and νῦν stand without the article. Brandis himself does not justify his addition of the article, nor do the others.

[317] On the term αἴτιον καθ' αὑτό see *Metaph.* B 1, 995b31-33: μάλιστα δὲ ζητητέον καὶ πραγματευτέον πότερον ἔστι τι παρὰ τὴν ὕλην αἴτιον καθ' αὑτό ἢ οὔ. Cf. also *Ph.* B 5, 196b24-29; Δ 13, 222b19-22.

[318] The conjecture by Luthe (1880: 198-99) οὐ τὸ καθ' αὑτό results from a misunderstanding of the apparatus in Bonitz's edition. The information in Bonitz's apparatus is indeed misleading, for he lists the lemmata οὐκ ἀΐδιον and οὐ τὸ αἴτιον separately from the lemma καθ' αὑτό. Brandis 1836: 592b27 refers to a scholium in E: γρ. "οὐ τὸ αἴτιον καθ' αὑτό," καὶ οὐ "τὸ ἀΐδιον."

states explicitly that he knows as a *varia lectio* the reading οὐ τὸ αἴτιον καθ' αὑτὸ, which we read in the α-version.[319]

Alexander comments on our passage at three different places in his commentary. Alexander approaches our sentence for the first time in his summary of book α ἔλαττον. He touches on it only briefly here, reporting Aristotle's position on "theoretical philosophy"[320] (139.3–5): ταύτης μόνης τέλος ἡ γνῶσις τῆς ἀληθείας καὶ τῆς κυρίως ἀληθείας· τὸ γὰρ ἀίδιον ἀληθὲς αὕτη θεωρεῖ ("Theoretical philosophy alone has the knowledge of truth as its end, and of truth that is such in the most proper sense; for it is eternal truth that this philosophy investigates"). Later on, at 993b21–23 Alexander offers his proper commentary on our passage:

Alexander, *In Metaph*. 145.8–12 Hayduck

τέλος γὰρ τῇ πρακτικῇ ἡ πρᾶξις, καὶ [9] οὐχ ἡ γνῶσις τέλος τῆς ἐν τοῖς πρακτοῖς ἀληθείας· καὶ γὰρ ἐν οἷς τὸ [10] πῶς ἔχει τὸ ὑποκείμενον ἀληθείας σκοποῦσιν οἱ πρακτικοί, οὐ περὶ ἀιδίου [11] τινὸς ἀλήθειαν σκοποῦσιν. ὡς δὲ τῆς περὶ τὰ ἀίδια ἀληθείας κυρίως καὶ [12] μάλιστα ἀληθείας οὔσης, οὐ τῆς ἐν τοῖς πρακτοῖς, τοῦτο προσέθηκεν.

For the end of practical science is action, and not knowledge of the truth [involved] in things to be done. For even in cases in which practical men do examine the truth in the subject [with which they are dealing], they are not looking to the truth of anything eternal. Aristotle adds this remark in the belief that truth in the proper and fullest sense is that which deals with eternal things, not the truth involved in things to be done.

8 καὶ **A P**ᵇ **S** : om. **O** ‖ 9 πρακτοῖς **A P**ᵇ **S** : πρακτικοῖς **O** ‖ 11 ἀλήθειαν **A S** : ἀληθείας **O P**ᵇ ‖ 12 πρακτοῖς **A P**ᵇ **S** : πρακτικοῖς **O**

Alexander's rendering of the *Metaphysics* passage indicates that he read οὐκ ἀίδιον in line b22 of his text. He does not mention anything like the cause in itself (οὐ τὸ αἴτιον καθ' αὑτὸ), which we know as the reading of the α-version. Also in the lines subsequent to this section (145.12–19), Alexander stresses that practical science aims at the truth in something of a particular time or place (τινὶ καὶ ποτέ, 145.13), that is to say, it aims at something that is not universally or eternally true (τοιαῦτα γὰρ τὰ πρακτά, οὐκ ἀεὶ οὐδὲ καθόλου, 145.14–15) but varies with circumstance.

A few lines later, Alexander continues in the following way:

---

[319] For this *Metaphysics* passage two Arabic translations are available. These two correspond here exactly to our two branches of the direct transmission. Walzer 1958: 223: "The Arabic translators were acquainted with both these old variants, Arᵘ following the tradition represented by Al and Aᵇ, Arⁱ siding with Alʸᵖ and E."

[320] Alex. *In Metaph*. 138.28–29. On the term θεωρητικὴ φιλοσοφία in Alexander and Aristotle see 5.2.4.

Alexander, *In Metaph.* 145.19–26 Hayduck

διὰ δὲ τοῦ εἰπεῖν οὐκ ἀίδιον [20] ἔδειξεν ὅτι μὴ τὸ τέλος τοῖς πρακτοῖς ἀλήθειά τε καὶ ἐπιστήμη· ἀιδίων [21] γὰρ αἱ ἐπιστῆμαι καὶ θεωρίαι. γράφεται δὲ ἔν τισιν ἀντιγράφοις <u>οὐ τὸ</u> [22] <u>αἴτιον καθ' αὑτὸ</u> ἀλλὰ πρός τι καὶ νῦν θεωροῦσιν. οὑ γεγραμ-[23]μένου εἴη ἂν λέγων ὅτι οὐ τὸ κυρίως καὶ καθ' αὑτὸ αἴτιον, ὃ τοῦ ἁπλῶς [24] ἀληθέσιν αὐτοῖς εἶναι αἴτιόν ἐστι, θεωροῦσιν οἱ πρακτικοί, ἀλλὰ τοῦ πρὸς [25] τόδε καὶ νῦν ἀληθὲς αὐτὸ εἶναι· τοιοῦτον γὰρ τό τε ἐν τοῖς πρακτοῖς [26] ἀληθὲς καὶ τὸ ὡς πρακτῶν αὐτῶν αἴτιον.

In saying, '[they do not study] what is eternal,' Aristotle points out that in things to be done the end is not truth or scientific knowledge, for the theoretical sciences deal with eternal objects. In certain manuscripts this reading occurs: '[practical men] do not study the cause in itself, but what is relative and in the present.' If the text was written thus, Aristotle would be saying that practical men do not consider the cause that is such in the proper sense and in itself, the one that is cause of the fact that things are true without qualification, but [the cause that explains why] something is true in relation to a particular thing, at this particular time. For this is the kind of truth [found] in things that are to be done, and is the cause [that explains them] as actions.

20 πρακτοῖς A P^b : πρακτικοῖς O S || 21 δὲ Bonitz Hayduck : om. codd. S || 24 πρακτικοί O P^b S : πρακτοί A || τοῦ S Bonitz : τὸ A O P^b || 25–26 αὐτὸ ... ἀληθὲς A O S : om. P^b || 25 πρακτοῖς A : πρακτικοῖς O || 26 πρακτῶν A P^b S : πρακτικῶν O

In this commentary section we once more gain insight into ω^{AL}: Alexander quotes the words οὐκ ἀίδιον (b22) and thereby confirms what his paraphrase showed, namely that ω^{AL} contains the β-reading. This commentary also shows that Alexander had access to an alternative reading found in other manuscripts. This *varia lectio* agrees with the α-reading. Alexander explicates the meaning of the variant reading thus: if Aristotle had said that practical science does not study *the cause in itself*, he would have meant to express that it does not search for the cause in the proper sense (κυρίως), that is, the cause through which something is true in an absolute sense, but in the sense in which something is true only for a particular time. At this point of his commentary, Alexander does not evaluate the alternative reading. Both possible readings seem to stand equally next to each other.

Alexander will evaluate the *varia lectio*, but before turning to his evaluation I would like to compare the two readings with each other. Arguments can be brought forward in support of both readings, so it is difficult to decide which of the two goes back to Aristotle.[321] According to the β-version, Aristotle excludes from practical science the study of what is eternal (οὐκ ἀίδιον) and attributes to it

---

[321] This is especially so because for Aristotle the causes are themselves eternal. See *Metaph*. E 1, 1026a17: ἀνάγκη δὲ πάντα μὲν τὰ αἴτια ἀίδια εἶναι. Cf. also *Metaph*. Θ 9, 1051a19–21. Alexander in the subsequent passage of his commentary also speaks about eternal causes (147.7): μάλιστα δὲ ἀληθῆ τὰ ἀίδια αἴτια.

the search for relative and temporary things. According to the α-version, Aristotle excludes from practical science the cause itself (οὐ τὸ αἴτιον καθ' αὐτὸ) and attributes to it the search for relative and temporary things, or, taking the two terms πρός τι and νῦν to be direct pendants to καθ' αὐτὸ, the search for the relative and temporary cause.

The following argument can be put forward in defense of the β-reading. Although the appearance of the eternal as subject of theoretical science might be unexpected in the context of α ἔλαττον, Aristotle declares in book E of the *Metaphysics* what is eternal (ἀΐδιον), what is unmoved (ἀκίνητον) and what is separate (χωριστόν) to be the subject of theoretical science.[322] This holds especially for the highest area of theoretical science, which is the subject of *Metaphysics* (πρώτη φιλοσοφία, 1026a24). The causes studied here are eternal to the highest degree.[323] And so it is fitting that 993b22 says that the practical science does not study the eternal.

Yet, in light of book A, where principles and *causes* were introduced as subjects of the inquiry, one could easily be led to a different view on the α 1 passage. In the first two chapters of book A, theoretical science (in contrast to productive sciences) was determined to be the science that deals with first principles and causes.[324] In addition to the promptings of book A, the α-reading (τὸ αἴτιον καθ' αὐτὸ) finds support from the sentence that follows directly upon the *Metaphysics* passage above. In α 1, 993b23-24, the cause is introduced as requirement for the knowledge of the truth: οὐκ ἴσμεν δὲ τὸ ἀληθὲς ἄνευ τῆς αἰτίας ("Now we do not know the truth without its cause").

Aristotle indeed further develops his treatment of the cause in lines 993b23-24. In these lines, however, the evidence shifts away from the α-reading and moves to

---

[322] *Metaph.* E 1, 1026a10-13: εἰ δέ τί ἐστιν ἀΐδιον καὶ ἀκίνητον καὶ χωριστόν, φανερὸν ὅτι θεωρητικῆς τὸ γνῶναι, οὐ μέντοι φυσικῆς γε (περὶ κινητῶν γάρ τινων ἡ φυσική) οὐδὲ μαθηματικῆς, ἀλλὰ προτέρας ἀμφοῖν. / "But if there is something which is eternal and immovable and separable, clearly the knowledge of it belongs to a theoretical science—not, however, to natural science (for natural science deals with certain movable things) nor to mathematics, but to a science prior to both."

[323] *Metaph.* E 1, 1026a13-18: ἡ μὲν γὰρ φυσικὴ περὶ χωριστὰ μὲν ἀλλ' οὐκ ἀκίνητα, τῆς δὲ μαθηματικῆς ἔνια περὶ ἀκίνητα μὲν οὐ χωριστὰ δὲ ἴσως ἀλλ' ὡς ἐν ὕλῃ· ἡ δὲ πρώτη καὶ περὶ χωριστὰ καὶ ἀκίνητα. ἀνάγκη δὲ πάντα μὲν τὰ αἴτια ἀΐδια εἶναι, μάλιστα δὲ ταῦτα. ταῦτα γὰρ αἴτια τοῖς φανεροῖς τῶν θείων. / "For natural science deals with things which are separable but not immovable, and some parts of mathematics deal with things which are immovable, but probably not separable, but embodied in matter; while the first science deals with things which are both separable and immovable. Now all causes must be eternal, but especially these; for they are the causes of so much of the divine as appears to us."

[324] *Metaph.* A 2, 982b7-12: ἐξ ἁπάντων οὖν τῶν εἰρημένων ἐπὶ τὴν αὐτὴν ἐπιστήμην πίπτει τὸ ζητούμενον ὄνομα· δεῖ γὰρ ταύτην τῶν πρώτων ἀρχῶν καὶ αἰτιῶν εἶναι θεωρητικήν· καὶ γὰρ τὸ ἀγαθὸν καὶ τὸ οὗ ἕνεκα ἓν τῶν αἰτίων ἐστίν. ὅτι δ' οὐ ποιητική, δῆλον καὶ ἐκ τῶν πρώτων φιλοσοφησάντων·. / "Judged by all the tests we have mentioned, then, the name in question falls to the same science; this must be a science that investigates (θεωρητικήν) the first principles and causes; for the good, i.e. that for the sake of which, is one of the causes. That it is not a science of production (ποιητική) is clear even from the history of the earliest philosophers."

favor the β-reading. The movement of thought in these lines is as follows. After having declared truth to be the aim of theoretical science (993b20–21) and after having indirectly introduced "what is eternal" (according to β) or the "cause in itself" (according to α) to be the subject of theoretical science (b22–23), Aristotle moves on to say that knowledge of the truth depends on knowledge of the cause (b23–24). He then describes in general what a cause is and states that the cause of the truth of a thing is itself true to the highest degree (b24–27).³²⁵ So he infers (b28–29): διὸ τὰς τῶν ἀεὶ ὄντων ἀρχὰς ἀναγκαῖον ἀεὶ εἶναι ἀληθεστάτας ("Therefore the principles of eternal things must be always most true"). This statement seems to be the conclusion of the thought that began in the passage of our concern, and so the following construction of Aristotle's thought suggests itself: theoretical science, which as we know already, aims at truth (book α ἔλαττον) by way of an investigation into causes (book A), is geared towards *what is eternal*, precisely because the causes of eternal things are in the highest sense true.³²⁶ This train of thought, then, would speak in favor of the β-reading (993b22).

Let us have a closer look at the *immediate* context of the line in question (993b22). Aristotle's specification of what is not a subject of practical science by either the words οὐκ ἀΐδιον (β) or the words οὐ τὸ αἴτιον καθ' αὑτό (α) is juxtaposed with the affirmation of the science's actual subject: ἀλλὰ πρός τι καὶ νῦν (993b22). How do the two possible juxtapositions compare? Following the β-reading, the ἀΐδιον ("what is eternal") pairs with νῦν ("what is in the present"), which follows after πρός τι ("what is relative"). This chiastic pairing could be accepted and justified as a stylistic *lectio difficilior*. Following the α-reading, by contrast, the term καθ' αὑτό pairs with πρός τι. This pairing is not only fitting, but it is also common in Aristotle's diction.³²⁷ The α-text, however, suggests taking the terms πρός τι and νῦν in the sense of <τὸ αἴτιον> πρός τι and <τὸ αἴτιον> νῦν. The object of practical science is not the "cause in itself," but the "cause in relation to something" and the "present cause." This pairing is unique: such a pairing does not occur elsewhere in Aristotle's writings.³²⁸

Each reading thus has merits of its own. And so one can clearly see how someone might have been motivated to adjust the text in the direction of one or the other of the readings. We cannot determine for certain which of the two was writ-

---

³²⁵ On the "Causal Resemblance Principle" see 5.3.3; p. 251.

³²⁶ Cf. again E 1, 1026a15–18.

³²⁷ Cf. *Metaph.* Γ 6, 1011a17–18: εἰ δὲ μὴ ἔστι πάντα πρός τι, ἀλλ' ἔνιά ἐστι καὶ αὐτὰ καθ' αὑτά … and *Metaph.* Α 9, 990b19–21: συμβαίνει γὰρ μὴ εἶναι τὴν δυάδα πρώτην ἀλλὰ τὸν ἀριθμόν, καὶ τὸ πρός τι τοῦ καθ' αὑτό… . Cf. also *Cat.* 6, 5b16–18: οὐδὲν γὰρ αὐτὸ καθ' αὑτὸ μέγα λέγεται ἢ μικρόν, ἀλλὰ πρὸς ἕτερον ἀναφέρεται.

³²⁸ Aristotle typically contrasts the αἴτιον καθ' αὑτό with the αἴτιον κατὰ συμβεβηκός. See *Ph.* Β 5, 196b24–27: ὥσπερ γὰρ καὶ ὄν ἐστι τὸ μὲν καθ' αὑτὸ τὸ δὲ κατὰ συμβεβηκός, οὕτω καὶ αἴτιον ἐνδέχεται εἶναι, οἷον οἰκίας καθ' αὑτὸ μὲν αἴτιον τὸ οἰκοδομικόν, κατὰ συμβεβηκὸς δὲ τὸ λευκὸν ἢ τὸ μουσικόν· and Β 6, 198a8–9: δῆλον ὅτι οὐδὲ τὸ κατὰ συμβεβηκὸς αἴτιον πρότερον τοῦ καθ' αὑτό.

ten by Aristotle, but we know both readings existed already by the time Alexander wrote his commentary. Thus we turn now to interpret the agreement between β and ω^AL as well as the agreement between α and Alexander's *varia lectio*. To prepare the way for this interpretation, I return to Alexander's commentary. As seen above, the sentence that in 993b23-24 follows directly upon our passage (οὐκ ἴσμεν δὲ τὸ ἀληθὲς ἄνευ τῆς αἰτίας) can be taken as evidence in support of the α-reading (οὐ τὸ αἴτιον καθ' αὑτὸ). In his commentary on this very sentence, Alexander refers back to our passage and to the variant reading (α-reading) he introduced there.

Alexander, *In Metaph.* 145.27–146.4 Hayduck

993b23 Οὐκ ἴσμεν δὲ τὸ ἀληθὲς ἄνευ αἰτίας·
[28] Τὸ λεγόμενον ἴσον ἐστὶ τῷ ἀλλὰ μὴν τὸ ἀληθὲς οὐχ οἷόν τε ἄνευ [29] αἰτίας εἰδέναι· ὥστε εἰ οἱ πρακτικοὶ μὴ κατὰ τὸ κυρίως αἴτιον τὴν γνῶσιν [146.1] περὶ τῶν προκειμένων ποιοῦνται, οὐδὲ τὸ ὄντως ἀληθὲς ἐν αὐτοῖς θεω-[2]ροῦσι. καὶ εἴη ἂν οὕτως μὲν λεγόμενον ἀκολούθως εἰρημένον <u>τῇ δευτέρᾳ</u> [3] <u>γραφῇ</u>· εἰ δὲ ἁπλῶς λέγοιτο, ὡς δεικτικὸν ἂν λέγοιτο τοῦ δεῖν τὸν περὶ [4] τὸ ἀληθὲς πραγματευόμενον τῶν αἰτιῶν εἶναι θεωρητικόν.

Now we do not know the truth without its cause.
This statement is equivalent to saying that it is indeed impossible to know the truth without its cause, so that if practical men do not base their knowledge of the actions before them on the cause that is such in the proper sense, neither do they consider the real truth in these actions. [Interpreted] thus, the statement might be a logical continuation of <u>the second reading</u> of the text. But if it is taken independently, it might be intended to show that one who devotes himself to the truth must have a theoretical knowledge of the causes.

27 ἄνευ τῆς αἰτίας *Metaph.* || 29 πρακτικοὶ O P^b S : πρακτοὶ A || 2 μὲν O P^b : μὴν A

Alexander offers two interpretations of lines 993b23-24, which are quoted in the lemma. According to the first interpretation, there is a direct connection between the quoted text and the *varia lectio* (= α-reading) in 993b22. Therefore, Alexander calls the present *Metaphysics* passage a logical continuation (ἀκολούθως) of the second reading (i.e. the *varia lectio*) and he takes it that Aristotle's words οὐκ ἴσμεν δὲ τὸ ἀληθὲς ἄνευ τῆς αἰτίας confirm the αἴτιον καθ' αὑτὸ (which he renders into τὸ κυρίως αἴτιον, 145.29)[329] as subject of theoretical science. Beside this interpretation Alexander gives a second (146.3-4), which holds that the present sentence is no continuation of the preceding sentence, and so entails no preference of the *varia lectio*.

I come now to the conclusion. The array of possibilities accounting for the present divergence between the α- and β-readings is similar to that discussed in the previous case. There is a decisive difference, however: in the present case

[329] Cf. Alexander's formulation in 145.23: τὸ κυρίως καὶ καθ' αὑτὸ αἴτιον.

the evaluation of the two possible readings of the *Metaphysics* is not as straightforward as in the former case. It is simply not clear whether we should take οὐ τὸ αἴτιον καθ' αὑτὸ (α) or οὐκ ἀίδιον (β) to be the original reading. With that said, the following scenarios seem possible.

(i) One could regard the α-version (οὐ τὸ αἴτιον καθ' αὑτὸ) as the original reading (see Jaeger's text), which was given in ω^{αβ} and corrupted to οὐκ ἀίδιον in ω^{AL}. Alexander knew (either via other commentators such as Aspasius or through marginal notes in his own copy) the ω^{αβ}-reading as a *varia lectio* (see 3.5.2). The β-text, initially reading (with ω^{αβ}) the original οὐ τὸ αἴτιον καθ' αὑτὸ, was changed to οὐκ ἀίδιον in order to be brought into accord with the ω^{AL}-text as presented in Alexander's commentary (cf. 5.2).[330] But the β-text would then have adopted not a suggestion made by Alexander, but the reading one can suppose to be in his *Metaphysics* copy.

(ii) Another viable option is that the β-version leads us to the original reading, οὐκ ἀίδιον, which was initially given in ω^{αβ} and is also preserved by Alexander's text (ω^{AL}). But there existed the alternative reading οὐ τὸ αἴτιον καθ' αὑτὸ in one of the versions that Alexander shows sporadic knowledge of (either via Aspasius or through marginal notes in his text): ω^{ASP2-n} or φ. The α-version adapted the variant reading οὐ τὸ αἴτιον καθ' αὑτὸ either (iia) from Alexander's commentary (cf. 5.3) or (iib) directly from the other version ω^{ASP2-n} or φ.

A decision between option (i) and option (ii) seems impossible as long as we do not know whether the α- or the β-reading is correct. If we regard the α-reading as original and go with option (i), it becomes likely that the β-reading is the result of contamination by Alexander's commentary, since there is no reason to assume that β adopted the reading—which, as far as we know, was only in ω^{AL}—by other means than through Alexander's commentary. If we instead regard the β-reading as correct and correspondingly opt for (ii)—on the grounds that the β-reading is well attested through the agreement of β and ω^{AL} and therefore most likely the reading of ω^{αβ}—then the contamination of the α-reading was triggered either (iia) directly by the version (ω^{ASP2-n} or φ) from which the alternative reading had come into Alexander's commentary, or (iib) by Alexander's commentary, in which the alternative reading appears as a *varia lectio*. Could anything in Alexander's commentary have invited the alteration of α? Perhaps it was the way in which Alexander, in one of his two interpretations of lines 993b23–24, presents the *varia lectio* as a plausible reading.

Since it remains an open question as to whether α or β leads to the original reading, I refrain from deciding conclusively between scenario (i) and (ii). Yet,

---

[330] Since in this scenario the β-reading was in ω^{AL} and we assume that β received it through contamination, we do not need to complicate the picture by including the (theoretical) possibility that the β-text incorporated the reading from a version other than ω^{AL}. The matter is different in scenario ii, for here we do not really know from where Alexander knew the reading that is in α. And so we should consider a variety of possible texts by which α could have been contaminated.

given the parallelism between this case and the case in 5.4.1, in which the contamination of α by Alexander's commentary was the slightly preferable option, one might perhaps be tempted to think that here again, β and ω^AL represent the older reading, while α adopted, perhaps from Alexander's commentary, a variant reading.

CHAPTER 6

Results

This study had two principle aims. The first aim has been to determine the relations between the text of the *Metaphysics* that Alexander used, ω^{AL}, which has to be reconstructed from Alexander's commentary, and the directly transmitted versions of the *Metaphysics*, α and β, as well as their common ancestor $\omega^{\alpha\beta}$. In performing the analyses that revealed these relations, I followed the basic rules of textual criticism and sought and examined peculiar errors that the various versions shared or did not share. The second aim has been to determine how Alexander's commentary influenced the tradition of the *Metaphysics* text.

The present study's results allow me to draw several conclusions about the ancient tradition of Aristotle's *Metaphysics* and the role Alexander's commentary played in it. First, we are now able to assess the textual situation of the *Metaphysics* at the time when Alexander wrote his commentary. From the first century BC edition of the *Metaphysics* that contained the fourteen books of our *Metaphysics* several copies were made. These copies developed into various versions of the text that differed from each other through errors that occurred in the text as well as through intentional changes made to the text. The oldest version of the *Metaphysics*, whose readings are at least partially reconstructible to us, is ω^{AL}, which is the exemplar Alexander used when writing his commentary around AD 200. Alexander himself had sporadic access to *variae lectiones* present in other *Metaphysics* versions either through notes in the margins of ω^{AL} or through other commentaries by previous scholars such as Aspasius. Among the variant readings known to Alexander are readings that are identical to the readings we can reconstruct for $\omega^{\alpha\beta}$, the ancestor of the directly transmitted texts α and β. Since some of these readings in $\omega^{\alpha\beta}$ are corrupt we are allowed to assume that Alexander knew indeed readings of, and hence had access to, $\omega^{\alpha\beta}$ or its ancestor (ω^{ASP1}), however limited that access may have been. Among the variant readings Alexander reports are also readings that differ not only from ω^{AL}, but also from $\omega^{\alpha\beta}$, and which therefore stem from one (or several) other version(s) of the *Metaphysics*.[1]

---

[1]This version or these versions of the *Metaphysics* text (ω^{ASP2-n}) I identified with any other texts that

We are, second, in a better position to reconstruct the text of the *Metaphysics*. Our manuscript evidence allows us to reconstruct two branches of the *Metaphysics* text, α and β. ω^{αβ} is their ancestor. ω^{αβ}'s *terminus ante quem* is the end of the fourth century AD. This text is independent of ω^{AL}, which itself is independent of ω^{αβ}. Therefore, if a reading in ω^{AL} agrees with either α or β against the other it is probably the reading of ω^{αβ}. The independence of ω^{AL} against ω^{αβ} makes it furthermore possible to correct corrupted readings in ω^{αβ} by means of the reading in ω^{AL}.

Third, the influence of Alexander's commentary on ω^{αβ}, which we are for the first time able to trace, allows us to give a more precise dating of ω^{αβ}. Alexander wrote his commentary on the *Metaphysics* around AD 200. The great success of this commentary as *the* commentary on Aristotle's work can explain the influence it had on the transmission of the *Metaphysics* text during the subsequent centuries. The present study demonstrates that Alexander's commentary influenced ω^{αβ} such that Alexander's reformulations or suggested corrections were incorporated into ω^{αβ} before its split into α and β. This means that we are now in the position to date the emergence of ω^{αβ} rather precisely to the time between AD 250 (i.e., the time when Alexander's commentary could first have established itself as an important commentary) and AD 400.

Fourth, we have a more comprehensive view of how Alexander's comments shaped parts of the β-version. Primavesi concluded for the first book of the *Metaphysics* that the β-version had had words and phrases from Alexander's commentary brought into it. His study of the character of this influence led him to conclude that the inclusion of these words and phrases was the result of a deliberate editorial revision of the *Metaphysics* text. The present study shows that such an influence occurred in several passages throughout books Α–Δ of the *Metaphysics*. The influence Alexander exerted on the text of β can be connected with the revision process that this version very likely underwent at some time before AD 400.[2] We do not need to suppose, however, that all changes in the β-version based on Alexander's commentary occurred simultaneously.

Fifth, we now see that Alexander's commentary even had an effect on the α-text. The types of influence are two: either Alexander's reformulations of an Aristotelian sentence were incorporated into the α-text, or his remarks about possible improvements to the *Metaphysics* text resulted in a change of the α-reading. The traces of contamination that one finds in the α-text are less extensive than those one finds in the β-text. The contamination of α is, as far as my evidence goes, mainly confined to book Α of the *Metaphysics*, with the exception of the contamination occurring in Δ 10.

Alexander's commentary thus influenced the text of the *Metaphysics* at all stages we can reconstruct. This means that Alexander's exegesis left clear footprints on the *Metaphysics* text as we know it. There are more instances of such an influence

---

Aspasius (first century AD) might have used, as Aspasius is the only textual source that Alexander names.

[2] Frede/Patzig 1988: 13–14 and Primavesi 2012b: 457–58.

than this study analyzed. Some instances are clearly determinable, while others are undetectable. For many cases of agreement between Alexander's paraphrase and the text of either $\omega^{\alpha\beta}$ or one of the descendants α and β we cannot determine whether this agreement is due to contamination by Alexander's comments, because the commentary does not offer sufficient evidence to securely determine the reading of ω^{AL}.

APPENDIX A: A DIAGRAM OF THE ANCIENT GREEK TRADITION OF THE *METAPHYSICS*

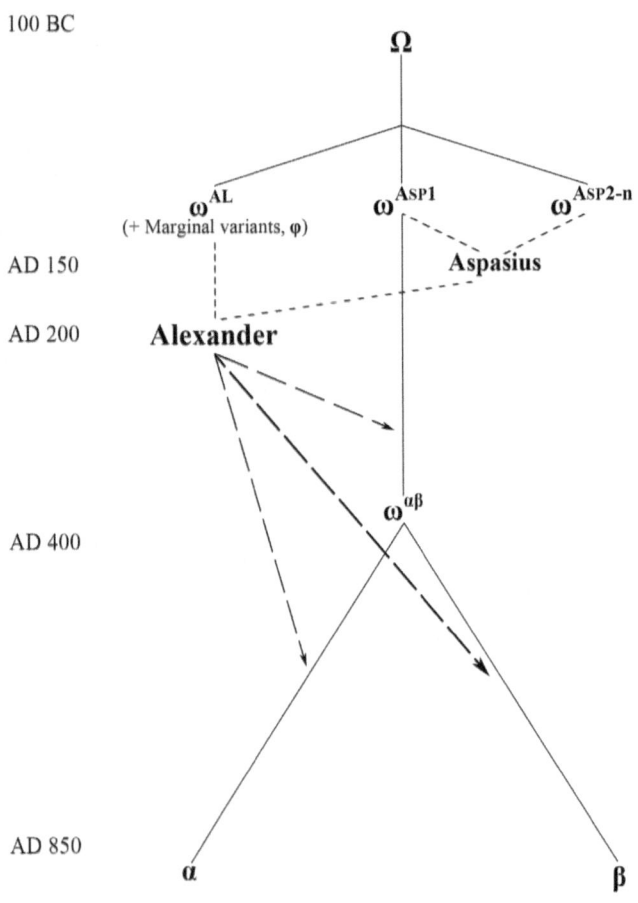

APPENDIX B: LEMMATA IN ALEXANDER'S COMMENTARY

This table lists all 296 lemmata in Alexander's commentary and depicts their relationship to the direct transmission of the *Metaphysics*. The 1st column provides the lines of the *Metaphysics*. The 2nd column lists the lemmata in Hayduck's edition. I put a mark in the 3rd column whenever the reading in Alexander's lemma agrees with the reading shared by α and β; in the 4th column, when the reading in Alexander's lemma agrees with the reading in α, rather than with the reading in β; and in the 5th column when the reading in Alexander's lemma agrees with the reading in β, rather than that in α. Finally, I put a mark in the 6th column whenever Alexander's lemma contains a reading that is peculiar to it, i.e., whenever it contains a reading not shared by either α or β. Some of the lemmata have marks in more than one of the four possible columns, because more than one feature applies to them. For example, lemma no. 24 agrees with α in reading ἀλλ' ἤ instead of β's ἀλλά (984a10), but also entirely omits the words καὶ διακρινόμενα that are contained in both α and β (984a10-11).

There is an inherent imprecision in the representation of agreements between witnesses in a list like this (see also appendix C), since a lemma or quotation often consists of several words. However, if there is a disagreement concerning *one* word between, for instance, Alexander's lemma or quotation and the text in ω^{αβ}, or between α and β (with Alexander siding with one of them), then the lemma or quotation will have marks that correspond to this difference; and the fact that the lemma or quotation contain other words that are in agreement across versions is ignored.

Please note that insignificant divergences, for example between δ' and δέ, and τὸ αὐτό and ταὐτό, have not been taken into consideration.

Lemmata: 296
Agreements with ω^{αβ}: 145; agreements with α: 61; agreements with β: 51; peculiar readings: 91.

Metaphysics	Alexander	ω^{αβ}	α	β	Peculiar reading
A 1					
980a21	(1) 1.3	x			
980a27–28	(2) 2.22	x			
980b25–26	(3) 4.12	x			
981a12–13	(4) 5.14–15			x	
981b13–14	(5) 6.13–14	x			
981b25–26	(6) 7.10–11	x			

981b27	(7) 8.6				X
A 2					
982a4	(8) 8.19	X			
982a6-7	(9) 9.17-18		X		
982a21-22	(10) 11.3-4				X
982a25-26	(11) 12.5			X	
982a26-27	(12) 12.15	X			
982b11	(13) 15.20-21				X
A 3					
983a24-26	(14) 19.21-23	X			
983a27	(15) 20.4				X
983a29	(16) 22.1		X		
983a31	(17) 22.4-6			X	
983b6	(18) 23.8			X	
983b32-33	(19) 25.11-12	X			
983b33-984a1	(20) 26.8	X			
984a4	(21) 26.14-15			X	
984a3-5	(22) 26.19-20				X
984a8-9	(23) 27.9-10		X		
984a10-11	(24) 27.13-14		X		
					X
984a11-14	(25) 27.26-27	X			
984a16-17	(26) 28.22				X
984a18	(27) 29.5	X			
984a27-28	(28) 29.9	X			
984a29-31	(29) 29.18-19	X			
984b1-3	(30) 30.12-13			X	
					X
984b3	(31) 31.6	X			
984b5	(32) 31.17			X	
984b8	(33) 31.27	X			
A 4					
984b29	(34) 33.6	X			
984b32-33	(35) 33.12	X			
985a21	(36) 35.5				X
			X		
985b19	(37) 36.19-20		X		
A 5					
985b23; 26-27	(38) 37.4-5	X			
985b26	(39) 37.17	X			
986a13	(40) 41.16	X			
986b8	(41) 42.18-19			X	
986b17-18	(42) 43.10	X			

APPENDIX B: LEMMATA   285

Bekker	Lemma					
987a2–3	(43) 45.10		x			
987a9	(44) 46.5–6		x			
987a27–28	(45) 49.16	x				
A 6						
987b9–10	(46) 50.18–19			x		
987b10–11	(47) 52.1		x			
987b14–15	(48) 52.9	x				
987b18	(49) 52.26	x				
987b22–23	(50) 53.12–13			x		
987b25–27	(51) 54.1–2		x			
987b29–31	(52) 54.20–22		x			
					x	
987b33–988a1	(53) 55.17–19	x				
988a1–2	(54) 58.1	x				
988a7–8	(55) 58.24	x				
988a11–12	(56) 59.9	x				
988a14–15	(57) 60.11–12					x
A 7						
988a18	(58) 60.27	x				
988a23–24	(59) 61.9	x				
988a28	(60) 61.17	x				
988a32	(61) 61.23	x				
988a34–35	(62) 62.1–2	x				
988b6	(63) 63.1–2	x				
988b16	(64) 63.32	x				
A 8						
988b22–24	(65) 64.13–15			x		
						x
989a18	(66) 66.15	x				
989a30–31	(67) 68.5					x
989b16–17	(68) 69.15	x				
989b21–22	(69) 70.10–11	x				
989b29–30	(70) 71.10–11					x
990a18	(71) 73.9–10					x
990a22–23	(72) 74.1–2	x				
990a24	(73) 75.18	x				
				x		
A 9						
990a34–b2	(74) 76.6–7			x		
990b11–12	(75) 79.1–2	x				
			x			
990b13–14	(76) 80.7	x				
990b14–15	(77) 81.23–24		x			

990b15–16	(78) 82.8–9		x		
		x			
990b18	(79) 85.13–14				x
				x	
990b21–22	(80) 87.1–2			x	
990b22	(81) 88.3–4			x	
990b31	(82) 90.3–4	x			
991a2–3	(83) 92.29	x			
991a9	(84) 95.3–4		x		
					x
991a19	(85) 99.1–2				x
991a20	(86) 101.11–12				x
991a23–24	(87) 102.1				x
991a27–28	(88) 104.19	x			
991b1	(89) 105.28–29		x		
991b3–4	(90) 106.7–8				x
					x
991b9	(91) 107.14				x
991b13	(92) 108.1	x			
991b21	(93) 110.3–4	x			
991b22–25	(94) 111.1–3		x		
				x	
				x	
991b27–8	(95) 112.17–18		x		
				x	
			x		
					x
991b31	(96) 113.23		x		
992a1–2	(97) 114.11	x			
992a2–3	(98) 114.20–21	x			
992a10–11	(99) 117.20–21			x	
992a20	(100) 119.13				x
992a24	(101) 120.18–19		x		
992b9–13	(102) 123.15–18			x	
			x		
				x	
			x		
					x
992b13–14	(103) 127.1	x			
992b19	(104) 128.10–11			x	
			x		
992b24	(105) 129.10	x			
993a1	(106) 131.12			x	

993a2-3	(107) 132.9				X
993a8	(108) 133.20-21				X
A 10					
993a11	(109) 134.15-16	X			
993a24	(110) 136.3		X		
α 1					
993a29-30	(111) 138.24-25	X			
993b4-5	(112) 140.10-11	X			
993b7-8	(113) 141.36	X			
993b11-13	(114) 143.3-4			X	
993b19-20	(115) 144.15-16			X	
			X		
993b23-4	(116) 145.27				X
993b24-5	(117) 147.1-2		X		
993b28-9	(118) 148.20-21	X			
α 2					
994a1-2	(119) 149.14-15				X
			X		
994a11-13	(120) 150.28-29			X	
				X	
					X
994a19-20	(121) 153.1-2			X	
					X
					X
994a25	(122) 155.12			X	
					X
994a30-31	(123) 156.23	X			
994b4-5	(124) 157.28	X			
994b6-8	(125) 158.1-3		X		
994b13	(126) 160.22	X			
994b16-18	(127) 160.28-29	X			
994b20-21	(128) 162.17-18			X	
			X		
994b21-2	(129) 163.15	X			
994b25-6	(130) 164.15				X
994b27-8	(131) 165.28-29	X			
α 3					
994b32	(132) 167.4	X			
B 1					
995a24-5	(133) 171.3-4		X		
995a29-30	(134) 172.23			X	
995a32-3	(135) 173.5	X			

995b4–5	(136) 174.5–6	x			
995b6–8	(137) 175.1–2	x			
995b10–11	(138) 175.15–16	x			
995b18–20	(139) 176.17–18				x
995b20–22	(140) 176.31–33	x			
995b25–6	(141) 177.15–16	x			
995b27–9	(142) 177.24–25	x			
995b31–3	(143) 178.3–4		x		
995b34–5	(144) 178.22–23	x			
996a1	(145) 179.6	x			
996a4–7	(146) 179.25–27				x
996a11–12	(147) 180.16–17				x
B 2					
996a21–2	(148) 181.24			x	
996b1–2	(149) 183.14				x
996b8–9	(150) 184.12–13		x		
				x	
996b13–14	(151) 184.28–29				x
996b18–19	(152) 185.21	x			
996b22	(153) 186.3	x			
996b26–7	(154) 187.14–15				x
996b35–997a1	(155) 188.7–8	x			
997a11–12	(156) 190.18	x			
997a14	(157) 191.1	x			
997a15–16	(158) 191,13–14	x			
997a25–6	(159) 194.8–9		x		
				x	
997b3–4	(160) 196.13–14	x			
997b5–7	(161) 196.29–30				x
997b12–14	(162) 197.29				x
	(162) 197.30				x
997b25–6	(163) 198.31–32			x	
998a7–9	(164) 200.32–34				x
B 3					
998a21–3	(165) 202.4–6		x		
			x		
998b4–6	(166) 203.1–2		x		
998b6–7	(167) 203.12–13			x	
					x
998b10	(168) 203.24–26		x		
					x
998b11–12	(169) 204.8				x

998b14-15	(170) 204.23-24					x
			x			
998b28-9	(171) 207.7-8	x				
999a1-2	(172) 208.4	x				
999a6-7	(173) 208.26-27					x
Β 4						
999a24-6	(174) 210.22-24	x				
999a32-4	(175) 211.18-19		x			
999b8-9	(176) 213.24-25					x
999b12-14	(177) 214.19-20	x				
						x
999b20-22	(178) 215.30-31					x
			x			
999b24-5	(179) 216.12-13		x			
999b27-8	(180) 217.26	x				
1000a5-6	(181) 218.18-19	x				
1000a27-8	(182) 220.1		x			
1000b28-9	(183) 222.4-5		x			
1001a4-5	(184) 223.6-7	x				
1001a29-30	(185) 225.33-34					x
1001b1-3	(186) 226.10-11		x			
1001b7-8	(187) 227.9-10	x				
Β 5						
1001b26-8	(188) 228.29-30					x
1002a28-9	(189) 231.26-27	x				
Β 6						
1002b12-14	(190) 233.1-3					x
				x		
1002b32-4	(191) 235.7-8	x				
Γ 1						
1003a21-2	(192) 239.4-5					x
Γ 2						
1003a33-4	(193) 240.31-32					x
1003b12-13	(194) 243.29-30	x				
1003b16	(195) 244.9	x				
1003b19-20	(196) 244.29-30	x				
1003b21-2	(197) 245.20-21		x			
						x
1003b22-5	(198) 246.25-27		x			
1003b32-3	(199) 249.1-2					x
1004a2-3	(200) 250.21	x				
1004a9-10	(201) 252.1-2	x				
						x

1004a10–11	(202) 252.17	x			
1004a20–21	(203) 254.16		x		
1004a22–3	(204) 255.3–4	x			
1004a30–31	(205) 256.19–20				x
					x
1004b1–3	(206) 257.17–18				x
1004b8–9	(207) 258.25	x			
1004b10–11	(208) 259.1	x			
1004b17	(209) 259.23				x
1004b25	(210) 260.21	x			
1004b28	(211) 260.30–31		x		
					x
1004b29–30	(212) 261.17–18			x	
		x			
1005a3–4	(213) 262.20–21	x			
Γ 3					
1005a19–21	(214) 264.28–30				x
					x
1005b2–4	(215) 266.29–31	x			
1005b5–7	(216) 267.22–23	x			
1005b8–10	(217) 268.7–8	x			
1005b17–18	(218) 269.18		x		
Γ 4					
1005b35–6a2	(219) 271.22–23	x			
1006a11–12	(220) 272.28–29	x			
1006a18–20	(221) 273.20–21				x
					x
1006a26–8	(222) 274.33–35			x	
1006a29–30	(223) 275.21–22				x
1006a31–2	(224) 276.1–2	x			
1006b11–13	(225) 279.15–16				x
			x		
1006b14–15	(226) 279.27–28	x			
1006b28–30	(227) 282.1–2				x
1006b34–a1	(228) 283.1			x	
1007a8–9	(229) 284.1	x			
1007a20–21	(230) 285.1–2		x		
1007a23	(231) 286.7	x			
1007a27–8	(232) 286.25		x		
1007a33–4	(233) 287.22–23	x			
1007b18–19	(234) 290.22–23	x			
1007b26–8	(235) 291.20–21	x			
1007b29–30	(236) 292.1–2	x			

APPENDIX B: LEMMATA   291

1008a2–4	(237) 292.22–23					x
1008a7–9	(238) 293.33–34					x
1008a16–17	(239) 294.22	x				
1008a25	(240) 295.29		x			
1008a28–9	(241) 296.3					x
1008a30–31	(242) 296.22					x
1008a34–5	(243) 297.7				x	
1008b2–3	(244) 297.27					x
1008b27–8	(245) 300.4	x				
1008b31–2	(246) 300.23	x				
Γ 5						
1009a6–7	(247) 301.27–28					x
						x
1009a38–b2	(248) 304.34–35	x				
1009b4–5	(249) 305.14	x				
1009b7–8	(250) 305.24–25		x			
					x	
					x	
1009b12–13	(251) 306.1–2		x			
1010a32–3	(252) 310.34–35	x				
1010b1–3	(253) 311.25	x				
1010b3–5	(254) 312.11–12	x				
1010b14–16	(255) 313.18–19	x				
1010b19–20	(256) 314.29		x			
1010b30–31	(257) 315.27–28					x
Γ 6						
1011a3–4	(258) 316.30–31	x				
1011a15–16	(259) 318.6–8		x			
1011a28–31	(260) 320.33–35					x
1011b7–9	(261) 323.11–12		x			
1011b13–14	(262) 326.20–21	x				
1011b15–17	(263) 326.28–29				x	
Γ 7						
1011b23–4	(264) 328.5–6	x				
1011b29–31	(265) 329.5–6					x
					x	
1012a2–3	(266) 330.17–18	x				
1012a5–6	(267) 331.7–8		x			
1012a9–10	(268) 332.1–2	x				
1012a12–13	(269) 332.16–17		x			
1012a15–16	(270) 333.18		x			
1012a17	(271) 334.4	x				
1012a21–2	(272) 335.20	x				

Γ 8						
1012a29-30	(273) 336.23-24					x
1012b5-6	(274) 338.23-24			x		
1012b8-9	(275) 339.1			x		
1012b11-12	(276) 340.8	x				
1012b13-14	(277) 340.19					x
			x			
1012b18-19	(278) 340.30-31	x				
1012b22-4	(279) 341.28-29	x				
1012b28	(280) 342.21					x
1012b29-30	(281) 342.35-36					x
Δ 2						
1013a24-5	(282) 348.25-26			x		
1013a32-3	(283) 350.4					x
1013b3-4	(284) 350.19	x				
1013b16-17	(285) 351.1-2		x			
1013b29-30	(286) 352.9	x				
1014a10-11	(287) 353.5-6	x				
1014a13-14	(288) 353.30	x				
Δ 3						
1014a26-7	(289) 354.26-27	x				
Δ 4						
1014b16-17	(290) 357.5-6			x		
Δ 5						
1015a20-21	(291) 360.17-18	x				
Δ 6						
1015b16	(292) 362.11					x
Δ 7						
1017a7-8	(293) 370.3-4	x				
Δ 8						
1017b10	(294) 373.1	x				
Δ 9						
1017b27	(295) 376.13					x
Δ 15						
1021a31-2	(296) 407.16	x				

APPENDIX C: QUOTATIONS FROM THE *METAPHYSICS* IN ALEXANDER'S COMMENTARY

This table lists the quotations of the *Metaphysics* text that Alexander provides in his commentary. The 1st column lists the quoted lines of the *Metaphysics*. The 2nd column gives the page and line numbers of the quotation in Hayduck's edition. I put a mark in the 3rd column whenever the reading in Alexander's quotation agrees with the reading shared by α and β; in the 4th column, when the reading in Alexander's quotation agrees with the reading in α, rather than the reading in β; and in the 5th column, when the reading in Alexander's quotation agrees with the reading in β, rather than that in α. The 6th column is marked whenever Alexander's quotation contains a reading that is peculiar to it, i.e., not shared by either α or β.[1]

Given that it is not always clear whether something is a quotation from the *Metaphysics* (see discussion in 3.3), one might worry whether some of the cases listed here are in fact quotations rather than paraphrases. This is especially relevant when the reading differs from our direct evidence. The instances of quotation that I present in the table agree mostly but not always with what Hayduck marks as a quotation in his edition of the commentary. My list does not contain quotations that Alexander cites as *varia lectio* from a *Metaphysics* copy other than ω^{AL} (for a complete list of the variant readings see 3.6).

Please note that insignificant divergences, for example between δ' and δέ, and τὸ αὐτό and ταὐτό, have not been taken into consideration.

Quotations 579:
Agreements with ω^{αβ}: 342; Agreements with α: 126; Agreements with β: 82; Peculiar readings: 187

Metaphysics	Alexander	ω^{αβ}	α	β	Peculiar reading
A 1					
980a21	1.8	x			
980a25–6	1.14–15	x			
980a27	1.21–22	x			
980a27–8	4.13–14	x			
980b28–9	4.21	x			
981a8	5.7				x
981b27–9	8.27–28; 10.22–23	x			
981b31–982a1	8.15				x
982a1	8.16–17	x			

[1] On the inherent imprecision of a list like this, see p. 283 (Appendix B).

A 2						
982a9	10.2	x				
982a13	10.7–8		x			
	10.14–15			x		
982a24	11.11–12	x				
982a32–b1	13.21–23			x		
982b5–6	14.7, 17–18					x
				x		
982b6	14.8–9	x				
982b9	15.10–11		x			
982b10	15.12–13					x
982b18	16.13		x			
983a2–3	18.2–3					x
983a4–5	18.4	x				
983a16–17	18.19–20	x				
					x	
983a18	19.8	x				
983a20–21	19.10–11		x			
983a22–23	19.12–13	x				
A 3						
983a24	19.24	x				
983a28	20.8–9; 21.2–3		x			
983a29	21.14–15, 22, 30	x				
983a32	22.10	x				
983b7	23.21	x				
983b8–10	23.17–18					x
983b11–13	23.23–24	x				
983b17	24.14–16	x				
				x		
983b18–21	24.18–20	x				
983b24	24.28–29					x
983b27	25.3–4					x
983b27–8	25.7			x		
984a 3	26.16				x	
984a16	31.4; 34.12–35.1				x	
984a27–8	29.11	x				
984a29	29.14				x	
984a31	30.6–7	x				
984b3	31.15	x				
984b16	32.11–12				x	
A 4						
985a17–18	34.5–6					x
985b21	37.2					x

APPENDIX C: QUOTATIONS 295

A 5					
985b23	37.7–8	x			
986a3	39.23–24				x
986a17	41.21	x			
986b12	43.2				x
987a3	45.11–12			x	
987a4	45.14	x			
987a5–6	45.21	x			
987a7–8	45.24–25	x			
987a11–13	46.20–22				x
					x
987a13	47.2–3	x			
987a26	48.21–23			x	
A 6					
987b8–9	50.19–20	x			
987b10	50.22–23			x	
987b17	52.21				x
987b21–22	53.5–6	x			
987b26	54.11			x	
987b33–988a1	57.1–3	x			
988a6–7	58.22–23				x
988a8–9	58.26–27			x	
988a10–11	59.14–15	x			
988a12–13	59.16–17				x
988a14	60.13–14				x
988a15–16	60.22–23			x	
				x	
A 7					
988a20	60.30–31				x
988a21–22	61.7	x			
988a29–31	61.25–26	x			
988a33	61.28	x			
988b2	62.12	x			
988b15	63.18			x	
988b18–19	64.3–4	x			
A 8					
988b22–3	64.16–17			x	
			x		
988b26	64.26–29				x
988b28	64.29				x
989a15–16	66.1			x	
989a20	66.17–67.1				x
989a21–22	67.6–7	x			

989a24	67.14–15	x				
989a33–4	68.15	x				
989a33	69.13	x				
989b6	69.4–5	x				
989b17	69.18	x				
989b19	69.22			x		
989b19–20	70.5–6; 28.12–13					x
		x				
			x			
990a7–8	72.20–21	x				
990a25–6	75.2–3		x			
Α 9						
990a33–4	76.1–2					x
						x
				x		
						x
990b1–2	95.6–7	x				
990b6–7	77.11–12	x				
990b7–8	77.17–19					x
	77.27–28, 31; 96.6					x
			x			
				x		
990b8–9	77.34–35		x			
			x			
						x
990b16–17	83.31–32		x			
990b34	91.11	x				
990b34	91.13, 17, 27	x				
990b34	91.17					x
991a1–2	91.17–18, 26; 94.10–11	x				
						x
991a3–5	93.15–17					x
					x	
			x			
991a9–10	96.2–3					x
					x	
991a18–19	98.23–24					x
	100.32–33		x			
991a19–20	99.6–7; 100.23–24, 33–34	x				
991a22–3	101.7–8, 22	x				
991a29–31	105.24–25					x
991b3	106.9					x

991b3–4	106.13–14	X			
991b19	109.17	X			
991b20	109.30			X	
991b23	112.14–16	X			
991b25	112.7		X		
		X			
991b29	113.9		X		
991b31	113.21	X			
992a2–3	115.5				X
992a6–7	117.8	X			
992a7–8	115.22–116.1	X			
992a8	116.15	X			
992a13–14	118.3–4		X		
992b3–4	122.15–16			X	
992b7	122.21–22				X
			X		
992b11	126.22	X			
992b25	129.13–14				X
992b31	131.6	X			
993a2	133.17	X			
Α 10					
993a11–16	63.27–31			X	
					X
					X
993a25–26	136.15–16				X
	137.8	X			
993a26–7	137.9	X			
				X	
α 1					
993a29	141.31	X			
993b1	139.19	X			
993b1–2	139.21	X			
993b2	141.2	X			
					X
993b2–4	141.27–29	X			
					X
					X
993b5	140.14	X			
993b6–7	140.19–20; 141.6–8		X		
	141.22			X	
993b14	143.14	X			
	143.16		X		
993b20–21	144.19–145.1	X			

993b22	145.19				X	
993b23	146.19	X				
993b24–5	147.15–16, 23					X
993b26–30	146.22–25; 148.32–149.3, 149.11–12					X
				X		
		X				
						X
				X		
						X
						X
993b30–31	149.7–8	X				
α 2						
994a5	150.2	X				
994a17	151.25	X				
994a18	152.2, 17–18	X				
994a20–22	153.12–13	X				
994a25–6	155.26–27	X				
						X
994a31–2	156.28–29	X				
994a32–994b1	156.32–33		X			
						X
994b2	157.16	X				
994b4	157.35	X				
994b5–6	157.33–34					X
994b6–8	159.6–7	X				
	159.8	X				
	159.10–11		X			
						X
994b9–10	159.29–160.1				X	
994b18	161.2; 162.10	X				
994b22–3	163.24	X				
994b24	164.4	X				
994b25	164.8	X				
994b26–7	165.6–7				X	
		X				
994b30–31	166.7–8		X			
α 3						
994b32	170. 4	X				
995a1–2	167.10–11					X
995a5–6	167.20–21		X			
995a10	168.5	X				
995a12–13	168.13–14				X	
				X		

APPENDIX C: QUOTATIONS   299

995a14–16	169.4–5				x
995a16–17	169.9				x
995a17	169.17–18				x
995a17–18	137.15–16;			x	
	169.20–21			x	
B 1					
995a25	172.1	x			
995a25–6	172.3–4	x			
995a30–31	172.31–32	x			
995b5–6	174.27		x		
		x			
995b8	175.8	x			
995b8	187.19	x			
995b15–16	176.12	x			
995b16–18	176.4–5	x			
995b26	177.17–18	x			
995b27–9	180.7–8	x			
995b35	178.30–31	x			
995b36	178.35–179.1	x			
996a1–2	179.18–20	x			
996a7	179.30	x			
B 2					
996a20–21	181.13–14	x			
996a24	182.5–6, 13–14				x
996a28–9	182.20	x			
996b4–5	184.8–10		x		
996b5–6	183.21–22				x
996b7	183.31	x			
996b14–16	185.1–3, 6	x			
996b19	185.24	x			
996b24	186.12	x			
996b24–26	186.14–15	x			
997a3	188.17–18				x
997a6–8	188.31–189.1, 5–6	x			
		x			
997a8–9	189.4–5	x			
997a9–10	189.11–12		x		
997a12–13	190.26, 28	x			
997a14	191.7	x			
997a18–19	192.4–5		x		
997a21	191.29	x			
997a22–4	192.6–7, 16; 193.21;			x	
	194.12	x			

997a24	192.11; 193.1–2; 193.32	x				
			x			
				x		
		x				
997a25–6	195.3–4		x			
						x
997a31	194.23–24					x
997a31	194.25–26					x
997b3	196.24	x				
997b6–7	196.31–32	x				
B 3						
998b20–21	204.33–34	x				
						x
998b24–5	206.6–7	x				
998b25	205.20				x	
				x		
998b25	206.9	x				
						x
998b27–8	207.5–6					x
998b29	207.28	x				
998b30	207.16, 17–18	x				
999a5–6	208.22		x			
				x		
999a14–16	210.11–12	x				
999a17–20	211.10–12		x			
						x
B 4						
999a32	211.9	x				
999a33–4	211.22	x				
999a34	215.22				x	
				x		
999b1	211.34–212.1				x	
999b4–5	212.10–11	x				
999b6	212.21–24			x		
		x				
			x			
						x
999b9–10	213.33–34	x				
999b12–13	214.22	x				
999b14	214.31	x				
999b15	215.5–6, 11–13					x
		x				

APPENDIX C: QUOTATIONS   301

999b16	215.8-9, 14	x			
					x
999b26-27	217.19-20	x			
1000a1-2	218.9-10	x			
				x	
			x		
1000a18-19	219.9			x	
1000a29-30	220.5				x	
1000b26-7	221.34-35					x
1000b27-8	221.35-222.1	x			
1001a2-3	222.24-26	x			
1001a6-7	223.23-24	x			
1001a7-8	223.33-34					x
1001a11-12	224.2-3					x
1001a20-21	224.18		x		
1001a22-3	224.23-24		x		
						x
1001a26-7	224.36			x	
1001a27-8	225.8, 23-24					x
						x
						x
1001a28-9	225.11, 29		x		
1001b4	226.27		x		
1001b4	226.29-30					x
1001b11	227.18-19		x		
1001b15-16	228.3					x
1001b20-21	228.12	x			
1001b23	228.24-25					x
B 5						
1001b29	229.3					x
1001b30-31	229.6		x		
1001b31-2	229.8	x			
1002a7	229.31	x			
1002a10-11	230.13-14	x			
1002a27	231.24-25	x			
1002b1	231.33-232.1					x
B 6						
1002b17-19	234.7-8					x
1002b24	233.21-22					x
1002b24-5	233.27-28		x		
1002b33-4	235.11	x			
1002b34	235.12	x			
1003a1-2	235.20-21	x			

1003a4–5	235.29–30	x			
1003a5	235.31	x			
1003a10–11	236.15–16	x			
1003a13	236.14–15, 20–21	x			
Γ 1					
1003a22–3	246.8–9	x			
Γ 2					
1003b4	242.3	x			
1003b9	242.30	x			
1003b9–10	243.7		x		
	243.15	x			
1003b17–19	250.33–251.1	x			
1003b20–22	251.4–5; 245.25				x
					x
1003b22–3	251.3–4		x		
1003b24–5	247.25	x			
1003b26	247.30	x			
1003b26	247.33–34			x	
1003b30	248.19	x			
1003b30–32	248.32–33				x
1003b33–34	249.18–19	x			
				x	
1003b35–6	249.34–35	x			
1004a1	250.13		x		
1004a1–2	252.3–4		x		
					x
1004a2–3	251.1–2, 6				x
		x			
1004a4–5	251.10		x		
1004a12–13	253.1–2				x
1004a13–14	253.10–11, 16–17	x			
1004a14–15	253.16	x			
1004a16–18	253.29–30	x			
1004a18–19	253.34–35; 254.7–8				x
					x
					x
1004a21–22	254.18–19		x		
1004a24	255.19–20	x			
1004a27	255.32	x			
1004a28–30	256.5–6	x			
1004b5–8	258.2–5, 15			x	
			x		
			x		

Ref	Page	1	2	3	4	5
1004b22-3	259.32-33					X
					X	
1004b23-5	259.35-260.1		X			
1004b27	261.10-11	X				
1004b27-8	261.14-15; 262.15			X		
				X		
1004b29-30	261.27-29; 262.13-14					X
		X				
1004b34-1005a1	262.18	X				
1005a6-7	263.8-9		X			
1005a10-11	263.20, 22-25	X				
		X				
1005a12-13	264.9					X
1005a14-15	264.17-18	X				
Γ 3						
1005a20	265.3-4	X				
1005a25-7	265.22-23					X
1005a31	265.28-29	X				
1005a34	265.40-266.1	X				
1005b1-2	266.15-16		X			
1005b2-3	267.15, 19-20					X
1005b4-5	267.7, 16					X
		X				
1005b5-8	267.17-19, 24-25, 28	X				
1005b8	267.21	X				
1005b19-22	269.23-25				X	
						X
						X
			X			
1005b21-2	269.31	X				
1005b23	269.33	X				
1005b23-4	269.35-36; 270.1-2	X				
						X
1005b24-5	270.4-5	X				
1005b26-7	270.15-16		X			
1005b27	270.27		X			
1005b30-32	270.38-271.1				X	
1005b32-4	271.5-7				X	
Γ 4						
1006a2-3	271.37-38	X				
1006a3-4	272.4-5	X				
1006a9	272.21-22, 26	X				

1006a14	273.1-3					x
				x		
1006a20-21	273.35	x				
1006a24-5	274.18-19	x				
						x
1006a26	274.27	x				
1006a28-30	275.3-4; 275.31-32	x				
		x				
1006a32-4	276.30-32, 34		x			
		x				
			x			
			x			
				x		
1006b9	278.16	x				
1006b13-16	279.29-32		x			
			x			
1006b19-20	280.35-36; 281.36				x	
		x				
1006b20-22	281.32-34			x		
1006b22-4	281.28-30					x
1006b24-5	281.20, 31, 34-35	x				
1006b29	282.11	x				
1006b31-2	282.29-30	x				
1007a20-21	285.33-34	x				
1007a21	285.12		x			
1007a22-3	285.32; 286.3	x				
						x
1007a23	285.34					x
1007a25-6	286.20-21			x		
1007a27-8	287.2-3		x			
			x			
						x
1007a29	287.4		x			
1007a34-5	288.17-18		x			
						x
1007b9-10	289.29-30				x	
1007b11	289.33				x	
1007b22-3	290.34			x		
1007b23-4	291.4-5		x			
				x		
1008a21	295.9-10	x				
1008a31-2	296.30	x				
1008a32-3	296.33	x				

APPENDIX C: QUOTATIONS 305

1008b7–8	298.18–19	x			
1008b10	298.23				x
1008b25–7	299.28–30	x			
1009a4–5	301.16–17	x			
Γ 5					
1009a9	302.9–10				x
1009a9–11	302.10–12				x
1009a38–1009b1	305.3	x			
1009b17–18	306.22–23	x			
1009b20–21	306.24–25	x			
1009b22–3	306.29–30, 35				x
1009b24–5	306.36–307.1			x	
					x
1010a6–7	308.11–12	x			
1010a19	309.7				x
1010a22–3	309.36–310.1	x			
1010a23–4	310.8–9				x
					x
1010b1	311.27	x			
1010b16	314.1	x			
1010b18–19	314.3–4		x		
1010b30	316.27–28				x
Γ 6					
1011a13–14	317.36	x			
1011a23–4	319.17–18	x			
1011a31	321.5				x
1011a33	321.10	x			
1011a34–1101b1	322.2–4				x
					x
1011b1–2	322.7	x			
1011b11–12	325.20–21			x	
1011b18–19	326.32–33, 35	x			
1011b19–20	327.8, 10–11, 14, 25–26		x		
				x	
Γ 7					
1011b24	328.19		x		
1011b27–8	328.25		x		
					x
1011b32–4	329.25–26		x		
1011b35	329.35	x			
1011b35–12a1	330.1–2				x
1012a2–3	330.33–34; 331.1	x			
					x

1012a7–8	331.17–18					x
1012a8–9	331.35–36	x				
1012a13–14	333.7–8	x				
			x			
1012a14–15	333.17	x				
Γ 8						
1012a29–30	336.29–30	x				
						x
1012a30–31	336.32–33					x
1012a31–33	337.4–5			x		
						x
						x
1012a33	337.8	x				
1012a33–4	337.8–9					x
						x
1012b1	337.30, 33	x				
1012b4	338.9–10; 339.14–15				x	
1012b14	340.20		x			
1012b21–22	341.25–26		x			
						x
1012b30	343.2	x				
1012b30–31	343.5		x			
			x			
Δ 1						
1012b34–5	345.23–24	x				
1013a4	346.3	x				
1013a7	347.28–29; 348.32	x				
1013a7–8	346.10–13		x			
						x
1013a16	346.25	x				
1013a17	346.33–34	x				
1013a18–19	346.35–36	x				
1013a20–21	347.6–7				x	
1013a21–3	347.20–21				x	
Δ 2						
1013a24–5	348.27	x				
1013a27–9	349.3–4		x			
						x
1013a27–8	349.16–17	x				
1013a29–30	349.28, 31–32	x				
1013a31–2	349.37–350.1					x
1013a35–6	350.7	x				
1013b4	350.20	x				

APPENDIX C: QUOTATIONS 307

1013b4-7	350.23-26				x
			x		
1013b8-9	350.27	x			
1013b17-18	351.5-6	x			
1013b21	351.3-4, 19, 23-24	x			
1013b22-3	351.26-27, 31	x			
1013b23-4	351.35-36	x			
1013b25-6	351.38			x	
1013b26-8	352.3-4				x
			x		
1013b30-1	352.11		x		
					x
1013b33-4	353.2-3				x
				x	
1014a10	353.8	x			
1014a12	353.14			x	
			x		
1014a15-16	353.34-354.1				x
1014a20-22	354.11-13, 17	x			
					x
			x		
Δ 3					
1014a26-7	354.29-30; 356.12-13	x			
1014a30	355.6		x		
1014a31-2	355.9-10	x			
1014b2-3	356.20-21, 28-29				x
1014b4-5	355.25-26	x			
1014b5-6	355.28-29	x			
1014b6-7	355.30-31				x
1014b8	355.34	x			
					x
1014b8-9	355.36-37		x		
					x
				x	
1014b10	356.6		x		
Δ 4					
1014b17-18	357.21	x			
1014b18-20	357.22-24, 31; 358.8	x			
1014b21	358.18, 26-27		x		
1014b27	359.4		x		
1014b32-3	359.5-6	x			
1015a11	359.30-31	x			

1015a13–15	360.1–3		x		
					x
1015a17–18	360.9–10		x		
1015a18–19	360.11				x
Δ 5					
1015a29–30	360.33–34	x			
1015a31	360.35	x			
Δ 6					
1015b23–4	362.31				x
1016a5–6	363.26–28				x
		x			
1016a16	364.16–17	x			
1016a18–19	364.20–21	x			
1016a20	365.35	x			
1016a30	365.22				x
1016a33	366.11–12		x		
					x
1016a34–5	366.12–13		x		
1016a35–6	366.21	x			
1016a6–1016b1	366.17–18			x	
1016b9	367.23–25	x			
1016b10–11	367.32–33		x		
1016b11–12	368.8–9, 14	x			
1016b19–20	368.20–21				x
1016b33	369.6		x		
1016b33	369.9	x			
1016b33–4	369.12	x			
1016b34–5	369.15	x			
1017a3–4	369.27–28				x
Δ 7					
1017a10–11	370.27–28			x	
1017b1	372.12				x
1017b1–2	372.14	x			
1017b8–9	372.27–28	x			
Δ 8					
1017b15–16	373.17–18	x			
1017b17	374.1–2		x		
1017b18–19	374.12–13				x
1017b19	374.18				x
1017b23–4	375.18–20				x
1017b24–6	375.24–26; 376.6	x			
			x		
Δ 9					

APPENDIX C: QUOTATIONS 309

1017b30–31	376.23–24		X		
1017b33–4	376.33–34				X
1018a8	377.29–30; 378.5	X			
1018a9–11	378.18, 22–23	X			
1018a12–13	378.30–31, 34		X		
		X			
1018a15	379.25				X
				X	
1018a17	380.5	X			
Δ 10					
1018a22–4	380.25–26, 30–31, 33; 381.1	X			
1018a25	381.3			X	
1018a30–31	381.37–383.1, 5–6	X			
1018a35–6	383.4–5	X			
1018b4	383.30	X			
1018b6–7	384.5–6	X			
Δ 11					
1018b9–10	385.2–3		X		
1018b10–12	385.12–13				X
1018b21	385.35, 38	X			
1019a2–3	387.4				X
1019a12–14	387.33–36; 388.4–5, 14–15	X			
Δ 12					
1019a16	389.16		X		
1019a19	389.19	X			
1019a20	389.29	X			
1019a23–6	395.18–21	X			
			X		
1019a26	390.9, 18	X			
1019a34	391.3				X
1019a35	391.5–6		X		
1019a35–1019b1	391.15	X			
1019b3	391.26–27	X			
1019b6–7	391.33–34		X		
					X
1019b9–10	392.16	X			
1019b11	392.20–21	X			
1019b32–3	394.28, 31–32	X			
			X		
1020a3–4	395.22	X			
Δ 13					
1020a7–8	396.2–3	X			

Aristotle	Alexander						
1020a19	397.24-25	X					
1020a23-4	397.29-30		X				
							X
1020a29-30	398.11-12	X					
			X				
Δ 14							
1020b4-5	399.26, 29	X					
1020b18-20	401.15-16	X					
Δ 15							
1020b33-4	402.17-19						X
1020b34	403.18		X				
1021a1-2	403.17, 19-20	X					
1021a4	404.18-19	X					
1021a5	404.22-23		X				
	404.3-4, 11-12	X					
			X				
				X			
1021a6-7	404.13-14		X				
					X		
					X		
1021a8	404.15		X				
1021a10	405.1		X				
1021a10-11	405.6	X					
1021a11	405.8		X				
				X			
1021a11-12	405.11		X				
1021a19-20	405.27-28	X					
1021b1-3	407.32-35						X
					X		
			X				
			X				
							X
Δ 16							
1021b15	410.34	X					
1021b16-17	410.35-36		X				
1021b22-3	411.6-7, 15-16		X				
1021b29-30	411.34-35	X					
Δ 17							
1022a4-5	412.26, 33	X					
1022a5-6	413.14-15, 22-23	X					
1022a7	413.34-35	X					
						X	
1022a8	414.4-6	X					

APPENDIX C: QUOTATIONS

1022a12	414.15	x				
1022a12–13	414.24	x				
Δ 18						
1022a16	414.33–34	x				
1022a35	416.37–417.1					x
						x
Δ 19						
1022b2–3	417.14–15				x	
Δ 20						
1022b8	417.33	x				
1022b9–10	417.37–418.1			x		
Δ 22						
1022b30–31	419.10–11, 18		x			
			x			
1022b32–3	419.22–24					x
1022b35	419.32		x			
Δ 23						
1023a8–9	420.26–28				x	
			x			
						x
Δ 24						
1023a27–28	421.31–32	x				
1023a34	422.15–16	x				
1023a36	422.33	x				
1023b3–4	423.9	x				
Δ 25						
1023b13	423.36	x				
1023b17	424.15		x			
						x
1023b19–20	424.22, 26–27, 31				x	
			x			
						x
1023b22–24	424.37–338	x				
Δ 26						
1023b34	425.29	x				
Δ 27						
1024a27–8	428.1		x			
						x
Δ 28						
1024a29–30	428.14–15	x				
1024b3–4	429.2–3	x				
1024b4–5	429.10		x			
1024b8	429.25	x				

		429.28		x		
1024b8–9		429.30–31	x			
Δ 29						
1024b17		431.1	x			
1024b18–19		431.22; 432.2–3	x			
1024b22–3		433.6–8	x			
1024b27–8		433.25–26	x			
1024b31–32		434.13–14	x			
Δ 30						
1025a23–4		438.8–9, 17	x			
1025a24–5		438.14–15				x
1025a31		439.10–11				x
1025a33		438.33–35	x			
1025a33–4		439.7	x			

APPENDIX D: ALEXANDER'S PARAPHRASE IN CASES OF α-/β-DIVERGENCES

This list gives an overview of a selection of paraphrases in Alexander's commentary. The selection includes those passages where the readings in α and β differ substantially. Not included are differences where the evidence in Alexander's paraphrase cannot be taken as secure evidence for him having found the one or the other in ω^AL, such as, for example, αὐτῆς rather than ἑαυτῆς or αἰτίας καὶ ἀρχὰς rather than ἀρχὰς καὶ αἰτίας.[1]

It is not possible to represent in a list like this the way in which Alexander's paraphrase relates to the *Metaphysics* text except with regard to the specific α-/β-divergences. To begin with, instances where Alexander's paraphrase 'agrees' with the *Metaphysics* text in ω^αβ are far too many—he is, after all, paraphrasing *the Metaphysics*. Moreover, instances where Alexander's paraphrase 'disagrees' with the *Metaphysics* text in ω^αβ are also far too many—he is, after all, only *paraphrasing the Metaphysics*.

Paraphrases 341:
Agreements with α: 198; Agreements with β: 143.

Metaphysics	Alexander	α	β
A 1			
980a26	1.16	x	
980b21	3.9–10, 19–20		x
981a4–5	5.11–13		x
981a11–12	4.13–5.13		x
981a20	5.25	x	
981b2–5	5.16–6.12		x
A 2			
982a4	8.26	x	
982a6	8.26; 9.26		x
982a8	9.29–30; 10.1–2		x
982a10	10.2–3		x
982b5	14.5	x	
982b6	15.15–16	x	
982b14	16.3–4		x

[1] See, e.g. in A 1, 982a2, where α's αἰτίας καὶ ἀρχὰς is confirmed by Al.^P 8.26, 28 and β's ἀρχὰς καὶ αἰτίας is confirmed by Al.^P 8.22–24; 9.1–2; 9.9.

982b27	17.9–10		x
982b32	17.22–23		x
983a10	18.11–12	x	
983a11	19.14		x
983a17	18.22		x
A 3			
984a9	27.12	x	
984a32–3	30.9–10		x
984b1	30.11		x
984b11	32.8	x	
A 4			
984b29–30	33.8–9	x	
985a10	33.26		x
985a19–20	35.1–4		x
985a26	35.11	x	
985a30	34.7	x	
985b6	35.27–36.1		x
985b7	36.1		x
985b16	36.6	x	
985b17	36.6–7		x
A 5			
985b25	37.13, 19	x	
985b27	37.22; 38.5–6		x
986a3	40.21	x	
986a4	40.21		x
986a6	40.24	x	
986a9–10	40.28		x
986a16	41.19	x	
986a20	41.30–31	x	
986b11	42.24		x
986b17	42.28		x
986b22	44.7		x
986b23	44.7	x	
986b24	44.9	x	
986b32	45.5		x
987a6	45.23		x
987a16	47.11		x
987a21	48.14	x	
987a23	49.4		x
A 6			
987a32	49.21–22		x
987b5	50.9	x	
987b6	50.9	x	

APPENDIX D: PARAPHRASES   315

987b6	50.12		x
987b12–13	52.3	x	
987b23	53.19	x	
987b27	54.13	x	
988a2	58.5		x
988a13–14	59.20–23	x	
A 7			
988a25	61.11–12		x
988a34	61.30	x	
988b2	62.16	x	
A 8			
988b25–6	64.23	x	
989a1	65.22		x
989a4	65.25	x	
989a4	65.25		x
989a5	65.25–27		x
989a8	65.32	x	
989a15–16	66.11–12		x
989a32	68.12		x
989a33	68.13	x	
989b8	69.7	x	
989b8	69.8	x	
989b11	69.8		x
990a28	75.12	x	
A 9			
990a34	76.8		x
990b4	77.11	x	
990b15	83.18; 85.6		x
990b21	86.7, 13–14	x	
990b29	89.8		x
991a6	94.3–4		x
991a7	94.7		x
991a15	97.3		x
991a22	102.11		x
991b11	107.20–21	x	
991b18–19	109.14–15	x	
991b24	111.13		x
991b25	112.5	x	
991b28	112.21	x	
991b31	114.3	x	
992a1	114.6–7		x
992a16	118.14, 21	x	
992a20	119.14–15	x	

992a26	121.3	x	
992a33	122.3	x	
992b7	122.19	x	
992b9	123.14		x
992b10	124.9	x	
992b12–13	126.30–31	x	
992b15	127.10–11		x
992b16	127.15; 128.6	x	
992b18	127.20–21; 128.8–9	x	
992b20	128.20		x
992b21	129.4	x	
992b23	129.9	x	
992b26	130.2–3	x	
993a20	135.22		x
993a20	135.22	x	
993a20	135.23	x	
993a24	136.4		x

α 1

993b12	143.11–12	x	
993b13	143.12		x
993b13	143.12		x
993b13	144.5	x	
993b20	144.17–19	x	

α 2

994a3	149.30	x	
994a13	151.5		x
994a15	151.7		x
994a20	153.6	x	
994a22	154.7–15	x	
994a25	155.16	x	
994a28	156.16		x
994a29–30	156.15–18	x	
994b2	157.10		x

α 3

995a12	167.6		x

Β 1

995a24	171.5	x	
995a25	171.13	x	
995a36	173.19		x
995b16	176.3	x	
995b33	178.14–16	x	
996a11	180.13–15	x	

APPENDIX D: PARAPHRASES   317

996a14	180.28	x	
B 2			
996a35–b1	182.37–38	x	
996b4	183.20	x	
996b9	184.14–15	x	
996b10	184.21–22	x	
997a9	189.13	x	
997a15	191.6	x	
997b10	197.15		x
997b35	200.9		x
B 3			
998a20	202.1	x	
998b2	202.28		x
998b8	203.17	x	
998b8	203.18	x	
998b10	203.29		x
998b17	204.29	x	
998b22	205.1	x	
998b22	205.5	x	
998b27	206.4		x
999a3	208.11		x
999a17–18	210.13–14	x	
B 4			
999b24–5	216.17	x	
1000a8	218.25		x
1000a14	218.34	x	
1000b1	220.7–8		x
1000b2	220.10	x	
1000b5	220.23	x	
1000b28	222.9	x	
1001a1	222.20		x
1001b5	226.30		x
1001b9	227.14–15		x
1001b12	227.20	x	
1001b13	227.32		x
1001b14	228.1–2		x
1001b17	228.6		x
B 5			
1001b28	229.1	x	
1002a19	230.28	x	
1002a25	231.16	x	
1002a30	231.29	x	
1002a30	231.29–30	x	

1002a30	231.29–31	x	
Β 6			
1002b20	234.9, 13–14		x
1002b26	233.31		x
1002b28	234.29		x
1002b31	235.2		x
1003a14	236.25	x	
1003a15	236.23		x
Γ 1			
1003a31	240.28–29	x	
Γ 2			
1003b2	241.35	x	
1003b15	244.1	x	
1003b21	245.24–25		x
1004a4	250.31	x	
1004a7	251.26		x
1004a25	255.16		x
1004a26	255.28	x	
1004b15–16	259.4, 20	x	
1004b25	260.15	x	
1004b28	260.35	x	
1005a5	263.1–2	x	
1005a8	263.9–17		x
Γ 3			
1005a25	265.12	x	
1005b1	266.6	x	
1005b11	268.24	x	
1005b15	269.8	x	
1005b16	269.11	x	
1005b27	270.17		x
1005b31	271.2		x
Γ 4			
1006a33	276.34		x
1006a34	277.11	x	
1006b10	278.17		x
1006b16	280.4, 17		x
1006b17	280.4, 17		x
1006b26	281.24	x	
1006b31	282.15, 16	x	
1007a15	284.27	x	
1007a29	286.29	x	
1007b33	292.15–16	x	
1008a1	292.13	x	

1008a7	293.21-22		x
1008a17	294.24	x	
1008a18	294.24	x	
1008a18	295.1	x	
1008a23	295.17	x	
1008a26	297.30	x	
1008a36	297.14		x
1008b4	297.34-298.2		x
1008b15	299.7-9		x
1008b15	299.10		x
1008b33-4	300.31	x	
Γ 5			
1009a9	301.35-36		x
1009a24	303.25		x
1009a26	303.27-28	x	
1009a34	304.20	x	
1009a37	304.31		x
1009b31	307.12	x	
1010a14	308.28-29	x	
1010a17	310.2	x	
1010a36	311.10-11	x	
1010a37	311.19	x	
1010b8	312.22	x	
1010b22	315.4		x
1010b32	315.35-316.2		x
Γ 6			
1011a8	317.21	x	
1011b5	322.23	x	
1011b10	324.3		x
1011b15	326.24-25	x	
1011b22	327.35-36		x
Γ 7			
1011b24	328.15	x	
1011b25	328.20	x	
1011b26	328.23		x
1011b27	328.21-22		x
1011b27	328.22		x
1011b34	329.18-19	x	
1012a6	331.12	x	
1012a12-13	332.19	x	
1012a15	333.19	x	
1012a16	333.21	x	
1012a18	334.8	x	

Γ 8			
1012a33	337.1		x
1012b8–9	339.2–8		x
1012b31	343.8–10	x	
Δ 1			
1013a14	346.24		x
Δ 2			
1013a25	349.1	x	
1013a28	349.21	x	
1013b12	350.31–32		x
1013b32	352.22	x	
Δ 4			
1014b21	358.17	x	
1015a17	360.8	x	
Δ 5			
1015a23	360.24		x
1015a27	360.30	x	
1015b10	361.21	x	
Δ 6			
1015b16–17	362.12–13	x	
1015b18–19	362.15–16	x	
1015b21	362.20	x	
1015b22–3	362.22–23	x	
1015b27	362.34	x	
1015b27	362.33–363.3		x
1016a1	363.17	x	
1016a17	364.19		x
1016a33	366.9	x	
1016b4	366.25–367.8		x
1016b11	367.36–37		x
1016b13	368.2	x	
1016b18	368.15	x	
1016b24	368.34		x
1016b31	369.4		x
Δ 7			
1017a14	370.36		x
1017a16	371.2	x	
1017a18	371.15–16		x
1017a19	371.17		x
1017a28	371.31	x	
1017b2	372.15	x	
Δ 8			
1017b17	372.26	x	

APPENDIX D: PARAPHRASES 321

1017b18	372.26	x	
Δ 10			
1018a21-2	380.14		x
Δ 11			
1018b28	386.14		x
1019a4	387.7		x
Δ 12			
1019a16	389.3	x	
1019b13	392.25	x	
1019b16	392.38		x
1019b17	393.10	x	
1019b19	393.14		x
1019b33	394.34		x
1020a3	395.12	x	
Δ 13			
1020a15	396.34		x
1020a17	397.2		x
1020a20	397.12		x
Δ 14			
1020a33	399.2	x	
1020b11	400.20	x	
Δ 15			
1020b26	402.4	x	
1020b29	402.6-7		x
1021a5	404.4	x	
1021b5	410.1	x	
1021b7	410.7		x
1021b10	410.11	x	
Δ 16			
1021b13	410.19		x
1021b15	410.33		x
1021b15	410.30	x	
1021b21	411.7	x	
1021b24	411.26		x
1021b27	411.30	x	
1021b28	411.29-30		x
Δ 18			
1022a18	415.2-3	x	
1022a26-7	416.3		x
1022a29	416.6	x	
1022a31	416.16		x
1022a33	416.22		x
Δ 19			

1022b1	417.6	x	
Δ 20			
1022b9–10	417.34–36	x	
Δ 21			
1022b21	418.31	x	
Δ 22			
1022b34	419.29	x	
1022b35	419.32–420.1		x
1022b36	420.2	x	
1022b36	420.3		x
Δ 23			
1023a13	421.3		x
1023a14	421.4	x	
1023a22	421.16		x
Δ 24			
1023a29–30	421.36–422.1		x
1023b6	423.24	x	
Δ 26			
1023b34	425.30	x	
Δ 27			
1024a12	426.29	x	
1024a14	427.4	x	
1024a21	427.19	x	
Δ 28			
1024a31	428.14	x	
1024a36	428.23	x	
1024b10	429.38	x	
Δ 29			
1024b31	434.7–8	x	
1025a5	436.21		x
1025a6	436.21–22	x	
1025a9	437.8	x	
Δ 30			
1025a20	437.31	x	
1025a22	437.33	x	

Bibliography

Abbreviations in the apparatus

Aristotle, *Metaphysics*:
Bekker = Bekker 1831
Schwegler = Schwegler 1847a
Bonitz = Bonitz 1848
Christ = v. Christ 1886a
Ross = Ross 1924
Jaeger = Jaeger 1957
Cassin/Narcy = Cassin/Narcy 1989
Hecquet-Devienne = Hecquet-Devienne 2008
Primavesi = Primavesi 2012c

Alexander, *Commentary*:
Hayduck = Hayduck 1891
Bonitz = Bonitz 1847

Adamson, P. 2012. "Aristotle in the Arabic Commentary Tradition." In Shields 2012: 645–64.
Adrados, F. R. 2002. *Geschichte der griechischen Sprache. Von den Anfängen bis heute*. Aus dem Spanischen übersetzt von H. Bertsch. Tübingen/Basel.
Alberti, A., and R. W. Sharples, eds. 1999. *Aspasius: The Earliest Extant Commentary on Aristotle's Ethics*. Berlin/New York.
Alexandru, S. 1999. "A New Manuscript of Pseudo-Philoponus' Commentary on Aristotle's 'Metaphysics' Containing a Hitherto Unknown Ascription of the Work." *Phronesis* 44: 347–52.
— . 2000. "Traces of Ancient *Reclamantes* Surviving in Further Manuscripts of Aristotle's Metaphysics." *ZPE* 131: 13–14.
— . 2014. *Aristotle's* Metaphysics *Lambda: Annotated Critical Edition Based upon a Systematic Investigation of Greek, Latin, Arabic and Hebrew Sources*. Leiden/Boston.
Annas, J. 1976. *Aristotle's Metaphysics Book M and N. Translated with Introduction and Notes*. Oxford.
Apelt, O. 1891. *Beiträge zur Geschichte der Philosophie*. Leipzig.
Balme, D. M., ed., and A. Gotthelf. 2002. *Aristotle, Historia Animalium*, Vol. I, Books I–X: Text. Cambridge.
Baltussen, H. 2008. *Philosophy and Exegesis in Simplicius: The Methodology of a Commentator*. London.
Barnes, J. 1997. "Roman Aristotle." In J. Barnes and M. Griffin, eds., *Philosophia Togata II. Plato and Aristotle at Rome*, 1–69. Oxford.
— . 1999. "An Introduction to Aspasius." In Alberti and Sharples 1999: 1–50.
— . 2003. *Porphyry: Introduction*. Oxford.

—, ed. 1984. *The Complete Works of Aristotle: The Revised Oxford Translation.* Vol. II. Princeton.
Barney, R. 2012. "History and Dialectic in *Metaphysics* A 3." In Steel 2012: 69–104.
Bassenge, F. 1960. "Das τὸ ἑνὶ εἶναι, τὸ ἀγαθῷ εἶναι etc. etc. und das τὸ τί ἦν εἶναι bei Aristoteles." *Philologus* 104: 14–47; 201–35.
Bastianini, G., and D. N. Sedley. 1995. "Commentarium in Platonis Theaetetum." *Corpus dei papiri filosofici greci e latini.* Vol III. *Commentari.* Florence.
Beere, J. 2009. *Doing and Being: An Interpretation of Aristotle's* Metaphysics *Theta.* Oxford.
Bekker, I., ed. 1831. *Aristoteles Graece.* Edidit Academia Regia Borussica. Volumen Prius; Volumen alterum. Berlin.
Berti, E. 1983. "La fonction de Métaph. Alpha elatton dans la philosophie d'Aristote." In Moraux and Wiesner 1983: 260–94.
—. 2009. "Aporiai 6–7." In Crubellier and Laks 2009: 105–33.
Bertolacci, A. 2005. "On the Arabic translations of Aristotle's *Metaphysics.*" *Arabic Science and Philosophy* 15: 241–75.
Betegh, G. 2012. "'The Next Principle': *Metaphysics* A 3–4, 984b8–985b22." In Steel 2012: 105–40.
Bloch, D. 2003. "Alexander of Aphrodisias as a Textual Witness: The Commentary on the *De Sensu.*" *Cahiers de l'institut du moyen-âge grec et latin* 74: 21–38.
Bluck, R. S. 1947. "Aristotle, Plato, and Ideas of Artefacta." *CR* 61: 75–76.
Bonitz, H., and H. Seidl, eds. 1989. *Aristoteles' Metaphysik. Erster Halbband: Bücher I (A) – VI (E). Neubearbeitung der Übersetzung von Hermann Bonitz. Mit Einleitung und Kommentar von Horst Seidl.* Hamburg.
Bonitz, H. 1842. *Observationes criticae in Aristotelis libros metaphysicos.* Berlin.
—, ed. 1847. *Alexandri Aphrodisiensis commentarius in libros Metaphysicos Aristotelis.* Berlin.
—, ed. 1848. *Aristotelis Metaphysica.* Pars prior. Bonn.
—, ed. 1849. *Aristotelis Metaphysica.* Pars posterior. Bonn.
—, ed. 1870. *Index Aristotelicus.* Berlin.
Borgia, A. 2007. "Commento al libro Δ (quinto). Presentazione, traduzione e note." In G. Movia, ed., *Alessandro di Afrodisia. Commentario alla 'Metafisica' di Aristotele*, 849–1127. Milan.
Bouyges, M. 1948. *Averroès, Tafsīr Mā ba'd at-Tabī'at. Troisième Volume: Livres ya' et lam. - Index alphabétiques.* Bibliotheca arabica scholasticorum, Tome VII. Beyrouth.
—. 1952. *Averroès, Tafsīr Mā ba'd at-Tabī'at. Notice.* Bibliotheca arabica scholasticorum, Tome V.1. Beyrouth.
Brandis, Ch. A. 1831. "Die Aristotelischen Handschriften der Vaticanischen Bibliothek." *Abhandlungen der Königlichen Preußischen Akademie der Wissenschaften zu Berlin*: 47–86. Berlin.
—. 1833. "Über die Reihenfolge der Bücher des Aristotelischen Organons und ihre Griechischen Ausleger, nebst Beiträgen zur Geschichte des Textes jener Bücher des Aristoteles und ihrer Ausgaben." *Abhandlungen der Königlichen Akademie der Wissenschaften.* Histor.-philol. Klasse: 249–99. Berlin.
—. 1836. *Scholia in Aristotelem.* Edidit Academia Regia Borussica. Berlin.
—, ed. 1823. *Aristotelis et Theophrasti Metaphysica* ad veterum codicum manuscriptorum fidem recensita indicibusque instructa in usum scholarum. Tomus Prior. Berlin.
Broadie, S. 2007. "Why no Platonistic Ideas of Artefacts?" In D. Scott, ed., *Maieusis: Essays*

on Ancient Philosophy in Honour of Myles Burnyeat, 232–53. Oxford.
— . 2009. "Aporia 8." In Crubellier and Laks 2009: 135–50.
— . 2012. "A Science of First Principles. Metaphysics A 2." In Steel 2012: 43–67.
Bülow-Jacobsen, A. 2009. "Writing Materials in the Ancient World." In R. S. Bagnall, ed., *The Oxford Handbook of Papyrology*, 3–29. Oxford.
Burkert, W. 1959. "ΣΤΟΙΧΕΙΟΝ. Eine semasiologische Studie." *Philologus* 103: 167–97.
Busse, A. 1990. "Ueber die in Ammonius' Kommentar erhaltene Ueberlieferung der aristotelischen Schrift περὶ ἑρμηνείας." In *Festschrift Johannes Vahlen zum siebenzigsten Geburtstag, gewidmet von seinen Schülern*, 71–85. Berlin.
Bydén, B. 2005. "Some Remarks on the Text of Aristotle's 'Metaphysics.'" *CQ* 55: 105–20.
Cacouros, M. 2000. "Le Laur. 85,1, témoin de l'activité conjointe d'un groupe de copistes travaillant dans la seconde moitié du XIIIe siècle." In G. Prato, ed., *I manoscritti greci tra riflessione e dibattito. Atti de V colloquio internazionale di paleografia greca* (Cremona, 4–10 ottobre 1998), Tom. I. Florence.
Cassin, B., and M. Narcy, eds. 1989. *La décision du sens. Le livre Gamma de la Métaphysique d'Aristote, introduction, texte, traduction et commentaire*. Paris.
Casu, M. 2007. "Commento al libro Γ (quatro). Presentazione, traduzione e note." In G. Movia, ed., *Alessandro di Afrodisia. Commentario alla 'Metafisica' di Aristotele*, 561–848. Milan.
Cavini, W. 2009. "Aporia 11." In Crubellier and Laks: 175–88.
Chaniotis, A. 2004. "Epigraphic Evidence for the Philosopher Alexander of Aphrodisias." *BICS* 47: 79–81.
Christ, W. (v.) 1853. *Studia in Aristotelis libros metaphysicos collata*. Berlin.
— , ed. 1886a. *Aristotelis Metaphysica*. Leipzig.
— , ed. 1886b. "Kritische Beiträge zur Metaphysik des Aristoteles." *Sitzungsberichte der philosophisch-philologischen und historischen Classe der königlich bayerischen Akademie der Wissenschaften zu München*. 1885, IV: 406–23. München.
Cooper, J. M. 1997. *Plato Complete Works*. Indianapolis.
— . 2012. "Conclusion – and Retrospect. Metaphysics A 10." In Steel 2012: 333–64.
Coroleu, A. 1995. "A philological analysis of Juan Ginés de Sepúlveda's Latin Translations of Aristotle and Alexander of Aphrodisias." *Euphrosyne* XXIII: 175–95.
— . 1996. "The Fortuna of Juan Ginés de Sepúlveda's Translations of Aristotle and of Alexander of Aphrodisias." *Journal of the Warburg and Courtauld Institutes* 59: 325–32.
Crubellier, M. 2009. "Aporiai 1–2." In Crubellier and Laks 2009: 47–72.
— . 2012. "The Doctrine of Forms under Critique – Part II." In Steel 2012: 297–334.
Crubellier, M., and A. Laks, eds. 2009. *Aristotle's* Metaphysics *Beta*. Oxford.
D'Ancona Costa, Ch. 2002. "Commenting on Aristotle: from Late Antiquity to the Arab Aristotelianism." In W. Geerlings and Ch. Schulze, eds., *Der Kommentar in Antike und Mittelalter. Beiträge zu seiner Erforschung*, 201–51. Leiden.
de Haas, F. 1997. *John Philoponus' New Definition of Prime Matter: Aspects of its Background in Neoplatonism and the Ancient Commentary Tradition*. Leiden.
Denniston, J. D. 1954. *The Greek Particles*. 2nd edition revised by K. J. Dover. Oxford.
Detel, W. 2009. *Aristoteles Metaphysik Bücher VII und VIII. Griechisch-deutsch*. Frankfurt am Main.
Di Giovanni, M., and O. Primavesi (forthcoming). "Who Wrote Alexander's Commentary on Metaphysics Λ? Issues in the Syro-Arabic Tradition."
Dickey, E. 2007. *Ancient Greek Scholarship: A Guide to Finding, Reading, and Understand-*

ing Scholia, Commentaries, Lexica, and Grammatical Treatises, from Their Beginnings to the Byzantine Period. Oxford.
Diels, H. 1882. "Zur Textgeschichte der Aristotelischen Physik." *Philos.-hist. Abhandlungen der königlichen Akademie der Wissenschaften zu Berlin*: 1–42.
—. 1899. *Elementum. Eine Vorarbeit zum griechischen und lateinischen Thesaurus.* Leipzig.
Diels, H., and W. Schubart. 1905. *Anonymer Kommentar zu Platons Theaitet.* Berliner Klassikertexte 2. Berlin.
Dietrich, A. 1964. *Die arabische Version einer unbekannten Schrift des Alexander von Aphrodisias über die Differentia specifica.* Göttingen.
Dillon, J. 2000. "Eudore d'Alexandrie." In R. Goulet, ed., *Dictionnaire des philosophes antiques*, 290–93. Paris.
Dodds, E. R. 1928. "The Parmenides of Plato and the Origin of the Neoplatonic 'One.'" *CQ* 22: 129–42.
Dooley, W. E. 1989. *Alexander of Aphrodisias, On Aristotle's Metaphysics 1.* Transl. by W. E. Dooley, S.J. Ithaca.
—. 1992. *Alexander of Aphrodisias, On Aristotle's Metaphysics 2.* Transl. by W. E. Dooley, S.J. Ithaca.
—. 1993. *Alexander of Aphrodisias, On Aristotle's Metaphysics 5.* Transl. by W. E. Dooley, S.J. Ithaca.
Dorandi, T. 1997. "Tradierung der Texte im Altertum; Buchwesen." In H.-G. Nesselrath, ed., *Einleitung in die griechische Philologie*, 3–16. Leipzig.
Dörrie, H., and M. Baltes, eds. 1996. *Die philosophische Lehre des Platonismus. Einige grundlegende Axiome / Platonische Physik (im antiken Verständnis) I*, Bausteine 101–24: Text, Übersetzung, Kommentar. Stuttgart.
Dover, K. 1997. "Textkritik." In H.-G. Nesselrath, ed., *Einleitung in die griechische Philologie.* 45–58. Stuttgart.
—, ed. 1980. *Plato, Symposium.* Cambridge.
Dreizehnter, L. 1962. *Untersuchungen zur Textgeschichte der Aristotelischen Politik.* Leiden.
Drossaart Lulofs, H. J., ed. 1965. *Nicolaus Damascenus on the Philosophy of Aristotle. Fragments of the first five books translated from the Syriac with an introduction and commentary.* Leiden.
Düring, I. 1957. *Aristotle in the Ancient Biographical Tradition.* Gothenburg.
Erbse, H. 1979. "Besprechung von Paul Maas: Textkritik." In *Ausgewählte Schriften zur Klassischen Philologie*, 547–54. Berlin.
Fazzo, S. 2004. "Aristotelianism as a Commentary Tradition." *Supplement of the Bulletin of the Institute of Classical Studies*: 1–19.
—. 2008. "Nicolas, L'auteur du 'Sommaire de la philosophie d'Aristote': doutes sur son identité, sa datation, son origine." *Revue des études grecques* 121: 99–126.
—. 2010. "Lo stemma codicum dei libri *Kappa* e *Lambda* della *Metafisica*: una revisione necessaria." *Aevum* 84: 339–59.
—. 2012a. "The *Metaphysics* from Aristotle to Alexander of Aphrodisias." *BICS* 55: 51–68.
—. 2012b. *Il libro Lambda della* Metafisica *di Aristotele.* Naples.
Fazzo, S., and M. Zonta. 2008. "Aristotle's Theory of Causes and the Holy Trinity. New Evidence about the Chronology and Religion of Nicolaus 'of Damascus.'" *Laval théologique et philosophique* 64: 681–90.

Fine, G. 1993. *On Ideas: Aristotle's Criticism of Plato's Theory of Forms.* Oxford.
Flannery, K. L. 2003. "Logic and Ontology in Alexander of Aphrodisias's Commentary on *Metaphysics* IV." In G. Movia, ed., *Alessandro di Afrodisia. Commentario alla 'Metafisica' di Aristotele*, 117–34. Milan.
Fränkel, S. 1885. "Anmerkungen." In Freudenthal 1885: 114–15.
Frede, D. 2012. "The Doctrine of the Forms Under Critique." In Steel and Primavesi 2012: 265–96.
Frede, M. 1994. "The Stoic Notion of a *lekton*." In S. Everson, ed., *Language.*, 109–28. Cambridge.
Frede, M., and D. Charles, eds. 2000. *Aristotle's Metaphysics Lambda. Symposium Aristotelicum.* Oxford.
Frede, M., and G. Patzig. 1988. *Aristoteles ,Metaphysik Z', Text, Übersetzung und Kommentar*, Erster Band: *Einleitung, Text und Übersetzung.* München.
Freudenthal, J. 1885. *Die durch Averroes erhaltenen Fragmente Alexanders zur Metaphysik des Aristoteles. Mit Beiträgen zur Erläuterung des arabischen Textes durch S. Fränkel.* Berlin: Verlag der Königlichen Akademie der Wissenschaften.
Gastgeber, Ch. 2003. "Die Überlieferung der griechischen Literatur im Mittelalter." In Pöhlmann 2003b: 1–46.
Genequand, Ch. 1986. *Ibn Rushd's Metaphysics: A translation with Introduction of Ibn Rushd's Commentary on Aristotle's Metaphysics, Book Lām.* Leiden.
Gercke, A. 1892. "Aristoteleum." *WS* 14: 146–8.
— . 1897. "Boethos von Sidon." *RE*, III 1, 603–604.
Gigon, O. 1983. "Versuch einer Interpretation von Metaphysik Alpha Elatton." In Moraux and Wiesner 1983: 193–220.
Golitsis, P. 2008. *Les Commentaires de Simplicius et de Jean Philopon à la Physique d'Aristote, Tradition et Innovation.* Berlin.
— . 2013. "Review: Silvia Fazzo, Il libro Lambda della Metafisica di Aristotele." *BMCR* 2013.06.32.
— . 2014a. "Trois annotations de manuscrits aristotéliciens au XIIe siècle: les Parisini gr. 1901 et 1853 et l'Oxoniensis Corporis Christi 108." In D. Bianconi, ed., *Paleografia e oltre (Supplemento del Bollettino dei Classici dell' Accademia Nazionale dei Lincei)*, 33–52. Rome.
— . 2014b. "La *recensio altera* du Commentaire d' Alexandre d' Aphrodise à la *Métaphysique* d' Aristote et le témoignage des manuscrits byzantins *Laurentianus plut.* 87,12 et *Ambrosianus* F 113 sup." In J. Signes Codoñer and I. Pérez Martín, eds., *Textual Transmission in Byzantium: Between Textual Criticism and Quellenforschung*, 201–32. Turnhout.
— . (forthcoming) 2016. "The manuscript tradition of Alexander of Aphrodisias' commentary on Aristotle's *Metaphysics*: towards a new critical edition." *Revue d'Histoire des Textes* 11.
Gottschalk, H. B. 1987. "Aristotelian Philosophy in the Roman World." *ANRW* 36.2: 1079–174. Berlin.
— . 1990. "The earliest Aristotelian commentators." In Sorabji 1990: 55–81.
Goulet, R. 1994. "Aspasios." In R. Goulet, ed., *Dictionnaire des philosophes antiques*, I Abam(m)on à Axiothéa. 635–36. Paris.
Gutas, D. 1993. "Aspects of Literary Form and Genre in Arabic Logical Works." In Ch. Burnett, ed., *Glosses and Commentaries on Aristotelian Logical Texts: The Syriac, Arabic*

and Medieval Latin Traditions, 29–76. London.
Hadot, I. 1987. "Recherches sur les fragments du commentaire de Simplicius sur la Métaphysique d'Aristote." In I. Hadot, ed., *Simplicius sa vie, sa œuvre, sa survie. Actes du colloque international de Paris* (28 Sept.–1er Oct. 1985), 225–45. Berlin.
—. 2002. "Der fortlaufende philosophische Kommentar." In W. Geerlings and Ch. Schulze, eds., *Der Kommentar in Antike und Mittelalter. Beiträge zu seiner Erforschung*, 183–99. Leiden.
Hankinson, R. J. 2006. *Simplicius On Aristotle On the Heavens 1.10–12*. Transl. by R. J. Hankinson. London.
Harlfinger, D. 1971. *Die Textgeschichte der Pseudo-Aristotelischen Schrift ΠΕΡΙ ΑΤΟΜΩΝ ΓΡΑΜΜΩΝ. Ein kodikologisch-kulturgeschichtlicher Beitrag zur Klärung der Überlieferungsverhältnisse im Corpus Aristotelicum*. Amsterdam.
—. 1975. "Edizione critica del testo del 'De ideis' di Aristotele." In Leszl 1975: 15–54.
—. 1979. "Zur Überlieferungsgeschichte der Metaphysik." In P. Aubenque, ed., *Études sur la Métaphysique d'Aristote. Actes du VIe Symposium Aristotelicum*, 7–33. Paris.
Hatzimichali, M. 2013. "The Text of Plato and Aristotle in the First Century BC." In M. Schofield, ed., *Aristotle, Plato and Pythagoreanism in the First Century BC. New Directions for Philosophy*, 1–27. Cambridge.
Hayduck, M. 1888, ed. *Asclepii in Aristotelis metaphysicorum libros A–Z Commentaria, consilio et auctoritate Academiae Litterarum Regiae Borussicae*. Berlin.
—. 1891, ed. *Alexandri Aphrodisiensis in Aristotelis metaphysica commentaria, consilio et auctoritate Academiae Litterarum Regiae Borussicae*. Berlin.
Heath, Th. L. 1949. *Mathematics in Aristotle*, Oxford.
Hecquet-Devienne, M. 2008. *Aristote*, Métaphysique Gamma. *Édition, traduction, études. Introduction, texte grec et traduction*. Louvain-la-Neuve.
Hein, Ch. 1985. *Definition und Einteilung der Philosophie. Von der spätantiken Einleitungsliteratur zur arabischen Enzyklopädie*. Frankfurt.
Hoffmann, Ph. 2009. "What was Commentary in Late Antiquity? The Example of the Neoplatonic Commentators." In M. L. Gill and P. Pellegrin, eds., *A Companion to Ancient Philosophy*, 597–622. Malden.
Hunger, H. 1989. *Schreiben und Lesen in Byzanz. Die byzantinische Buchkultur*. München.
Hunger, H., et al. 1961. *Geschichte der Textüberlieferung der antiken und mittelalterlichen Literatur*, Bd. 1. *Antikes und mittelalterliches Buch- und Schriftwesen, Überlieferungsgeschichte der antiken Literatur*. Zürich.
Ilberg, J. 1890. "Die Hippokratesausgaben des Artemidoros Kapiton und Dioskurides." *RhM* 45: 111–37.
Irigoin, J. 1954. "Stemmas bifides et états de manuscrits." *RPh* 80: 211–17.
—. 1962. "Survie et renouveau de la littérature antique à Constantinople (IXe siècle)." *Cahiers de civilisation médiévale* 19: 287–302.
Jaeger, W. 1912. *Studien zur Entstehungsgeschichte der Metaphysik des Aristoteles*. Berlin.
—. 1917. "Emendationen zur Aristotelischen Metaphysik Α–Δ." *Hermes* 52: 481–519.
—. 1923. "Emendationen zur Aristotelischen Metaphysik (Zweiter Teil)." In *Sitzungsberichte der Preuss. Akad. der Wissensch. Phil.-Hist. Kl.*, XXXIV: 263–79 [= Jaeger, W. 1960. *Scripta Minora I*. Rome, 257–80].
—. 1923. *Aristoteles. Grundlegung einer Geschichte seiner Entwicklung*, Berlin.
—. 1957. *Aristotelis Metaphysica*. Oxford.
—. 1965. "'We say in the *Phaedo*.'" In *Harry Austryn Wolfson Jubilee Volume*, Vol. 1,

407–21. Jerusalem.
Judson, L. 2000. "Formlessness and the Priority of Form: *Metaphysics*: Z 7–9 and Λ 3." In Frede and Charles 2000: 111–35.
Kenny, A. 1983. "A Stylometric Comparison between Five Disputed Works and the Remainder of the Aristotelian Corpus." In Moraux and Wiesner 1983: 345–66.
Kirwan, Ch. 1971. *Aristotle's Metaphysics, Books Γ, Δ, E*, Transl. with Notes. Oxford.
Kotwick, M. E. 2015. "On Aristotle's *Metaphysics* A 4, 985a18–21: A Platonic Interpolation from Asclepius of Tralles' commentary?" In V. Gysembergh and A. Schwab, eds., *Le Travail du Savoir / Wissensbewältigung. Philosophie, sciences exactes et sciences appliquées dans l'Antiquité*. AKAN-Einzelschriften, Band 10, 215–31. Trier.
— . forthcoming 2016. "The *Entwicklungsgeschichte* of a Text: On Werner Jaeger's Edition of Aristotle's *Metaphysics*." In C. Guthrie King and R. Lo Presti, eds., *Werner Jaeger. Wissenschaft, Bildung, Politik. Philologus. Supplemente*. Berlin.
Krämer, H. 2004. "Die Ältere Akademie." In H. Flashar, ed., *Die Philosophie der Antike*, Band 3, *Ältere Akademie, Aristoteles, Peripatos*, 2., durchgesehene u. erweiterte Auflage, 1–165. Basel.
Kühner, R., and B. Gerth. 1898. *Ausführliche Grammatik der griechischen Sprache*, II. Teil: Satzlehre. 1. Band. Hannover.
— . 1904. *Ausführliche Grammatik der griechischen Sprache*, II. Teil: Satzlehre. 2. Band. Hannover.
Kupreeva, I. 2010. "Alexander of Aphrodisias on Form: A Discussion of Marwan Rashed, *Essentialisme*." *OSAPh* 38: 211–49.
— . 2012. "Alexander of Aphrodisias and Aristotle's *De anima*: What's in a commetary?" *BICS* 55: 109–29.
Lai, P. 2007. "Commento al libro B (terzo). Presentazione, traduzione e note." In G. Movia, ed., *Alessandro di Afrodisia. Commentario alla 'Metafisica' di Aristotele*. 359–557. Milan.
Laks, A. 2009. "Aporia Zero (*Metaphysics* B 1, 995a24–995b4)." In Crubellier and Laks 2009: 25–46.
Lamberz, E. 1987. "Proklos und die Form des philosophischen Kommentars." In J. Pépin and H. D. Saffrey, eds., *Proclus. Lecteur et interprète des anciens. Actes du colloque international du CNRS*, 1–20. Paris.
Lapini, W. 1997. "I libri dell' Ephemeris di Ditti-Settimio." *ZPE* 117: 85–89.
Lennox, J. G. 2011. "Aristotle on Norms of Inquiry." *HOPOS* 1: 23–46.
Leszl, W. 1975. *Il ,De Ideis' di Aristotele e la teoria platonica delle idee*. Accademia Toscana di scienze e lettere „La colombaria", Studi 40. Florence.
Lewis, F. A. 2004. "Aristotle on the Homonymy of Being." *Ph&PhenR* 68: 1–36.
Lloyd, A. C. 1976. "The Principle that the Cause is Greater than its Effect." *Phronesis* 21: 146–56.
Loredana Cardullo, R. 2012. *Asclepio di Tralle. Commentario al libro* Alpha Meizon *(A) della* Metafisica *di Aristotele. Introduzione, testo greco, traduzione e note di commento*. Rome.
Lumpe, A. 1955. "Der Terminus 'Prinzip' (ἀρχή) von den Vorsokratikern bis auf Aristoteles." *ABG* 1: 104–16.
Luna, C. 2001. *Trois études sur la tradition des commentaires anciens à la Métaphysique d'Aristote*. Leiden.
— . 2003. "Le commentaires grecs à la Métaphysique." In R. Goulet et al., eds., *Diction-*

naire des Philosophes Antiques. Supplément. 249-58. Paris.
— . 2005. "Observations sur le texte des livres M-N de la Métaphysique d'Aristote." *DSTradF* XVI: 553-93.
Luthe, W. 1880. "Zur Kritik und Erklärung von Aristoteles Metaphysik und Alexanders Commentar." *Hermes* 15: 189-210.
Maas, P. 1927. "Textkritik." In A. Gercke and E. Norden, eds., *Einleitung in die Altertumswissenschaft*, I. Band/ Zweites Heft, 1-18. Leipzig.
— . 1958. *Textual Criticism*. Translated by Barbara Flower. Oxford.
— . 1960. *Textkritik*. Fourth edition. Leipzig.
Madigan, A. 1992. *Alexander of Aphrodisias, On Aristotle's Metaphysics 3*. Transl. by A. Madigan. Ithaca.
— . 1993. *Alexander of Aphrodisias, On Aristotle's Metaphysics 4*. Transl. by A. Madigan. Ithaca.
— . 1999. *Aristotle Metaphysics Book B and Book K 1-2. Transl. with Commentary*. Oxford.
Makin, S. 1990. "An Ancient Principle about Causation." *PAS* 91: 135-52.
Mansfeld, J., and O. Primavesi, eds. 2012. *Die Vorsokratiker*. Griechisch/Deutsch. Stuttgart.
Martin, A. 1984. *Averroès, Grand commentaire de la Métaphysique d'Aristote (Tafsīr mā ba'd at-tabī'at)*. Livre lam-lambda. Paris.
Martin, A., and O. Primavesi, eds. 1999. *L'Empédocle de Strasbourg (P. Strasb. gr. Inv. 1665-1666)*. Berlin.
McNamee, K. 1977. *Marginalia and commentaries in Greek literary papyri*. PhD diss., Duke University.
Menn, S. 1995. "The Editors of the 'Metaphysics.'" *Phronesis* 40: 202-208.
Mercken, H. P. F. 1990. "The Greek Commentators on Aristotle's Ethics." In Sorabji 1990: 407-43.
Mondrain, B. 2000. "La constitution de corpus d'Aristote et de ses commentateurs aux XIIIe-XIVe siècles." *CodMan* 29: 11-33.
Moraux, P. 1951. *Les listes anciennes des ouvrages d'Aristote*. Louvain.
— . 1965. *Aristote Du ciel*. Paris.
— . 1967. "Aristoteles, der Lehrer Alexanders von Aphrodisias." *AGPh* 49: 169-82.
— . 1968. "Einige Aspekte des Aristotelismus von Andronikos bis Alexander von Aphrodisias." In J. Burian and L. Vidman, eds., *Antiquitas graeco-romana ac tempora nostra. Acta congressus internationalis habiti Brunae diebus 12-16 mensis Aprilis MCMLXVI*, 203-208. Prag.
— . 1969. "Eine Korrektur des Mittelplatonikers Eudoros zum Text der Metaphysik des Aristoteles." In R. Stiehl and H. E. Stier, eds., *Beiträge zur Alten Geschichte und deren Nachleben. Festschrift für Franz Altheim zum 6. 10. 1968*. Erster Band, 492-504. Berlin.
— . 1973. *Der Aristotelismus bei den Griechen. Von Andronikos bis Alexander von Aphrodisias*. Erster Band: *Die Renaissance des Aristotelismus im 1. Jh. v. Chr.* Berlin.
— . 1976. *Aristoteles Graecus. Die Griechischen Manuskripte des Aristoteles*. Erster Band *Alexandrien - London*. Berlin.
— . 1984. *Der Aristotelismus bei den Griechen. Von Andronikos bis Alexander von Aphrodisias*. Zweiter Band: *Der Aristotelismus im 1. und 2. Jh. n. Chr.* Berlin.
— . 1985. "Ein neues Zeugnis über Aristoteles, den Lehrer Alexanders von Aphrodisias." *AGPh* 67: 266-9.
— . 2001. *Der Aristotelismus bei den Griechen. Von Andronikos bis Alexander von Aph-*

rodisias. Dritter Band: *Alexander von Aphrodisias* von Paul Moraux and J. Wiesner. Berlin.
Moraux, P., and J. Wiesner, eds. 1983. *Zweifelhaftes im Corpus Aristotelicum. Studien zu einigen Dubia, Akten des 9. Symposium Aristotelicum*. Berlin.
Mueller-Goldingen, Ch. 1991. "Politik und Philosophie bei Aristoteles und im frühen Peripatos." *AGPh* 73: 1–19.
Mueller, I. 2009. "Aporia 12 (and 12 bis)." In Crubellier and Laks 2009: 189–209.
Nussbaum, M. 1978. *Aristotle's De Motu Animalium. Text with Translation, Commentary, and Interpretative Essays*. Princeton.
Owens, J. 1978. *The Doctrine of Being in the Aristotelian 'Metaphysics.' A Study in the Greek Background of Mediaeval Thought*. Third, revised edition. Toronto.
Palmer, J. A. 2000. "Aristotle on the Ancient Theologians." *Apeiron* 33: 181–205.
Pasquali, G. 1929. "Paul Maas: Textkritik." *Gnomon* 5: 417–435 and 498–521.
— . 1962. *Storia della tradizione e critica del testo*. Seconda edizione. Firenze.
Perilli, L. 1996. *La teoria del vortice nel pensiero antico*. Pisa.
Pfeiffer, R. 1968. *History of Classical Scholarship. Form the Beginnings to the End of the Hellenistic Age*. Oxford.
Pöhlmann, E. 2003a. *Einführung in die Überlieferungsgeschichte und die Textkritik der antiken Literatur*, Band 1, *Altertum*, 2. durchgesehene Auflage. Darmstadt.
— . 2003b. *Einführung in die Überlieferungsgeschichte und die Textkritik der antiken Literatur*, Band 2, *Mittelalter und Neuzeit*. Mit Beiträgen von Ch. Gastgeber, P. Klopsch u. G. Heldmann. Darmstadt.
Praechter, K. 1906. "Michaelis Ephesii" [Review of CAG XXII 2]. *GGA* 11: 861–907.
— . 1909. "Die griechischen Aristoteleskommentare." *ByzZ* 18: 516–38.
Primavesi, O. 1998. "Neues zur Aristotelischen Vorsokratiker-Doxographie." In K. Döring, B. Herhoff, and G. Wöhrle, eds., *Antike Naturwissenschaft und ihre Rezeption*, Bd. 8, 25–41. Trier.
— . 2005. "Aristoteles." In H. H. Schmitt and E. Vogt, eds., *Lexikon des Hellenismus*, 133–41. Wiesbaden.
— . 2006. "Empedokles in Florentiner Aristoteles-Scholien." *ZPE* 157: 27–40.
— . 2007. "Ein Blick in den Stollen von Skepsis: Vier Kapitel zur frühen Überlieferung des Corpus Aristotelicum." *Philologus* 151: 51–77.
— . 2008. *Empedokles Physika I. Eine Rekonstruktion des zentralen Gedankengangs*. APF Beiheft 22. Berlin.
— . 2011a. "Vorsokratiker im Lateinsichen Mittelalter I: Helinand, Vincenz, der Liber de vita et moribus und die Parvi flores." In Primavesi and Luchner 2011: 157–96.
— . 2011b. "Werk und Überlieferung." In Rapp and Corcilius 2011: 57–64.
— . 2011c. "Philosophische Fragmente." In Rapp and Corcilius 2011: 170–75.
— . 2012a. "Second Thoughts on Some Presocratics: *Metaphysics* A 8, 989a18–990a32." In Steel 2012: 225–63.
— . 2012b. "Introduction: The Transmission of the Text and the Riddle of the Two Versions." In Steel 2012: 388–464.
— . 2012c. "Text of *Metaphysics* A (and of the corresponding parts of M 4–5)." In Steel 2012: 465–516.
— . 2013. "Empedokles." In H. Flashar, D. Bremer, and G. Rechenauer, eds., *Grundriss der Geschichte der Philosophie. Philosophie der Antike*. Bd. 1, *Frühgriechische Philosophie*, 667–739. Basel.

—. 2014. "Aristotle on the 'so-called Pythagoreans': From Lore to Principles." In C. A. Huffmann, ed., *A History of Pythagoreanism*, 227–49. Cambridge.
Primavesi, O., and K. Luchner, eds. 2011. *The Presocratics from the Latin Middle Ages to Hermann Diels. Akten der 9. Tagung der Karl und Gertrud Abel-Stiftung*. Stuttgart.
Radice, R., and R. Davies. 1997. *Aristotle's Metaphysics. Annotated Bibliography of the Twentieth-Century Literature*. Leiden.
Rapp, Ch. 1992. "Ähnlichkeit, Analogie und Homonymie bei Aristoteles." *ZPhF* 46: 526–44.
—. 1993. "Aristoteles über die Rechtfertigung des Satzes vom Widerspruch." *ZPhF* 47: 521–41.
—. 2002. *Aristoteles Rhetorik*. Berlin.
Rapp, Ch., and K. Corcilius, eds. 2011. *Aristoteles Handbuch. Leben – Werk – Wirkung*. Stuttgart.
Rapp, Ch., and Ph. Brüllmann, eds. 2008. *Focus: The Practical Syllogism – Schwerpunkt: der praktische Syllogismus*. Philosophiegeschichte und logische Analyse 11. Paderborn.
Rashed, M. 2001a. *Die Überlieferungsgeschichte der aristotelischen Schrift De generatione et corruptione*. Serta Graeca, Beiträge zur Erforschung griechischer Texte, Band 12. Wiesbaden.
—. 2001b. "Le chronographie du système d'Empédocle: Documents byzantins inédits." *Aevum Antiquum* 1: 237–59.
—. 2007. *Essentialisme. Alexandre d'Aphrodise entre logique, physique et cosmologie*. Berlin.
—. 2011. *Alexandre d'Aphrodise, Commentaire perdu à la* Physique *d'Aristote* (Livres IV–VIII). Berlin.
Rashed, M., and Th. Auffret. 2014. "Aristote, *Métaphysique* A 6, 988a 7–14, Eudore d'Alexandrie et l'histoire ancienne du texte de la *Métaphysique*." In Chr. Brockmann et al., eds., *Handschriften- und Textforschung heute. Zur Überlieferung der griechischen Literatur. Festschrift für Dieter Harlfinger aus Anlass seines 70. Geburtstages*, 55–84. Wiesbaden.
Ravaisson, F. 1837. *Essai sur le Métaphysique d'Aristote*. Tome I. Paris.
Reeve, M. D. 2007. "Reconstructing Archetypes: A New Proposal and an Old Fallacy." In P. Finglass, C. Collard, and N. J. Richardson, eds. *Hesperos: Studies in Ancient Greek Poetry Presented to M. L. West on his Seventieth Birthday*, 326–340. Oxford.
—. 2011. "Stemmatic Method: «qualcosa che non funziona»?" In *Manuscripts and Methods. Essays on Editing and Transmission*, 27–44. Rome: Edizioni di storia e letteratura. Originally published in P. Ganz, ed. 1986. *The Role of the Book in Medieval Culture. Proceedings of the Oxford International Symposium 26 September – 1 October 1982*. Vol. I: 57–69. Turnhout.
Reis, B. 1999. *Der Platoniker Albinos und sein sogenannter Prologos. Prolegomena, Überlieferungsgeschichte, kritische Edition und Übersetzung*. Wiesbaden.
Resigno, A. 2004. *Alessandro di Afrodisia. Commentario al de Caelo di Aristotele: frammenti del primo libro*. Amsterdam.
Reynolds, L. D., and N. G. Wilson. 2013. *Scribes and Scholars: A Guide to the Transmission of Greek and Latin Literature*. Fourth edition. Oxford.
Rose, V. 1854. *De Aristotelis librorum ordine et autoritate commentatio*. Berlin.
Rosemann, Ph. W. 1989. "Averroes and Aristotle's Philosophical Dictionary." *ModSch* 66:

95–111.
Ross, W. D. 1908. *The Works of Aristotle: Translated into English under the Editorship of W. D. Ross.* Vol. VIII, *Metaphysica.* Oxford.
— . 1924. *Aristotle's Metaphysics. A Revised Text with Introduction and Commentary.* Volume I–II. Oxford [Ross 1924 = Vol. I, Ross 1924, II = Vol. II].
— . 1936. *Aristotle's Physics. A Revised Text with Introduction and Commentary.* Oxford.
— . 1984. "Metaphysics." In Barnes 1984: 1552–728.
Schironi, F. 2010. ΤΟ ΜΕΓΑ ΒΙΒΛΙΟΝ: *Book-Ends, End-Titles, and* Coronides *in Papyri with Hexametric Poetry.* Durham, NC.
— . 2012. "Greek Commentaries." *Dead Sea Discoveries* 19: 399–441.
Schneider, J.-P. 1994. "Boéthos de Sidon." In R. Goulet, ed., *Dictionnaire des philosophes antiques.* II *Babélyca d'Argos à Dyscolius,* 126–30. Paris.
Schofield, M. 2012. "Pythagoreanism: Emerging from the Presocratic Fog." In Steel 2012: 141–66.
Schwegler, A. 1847a. *Die Metaphysik des Aristoteles. Grundtext, Übersetzung und Commentar.* Erster Band: *Grundtext und kritischer Apparat.* Tübingen.
— . 1847b. *Die Metaphysik des Aristoteles. Grundtext, Übersetzung und Commentar.* Zweiter Band: *Übersetzung.* Tübingen.
— . 1847c. *Die Metaphysik des Aristoteles. Grundtext, Übersetzung und Commentar.* Dritter Band: *Des Commentars erste Hälfte.* Tübingen.
— . 1848. *Die Metaphysik des Aristoteles. Grundtext, Übersetzung und Commentar.* Vierter Band: *Des Commentars zweite Hälfte.* Tübingen.
Schwyzer, E., and A. Debrunner. 1950. *Griechische Grammatik. Auf der Grundlage von Karl Brugmanns Griechischer Grammatik von Eduard Schwyzer.* Zweiter Band: *Syntax und syntaktische Stilistik.* München.
Sedley, D. 1998. "Platonic Causes." *Phronesis* 43: 114–32.
Sepúlveda, J. G. 1527. *Alexandri Aphrodisiei commentaria in duodecim Aristotelos libros de prima Philosophia.* Rome.
Sharples, R. W. 1987. "Alexander of Aphrodisias: Scholasticism and Innovation." *ANRW* II, 36.2: 1176–243.
— . 1990. "The School of Alexander?" In Sorabji 1990: 83–111.
— . 2005. "Implications of the New Alexander of Aphrodisias Inscription." *BICS* 48: 47–56.
Shields, Ch. 1999. *Order in Multiplicity: Homonymy in the Philosophy of Aristotle.* Oxford.
— , ed. 2012. *The Oxford Handbook of Aristotle.* Oxford.
Sluiter, I. 1999. "Commentaries and the Didactic Tradition." In G. W. Most, ed., *Commentaries – Kommentare,* 173–205. Göttingen.
Smyth, H. W. 1920. *Greek Grammar for Colleges.* New York.
Sorabji, R., ed. 1990. *Aristotle Transformed: The Ancient Commentators and Their Influence.* Ithaca.
Steel, C., ed. 2012. *Aristotle's Metaphysics Alpha. With a new critical edition of the Greek Text by Oliver Primavesi.* Oxford.
Szlezák, Th. A. 1983. "Alpha Elatton: Einheit und Einordnung in die Metaphysik." In Moraux and Wiesner 1983: 221–59.
Timpanaro, S. 2005. *The Genesis of Lachmann's Method.* Transl. and ed. by G. Most. Chi-

cago.

Usener, H. 1892. "Rezension zu *Commentaria in Aristotelem Graeca.*" *GGA* 26: 1001–22 [= *Kleine Schriften*, Dritter Band, Leipzig/Berlin 1914, 193–214].

Vuillemin-Diem, G. 1970. *Metaphysica lib. I–IV.4: Translatio Iacobi sive 'Vetustissima' cum Scholiis et Translatio Composita sive 'Vetus.'* Aristoteles Latinus XXV 1–1a, Bruxelles.

—. 1976. *Metaphysica lib. I–X, XII–XIV: Translatio Anonyma sive 'Media.'* Aristoteles Latinus XXV 2. Bruxelles.

—. 1983. "Anmerkungen zum Pasikles-Bericht und zu Echtheitszweifeln am größeren und kleineren Alpha in Handschriften und Kommentaren." In Moraux and Wiesner 1983: 157–92.

—. 1995a. *Metaphysica lib. I–XIV: Recensio et Translatio Guillelmi de Moerbeka. Praefatio.* Aristoteles Latinus XXV 3.1. Bruxelles.

—. 1995b. *Metaphysica lib. I–XIV: Recensio et Translatio Guillelmi de Moerbeka. Editio textus.* Aristoteles Latinus XXV 3.2. Bruxelles.

Wagner, T., and Ch. Rapp. 2004. *Aristoteles, Topik.* Stuttgart.

Walzer, R. 1958. "On the Arabic Versions of Books A, α, and Λ of Aristotle's Metaphysics." *HSPh* 63: 217–31.

Weidemann, H. 1980. "In Defence of Aristotle's Theory of Predication." *Phronesis* 25: 76–87.

—. 1996. "Zum Begriff des *ti ên einai* und zum Verständnis von Met. Z 4, 1029b22–1030a6." In Ch. Rapp, ed., *Aristoteles: Metaphysik. Die Substanzbücher (Z, H, Θ)*, 75–103. Berlin.

West, M. L. 1966. *Hesiod, Theogony. Edited with Prolegomena and Commentary.* Oxford.

—. 1973. *Textual Criticism and Editorial Technique.* Stuttgart.

Wildberg, Ch. 1993. "Simplicius und das Zitat. Zur Überlieferung des Anführungszeichens." In F. Berger et al., eds., *Symbolae Berolinenses. Für Dieter Harlfinger*, 187–99. Amsterdam.

—. 2009. "Aporiai 9–10." In Crubellier and Laks 2009: 151–74.

Wilpert, P. 1940. "Reste verlorener Aristotelesschriften bei Alexander von Aphrodisias." *Hermes* 75: 369–96.

Wilson, N. G. 1983. *Scholars of Byzantium.* London.

Wittwer, R. 1999. "Aspasian Lemmatology." In Alberti and Sharples 1999: 51–84.

INDEX LOCORUM

This index includes only those passages that are themselves discussed in the study or are directly related to or quoted in the discussion of a passage. This index does not include all those passages that play only an illustrative role or are referred to in lists in footnotes or the main text (except the lists A and B on pp. 90–95), or that appear in the appendices.

Alexander of Aphrodisias
 In Metaph., *11.3–6*, 92, 124–26; *26.14–18*, 242–45; *31.27–32.9*, 230–34; *33.17–19*, 249–54; *33.23–26*, 249–54; *36.12–13*, 90; *37.20–21*, 92; *37.21–38.1*, 247–48; *38.5–7*, 245–49; *41.26–27*, 61–62; *46.20–23*, 93; *46.23–24*, 67, 90; *58.31–59.8*, 62–64, 90; *59.23–27*, 90; *67.20–68.4*, 254–59; *68.3–4*, 93; *70.7–8*, 93; *75.26–28*, 67, 90; *91.5–6*, 90; *100.25–27*, 67; *104.19–22*, 67–69, 90; *114.22*, 93; *116.25–27*, 93; *137.2–5*, 78–83; *137.5–7*, 81n245; *138.24–28*, 78–83, 90; *139.3–5*, 272; *141.11–13*, 67, 93; *141.19–21*, 93; *141.24–26*, 93; *144.15–145.8*, 224–30; *145.8–12*, 271–78; *145.12–19*, 272; *145.19–146.4*, 271–78; *145.21–25*, 90; *156.14–18*, 57–58; *162.10–16*, 67; *163.6–7*, 67; *164.15–165.5*, 48, 198–200; *164.22–25*, 67, 93; *166.19–167.1*, 66; *167.7–14*, 93, 126–30; *169.4–11*, 83–90; *169.11–15*, 86n264; *172.13–15*, 93; *172.20–22*, 67; *174.5–6*, 101–105; *174.25–27*, 67, 93, 101–105; *177.10*, 67; *182.32–38*, 161–64; *184.12–13*, 48–49; *185.22–24*, 93; *186.11–187.6*, 257n282; *186.31–33*, 93; *193.32–33*, 93; *194.3–4*, 90; *204.23–31*, 121–24; *205.28–33*, 182–83; *206.9–12*, 178–87; *206.13–19*, 186n32; *219.15–37*, 258–59; *220.1–4*, 112–21; *224.18–19*, 94; *228.29–229.1*, 47, 134–38; *233.26*, 94; *243.31–32*, 220; *244.31–32*, 94; *251.2–5*, 94; *251.21*, 90; *257.7–16*, 153–57; *262.37–263.5*, 219–24; *264.17–18*, 94; *264.28–35*, 48, 105–12; *265.6–9*, 105–12; *267.14–21*, 94; *270.15–17*, 94; *273.20–26*, 130–34; *273.34–36*, 96; *273.34–274.2*, 94, 130–34, 90; *285.32–36*, 94, 212–19; *286.2–6*, 212–19; *288.9–11*, 94; *292.13–16*, 157–60; *295.29–32*, 235–39; *299.5–9*, 140–46; *303.23–29*, 164–67; *321.1*, 94; *329.33–330.8*, 167–76; *330.1–3*, 94; *339.18–20*, 90; *341.30*, 91; *345.4–6*, 68; *347.19–25*, 266–70; *348.5–8*, 266–70; *91;* *349.5–6*, 94; *354.28–355.5*, 70–75; *354.31–32*, 91; *356.34–35*, 91; *357.24*, 95; *360.9–12*, 59–60; *368.7–15*, 95; *372.10–17*, 191–98; *378.28–379.8*, 64; *380.25–30*, 259–65; *381.1–4*, 259–65; *417.2–3*, 91; *419.25–420.3*, 146–53; *421.7–15*, 208–12; *433.15–16*, 95; *438.14–17*, 187–91; *439.3–5*, 91; *Fr. 4b* Freudenthal, 91; *Fr. 10a* Freudenthal, 68; *Fr. 12* Freudenthal, 75–78, 91, 95, 200–205; *Fr. 13b* Freudenthal, 91
 De fato, I, *164.3–6*, 18n100
 In Sens. *9.24–25*, 63n162
 De diff. *136a–b*, 185–86
Aristotle
 Categories, *1, 1a1–4*, 220–21; *10, 11b16–19*, 260–61; *10, 11b32–12a25*, 261
 De anima, *B 10, 422a20–31*, 151–53
 De caelo, *B 1, 284a18–20*, 209n110; *1, 284a18–26*, 211; *Γ 2, 301a14–20*, 115–16, 119–20; *3, 302a15–18*, 74n213
 EE, *A 6, 1216b35–1217a10*, 87; *8, 1218b4–6; 9–12*, 267n306
 Int., *2, 16a16–18*, 130n106
 MA, *4, 699 b17–b21*, 151–53; *6, 700b17–b19*, 141n137; *6, 700b25–b28*, 141n138; *7, 702a13–15*, 145; *7, 701a23–25*, 146n157
 Metaphysics A, *2, 982a19–25*, 124–26; *2, 982a25–28*, 85; *2, 982b7–12*, 274n324; *3, 983b33–984a3*, 242–45; *3, 984a21–22*, 232–33; *3, 984b8–13*, 230–34; *4, 985a4–10*, 249–54; *4, 985b12–13*, 90; *5, 985b23–29*, 245–49; *5, 986a15–18*, 61–62; *5, 987a10*, 90; *6, 988a2–4*, 76; *6, 988a9–11*, 62–64, 90; *6, 988a12–13*, 90; *8, 989a22–26*, 254–59; *8, 990a24*, 90; *9, 990b30–31*, 90; *9, 991a27–b1*, 68–69; *9, 991b3*, 48, 89, 100; *9, 993a5*, 101
 Metaphysics α, *1, 993a29–b2*, 78–83, 90; *1, 993b19–23*, 90, 224–30, 271–78; *2, 994b21–27*, 48, 198–200; *2, 994a27–30*, 57–58; *3, 994b32–995a3*, 126–30; *3, 995a16–17*, 90; *3,*

995a12-20, 83-89, 101-105
Metaphysics B, 1, 995b6-8, 105; 2, 996a
  29-996b1, 161-64; 2, 996b 26-27, 31-33,
  105; 2, 997a24, 90; 3, 998b14-19, 121-24;
  3, 998b23-28, 178-87; 3, 998b14-17, 179;
  4, 1000a5-1001a3, 258-59; 4, 1000a26-32,
  112-21; 5, 1001b26-28, 134-38; 5, 1002a4-6,
  136; 5, 1002a15-16, 136; 5, 1002a23-25, 136;
  5, 1002a32-33, 136; 5, 1002b8-9, 136
Metaphysics Γ, 2, 1003a33-35, 219-21;
  2, 1003b5-10, 219-21; 2, 1003b11-12,
  b14-15, 221; 2, 1004a5, 90; 2, 1004a24-25,
  221; 2, 1004a31-b3, 110, 153-57; 2,
  1005a2-8, 219-24; 3, 1005a19-23, 105-12;
  3, 1005b5-8, 110; 4, 1006a18-24, 90, 130-34;
  4, 1007a20-23, 212-19; 4, 1007b 29-1008a2,
  157-60; 4, 1008a17, 238n211; 4, 1008a18-19,
  238n212, 239n213; 4, 1008a18-27, 235-39;
  4, 1008a23, 239n214; 4, 1008b11-12, 101; 4,
  1008b12-19, 140-46; 5, 1009a22-28, 164-67;
  6, 1011a17-18, 275n327; 7, 1011b35-1012a1,
  167-75; 8, 1012b8-10, 90; 8, 1012b22-28, 91
Metaphysics Δ, 1, 1012b34-1013a23, 266; 1,
  1013a17-23, 91, 266-70; 3, 1014a26-31,
  70-75, 91; 3, 1014b2-3, 91; 4, 1015a17-19,
  59-60; 7, 1017a35-b6, 191-98; 9, 1018a12-
  13, 64; 10, 1018a20-25, 259-66; 18,1022a35-
  6, 91; 22, 1022b32-36, 140-53; 23, 1023a
  17-21, 208-12; 30, 1025a21-25, 187-91;
  30,1025a32-33, 91
Metaphysics E, 1, 1026a10-13, 274n322;
  1, 1026a13-18, 274n323; 1, 1026a18-19,
  225-26
Metaphysics K, 7, 1064b1-3, 225n169
Metaphysics Λ, 1, 1069a32, 91; 3, 1070a13-
  19, 200-205; 3, 1070a18-19, 75-78, 91
Metaphysics M, 1, 1076a9, 86-87n267; 3,
  1078a31-34, 163-64; 3, 1078a36-b1,
  163n212
Physics, B 3, 194b32-35, 144; Γ 6, 207a21-26,
  199-200; Θ 8, 263a23-b9, 199-200
Po., 20, 1456b20-24, 71n202
Topics, Z 6, 144a31-b11, 180-81
Asclepius
  In Metaph., 114.1-2, 81n246; 140.22-27,
  105n20; 275.14-17, 166n218; 294.23-25,
  169n231; 318.32-34, 196; 322.4-6, 261n294

Empedocles
  Physica I, 265-72, 114-15; B 21.9 DK, 114

Galen
  In Hippocr. libr. de officina med. comm. XVIII,
  II, 35n14

Hesiod
  Theogony, 517, 209-11

Plato
  Alcibiades, I 107a7, 189
  Parmenides, 166a1-2, 189
  Gorgias, 468b1-4, 145
  Phaedrus, 227d2-5, 143-44

Simplicius
  In Cael.
  336.29-31, 40
  In Phys.
  159.12, 114; 377.24-26, 36; 1317.6-7, 36
Syrianus
  In Metaph. 32.34-36, 184

GENERAL INDEX

Academy, 76, 78, 100, 164
Ammonius Hermiae, 24n36, 34–36, 53–54, 196
Anaxagoras, 165, 231, 250n256, 254–55, 258
Ancestor, 1–2, 13–14, 19, 70, 78, 87–88, 130, 133–35, 138, 191, 197, 204–205, 230, 270, 279–80
Andronicus of Rhodes / Andronican, 15–18, 137
Aporia, 103–107, 109–12, 121, 123, 134–37, 153–56, 161, 178–80, 185, 220, 258–59
Aristippus, 161–63
Asclepius (of Tralles), *passim*, 2, 7n34, 15, 22–25, 28–29, 80–81, 104–105, 122n79, 126, 128–29, 142n143, 152n174, 159n198, 162n209, 166–69, 191n50, 196–97, 209n111, 223n159, 229n185, 232n193, 237n207, 252–54, 261n294, 270n313
Aspasius, 8, 34, 36–37, 39n37, 41n50, 43, 45, 56n130, 60–65, 69, 88, 138, 238, 269, 277, 279–80
Asyndeton / asyndetic, 41, 43, 52, 94
Averroes, 2, 22, 29–31, 45, 75–77, 192n52, 200–202, 205n95
Axioms, 106–12

Being
 actually, 59, 165, 192–200
 potentially, 59, 165, 192–200
Bekker, Immanuel, *passim*, 9–10, 141, 147, 154, 224
Boethus of Sidon, 65–66
Bonitz, Hermann, *passim*, 3n7, 10–11, 22–27, 29, 43n61, 46, 80n243, 99–101, 116, 141, 147–50, 154, 172–74, 193, 213n128, 217n137, 220n147, 232, 256
Byzantine (scholarship / scholars), 4, 6n28, 21

Cause
 definite, 188–91
 efficient, 116n58, 23–34, 254
 final, 161–63, 231–32, 267
 formal, 258n283
 four cause theory, 144, 245
 material, 230–31, 233, 242, 249
 of generation / destruction, 112–13
Conjecture, 13n65, 28n63, 48, 61–64, 89, 93n316, 95–98, 138n129, 170, 172n241, 199, 204, 252n262, 257–58, 269
Contamination, 2, 6n31, 11–13, 33, 49, 52–53, 57, 138–39, 164, 178, 206–209, 219, 224, 239–41, 248n252, 265–66, 270, 277–78, 280–81
Contradiction, 147n160, 169–72, 174, 260
Contraries, 164–68, 221, 260–62
Correction (of a text), 56n130, 60, 74n214, 87, 89, 104, 112, 122, 124, 133–34, 137, 152n173, 164, 172, 197, 219, 245, 247n249, 249, 257, 269
Corruption (textual), 36n19, 41, 47, 49n96, 53, 56n129, 58, 78, 99–100, 124, 126n90, 139, 142, 169, 175n250, 191, 199, 205n95, 218–19, 227, 256, 265

Deletion, 96–97, 129–30, 133, 193
Diogenes Laertius, 16–17

Empedocles, 44, 112–20, 209, 211, 230–31, 246n242, 249–59
Error
 conjunctive 26, 99, 133, 137n126, 139, 175
 indicative, 13, 59n138, 99
 saut du même au même, 47, 54, 58, 114n47, 139n135, 147, 165, 217–18
 scribal, 139n135
 separative, 4, 13–14, 99, 178
 α against β + ω^{AL}, 140–57
 β against α + ω^{AL}, 157–67
 ω^{αβ} against ω^{AL}, 99–124
 ω^{AL} against ω^{αβ}, 124–38
Essence, 54n119, 62n155–58, 95n336, 212–217
Eudorus of Alexandria, 62–63, 139n135
Exegetical / exegesis, 40, 60, 66, 97, 238, 280

Freudenthal, Jacob, *passim*, 3, 22, 25, 30–31, 76–78, 201–205
Forms (Ideas)
 Aristotle's criticisms, 207
 paradigms, 68
 theory of, 42–43, 62, 67–69, 75–78, 100, 200–204

Galen, 35–36, 257n279

Genus, 64, 106–107, 179–87, 220, 223
God(s), 113, 115, 117–20, 236, 242
Gloss, 7, 9, 36n19, 89, 89n276, 137n126, 154–55, 197–98, 224, 227, 230, 234, 238, 240–41, 248n252,

Hermias of Alexandria, 144
Herminus, 61n143, 66n177
Hesiod, 112, 209, 209–211, 231, 242–43, 249
Hesychius, 16–17
Homer, 16n80, 209n112, 211n117, 242–43
Homonymous / Homonymy, 66n175, 219–20

Infinite, 66, 198–200
Influence (on a text), 7, 12–14, 28, 33, 47, 50, 53, 57, 75n215, 98, 105, 178, 198, 205–206, 208, 211–12, 218–219, 223n160, 230, 240, 241, 248n249, 253n268, 265–66, 269, 279–80
Intermediate, 167–74, 261–65
Interpolation 3, 103–104, 111, 157, 160, 175n251, 204, 207n105, 210, 234, 248n252
Isḥāq 84n257

Jaeger, Werner, *passim*, 2n3, 5–6, 10–12, 18n98, 23n31, 37n23, 78, 81, 93n314, 94–95, 100n5, 103, 116n55, 142, 147, 154, 160, 178, 211–13, 224, 232, 236n205, 260n288

Knowledge, 124–26, 198–99
 of truth, 224, 226–27, 230, 274–75
 practical 226–27, 271
 scientific, 273
 theoretical 226–30, 271–72

Latin translation of Alexander's commentary (by Sepúlveda), 24–28, 62, 131n109, 253n267, 257n277
Latin translation of Averroes' commentary (by Michael Scotus), 30, 77n231, 122n78, 133n112, 191n50
lectio difficilior, 7n32, 87, 107n25, 137, 180n6, 236, 275
Lemma, 9–10, 14, 27–28, 31, 33, 34n3, 38–50, 52–53, 56–60, 68, 76–77, 80, 108, 112, 116–17, 125–26, 131, 134–35, 202, 207n105, 214, 225, 228, 237, 243–45, 276, 283–92
Love and Strife (Empedocles), 112–16, 119–20, 231, 249–52, 255–59

Maas, Paul, 13, 99, 137–38
Mathematical / mathematics, 83–87, 101–102, 106–107, 134–36, 161–63, 226, 245–46, 271, 274

Abū Bišr Mattā, 30–31, 76–78, 201n83
Megara, 140–46
Michael of Ephesus, 2, 11n58, 15, 21–25, 92n304
Michael Scotus, *passim*, 2n3, 30, 68, 76–77, 105n20, 108n29, 118n65, 128n96, 122n78, 133n112, 135n120, 141n139, 147n162, 154n183, 158, 162n209, 166n217, 169n231, 183n24, 191n50, 194n62, 201–204, 223n159, 224n163, 237n207
Movement, 59, 141, 256, 267–68

Natural philosophy / science, 78n238, 84–86, 102–103, 110, 209–211, 230, 274
Naturally / by nature, 59, 75–76, 86, 146–52, 200–204, 246–47
Neoplatonic / Neoplatonism, 30, 36, 41n50, 55n126, 113, 144, 183
Nicolaus of Damascus, 18

Omission, 131n108, 137, 197, 252
Opposites, 168, 174, 259–60, 262

Particle (grammatical), 41n50, 51–52, 57–58, 96, 98n347, 117, 132, 169, 188, 191, 236–38, 247n249
Peripatetic, 11, 21, 30, 61, 65, 69
Philosophy / philosopher, 153, 155–56, 209–10, 224–31, 233, 246, 254, 271–72
Plato / Platonic, 42, 62n157, 75–78, 100, 143–45, 179, 200–201, 250–51, 267
Platonist, 62, 82, 139n135
Principle
 of excluded middle, 167, 261n292
 of non-contradiction, 42, 130, 140, 146, 157, 164, 167, 212, 214, 216, 235
 utrum in alterum, 82, 111, 247
Privation, 146–50
πρὸς ἕν ("focal meaning"), 153, 192n52, 219–22
Protagoras, 157, 164
Ptolemy al-Gharīb, 16–17
Pythagorean, 61, 63n161, 67, 245–47, 254

recensio altera of Alexander's commentary, 2n2, 23n26, 26, 26n56, 28–29, 33n1, 81n245, 126n92, 129n102, 229n184, 240–41, 257n277
Reception, 33, 39, 265
reclamantes 4–5, 212n120, 240n219
Revision of the β-text, 3, 5–7, 206–208, 212, 217n139, 219n141, 238–41, 265, 280
Ross, David, *passim*, 11–12, 73n208, 77n233, 80n243, 116n55, 126n93, 130n104, 141–42, 160, 168n227, 170n235, 193n54–57, 196,

198, 225n169, 231, 235n201, 257

Scribe, 4n10, 8, 13, 56, 135, 147, 153, 160, 162n210, 187, 197, 234, 248-49, 265
Scroll(s) (papyrus), 5n18, 34n5, 36n14, 37, 240n219
Simplicius, 9, 36-37, 40-41, 47, 53, 65, 92n304, 114, 190n48, 211n117
Sepúlveda, Juan Ginés de, *passim*, 20-28, 28n63, 45n70, 62, 80n242, 84n259, 194n63, 228n179, 243n235, 253n267, 257n277
Socrates, 250n259, 143-45
Sosigenes, 66n177
Species, 64, 179-87, 220, 222
Strabo, 15n72, 38n29
Synonymy / synonyms, 220, 222-23
Syrianus, 2, 15, 22-23, 25, 142n143, 183-84, 187

Textual criticism, 99, 137, 279
Thales, 242-44
Theologians, theology, 112, 210n113, 226, 271
Theophrastus, 15
Themistius, 30n77, 75, 77-78, 91n300
Tradition
　Arabic, 2n3, 29n72, 45, 91n300, 224n163, 128n96, 134-35, 209n111, 223n159, 224n163, 237n207
　commentary, 36, 37n26, 65, 69
　of the *Metaphysics*, 1, 47, 50, 53, 75, 78, 83, 87-89, 104-107, 130, 133, 137, 154, 159n198, 162n209, 169n231, 185, 187, 196, 205, 217n137, 279
　scholiastic, 238
Transliteration, 1, 4-5, 15, 240

Ustāth, 31, 84n257, 108n29, 133n112, 147, 154

varia lectio / variant (reading), 9-11, 13-14, 28n63, 34-38, 41n50, 48, 50n103, 60-65, 67-70, 72-75, 77-78, 80-97, 99, 133, 194-97, 214n29, 224, 257-58, 266-79

Witness
　independent 13, 28n65, 137-40, 175
　indirect 1, 3, 13, 15
　textual, 2-3, 8-11, 13-14, 29, 33, 37n26, 99-100
World, 115-16, 120, 165, 211n115, 231, 242, 249

Xenocrates, 76, 179n3

Zeno's paradox, 199

www.ingramcontent.com/pod-product-compliance
Lightning Source LLC
Chambersburg PA
CBHW021817300426
44114CB00009BA/213